# Problem Books in Mathematics

Series Editor

Peter Winkler
Department of Mathematics
Dartmouth College
Hanover, NH 03755
USA

For further volumes:
http://www.springer.com/series/714

Albert N. Shiryaev

# Problems in Probability

Translated by Andrew Lyasoff

 Springer

Albert N. Shiryaev
Chair, Department of Probability Theory
M.V. Lomonosov Moscow State Univeristy
Moscow
Russia

V.A. Steklov Institute of Mathematics
Russian Academy of Sciences
Moscow
Russia

Andrew Lyasoff (Translator)
School of Management
Boston University
Boston, MA
USA

Department of Mathematics and Statistics
Boston University
Boston, MA
USA

ISSN 0941-3502
ISBN 978-1-4899-9941-2        ISBN 978-1-4614-3688-1 (eBook)
DOI 10.1007/978-1-4614-3688-1
Springer New York Heidelberg Dordrecht London


Mathematics Subject Classification (2010): 60-01; 28-01

© Springer Science+Business Media New York 2012
Softcover reprint of the hardcover 1st edition 2012
This work is subject to copyright. All rights are reserved by the Publisher, whether the whole or part of
the material is concerned, specifically the rights of translation, reprinting, reuse of illustrations, recitation,
broadcasting, reproduction on microfilms or in any other physical way, and transmission or information
storage and retrieval, electronic adaptation, computer software, or by similar or dissimilar methodology
now known or hereafter developed. Exempted from this legal reservation are brief excerpts in connection
with reviews or scholarly analysis or material supplied specifically for the purpose of being entered
and executed on a computer system, for exclusive use by the purchaser of the work. Duplication of
this publication or parts thereof is permitted only under the provisions of the Copyright Law of the
Publisher's location, in its current version, and permission for use must always be obtained from Springer.
Permissions for use may be obtained through RightsLink at the Copyright Clearance Center. Violations
are liable to prosecution under the respective Copyright Law.
The use of general descriptive names, registered names, trademarks, service marks, etc. in this publication
does not imply, even in the absence of a specific statement, that such names are exempt from the relevant
protective laws and regulations and therefore free for general use.
While the advice and information in this book are believed to be true and accurate at the date of
publication, neither the authors nor the editors nor the publisher can accept any legal responsibility for
any errors or omissions that may be made. The publisher makes no warranty, express or implied, with
respect to the material contained herein.

Printed on acid-free paper

Springer is part of Springer Science+Business Media (www.springer.com)

# Preface

In the first two editions of the book "Probability," which appeared in 1980 and 1989 (see [118]), and were translated into English in 1984 and 1990 (see [119]), all chapters were supplemented with a fairly comprehensive and diverse set of relevant exercises. The next two (considerably revised and expanded) editions appeared in 2004 and 2007 (see [121]) in two volumes entitled "Probability 1" and "Probability 2." While the work on the third edition was still in progress, it was determined that it would be more appropriate to publish a separate book that includes all "old" exercises, i.e., exercises included in the previous two editions, and many "new" exercises, i.e., exercises, which, for one reason or another, were not included in any of the previous editions (the main reason for that was the constrain on the size of the volume that could go to print). This is how the present volume "Problems in Probability" came to life. On the most part, this book includes problems and exercises that I have created, collected and compiled over the course of many years, while working on topics and subjects that interested me the most. These problems derive from a rather diverse set of sources: textbooks, lecture notes, exercise manuals, monographs, research papers, private communications and such. Some of the problems came out of discussions that took place during special seminars for graduate and undergraduate students in which I was involved.

It is impossible to cite here with complete accuracy all of the original sources from which the problems and the exercises are derived. The bibliography included at the end of the book and the citations throughout the main text are simply the result of my best effort to give credit where credit is due.

I would like to draw the reader's attention to the appendix included at the end of the book. I strongly recommend that anyone using this book becomes familiar—at least in passing—with the material included in the Appendix. There are two reasons for this recommendation. First, the appendix contains a summary of the main results, notation and terminology from probability theory, that are used not only throughout this book, but also throughout the books "Probability." Second, the appendix contains additional material from combinatorics, potential theory and

Markov chains, which is not covered in these books, but is nevertheless needed for many of the exercises included here.

The following referencing conventions are adopted throughout the book:

(a) All references to the books "Probability" (see [121]) start with the token P . For example, "[ P §1.1, **3** ]" points to Part 3 of Sect. 1 in Chap. 1 in [121], "[ P §2.6, (6)]" and "[ P §2.6, Theorem 6]" point, respectively, to (6) and Theorem 6 from Sect. 6 in Chap. 2 in [121]—and so on.

(b) Problems included in this book are referenced, for example, as "Problem 1.4.9," which points to Problem 1.4.9 from Sect. 1.4 in Chap. 1 below.

The reader must be forewarned that the problems and the exercises collected in the present volume differ from each other in nature:

(a) Some problems are simply meant to test the reader's understanding of the basic concepts and facts from the books "Probability." For example, the exercises from Sects. 1.1 and 1.2 in Chap. 1 relate to the various combinatorial methods for counting the favorable outcomes of an event and illustrate the important notions of partial factorial $(N)_n$, combinations $C_N^n$ and $C_{N+n-1}^n$, Catalan numbers $C_n$, Stirling numbers of the first and second kind $s_N^n$ and $S_N^n$, Bell numbers $B_N$, Fibonachi numbers $F_n$, etc.

(b) Other problems are of a medium-to-high degree of difficulty and require more creative thinking. A good example is Problem 7.4.3, which is asking for a unified proof of Lebesgue's dominated convergence theorem and Levy's theorem of convergence of conditional expectations.

(c) Some of the problems are meant to develop additional theoretical concepts and tools that supplement the material covered in the books "Probability," or simply to familiarize the reader with various facts that, typically, are not covered in the mainstream texts in probability theory, but are nevertheless "good to know"—or at least good to know that such results exist and be aware of the respective sources. One such example is M. Suslin's result (see Problem 2.2.27 below), which states that the projection of a Borel set in the plane onto one of the coordinate axes may not be a Borel set inside the real line, or the result describing the set-operations that allow one to produce the smallest algebra or $\sigma$-algebra that contains a given collection of sets—see Problems 2.2.25, 2.2.26 and 2.2.32. One must realize that, in fact, many problems of this type represent fairly difficult theorems. The formulation of such theorems in the form of exercises has the goal of inviting the reader to think and to ask questions like: how does one construct a $\sigma$-algebra anyway? The answer to this and similar questions is of paramount importance in the study of models and phenomena that pertain to what one may call "non-elementary probability theory."

(d) Some of the problems are related to the passage from random walks to Brownian motions and Brownian bridges—see Sect. 3.4, for example. The statements in these problems are intimately related to what is known as the "invariance principle" and may be viewed as some sort of a prelude by way of problems and exercises to the general theory of stochastic processes in continuous time and, in particular, to the functional limit theorems.

Many (but not all, by far) of the problems included in this book contain hints and other relevant comments. I very much hope that these hints and comments will be helpful not only for deriving the solution, but also for learning how to think about the related concepts and problems.

Over nearly 50 years several of my colleagues at MSU have published exercise manuals in probability theory that have been in continuous use in courses offered at MSU, as well as in other institutions of higher education. I would like to mention them:

1963 – L. D. Meshalkin. Exercise Manual in Probability Theory. Moscow University Press, Moscow;

1980 – B. A. Sevastyanov, V. P. Chistyakov, A. M. Zubkov. Exercise Manual in Probability Theory. Nauka, Moscow;

1986 – A. V. Prohorov, V. G. Ushakov, N. G. Ushakov. Exercises in Probability Theory: Basic Notions, Limit Theorems, Random Processes. Nauka, Moscow;

1989 – A. M. Zubkov, B. A. Sevastyanov, V. P. Chistyakov. Exercise Manual in Probability Theory, a-ed.. Nauka, Moscow;

1990 – M. V. Kozlov. Elements of Probability Theory Through Examples and Exercises. Moscow University Press, Moscow.

Since this last book was published nearly 15 years ago, the curriculum in most graduate-level courses in probability theory has changed considerably. Some new directions have emerged, new areas of research were developed, and new problems were formulated. An earnest effort was made to adequately reflect these changes in the books "Probability" and, naturally, in the present volume, which first appeared in 2006. At the same time, the traditional coverage of all classical domains of probability theory was kept intact. At the end more than 1,500 problems (counting the various parts of problems) found their way into the present volume.

As was the case with the books "Probability," the final edit, arrangement and proof-reading of the text was done by Tatyana Borisovna Tolozova, to whom I am deeply indebted.

Finally, I would like to express my gratitude to Andrew Lyasoff not only for translating the present volume into English, but also for making a number of corrections in the original and for enriching the text with many comments and clarifications.

Moscow                                                          Albert N. Shiryaev

# Contents

# Chapter 1
# Elementary Probability Theory

## 1.1 Probabilistic Models for Experiments with Finitely Many Outcomes

**Problem 1.1.1.** Verify the following relations involving the operations $\cap$ (intersection) and $\cup$ (union):

$A \cup B = B \cup A$, $A \cap B = B \cap A$ (commutativity),

$A \cup (B \cup C) = (A \cup B) \cup C$, $A \cap (B \cap C) = (A \cap B) \cap C$ (associativity),

$A \cap (B \cup C) = (A \cap B) \cup (A \cap C)$, $A \cup (B \cap C) = (A \cup B) \cap (A \cup C)$ (distributivity),

$A \cup A = A$, $A \cap A = A$ (idempotent property of $\cap$ and $\cup$).

Then prove that

$$\overline{A \cup B} = \overline{A} \cap \overline{B} \quad \text{and} \quad \overline{A \cap B} = \overline{A} \cup \overline{B},$$

where ⁻ stands for the operation "complement of a set."

**Problem 1.1.2.** (*Various interpretations of the partial factorial* $(N)_n \equiv N(N-1)$ $\ldots (N-n+1)$, *i.e., the number of permutations $N$ take $n$—see Sect. A.1.*) Prove that:

(a) The number of all *ordered* samples (...) without replacement (equivalently, samples without repetition) of size $n$ drawn from any finite set $A$ of size $|A| = N$, $1 \leq n \leq N$, equals $(N)_n$.

(b) The number of all words of length $n$ composed from *different* letters selected from an alphabet that consists of $N$ letters, $1 \leq n \leq N$, equals $(N)_n$.

(c) Given a finite set $X$ of size $|X| = n$ and a finite set $Y$ of size $|Y| = N$, $n \leq N$, the number of all functions $f : X \mapsto Y$ such that if $x_1, x_2 \in X$ and $x_1 \neq x_2$ then $f(x_1) \neq f(x_2)$ (i.e., the number of all injections from $X$ to $Y$) equals $(N)_n$.

**Problem 1.1.3.** (*Various interpretations of the binomial coefficients* $C_N^n \equiv \frac{N!}{n!(N-n)!}$—*see Sect. A.1.*) Prove that:

A.N. Shiryaev, *Problems in Probability*, Problem Books in Mathematics,
DOI 10.1007/978-1-4614-3688-1_1,
© Springer Science+Business Media New York 2012

(a) The number of all *unordered* samples [...] without replacement (equivalently, samples without repetition) of size $n$, drawn from any finite set $A$ of size $|A| = N$, $1 \leq n \leq N$, equals $C_N^n$.

(b) The number of all *ordered* finite 0-1-sequences (...) of length $N$ that contain exactly $n$ 1's and exactly $(N - n)$ 0's, $1 \leq n \leq N$, equals $C_N^n$.

(c) The number of all possible placements of $n$ *indistinguishable* particles into $N$ *distinguishable* cells, $1 \leq n \leq N$, in such a way that each cell can contain at most one particle (the so called "placement with locks"), equals $C_N^n$.

(d) The number of all possible *nondecreasing* paths on the two-dimensional lattice $\mathbb{Z}_+^2 \equiv \{(i, j) : i, j = 0, 1, 2, \ldots\}$, that start from the point $(0, 0)$ and end at the point $(n, N - n)$, $0 \leq n \leq N$, equals $C_N^n$ (a path on the two-dimensional lattice is said to be nondecreasing if at each step the path moves either up by $+1$ or to the right by $+1$—notice that $C_N^0 = 1$).

(e) The number of all *different* subsets $D$ of size $|D| = n$ that are contained in some finite set $A$ of size $|A| = N$, $n \leq N$, equals $C_N^n$.

*Hint.* Assuming that (a) has already been established, then one can establish (b), (c), (d) and (e) by proving the equivalence relations $(a) \Longleftrightarrow (b)$, $(a) \Longleftrightarrow (c)$, ...—exactly as this is done in [ $\underline{P}$ §1.1, Example 6].

**Problem 1.1.4.** Similarly to Part (d) in the previous problem, consider the class of all *nondecreasing* paths on the lattice $\mathbb{Z}_+^2 \equiv \{(i, j) : i, j = 0, 1, 2, \ldots\}$, that start from the point $(0, 0)$ and end at the point $(n, n)$, while never moving above the diagonal, i.e. all paths that go from $(0, 0)$ to $(n, n)$ and remain in the set $\{(i, j) \in \mathbb{Z}_+^2, 0 \leq j \leq i \leq n\}$. Prove that the number of paths in this class is given by the $(n + 1)^{\text{st}}$ Catalan number $C_{n+1}$, the $n^{\text{th}}$ Catalan number, $n \geq 1$, being defined as

$$C_n \equiv \frac{1}{n} C_{2(n-1)}^{n-1}.$$

*Note:* Sometimes the Catalan numbers are defined as $c_n \equiv \frac{1}{n+1} C_{2n}^n$ $(= C_{n+1})$, $n \geq 0$ (see, for example, [6]).

Prove that $C_1, \ldots, C_9$ equal, respectively, 1, 1, 2, 5, 14, 42, 132, 429, 1430.

**Problem 1.1.5.** The Catalan numbers $C_n$, $n \geq 1$, show up in many combinatorial problems. Consider, for example, the number of *binary bracketings* of $n$ letters— this is the number of all possible ways in which one can compute the sum of $n$ numbers that are arranged in a row by adding only 2 *neighboring* numbers at the time. For instance, one can compute the sum $a + b + c$ either as $((a + b) + c)$ or as $(a + (b + c))$. Thus, the number of binary bracketings of three letters equals 2. It is not hard to see that there is a total of 5 binary bracketings of 4 letters:

$$a + b + c + d = ((a + b) + (c + d))$$
$$= (((a + b) + c) + d)$$

$$= ((a + (b + c)) + d)$$
$$= (a + ((b + c) + d))$$
$$= (a + (b + (c + d))).$$

(a) Prove that for any integer $n \geq 3$, the number of all binary bracketings of $n$ letters is given by the Catalan number $C_n$.

(b) Consider *Euler's polygon divison problem:* In how many different ways can a plane convex polygon of $n$ sides, $n \geq 4$, be divided into triangles by non-intersecting diagonals? Prove that the answer to Euler's polygon divison problem is given by the Catalan number $C_{n-1}$.

*Hint.* If a convex $n$-gone is divided into (non-intersecting) triangles whose vertices are also vertices of the $n$-gone, then there would be exactly $(n-2)$ triangles and any such division corresponds to the choice of $(n-3)$ non-intersecting diagonals.

(c) Consider the numbers $C_n^*$, $n \geq 0$, defined recursively by the relations:

$$C_0^* = 0, \quad C_1^* = 1 \quad \text{and} \quad C_n^* = \sum_{i=1}^{n-1} C_i^* C_{n-i}^*, \quad \text{for } n > 1. \qquad (*)$$

Prove that, for any $n \geq 1$, the number $C_n^*$ coincides with the $n^{\text{th}}$ Catalan number $C_n$; in other words, prove that the Catalan numbers can be defined equivalently by way of the recursive relation $(*)$.

(d) Prove that the generating function $F^*(x) = \sum_{n \geq 1} C_n^* x^n$, associated with the sequence $(C_n^*)_{n \geq 1}$ and defined by the recursive relation $(*)$ above, satisfies the following relation:
$$F^*(x) = x + (F^*(x))^2.$$

(e) By taking into account that $F^*(0) = 0$, prove that

$$F^*(x) = \frac{1}{2}(1 - (1 - 4x)^{1/2}), \qquad |x| < \frac{1}{4},$$

and conclude that, just as one would expect, the coefficients $C_n^*$ in the expansion of the function $F^*(x)$ coincide with the Catalan numbers $C_n$:

$$C_n^* = -\frac{1}{2} C_{1/2}^n (-4)^n = \frac{1}{n} C_{2(n-1)}^{n-1} = C_n.$$

(For the definition of the quantity $C_{1/2}^n$ see Problem 1.2.22.)

**Problem 1.1.6.** (*Various interpretations of the binomial coefficients* $C_{N+n-1}^n$.) Prove that:

(a) The number of all *unordered* samples with replacement [...] of size $n$ drawn from any finite set $A$ of size $|A| = N$ equals $C_{N+n-1}^n$.

(b) The number of all *ordered* lists $(n_1, \cdots, n_N)$, $N \geq 1$, whose entries $n_i$, $i = 1, \ldots, N$, are non-negative integer numbers that satisfy the relation $n_1 + \cdots + n_N = n$, for some fixed $n \geq 1$, equals $C^n_{N+n-1}$.

(c) The number of all possible placements of $n \geq 1$ *indistinguishable* particles into $N \geq 1$ *distinguishable* cells without a restriction on the number of particles in each cell (the so-called "placement without locks"), equals $C^n_{N+n-1}$.

*Hint.* Follow the hint to Problem 1.1.3.

**Problem 1.1.7.** (*Continuation of Part (b) in the previous problem.*) Given some fixed integers, $N \geq 1$ and $n \geq 1$, consider the collection of all *unordered* solutions $[n_1, \ldots, n_N]$ to the equation $n_1 + \cdots + n_N = n$, in terms of some non-negative integers $n_i \geq 0$, $i = 1, \ldots, N$. What is the total number of solutions in this collection? What is the total number of all—still unordered—strictly positive solutions $n_i > 0$, $i = 1, \ldots, N$? What is the total number of all *ordered* solutions $(n_1, \ldots, n_N)$ to the same equation, $n_1 + \cdots + n_N = n$, in terms of positive integers $n_i > 0$, $i = 1, \ldots, N$?

**Problem 1.1.8.** (*Continuation of Part (b) in Problem 1.1.6 and Problem 1.1.7.*) Given some fixed integers, $n \geq 1$ and $N \geq 1$, consider the *inequality* $n_1 + \cdots + n_N \leq n$. Count the total number of all ordered solutions $(n_1, \ldots, n_N)$ and the total number of all unordered solutions $[n_1, \ldots, n_N]$ to this inequality, in terms of non-negative or strictly positive integers $n_i$, $i = 1, \ldots, N$.

**Problem 1.1.9.** Prove that:

(a) The maximal number of disjoint regions in the plane $\mathbb{R}^2$, determined by $n$ different lines that are placed arbitrarily in the plane $\mathbb{R}^2$, equals

$$1 + \frac{n(n+1)}{2}.$$

(b) The maximal number of disjoint regions in the space $\mathbb{R}^3$, determined by $n$ different planes that are placed arbitrarily in the space $\mathbb{R}^3$, equals

$$\frac{1}{6}(n^3 + 5n + 6).$$

**Problem 1.1.10.** Suppose that $A$ and $B$ are any two subsets of the set $\Omega$. Prove that the algebra, $\alpha(A, B)$, generated by these two sets—i.e., following the terminology introduced in [P §1.1, **3**], the algebra generated by the system $\mathscr{A}_0 = \{A, B\}$—consists of the following $N(2) = 16$ subsets of $\Omega$:

$$\{A, B, \overline{A}, \overline{B}, A \cap B, \overline{A} \cap \overline{B}, A \setminus B, B \setminus A,$$

$$A \cup B, A \cup \overline{B}, \overline{A} \cup B, \overline{A} \cup \overline{B}, A \triangle B, \overline{A \triangle B}, \Omega, \varnothing\},$$

where $A \bigtriangleup B = (A \setminus B) \cup (B \setminus A)$ is the so called "symmetric difference" of the sets $A$ and $B$ (see [P §2.1, Table 1]).

Find those partitions $\mathscr{D}$ (see [P §1.1, 3]) of the set $\Omega$, for which the algebra $\alpha(\mathscr{D})$, i.e., the algebra generated by $\mathscr{D}$, coincides with $\alpha(A, B)$.

Finally, prove that the algebra $\alpha(A_1, \ldots, A_n)$, generated by the system $\mathscr{A}_0 = \{A_1, \ldots, A_n\}$, where $A_i \subseteq \Omega$, $i = 1, \ldots, n$, consists of $N(n) = 2^{2^n}$ different subsets of $\Omega$, so that $N(2) = 16$, $N(3) = 256$, and so on.

**Problem 1.1.11.** Prove *Boole's inequalities*

$$\text{(a)} \quad \mathsf{P}\left(\bigcup_{i=1}^{n} A_i\right) \le \sum_{i=1}^{n} \mathsf{P}(A_i), \quad \mathsf{P}\left(\bigcap_{i=1}^{n} A_i\right) \ge 1 - \sum_{i=1}^{n} \mathsf{P}(\overline{A}_i).$$

Prove that for any integer $n \ge 1$ the following inequality is in force

$$\text{(b)} \quad \mathsf{P}\left(\bigcap_{i=1}^{n} A_i\right) \ge \sum_{i=1}^{n} \mathsf{P}(A_i) - (n-1).$$

Prove the *Kounias inequality*

$$\text{(c)} \quad \mathsf{P}\left(\bigcup_{i=1}^{n} A_i\right) \le \min_{k}\left\{\sum_{i=1}^{n} \mathsf{P}(A_i) - \sum_{i \neq k} \mathsf{P}(A_i \cap A_k)\right\}.$$

Prove the *Chung-Erdös inequality*

$$\text{(d)} \quad \mathsf{P}\left(\bigcup_{i=1}^{m} A_i\right) \le \frac{\left(\sum_{i=1}^{n} \mathsf{P}(A_i)\right)^2}{\sum_{i,j=1}^{n} \mathsf{P}(A_i \cap A_j)}.$$

*Hint.* With $n = 3$ Part (b) comes down to the inequality $\mathsf{P}(A_1 \cap A_2 \cap A_3) \ge \mathsf{P}(A_1) + \mathsf{P}(A_2) + \mathsf{P}(A_3) - 2$, which can be established with elementary considerations. The general case can then be handled by using induction with respect to $n$. To prove (c), it is enough to establish that the following inequality holds:

$$\mathsf{P}\left(\bigcup_{i=1}^{n} A_i\right) \le \sum_{i=1}^{n} \mathsf{P}(A_i) - \sum_{2 \le i \le n} \mathsf{P}(A_1 \cap A_i).$$

This inequality, too, can be established by induction.

**Problem 1.1.12.** Prove the "inclusion–exclusion formulas" (also known as *Poincaré's formulas, Poincaré's theorems, Poincaré's identities*) for the probability of a union of events and the probability of an intersection of events; namely, prove that, for any $n \geq 1$ and for any choice of the events $A_1, \ldots, A_n$, one has:

(a) $P(A_1 \cup \ldots \cup A_n) = \displaystyle\sum_{m=1}^{n} (-1)^{m+1} \sum_{1 \leq i_1 < \ldots < i_m \leq n} P(A_{i_1} \cap \ldots \cap A_{i_m})$

$\equiv \displaystyle\sum_{1 \leq i_1 \leq n} P(A_{i_1}) - \sum_{1 \leq i_1 < i_2 \leq n} P(A_{i_1} \cap A_{i_2}) + \sum_{1 \leq i_1 < i_2 < i_3 \leq n} P(A_{i_1} \cap A_{i_2} \cap A_{i_3})$

$+ \ldots + (-1)^{n+1} P(A_1 \cap \ldots \cap A_n)$

and

(b) $P(A_1 \cap \ldots \cap A_n) = \displaystyle\sum_{m=1}^{n} (-1)^{m+1} \sum_{1 \leq i_1 < \ldots < i_m \leq n} P(A_{i_1} \cup \ldots \cup A_{i_m})$

$\equiv \displaystyle\sum_{1 \leq i_1 \leq n} P(A_{i_1}) - \sum_{1 \leq i_1 < i_2 \leq n} P(A_{i_1} \cup A_{i_2}) + \sum_{1 \leq i_1 < i_2 < i_3 \leq n} P(A_{i_1} \cup A_{i_2} \cup A_{i_3})$

$+ \ldots + (-1)^{n+1} P(A_1 \cup \ldots \cup A_n).$

*Note 1.* The formula in Part (a) is often written in the form

$$P\left( \bigcup_{i=1}^{n} A_i \right) = S_1 - S_2 + \ldots + (-1)^{n+1} S_n,$$

where

$$S_m = \sum_{1 \leq i_1 < \ldots < i_m \leq n} P(A_{i_1} \cap \ldots \cap A_{i_m}),$$

while the formula in Part (b) is often written in the form

$$P\left( \bigcap_{i=1}^{n} A_i \right) = \tilde{S}_1 - \tilde{S}_2 + \ldots + (-1)^{n+1} \tilde{S}_n,$$

where

$$\tilde{S}_m = \sum_{1 \leq i_1 < \ldots < i_m \leq n} P(A_{i_1} \cup \ldots \cup A_{i_m}).$$

*Note 2.* Although the inclusion–exclusion formulas are considered here in the context of [ P Chap. 1], which deals only with *finite* probability spaces $(\Omega, \mathscr{A}, P)$, it is important to recognize that, in fact, these formulas are valid on *any* (finite or infinite, countable or uncountable) probability space $(\Omega, \mathscr{F}, P)$, regardless of its nature (see [ P Chap. 2]).

Nevertheless, in order to use the inclusion–exclusion formulas in concrete situations, one must be able to compute somehow the quantities $S_m$, or, which amounts to the same, the probabilities $P(A_{i_1} \cap \ldots \cap A_{i_m})$. Usually, such computations take into account the concrete probabilistic structure of the model encoded in the space $(\Omega, \mathscr{A}, P)$—the models associated with the Bose–Einstein statistics and Fermi–Dirac statistics illustrate this point rather well.

*Hint.* The formula in Part (a) can be established by induction with respect to the number of events $n$, after showing first that for $n = 2$ one has

$$P(A_1 \cup A_2) = (P(A_1) + P(A_2)) - P(A_1 \cap A_2).$$

(See also Problem 1.4.9).

To prove the formula in Part (b), notice first that

$$P\left(\bigcap_{i=1}^{n} A_i\right) = P\left(\overline{\bigcup_{i=1}^{n} \overline{A_i}}\right) = 1 - P\left(\bigcup_{i=1}^{n} \overline{A_i}\right).$$

and then apply the formula from Part (a) to the events $\overline{A_1}, \ldots, \overline{A_n}$, instead of the events $A_1, \ldots, A_n$.

*Note 3.* Anticipating the use of the inclusion–exclusion formulas later in this book, notice that $P\left(\bigcap_{i=1}^{n} \overline{A_i}\right)$ is the probability that neither of the events $A_1, \ldots, A_n$ occurs.

**Problem 1.1.13.** Let $B_m$ denote the event that *exactly m* of the events $A_1, \ldots, A_n$ occur, the integers $n \geq m \geq 0$ being fixed. Assuming that the quantities $S_1, \ldots, S_n$ are defined as in Problem 1.1.12, prove that

$$P(B_m) = \sum_{k=m}^{n} (-1)^{k-m} C_k^m \, S_k \,,$$

which can be written also as

$$P(B_m) = S_m - C_{m+1}^1 S_{m+1} + \ldots + (-1)^{n-m} C_n^{n-m} \, S_n \,,$$

and conclude that the probability $P(B_{\geq m})$ of the event $B_{\geq m}$ that *at least m* of the events $A_1, \ldots, A_n$ occur, is given by

$$P(B_{\geq m}) \equiv P(B_m) + \ldots + P(B_n) = \sum_{k=m}^{n} (-1)^{k-m} C_{k-1}^{m-1} S_k,$$

which can be written also as

$$P(B_{\geq m}) = S_m - C_m^1 S_{m+1} + \ldots + (-1)^{n-m} C_{n-1}^{n-m} \, S_n \,.$$

*Hint.* The formula for $P(B_m)$ is known as *Waring's formula* and can be proved by using the method of "inclusion–exclusion," just as in Problem 1.1.12. Such a proof can be found, for example, in W. Feller's book [39, vol. 1, Chap. IV, § 3]. However, readers familiar with the notions of random variables and expected values (see [ P §1.4] and [ P §2.6, **4** ]) may follow these steps:

Given any $i = 1,\ldots,n$, let $X_i \equiv I_{A_i}$ denote the indicator of the event $A_i$ and consider the sum

$$\sum X_{i_1} \ldots X_{i_m}(1 - X_{j_1})\ldots(1 - X_{j_{n-m}}), \qquad (*)$$

where the summation is taken over all $C_n^m$ possible choices of the (unordered) list $[i_1,\ldots,i_m]$ from the list $[1,\ldots,n]$ and

$$[j_1,\ldots,j_{n-m}] \equiv [1,\ldots,n] \setminus [i_1,\ldots,i_m].$$

When evaluated at a particular outcome $\omega$, the sum in $(*)$ is equal to 1 precisely when $\omega$ belongs to exactly $m$ of the events $A_1,\ldots,A_n$ and is equal to 0 in all other cases. Consequently, the quantity $P(B_m)$ is nothing but the expected value of the sum $(*)$. The remaining steps are similar to those described in the hint to Problem 1.4.9. (See also Problem 2.6.31.)

**Problem 1.1.14.** By using the formulas for $P(B_m)$ and $P(B_{\geq m})$ obtained in Problem 1.1.13, derive *Bonferroni formulas*: for any even integer number $r \geq 2$ one has

$$S_m + \sum_{k=1}^{r+1}(-1)^k C_{m+k}^k S_{m+k} \leq P(B_m) \leq S_m + \sum_{k=1}^{r}(-1)^k C_{m+k}^k S_{m+k},$$

$$S_m + \sum_{k=1}^{r+1}(-1)^k C_{m+k-1}^k S_{m+k} \leq P(B_{\geq m}) \leq S_m + \sum_{k=1}^{r}(-1)^k C_{m+k-1}^k S_{m+k},$$

where the quantities $S_1,\ldots,S_n$ are defined as in Problem 1.1.12.

*Hint.* One possibility is to prove first the following (also very useful) identities:

$$S_m = \sum_{r=m}^{n} C_r^m P(B_m), \quad S_m = \sum_{r=m}^{n} C_{r-1}^{m-1} P(B_{\geq m}).$$

**Problem 1.1.15.** By using the definition of the quantities $S_1,\ldots,S_n$ given in Problem 1.1.12, derive:

(a) *Bonferroni inequalities* (this is a special case of the formulas obtains in the previous problem): for any integer $k \geq 1$ with the property $2k \leq n$, one has

$$S_1 - S_2 + \ldots - S_{2k} \leq P\left(\bigcup_{i=1}^{n} A_i\right) \leq S_1 - S_2 + \ldots + S_{2k-1}.$$

(b) *Fréchet inequality*: for any integer $0 \leq r \leq n - 1$ one has

$$\frac{S_{r+1}}{C_n^{r+1}} \leq \frac{S_r}{C_n^r}.$$

(c) *Gumbel inequality*: for any integer $1 \leq r \leq n - 1$ one has

$$\frac{C_n^{r+1} - S_{r+1}}{C_{n-1}^r} \leq \frac{C_n^r - S_r}{C_{n-1}^{r-1}}.$$

**Problem 1.1.16.** (*"The matching problem."*) Given some fixed integer $n \geq 1$, consider the set of all possible permutations of the list $(1, \ldots, n)$, suppose that one permuatation is chosen at random from that set and denote this randomly chosen permutation by $(i_1, \ldots, i_n)$. Assuming that all permutations are equally likely to occur, i.e., each permutation is chosen with probability $1/n!$, prove that:

(a) The probability $P_{(m)}$ that *exactly* $m$ of the numbers $1, \ldots, n$, $1 \leq m \leq n$, appear in the permutation $(i_1, \ldots, i_n)$ in their own positions (i.e., in the same positions in which they appear in the list $(1, \ldots, n)$) is given by

$$\frac{1}{m!}\left(1 - \frac{1}{1!} + \frac{1}{2!} - \frac{1}{3!} + \ldots + (-1)^{n-m}\frac{1}{(n-m)!}\right) \quad \left(\approx \frac{e^{-1}}{m}, \ n \to \infty\right).$$

(b) The probability $P_{(\geq 1)}$ that *at least* one of the numbers $1, \ldots, n$ appears in the permutation $(i_1, \ldots, i_n)$ in its own position is given by

$$1 - \frac{1}{2!} + \frac{1}{3!} - \ldots + (-1)^{n-1}\frac{1}{n!} \quad (\approx 1 - e^{-1}, \ n \to \infty),$$

and, consequently, the probability for a complete "disorder" (i.e., a situation where none of the numbers $1, \ldots, n$ appears in its own position in the list $(i_1, \ldots, i_n)$) is given by $1 - P_{(\geq 1)} = \sum_{j=0}^{n} \frac{(-1)^j}{j!}$ ($\approx e^{-1}$ when $n \to \infty$).

*Hint.* For any $1 \leq i \leq n$ let $A_i$ denote the event that the number $i$ is located in the $i^{\text{th}}$ position of the list $(i_1, \ldots, i_n)$. The probability $P_{(m)}$ is then the same as the probability $P(B_m)$ in the previous problem, so that

$$P_{(m)} = S_m - C_{m+1}^1 S_{m+1} + \ldots + (-1)^{n-m}C_n^{n-m} S_n.$$

Showing that in the present setting one has $S_k = 1/k!$ for any $m \leq k \leq n$ would complete the proof.

In order to establish the formula for $P_{(\geq 1)}$, it is enough to notice that, using again the results from the previous problem, one has

$$P_{(\geq 1)} = \mathsf{P}(B_{\geq 1}) = S_1 - S_2 + S_3 - \ldots + (-1)^{n-1} S_n \, ,$$

where, in this case, $S_k = \frac{(n-k)!}{n!}$.

**Problem 1.1.17.** (*"The absent-minded secretary problem."*) There are $n$ different letters and $n$ envelopes addressed to the respective recipients of the letters. The secretary who prepares the letters is absent-minded and stuffs the letters into the envelopes at random. Assume the "classical," i.e., equal-likelihood-of-outcomes, definition of the probabilities involved (see [P §1.1, **5**]), and let $P_{(m)}$ denote the probability that *exactly m* of the letters will reach their (correct) recipients.
    Prove that

$$P_{(m)} = \frac{1}{m!} \left( 1 - \sum_{j=0}^{n-m} \frac{(-1)^j}{j!} \right).$$

*Hint.* 1. First, one must clarify the assumption that "the secretary stuffs the letters into the envelopes at random." If we are to assume that the secretary chooses at random one of the $n$ envelopes and stuffs the first letter into that envelope, then chooses at random one envelope from the remaining $n - 1$ envelopes and stuffs the second letter into that envelope, and so on, then the entire procedure would be tantamount to taking an ordered sample of size $n$ without repetion from the set of symbols $(a_1, \ldots, a_n)$ that represent the different envelopes and then making the assumption that any such sample is equally likely to occur, according to the principles described in [P §1.1, **5**]. This means that we have an experiment with $(n)_n = n!$ possible outcomes, every one of which occurs with probability $1/n!$.
    2. Denote by $A_i$ the event that the $i$-th letter is placed in its own envelope. Then $P_{(m)} = \mathsf{P}(B_m)$ (see Problem 1.1.13) and, consequently,

$$P_{(m)} = \sum_{k=m}^{n} (-1)^{k-m} C_k^m \, S_k.$$

After noticing that in this setting $S_k = 1/k!$, $1 \leq k \leq n$, one obtains the desired formula for $P_{(m)}$. Notice that the probability $P_{(0)}$ that none of the letters reaches its recipient equals $\sum_{k=1}^{n} (-1)^{k+1} \frac{1}{k!}$, which is close to $1 - e^{-1}$ even for relatively small values for $n$—for example, with $n = 5$ this sum equals $0.633333$, while $1 - e^{-1} \approx 0.632121$.

**Problem 1.1.18.** There are $n$ children in a given kindergarten. When leaving the kindergarten each child chooses at random one left and one right shoe. Prove that:
    (a) The probability $P_a$ that none of the children will bring home his or her own *pair* of shoes is given by

$$P_a = \sum_{k=1}^{n} (-1)^k \frac{(n-k)!}{k!\,n!}.$$

(b) The probability $P_b$ that none of the children will bring home at least one of his or her own shoes is given by

$$P_b = \left( \sum_{k=2}^{n} (-1)^k \frac{1}{k!} \right)^2.$$

*Hint.* First, one must give meaning to the phrase "each of the $n$ children chooses *at random* one left and one right shoe"—this can be done by following the principle outlined in the hint to Problem 1.1.17.

(a) Let $A_i$ denote the event that the $i^{\text{th}}$ child takes both his or her left and right shoes. According to the inclusion–exclusion formula, we have

$$P_a = P\left( \bigcap_{i=1}^{n} \overline{A_i} \right) = 1 - P\left( \bigcup_{i=1}^{n} A_i \right) = 1 - S_1 + S_2 - \ldots + (-1)^n S_n,$$

and one must show that in this case $S_k = \frac{(N-k)!}{k!\,n!}$, which gives the desired formula for $P_a$.

(b) In order to established the formula for $P_b$, it is enough to notice that $P_b$ is simply the product of the probability that none of the children brings home his or her left shoe and the probability none of the children brings home his or her right shoe, after which the statement in (b) follows with a straight-forward application of the result established in Problem 1.1.17.

**Problem 1.1.19.** There are $n$ particles that are distributed in $M$ boxes according to the *Maxwell–Boltzmann* statistics (placement without locks of distinguishable particles in distinguishable cells). By following the classical method of Laplace for counting probabilities (see [$\underline{P}$ §1.1, (10)]), which encodes, so to speak, "the random nature" of the placement of the particles, prove that the probability, $P_k(n; M)$, that exactly $k$ particles appear in any fixed cell is given by

$$P_k(n; M) = C_n^k \frac{(M-1)^{n-k}}{M^n}.$$

Conclude from the above formula that when $n \to \infty$ and $M \to \infty$ in such a way that $n/M \to \lambda > 0$, then

$$P_k(n; M) \to e^{-\lambda} \frac{\lambda^k}{k!}.$$

(Comp. with the Poisson distribution—see [$\underline{P}$ §1.6] and [$\underline{P}$ §2.3, Table 2].)

**Problem 1.1.20.** (*Continuation of Problem 1.1.19.*) Let $R_m(n; M)$ stand for the probability that exactly $m$ cells remain empty. Prove that

$$R_m(n; M) = C_M^n \sum_{k=0}^{M-m} (-1)^k C_{M-m}^k \left(1 - \frac{m+k}{M}\right)^n,$$

and conclude that if $n \to \infty$ and $M \to \infty$ in such a way that $Me^{-n/M} \to \lambda > 0$, then

$$R_m(n; M) \to e^{-\lambda} \frac{\lambda^m}{m!}.$$

**Problem 1.1.21.** Consider again a random placement of $n$ particles into $M$ cells, but according to the *Bose–Einstein* statistics (placement without locks of indistinguishable particles in distinguishable cells). Denote by $Q_k(n; M)$ the probability that there are exactly $k$ particles in any fixed cell. Prove that

$$Q_k(n; M) = \frac{C_{M+n-k-2}^{n-k}}{C_{M+n-1}^n}$$

and conclude that when $n \to \infty$ and $M \to \infty$ in such a way that $n/M \to \lambda > 0$, then

$$Q_k(n; M) \to p(1-p)^k, \quad \text{where } p = \frac{1}{1+\lambda}.$$

(Compare with the geometric distribution—see [P §2.3, Table 2].)

**Problem 1.1.22.** A box contains $N$ balls labeled $1, \ldots, N$. A ball is sampled $n$ times from the box *randomly* and *with repetition* (i.e., the ball is returned to the box after each sample). Given any fixed $k \in \{1, \ldots, N\}$, let $A_k$ denote the event that the largest label found among the sampled balls equals $k$. Prove that

$$P(A_k) = \frac{k^n - (k-1)^n}{N^n}.$$

In addition, prove that if the balls are sampled *randomly* but *without repetition*, then for any $n \le k \le N$ one has

$$P(A_k) = \frac{C_{k-1}^{n-1}}{C_N^k}.$$

**Problem 1.1.23.** Verify the *Leibniz formula* for the $N^{\text{th}}$ derivative of the product of two functions $f$ and $g$:

$$D^N(fg) = \sum_{n=0}^{N} C_N^n (D^n f)(D^{N-n} g).$$

*Hint.* Consider using induction and the property $C_{N+1}^n = C_N^n + C_N^{n-1}$, i.e., the so called "Pascal triangle property."

## 1.2 Some Classical Models and Distributions

**Problem 1.2.1.** Prove that:

$$(x+y)^n = \sum_{k=0}^n C_n^k \, x^k \, y^{n-k} \quad \textit{(binomial identity)},$$

$$(x+y)_n = \sum_{k=0}^n C_n^k \, (x)_k \, (y)_{n-k} \quad \textit{(Vandermonde's identity)},$$

$$[x+y]_n = \sum_{k=0}^n C_n^k \, [x]_k \, [y]_{n-k} \quad \textit{(Nørlund's identity)},$$

where

$$(x)_n = x(x-1)(x-2)\dots(x-n+1),$$
$$[x]_n = x(x+1)(x+2)\dots(x+n-1).$$

*Hint.* Consider using Taylor's expansion for polynomials.

**Problem 1.2.2.** By using probabilistic, combinatorial, or geometric arguments (say, by counting the number of favorable outcomes, or, counting the number of paths that connect one point with another), or some other type of reasoning (say, by way of some algebraic argument analogous to identifying the coefficients for $x^n$ in identities of the form $(1+x)^a(1+x)^b = (1+x)^{a+b}$)), verify the following claims about the binomial coefficients (below $\lfloor x \rfloor$ denotes the integer part of the real number $x$, i.e., the largest integer number which is not greater than $x$, while $\lceil x \rceil$ denotes the smallest integer number which is not smaller than $x$):

- $1 = C_n^0 < C_n^1 < \dots < C_n^{\lfloor n/2 \rfloor} = C_n^{\lceil n/2 \rceil} > \dots > C_n^{n-1} > C_n^n = 1$

  (*symmetry* and *unimodality*);

- $C_N^{k-1} + C_N^k = C_{N+1}^k \quad$ (*this is the Pascal triangle rule*)

$$
\begin{array}{ccccccccccccc}
 & & & & & & 1 & & & & & & \\
 & & & & & 1 & & 1 & & & & & \\
 & & & & 1 & & 2 & & 1 & & & & \\
 & & & 1 & & 3 & & 3 & & 1 & & & \\
 & & 1 & & 4 & & 6 & & 4 & & 1 & & \\
 & 1 & & 5 & & 10 & & 10 & & 5 & & 1 & \\
1 & & 6 & & 15 & & 20 & & 15 & & 6 & & 1
\end{array} \quad ;
$$

... ... ... ... ... ... ... ... ... ... ...

- $C_n^k = C_{n-2}^k + 2C_{n-2}^{k-1} + C_{n-2}^{k-2}$ ;

- $\displaystyle\sum_{k=0}^{N} C_N^k = 2^N, \qquad \sum_{k=0}^{N}(C_N^k)^2 = C_{2N}^N$ ;

- $\displaystyle\sum_{k=0}^{N} 2^k C_N^k = 3^N, \qquad \sum_{k=1'}^{N} k C_N^k = N 2^{N-1}$ ;

- $\displaystyle\sum_{k=0}^{N}(-1)^{N-k} C_M^k = C_{M-1}^N, \quad M \geq N+1$ ;

- $\displaystyle\sum_{k=0}^{N} k(k-1) C_N^k = N(N-1) 2^{N-2}, \quad N \geq 2$ ;

- $kC_N^k = N C_{N-1}^{k-1}, \qquad \displaystyle\sum_{m=k}^{N} C_m^k = C_{N+1}^{k+1}$ ;

- $C_N^k C_k^l = C_N^l C_{N-l}^{k-l}, \quad l \leq k \leq N$ ;

- $\displaystyle\sum_{j=0}^{k} C_N^j = \sum_{j=0}^{k} 2^j C_{N-1-j}^{k-j}, \quad k \leq N-1$ ;

- $\displaystyle\sum_{j=0}^{k} C_{N+j}^j = C_{N+k+1}^k$ ;

- $C_{N-k}^k + C_{N-k-1}^{k-1} = \dfrac{N}{N-k} C_{N-k}^k, \quad 0 \leq k \leq N$ ;

- $C_{N_1+N_2}^n = \displaystyle\sum_{k=0}^{N_1} C_{N_1}^k C_{N_2}^{n-k}$   (*Vandermonde's binomial convolution*) ;

- $C_N^{n-1} \leq C_N^n, \quad 1 \leq n \leq \dfrac{N+1}{2}$ ;

- $C_N^{n-1} C_N^{n+1} \leq (C_N^n)^2, \quad n \leq N-1$ ;

- $C_{M+N}^N \leq \left(1 + \dfrac{M}{N}\right)^N \left(1 + \dfrac{N}{M}\right)^M$ ;

  or, equivalently,   $\dfrac{(M+N)!}{(M+N)^{M+N}} \leq \dfrac{M!\,N!}{M^M N^N}$ ;

- $$\frac{N^k}{k^k} \le C_N^k \le \frac{N^k}{k!};$$

- $$\sum_{k=0}^{N} C_N^k (-1)^{N-k} 2^k = 1;$$

- $$C_N^n = \sum_{k=n}^{N} (-1)^{N+k} C_N^k C_k^n 2^{k-n}, \quad 1 \le n \le N;$$

- $$\sum_{k=0}^{N} C_N^k C_k^n = 2^{N-n} C_N^n, \quad \sum_{k=0}^{N} C_{2N+1}^{2k+1} C_{2N}^{n+k} = C_{2N}^{2n}, \quad 1 \le n \le N;$$

- $$C_{N-1}^{M-1} = C_N^M - C_N^{M+1} + \cdots \pm C_N^N, \quad M \le N;$$

- $$\sum_{k=0}^{M} (-1)^k C_N^k = (-1)^M C_{N-1}^M;$$

- $$\sum_{k=0}^{N} (-1)^k (C_N^k)^2 = \begin{cases} (-1)^m C_{2m}^m, & \text{if } N = 2m \\ 0 & \text{if } N \ne 2m \end{cases};$$

- $$\sum_{k=0}^{N} (-1)^k (C_N^k)^3 = \begin{cases} (-1)^m (3m)! \, (m!)^{-3}, & \text{if } N = 2m \\ 0 & \text{if } N \ne 2m \end{cases};$$

- $$\sum_{k=1}^{N} C_N^k \frac{(-1)^{k-1}}{k} = \sum_{k=1}^{N} \frac{1}{k};$$

- $$\sum_{k=0}^{N} (-1)^{N-k} k^l C_N^k = \begin{cases} 0, & \text{if } l < N \\ N!, & \text{if } l = N \end{cases};$$

- $$\sum_{k=0}^{N} C_{2(N-k)}^{N-k} C_{2k}^k = 2^{2N}.$$

(See also Problem 1.2.22.)

**Problem 1.2.3.** Prove that if $p$ is a prime number and $1 \le k \le p - 2$, then $p$ divides $C_p^k$ and one has $C_{2p}^p = 2 \pmod{p}$.

**Problem 1.2.4.** Prove that the number of different ways in which a set of $N$ objects can be split into no more than two disjoint sets, the order in the sets being irrelevant, equals $\lfloor N/2 \rfloor + 1$, where $\lfloor x \rfloor$ is the integer part of the real number $x$.

**Problem 1.2.5.** Given $N \ge n \ge 1$, the *Stirling number of the second kind*, $S_N^n$, is defined as the number of all possible partitions of a set of $N$ objects, say, the set

$\{1, 2, \ldots, N\}$, into $n$ disjoint and non-empty sets with no regard of the order in the sets.[1]

Prove that the following relations are in force:

(a)   $S_N^1 = S_N^N = 1$,     $S_N^2 = 2^{N-1} - 1$,     $S_N^{N-1} = C_N^2$;

(b)   $S_{N+1}^n = S_N^{n-1} + n S_N^n$,        $1 \le n \le N$;

(c)   $S_{N+1}^n = \sum_{k=0}^{N} C_N^k S_k^{n-1}$    $(S_k^p = 0 \ p > k)$;

(d)   $S_N^n = \dfrac{1}{n!} \sum_{k=0}^{N-1} (-1)^k C_n^k (n-k)^N$;

(e)   $S_N^n = \dfrac{1}{n!} \sum_{k=0}^{n} (-1)^{n-k} C_n^k k^N$;

(f)   $S_N^n < n^{N-n} C_{N-1}^{n-1}$.

*Hint.* To prove (b), which is the key to deriving (c) and (d), use the relation $x^N = \sum_{n=0}^{N} S_N^n (x)_n$ (see page 376 in the Appendix) and the relation $(x)_{n+1} = (x - n)(x)_n$.

**Problem 1.2.6.** By using the relation $x^N = \sum_{n=0}^{N} S_N^n (x)_n$, it is shown in Sect. A.3 that the exponential generating function

$$E_{S_\cdot^n}(x) \equiv \sum_{N \ge 0} S_N^n \frac{x^N}{N!},$$

associated with the sequence $S_\cdot^n = (S_N^n)_{n \ge 0}$, consisting of Stirling numbers of the second kind, has the property

$$E_{S_\cdot^n}(x) = \frac{(e^x - 1)^n}{n!}.$$

Prove the above identity by using property (e) in the previous problem.

---

[1] The definitions and some basic facts concerning the Stirling numbers (of the first and the second kind), and also of the Bell numbers, can be found in Sect. A.1.

**Problem 1.2.7.** According to one of the definitions of *the Stirling numbers of the first kind*, $s_N^n$, (see page 377 in Sect. A.3), the number $(-1)^{N-n} s_N^n$ gives the total number of permutations of the set $\{1, 2, \ldots, N\}$ with exactly $n$ cycles (note that $s_N^0 \equiv 0$).

Prove that

(a) $\quad s_N^1 = s_N^N = 1,$

(b) $\quad s_{N+1}^n = s_N^{n-1} - N s_N^n, \quad 1 \le n \le N,$

(c) $\quad \displaystyle\sum_{n=1}^{N} (-1)^{N-n} s_N^n = \sum_{n=1}^{N} |s_N^n| = N!.$

In addition, prove that the numbers $s_N^n$, $0 \le n \le N$, satisfy the following algebraic relation (see page 377)

(d) $\quad \displaystyle (x)_N = \sum_{n=0}^{N} s_N^n x^n,$

where $(x)_N \equiv x(x-1) \ldots (x - N + 1)$.

*Hint.* The recursive relation (b) may be established by way of combinatorial reasoning. Alternatively, it may be derived directly from the algebraic relation (d).

**Problem 1.2.8.** Prove the following *duality* property of the Stirling numbers of the first and second kinds:

$$\sum_{n \ge 0} S_N^n s_n^M = \delta_{NM},$$

where $\delta_{ab}$ is the Kronecker symbol associated with the quantities $a$ and $b$, i.e., $\delta_{ab} = 1$ if $a = b$ and $\delta_{ab} = 0$ if $a \ne b$.

**Problem 1.2.9.** Prove that the exponential generating function

$$E_{s_{\cdot}^n}(x) \equiv \sum_{N \ge 0} s_N^n \frac{x^N}{N!},$$

associated with the sequence $s_{\cdot}^n = (s_N^n)_{N \ge 0}$, comprised of Stirling numbers of the first kind, is given by the following formula

$$E_{s_{\cdot}^n}(x) = \frac{(\ln(1+x))^n}{n!}.$$

**Problem 1.2.10.** Given any $N \ge 1$, the Bell number $B_N$ is defined as (see page 362 in the Appendix) the number of all possible partitions of the set $\{1, 2, \ldots, N\}$, or, which amounts to the same,

$$B_N \equiv \sum_{n=1}^{N} S_N^n \,,$$

where $S_N^n$, $1 \leq n \leq N$ are the Stirling numbers of the second kind.

Setting $B_0 \equiv 1$, prove that:

(a) The following recursive relation is in force

$$B_N = \sum_{k=1}^{N} C_{N-1}^{k-1} \, B_{N-k} \,.$$

(b) The exponential generating function $E_B(x) \equiv \sum_{N \geq 0} B_N \frac{x^N}{N!}$ is given by the formula

$$E_B(x) = \exp\{e^x - 1\} \,.$$

(c) $B_N < N!$ and $\lim_{N \to \infty} (B_N / N!)^{1/N} = 0$.

Finally, verify that the numbers $B_1, \dots, B_5$ equal, respectively, 1, 2, 5, 15, 52.

*Hint.* To prove (b), use (a) and check that the function $x \longrightarrow E_B(x)$ satisfies the following first-order equation

$$\frac{dE_B(x)}{dx} = e^x E_B(x) \,,$$

with boundary condition $E_B(0) = 1$. In order to prove the second property in (c), consider the radius of convergence $R = 1 / \overline{\lim} \left( \frac{B_N}{N!} \right)^{1/N}$ for the series $\sum_{N \geq 0} B_N \frac{x^N}{N!}$, which, as is easy to see, converges for all real $x$.

**Problem 1.2.11.** (*Fibonacci numbers.*) Given any integer $n \geq 1$, let $F_n$ denote the number of all possible representations of the number $n$ as the sum of an *ordered* list of 1's and 2's. Thus, one has $F_1 = 1$, $F_2 = 2$ (since $2 = 1 + 1 = 2$), $F_3 = 3$ (since $3 = 1 + 1 + 1 = 1 + 2 = 2 + 1$), $F_4 = 5$ (since $4 = 1 + 1 + 1 + 1 = 2 + 1 + 1 = 1 + 2 + 1 = 1 + 1 + 2 = 2 + 2$), and so on.

(a) Setting $F_0 \equiv 1$, prove that for any $n \geq 2$ the Fibonacci numbers $F_n$ satisfy the following recursive relation

$$F_n = F_{n-1} + F_{n-2} \,, \quad n \geq 2. \tag{$*$}$$

(b) By using the above relation, prove that

$$F_n = \frac{1}{\sqrt{5}} \left[ \left( \frac{1 + \sqrt{5}}{2} \right)^{n+1} - \left( \frac{1 - \sqrt{5}}{2} \right)^{n+1} \right]. \tag{$**$}$$

Notice that $\frac{1+\sqrt{5}}{2} \approx 1.6180339887\dots$ and $\frac{1-\sqrt{5}}{2} \approx -0.6180339887\dots$ .

(c) By using $(*)$, prove that the generating function $F(x) \equiv \sum_{n\geq 0} F_n x^n$, associated with the sequence $(F_n)_{n\geq 0}$, is given by the formula

$$F(x) = \frac{1}{1 - x - x^2} . \qquad (***)$$

(d) The Fibonacci numbers[2] have many interesting properties. For example, setting $F_{-1} \equiv 0$, for any choice of the integers $m, n \geq 0$ one can write:

$$F_0 + F_1 + \ldots + F_n = F_{n+1} - 1, \qquad F_{n-1}^2 + F_n^2 = F_{2n},$$

$$F_{n-1} F_n + F_n F_{n+1} = F_{2n+1}, \qquad F_m F_n + F_{m-1} F_{n-1} = F_{m+n} ,$$

Verify the last four identities.

(e) Prove that for any $n \geq 0$ the $n^{\text{th}}$ Fibonacci number is given by

$$F_n = \sum_{k=0}^{\lfloor n/2 \rfloor} C_{n-k}^k .$$

For example, convince yourself that the list of the first 18 Fibonacci numbers $\{F_0, F_1, \ldots, F_{17}\}$ is given by

$$\{1, 1, 2, 3, 5, 8, 13, 21, 34, 55, 89, 144, 233, 377, 610, 987, 1597, 2584\}.$$

(f) Prove that for any $n \leq 9$ one has

$$\frac{F_n}{\lceil e^{(n-1)/2} \rceil} = 1,$$

but for $n = 10$ one has $F_{10} / \lceil e^{9/2} \rceil < 1$ and

$$\lim_{n \to \infty} \frac{F_n}{\lceil e^{(n-1)/2} \rceil} = 0. \qquad (****)$$

*Hint.* (b) To prove $(**)$, start by looking for a sequences of the form $(F_n \equiv \alpha^n)_{n\geq 1}$ that satisfies the recursive relation $F_n = F_{n-1} + F_{n-2}$. The formula $(**)$ may be obtained also by considering the coefficients for $x^n$ in Taylor's expansion of the function $(***)$. In this context it is useful to notice that $1 - x - x^2 = (1-ax)(1-bx)$, for $a = (1 + \sqrt{5})/2$ and $b = (1 - \sqrt{5})/2$.

---

[2]Tradtionally linked to the population growth of a colony of rabbits, and described as early as the thirteenth century AD, by *Leonardus Pisanus de filiis Bonaccii*, widely known under the nickname "*Fibonacci*," in his book "Liber Abaci," probably written around 1202 CE.

(f) To prove (****), notice that (**) implies $F_n \approx c_1 (1.618\ldots)^n$, while $\lceil e^{(n-1)/2} \rceil \approx c_2 (1.648\ldots)^n$, with some appropriate constants $c_1$ and $c_2$. Try to find these two constants.

**Problem 1.2.12.** Prove that the multinomial (polynomial) coefficients

$$C_N(n_1,\ldots,n_r) = \frac{N!}{n_1!\ldots n_r!}, \quad n_1 + \ldots + n_r = N, \quad n_i \geq 0,$$

satisfy the following formula, known as *Vandermonde's multinomial convolution formula*:

$$C_{N_1+N_2}(n_1,\ldots,n_r) \equiv \sum C_{N_1}(k_1,\ldots,k_r) C_{N_2}(n_1-k_1,\ldots,n_r-k_r),$$

the summation being taken over all possible choices of the integers $\{k_i ; i = 1,\ldots,r\}$, so that $0 \leq k_i \leq n_i$, for any $i = 1,\ldots,r$, $\sum_{i=1}^{r} k_i = N_1$, and $n_1 + \ldots + n_r = N_1 + N_2$.

**Problem 1.2.13.** Prove that

$$(x_1 + \ldots + x_r)^N = \sum C_N(n_1,\ldots,n_r) x_1^{n_1} \ldots x_r^{n_r},$$

the summation being taken over all possible choices for the integers $n_1,\ldots,n_r$, in such a way that $n_i \geq 0$, for any $i = 1,\ldots,r$, and $\sum_{i=1}^{r} n_i = N$.

**Problem 1.2.14.** Prove that the number of *nondecreasing* paths on the integer lattice $\mathbb{Z}_+^r = \{(i_1,\ldots,i_r) : i_1,\ldots,i_r = 0,1,2,\ldots\}$ that start at the origin $(0,\ldots,0)$ and end at some point $(n_1,\ldots,n_r)$ with $\sum_{i=1}^{r} n_i = N$, equals $C_N(n_1,\ldots,n_r)$. (A path on the lattice $\mathbb{Z}_+^r$ is said to be nondecreasing if at every step only one of the coordinates changes by $+1$.)

**Problem 1.2.15.** Consider the sets $A$ and $B$, chosen so that the numbers of their elements, resp. $N \equiv |A|$ and $M \equiv |B|$, are both finite, and:

let $\mathbb{F}: A \mapsto B$ denote any *function* from $A$ to $B$, i.e., any rule that assigns a unique $b \in B$ to any $a \in A$ (one and the same $b \in B$ can be assigned to many $a \in A$);

let $\mathbb{I}: A \mapsto B$ denote any *injection* of $A$ into $B$, i.e., any rule that assigns to different elements of $A$ different elements of $B$, so that no two elements of $A$ are assigned one and the same element from $B$ (for this to be possible one must have $|A| \leq |B|$);

let $\mathbb{S}: A \mapsto B$ denote any *surjection* of $A$ into $B$, i.e., any function from $A$ into $B$ with the property that for every $b \in B$ there is at least one $a \in A$ with $\mathbb{S}(a) = b$ (for this to be possible one must have $|A| \geq |B|$);

and, finally, let $\mathbb{B}: A \to B$ denote any *bijection* from $A$ into $B$, i.e., any function from $A$ into $B$ which is both surjection and injection (for this to be possible one must have $|A| = |B|$);

Prove that the total number of: all functions from $A$ into $B$, of all injections of $A$ into $B$, of all surjections from $A$ onto $B$, and of all bijections between $A$ and $B$, are given, respectively, by:

$$N(\mathbb{F}) = M^N, \quad N(\mathbb{I}) = (M)_N, \quad N(\mathbb{S}) = M! \, S_N^M, \quad \text{and} \quad N(\mathbb{B}) = N!.$$

**Problem 1.2.16.** Prove that the numbers $P_N = \sum_{n=0}^{N}(N)_n$, $N \geq 0$, with $(N)_0 = 1$ and $(N)_n = N(N-1)\ldots(N-n+1)$, satisfy the following recursive relation:

$$P_N = N P_{N-1} + 1, \qquad N \geq 1.$$

In addition, prove that

$$P_N = N! \sum_{n=0}^{N} \frac{1}{n!},$$

and that $P_N$ is the nearest integer to $e N!$.

**Problem 1.2.17.** Prove that the exponential generating function

$$E_P(x) = \sum_{N=0}^{\infty} P_N \frac{x^N}{N!},$$

associated with the sequence $P = (P_N)_{N \geq 0}$, which is defined in the previous problem, satisfies the relation

$$E_P(x) = \frac{e^x}{1-x}.$$

**Problem 1.2.18.** An urn contains $M$ balls labeled $1, 2, \ldots, M$. Each ball is painted in either red or blue. Let $M_1$ denote the number of red balls in the urn and let $M_2$ denote the number of blue balls in the urn ($M_1 + M_2 = M$). Consider an unordered sample from the urn without a replacement of size $n = n_1 + n_2 < M$ and let $B_{n_1,n_2}$ denote the random event that there are exactly $n_1$ red and $n_2$ blue balls in the sample. Suppose that $M \to \infty$, $M_1 \to \infty$ and $M_2 \to \infty$ in such a way that, for some finite number $0 < p < 1$, one has $M/M_1 \to p$ and $M/M_2 \to 1 - p$. Prove that

$$P(B_{n_1,n_2}) \to C_{n_1+n_2}^{n_2} \, p^{n_1} (1 - p)^{n_2}.$$

*Hint.* Use the identity

$$P(B_{n_1,n_2}) = \frac{C_n^{n_1} M_1^{n_1} M_2^{n_2}}{M^n}.$$

**Problem 1.2.19.** Prove that in the multinomial distribution $\{P(A_{n_1,\ldots,n_r})\}$ the probability $\{P(A_{n_1,\ldots,n_r})\}$ is the largest when the list $(k_1, \ldots, k_r)$ is chosen so that $np_i - 1 < k_i \leq (n + r - 1)p_i$, $i = 1, \ldots, r$.

**Problem 1.2.20.** (*One-dimensional Ising model.*) There are $n$ particles placed at locations $1, \ldots, n$. Each particle is either type-1 particle or type-2 particle. The total number of type-1 particles is $n_1$, the total number of type-2 particles is $n_2$ ($n_1 + n_2 = n$) and all $n!$ placements of the particles are equally likely.

Describe the associated probabilistic model and compute the probability of the event $A_n(m_{11}, m_{12}, m_{21}, m_{22}) = \{v_{11} = m_{11}, \ldots, v_{22} = m_{22}\}$, where $v_{ij}$ denotes the total number of type-$i$ particles that are placed immediately after a type $j$ particle $(i, j = 1, 2)$.

**Problem 1.2.21.** Suppose that one must estimate the size $N$ of a certain population and that the estimation effort must be "minimal"; in particular, straight counting of all individuals in the population cannot be used as a method. Such problems are of interest when one must estimate, for example, the total number of citizens in a given country, large city, etc.

In 1786 Pierre-Simon Laplace proposed the following method for estimating the total number $N$ of all French citizens:

Take some number, say, $M$, of French citizens and record their names. Then return those citizens back in the general population so that they are "perfectly mixed" with unrecorded individuals. Then choose a "perfectly random" sample of $n$ individuals and denote by $X$ the total number of recorded individuals in that sample.

(a) Given some fixed $N$, $M$ and $n$, prove that the probability $\mathsf{P}_{N,M;n}\{X = m\}$, i.e., the probability that the number of recorded individuals in the sample is exactly equal to $m$, is given by the formula for the hyper-geometric distribution (see [P §1.2, (4)]):

$$\mathsf{P}_{N,M;n}\{X = m\} = \frac{C_M^n \, C_{N-M}^{n-m}}{C_N^n}.$$

(b) For some fixed $M$, $n$ and $m$ find the maximum of $\mathsf{P}_{N,M;n}\{X = m\}$ for various choices of $N$. If $\widehat{N}$ denotes the value for $N$ at which that maximum is achieved, i.e., if $\widehat{N}$ is the "most likely" size of the entire population, given that the number of recorded individuals in the sample is $m$ (this is also known as *the maximim likelihood estimate of $N$*), prove that

$$\widehat{N} = \left\lfloor \frac{Mn}{m} \right\rfloor,$$

where $\lfloor \cdot \rfloor$ is the "integer part" function. (This problem continues in Problem 1.7.4.)

**Problem 1.2.22.** In the (elementary) combinatorial theory the binomial coefficients $C_M^n \equiv \frac{(M)_n}{n!} = \frac{M!}{n!(M-n)!}$ (denoted equivalently by $\binom{M}{n}$) and the number of ordered samples $(M)_n \equiv M(M-1) \ldots (M-n+1)$ are defined usually for *integer* numbers $n, M \in \mathbb{N} = \{1, 2, \ldots\}$. In some areas of analysis it is often useful to define "the number of ordered samples $(M)_n$" and "the binomial coefficient $C_M^n$," with $M$ replaced by some arbitrary $X \in \mathbb{R}$. Assuming that $n \in \{0, \pm 1, \pm 2, \ldots\}$ define $0! = 1$, $(X)_0 = 1$, $C_X^0 = 1$, and the define

$$(X)_n = X(X-1)\ldots(X-n+1), \quad C_X^n = \frac{(X)_n}{n!}, \quad \text{for any } n > 0,$$

and $C_X^n = 0$, for any $n < 0$. In conjunction with the above definitions (and some of the relations established in Problem 1.2.2) prove the following identities for arbitrary $X, Y \in \mathbb{R}$ and $n \in \mathbb{Z} = \{0, \pm 1, \pm 2, \ldots\}$):

$$C_X^{n-1} + C_X^n = C_{X+1}^n \quad \text{(Pascal triangle property)};$$

$$C_{X+Y}^n = \sum_{k=0}^n C_X^k C_Y^{n-k} \quad \left(\begin{array}{l}\textit{Vandermonde's binomial}\\ \textit{convolution}\end{array}\right),$$

$$C_{X-1}^n = \sum_{k=0}^n (-1)^{n-k} C_X^k;$$

$$C_{n-X}^{n-m} = \sum_{k=0}^{n-m} (-1)^k C_X^k C_{n-k}^m;$$

$$C_{X+Y+n-1}^n = \sum_{k=0}^n C_{X+n-k-1}^{n-k} C_{Y+k-1}^k;$$

$$C_{-X}^n = (-1)^n C_{X+n-1}^n.$$

**Problem 1.2.23.** Consider ordered samples without repetition of size $M$, taken from an urn that has $N \geq 2$ balls, of which $n \geq 2$ are white and $N - n$ are black. Let $A_{i,j}$ be the event that the $i^{\text{th}}$ and the $j^{\text{th}}$ balls in the sample are white, $i < j \leq M$, and let $A_{i,j,k}$ be the event that the $i^{\text{th}}$, the $j^{\text{th}}$ and the $k^{\text{th}}$ balls in the sample are white, $i < j < k \leq M$. Compute the probability of the events $A_{i,j}$ and $A_{i,j,k}$.

**Problem 1.2.24.** Find a formula for the probability $P_n$ of having $n$ spades in a hand of 13 cards, taken at random from a full deck of 52 playing cards.

**Problem 1.2.25.** Consider $n \geq 3$ different points on a circle and suppose that 2 of these points are chosen at random. What is the probability that these two points are "neighbors"?

**Problem 1.2.26.** (*"The married couples problem," a.k.a. "problème des ménage."*) In how many different ways can $n$ married couples ($n \geq 3$) be seated at a round table in such a manner that men and women alternate, i.e., there are no two men or two women sitting next to each other, and, at the same time, there are no husband and wife sitting next to each other?

  *Hint.* Suppose that the seats around the table are labeled (say, clockwise) $1, \ldots, 2n$, and that seat 1 is always occupied by a woman. Given some $1 \leq k \leq 2n$, let $A_k$ denote the event that seats $k$ and $k+1$ are occupied by some married couple, with the understanding that seat $2n+1$ is identified with seat 1. Then the event that there are no husband and wife sitting next to each other can be expressed as

$\bigcap_{k=1}^{2n} \overline{A}_k$. By the inclusion–exclusion formula (see Part (b) of Problem 1.1.12) one can write

$$P\left( \bigcap_{k=1}^{2n} \overline{A}_k \right) = 1 - P\left( \bigcup_{k=1}^{2n} A_k \right) = 1 - \sum_i P(A_i) + \sum_{i<j} P(A_i \cap A_j) - \dots .$$

A straight-forward calculation shows that, for any $1 \le i \le 2n$, one has

$$P(A_i) = n\left( \frac{(n-1)!}{n!} \right)^2 ,$$

for any $1 \le i < j \le 2n$ one has

$$P(A_i \cap A_j) = \begin{cases} n(n-1)\left( \frac{(n-2)!}{n!} \right)^2 , & \text{if } |i-j| \ne 1, \\ 0, & \text{if } |i-j| = 1, \end{cases}$$

where $P(A_1 \cap A_{2n}) = 0$, and, in general, for any $i_1 < \dots < i_k$ one has

$$P(A_{i_1} \cap \dots \cap A_{i_k}) = \begin{cases} \frac{n!}{(n-k)!}\left( \frac{(n-k)!}{n!} \right)^2 , & \text{if } |i_{j+1} - i_j| \ge 2 \text{ for } 1 \le k \le k, \\ & \text{and } 2n + i_1 - i_k \ge 2, \\ 0 & \text{in all other cases.} \end{cases}$$

Consequently,

$$P\left( \bigcap_{k=1}^{2n} \overline{A}_k \right) = \sum_{k=0}^{n} (-1)^k \frac{(n-k)!}{n!} d_n^k ,$$

where $d_n^k$ denotes the number of all possible choices of $k$ non-intersecting pairs of neighboring seats (the pairs $(i, i+1)$ and $(j, j+1)$ are said to be non-intersecting if, either $i+1 < j$, or $j+1 < i$). After showing that

$$d_n^k = C_{2n-k}^k \frac{2n}{2n-k} ,$$

one arrives at the following conclusion: the probability that no married couple is seated on two neighboring seats is given by

$$\frac{1}{n!} \sum_{k=0}^{n} (-1)^k (n-k)! \frac{2n}{2n-k} C_{2n-k}^k .$$

**Problem 1.2.27.** (*Latin squares.*) A Latin square of size $n \times n$ is simply a square matrix of size $n \times n$ which is filled with the numbers $1, 2, \dots, n$ in such a way that each of these numbers appears precisely once in every column and precisely once in every row. For example,

$$\boxed{\begin{array}{cc}1 & 2\\ 2 & 1\end{array}} \quad \text{and} \quad \boxed{\begin{array}{cc}2 & 1\\ 1 & 2\end{array}}$$

are Latin squares of size $2 \times 2$, while

$$\boxed{\begin{array}{ccc}1 & 2 & 3\\ 2 & 3 & 1\\ 3 & 1 & 2\end{array}} \quad \text{and} \quad \boxed{\begin{array}{ccc}1 & 2 & 3\\ 3 & 1 & 2\\ 2 & 3 & 1\end{array}}$$

are Latin squares of size $3 \times 3$. If $L_n$ stands for the total number of all Latin squares of size $n \times n$, prove that

$$L_n \geq n!\,(n-1)!\ldots 1! \quad \left(= \prod_{k=1}^{n} k!\right).$$

*Remark.* One can show, for example, that $L_2 = 2$, $L_3 = 12$, $L_4 = 576$, etc.; however, an exact general formula for $L_n$ is rather difficult to obtain. Nevertheless, the following asymptotic result is well known:

$$\ln L_n = n^2 \ln n + O(n^2), \qquad \text{as} \quad n \to \infty.$$

**Problem 1.2.28.** (*G. Pólya's urn scheme.*) Suppose that an urn contains $r$ red and $b$ black balls and consider the following trial: one ball is drawn at "random" from the urn, after which that ball and a *new* ball of the same color are placed back into the urn. Suppose that this trial is repeated many times and let $S_n$ denote the number of red balls that have been drawn from the urn during the first $n$ trials. Prove that

$$P\{S_n = x\} = \frac{C_{r+x-1}^{r-1}\, C_{b+n-x-1}^{b-1}}{C_{r+b+n-1}^{n}}, \qquad 0 \leq x \leq n.$$

**Problem 1.2.29.** In the context of Pólya's urn scheme described in the previous problem, set

$$p = \frac{r}{r+b}, \qquad q = \frac{b}{r+b}, \qquad \gamma = \frac{1}{r+b}.$$

and suppose that when $n \to \infty$ one has $p \to 0$ and $\gamma \to 0$ in such a way that $np \to \lambda$ and $n\gamma \to 1/\rho$. Prove that for any fixed $x$ one has

$$P\{S_n = x\} \to C_{\lambda\rho+x-1}^{x} \left(\frac{\rho}{1+\rho}\right)^{\lambda\rho} \left(\frac{1}{1+\rho}\right)^{x} \quad \text{as } n \to \infty.$$

**Problem 1.2.30.** Consider the random placement of $2n$ balls, of which $n$ are white and $n$ are black, into $m$ boxes, labeled $1, \ldots, m$. The probability for a black ball to be placed in the $j^{\text{th}}$ box is $p_j$ $(p_1 + \cdots + p_m = 1)$ and the probability for a white ball to be placed in the $j^{\text{th}}$ box is $q_j$ $(q_1 + \cdots + q_m = 1)$. Let $\nu$ denote the number of boxes that contain *exactly* one white and one black balls. Calculate the probability $P\{\nu = k\}, k = 0, 1, \ldots, m$, and the expected value $E\nu$.

**Problem 1.2.31.** (*On Stirling's formula—see also Problem 1.3.16 and Problem 8.8.1.*) By the well known asymptotic series expansion for the gamma function, one has

$$n! = \sqrt{2\pi n} \left(\frac{n}{e}\right)^n \left(1 + \frac{1}{12n} + \frac{1}{288n^2} - \frac{139}{5140n^3} + O\left(\frac{1}{n^4}\right)\right).$$

By using the relations

$$\ln n! = \sum_{k=2}^{n} \ln k \quad \text{and} \quad \ln(n-1)! < \int_{1}^{n} \ln t \, dt < \ln n!,$$

in which $\int_{1}^{n} \ln t \, dt = n \ln n - n + 1$, derive the following (rough) lower and upper bounds for $n!$:

$$e\left(\frac{n}{e}\right)^n < n! < en\left(\frac{n}{e}\right)^n, \tag{$*$}$$

which leads to *Stirling's formula*:

$$n! \sim \sqrt{2\pi n} \left(\frac{n}{e}\right)^n.$$

**Problem 1.2.32.** (*On the asymptotic decomposition of harmonic numbers.*) A harmonic number is a number of the form $H_n = \sum_{k=1}^{n} \frac{1}{k}, n \geq 1$. From the well known asymptotic expansion of the digamma function, one has

$$H_n = \ln n + \gamma + \frac{1}{2n} - \frac{1}{12n^2} + \frac{1}{120n^4} + O\left(\frac{1}{n^6}\right),$$

where $\gamma = 0.5772\ldots$ is the Euler constant (a.k.a. the Euler-Mascheroni constant).

By using the method developed in the previous problem, for estimating certain sums in terms of integrals, prove that for any $n \geq 1$ one has

$$\ln n + \frac{1}{n} \leq H_n \leq \ln n + 1,$$

and conclude that $\lim_n (H_n / \ln n) = 1$.

## 1.3 Conditional Probability: Independence

**Problem 1.3.1.** Prove by way of example that, in general, the following identities do not hold:

$$P(B \mid A) + P(B \mid \overline{A}) = 1,$$

$$P(B \mid A) + P(\overline{B} \mid \overline{A}) = 1.$$

**Problem 1.3.2.** An urn contains $M$ balls of which $M_1$ are white. Consider a random sample of size $n$ and let $B_j$ denote the event that the ball taken at the $j^{th}$ drawing in the sample is white. Let $A_k$ denote the event that there are exactly $k$ white balls in the entire sample of size $n$. Prove that, regardless of whether the sampling is with replacement or without replacement, one must have

$$P(B_j \mid A_k) = k/n.$$

*Hint.* Prove that in the case of sampling with replacement one must have

$$P(B_j \cap A_k) = \frac{C_{n-1}^{k-1} M_1^k (M - M_1)^{n-k}}{M^n},$$

$$P(A_k) = \frac{C_n^k M_1^k (M - M_1)^{n-k}}{M^n},$$

while in the case of sampling without replacement one must have

$$P(B_j \cap A_k) = \frac{C_{n-1}^{k-1} (M_1)_k (M - M_1)_{n-k}}{(M)_n},$$

$$P(A_k) = \frac{C_n^k (M_1)_k (M - M_1)_{n-k}}{(M)_n}.$$

**Problem 1.3.3.** Let $A_1, \ldots, A_n$ be independent events with $P(A_i) = p_i$.
(a) Prove that

$$P\left(\bigcup_{i=1}^{n} A_i\right) = 1 - \prod_{i=1}^{n} P(\overline{A}_i). \tag{$*$}$$

(b) Let $P_0$ be the probability that *none* of the events $A_1, \ldots, A_n$ occurs. Prove that

$$P_0 = \prod_{i=1}^{n} (1 - p_i).$$

*Hint.* Give a direct proof of the identity in $(*)$, i.e., a proof that makes no use of the inclusion–exclusion formula (see Problem 1.1.12), by showing that if $A_1, \ldots, A_n$ are independent events, then any events of the form $\widetilde{A}_1, \ldots, \widetilde{A}_n$, where $\widetilde{A}_i$ is taken to be either $A_i$, or $\overline{A}_i$, are also independent.

**Problem 1.3.4.** Suppose that the events $A$ and $B$ are independent. Calculate the probabilities that *exactly* $k$, *at least* $k$ and *at most* $k$ of the events $A$ and $B$ occur, $k = 0, 1, 2$—comp. with Problem 1.1.13.

**Problem 1.3.5.** Suppose that the event $A$ is such that it is independent from itself; in other words, one can claim that the events $A$ and $A$ are independent. Prove that $P(A)$ equals either 0 or 1. In addition, prove that if the events $A$ and $B$ are independent and $A \subseteq B$, then either $P(A) = 0$, or $P(B) = 1$.

**Problem 1.3.6.** Suppose that the event $A$ is such that either $P(A) = 1$ or $P(A) = 0$. Prove that, given *any* event $B$, one can claim that $A$ and $B$ are independent events.

**Problem 1.3.7.** Consider the electric circuit from [P §1.4, Fig. 4]. Each of the relays $A, B, C, D$ and $E$ function independently, and can be either *off* (i.e., not allow electric current to pass through), or *on* (i.e., allow electric current to pass through), respectively, with probabilities $p$ and $q$. What is the probability for a signal submitted at the input to eventually get transmitted through the circuit all the way to the output? What is the conditional probability for the relay $E$ to have been on, given that the signal has been transmitted through the circuit and has reached the output?

*Hint.* (a) Let $S$ denote the event that the signal submitted at the input has been received at the output. Then

$$P(S \mid E) = 1 - 2p^2 + p^4, \quad P(S \mid \overline{E}) = 2q^2 - q^4,$$

and, according to the total probability formula, one has

$$P(S) = q(1 - p^2)^2 + pq^2(2 - q^2),$$

while the Bayes formula implies that

$$P(E \mid S) = \frac{(1 - p^2)^2}{(1 - p^2)^2 + pq(2 - q^2)}.$$

**Problem 1.3.8.** Suppose that $P(A + B) > 0$. Prove that

$$P(A \mid A + B) = \frac{P(A)}{P(A) + P(B)}.$$

**Problem 1.3.9.** Suppose that the event $A$ is independent from each of the events $B_n$, $n \geq 1$, chosen so that $B_i \cap B_j = \varnothing$, $i \neq j$. Argue that the events $A$ and $\bigcup_{n=1}^{\infty} B_n$ are independent.

**Problem 1.3.10.** Prove that if $P(A \mid C) > P(B \mid C)$ and $P(A \mid \overline{C}) > P(B \mid \overline{C})$, then $P(A) > P(B)$.

**Problem 1.3.11.** Prove that

$$P(A \mid B) = P(A \mid BC) P(C \mid B) + P(A \mid B\overline{C}) P(\overline{C} \mid B).$$

**Problem 1.3.12.** Suppose that $X$ and $Y$ are two independent binomial random variables with parameters $n$ and $p$. Prove that

$$P(X = k \mid X + Y = m) = \frac{C_n^k \, C_n^{m-k}}{C_{2n}^m}, \qquad \text{for any } k = 0, 1, \dots, \min(m, n).$$

**Problem 1.3.13.** Suppose that $A$, $B$ and $C$ are pair-wise independent events, with $A \cap B \cap C = \emptyset$. Find the largest possible value for $P(A)$.

**Problem 1.3.14.** Consider an urn which already contains one white ball. One randomly chosen ball—either white or black, with equal probability—is added to the urn, after which one ball is taken from the urn at random. Assuming that this last ball happens to be white, what is the probability that the ball left in the urn is also white?

**Problem 1.3.15.** If the events $A$ and $B$ are independent, then, just by definition, one has $P(AB) = P(A)P(B)$. What conditions for $A$ and $B$ would gurantee that $P(AB) \le P(A)P(B)$, or that $P(AB) \ge P(A)P(B)$?

**Problem 1.3.16.** In conjunction with the generalization of Stirling's formula (in the form $n! \sim \sqrt{2\pi n}\, n^n e^{-n}$, $n \to \infty$), prove that the gamma-function $\Gamma(v) = \int_0^\infty u^{v-1} e^{-u}\, du$, $v > 0$ has the property:

$$\Gamma(v) \sim \sqrt{2\pi n}\, v^v e^{-v}, \qquad v \to \infty.$$

## 1.4   Random Variables and Their Characteristics

Recall that in the present chapter the underlying sample space, $\Omega$, is assumed to be *finite* and, therefore, all random variables under consideration can take only finitely many values.

**Problem 1.4.1.** Verify the following properties of the indicators $I_A = I_A(\omega)$:

$$I_\emptyset = 0, \quad I_\Omega = 1, \quad I_{\overline{A}} = 1 - I_A,$$

$$I_{AB} = I_A \cdot I_B, \quad I_{A \cup B} = I_A + I_B - I_{AB},$$

$$I_{A \setminus B} = I_A(1 - I_B), \quad I_{A \triangle B} = (I_A - I_B)^2 = I_A + I_B \quad (\text{mod } 2),$$

$$I_{\bigcup_{i=1}^n A_i} = 1 - \prod_{i=1}^n (1 - I_{A_i}), \quad I_{\overline{\bigcup_{i=1}^n A_i}} = \prod_{i=1}^n (1 - I_{A_i}), \quad I_{\sum_{i=1}^n A_i} = \sum_{i=1}^n I_{A_i},$$

where $A \triangle B$ is the *symmetric difference* of the sets $A$ and $B$, i.e., the set $(A \setminus B) \cup (B \setminus A)$, and the summation symbol $\sum$ stands for union ($\bigcup$) of *non-intersecting* events.

**Problem 1.4.2.** Conclude from the statement in Problem 1.4.1 that the following "inclusion–exclusion" formula for the indicators of the events $A$, $B$ and $C$ is in force:

$$I_{A \cup B \cup C} = I_A + I_B + I_C - [I_{A \cap B} + I_{A \cap C} + I_{B \cap C}] + I_{A \cap B \cap C}.$$

Find the analogous representation for the indicator $I_{A_1 \cup ... \cup A_n}$ of the union of $A_1, \ldots, A_n$.

**Problem 1.4.3.** Suppose that $\xi_1, \ldots, \xi_n$ are Bernoulli random variables with

$$P\{\xi_i = 0\} = 1 - \lambda_i \Delta,$$
$$P\{\xi_i = 1\} = \lambda_i \Delta,$$

for some small number $\Delta > 0$, and for some choice of $\lambda_i > 0$. Prove that

$$P\{\xi_1 + \ldots + \xi_n = 1\} = \left(\sum_{i=1}^{n} \lambda_i\right) \Delta + O(\Delta^2),$$

$$P\{\xi_1 + \ldots + \xi_n > 1\} = O(\Delta^2).$$

**Problem 1.4.4.** Prove that $\inf_{-\infty < a < \infty} E(\xi - a)^2$ is achieved with $a = E\xi$ and that, consequently,

$$\inf_{-\infty < a < \infty} E(\xi - a)^2 = D\xi.$$

*Hint.* Assuming that $E\xi = 0$, prove that $E(\xi - a)^2 = D\xi + a^2 \geq D\xi$.

**Problem 1.4.5.** Let $\xi$ be any random variable with distribution function $F_\xi(x) = P\{\xi \leq x\}$ and with *median* $\mu = \mu(\xi) \equiv \mu(F_\xi)$, defined as the only $\mu \in \mathbb{R}$ with

$$F_\xi(\mu-) \leq \frac{1}{2} \leq F_\xi(\mu).$$

(For an alternative definitions of the notion of median see Problem 1.4.23 below.)
    Prove that

$$\inf_{-\infty < a < \infty} E|\xi - a| = E|\xi - \mu|.$$

*Hint.* Assuming that $\mu = 0$, prove that for $a > 0$ one has

$$E|\xi - a| = E|\xi| + Ef(\xi),$$

where

$$f(x) = \begin{cases} a, & x \leq 0, \\ a - 2x, & 0 < x < a, \\ -a, & x \geq a. \end{cases}$$

Since $f(x) \geq 0$, we have $\mathsf{E} f(\xi) \geq 0$ and $\mathsf{E}|\xi - a| \geq \mathsf{E}|\xi|$. Analogous statement can be made for the case $a < 0$.

**Problem 1.4.6.** Let $P_\xi(x) = \mathsf{P}\{\xi = x\}$ and $F_\xi(x) = \mathsf{P}\{\xi \leq x\}$. Prove that for $a > 0$ and $-\infty < b < \infty$ one has

$$P_{a\xi+b}(x) = P_\xi\left(\frac{x-b}{a}\right),$$

$$F_{a\xi+b}(x) = F_\xi\left(\frac{x-b}{a}\right).$$

In addition, prove that for $y \geq 0$ one has

$$F_{\xi^2}(y) = F_\xi(+\sqrt{y}) - F_\xi(-\sqrt{y}) + P_\xi(-\sqrt{y})$$

and, with $\xi^+ = \max(\xi, 0)$, one has

$$F_{\xi^+}(x) = \begin{cases} 0, & x < 0, \\ F_\xi(x), & x \geq 0. \end{cases}$$

*Hint.* Use the following relations:

$$\{a\xi + b = x\} = \left\{\xi = \frac{x-b}{a}\right\}, \qquad \{a\xi + b \leq x\} = \left\{\xi \leq \frac{x-b}{a}\right\},$$

$$\{\xi^2 \leq y\} = \{\xi = -\sqrt{y}\} \cup (\{\xi \leq +\sqrt{y}\} \setminus \{\xi \leq -\sqrt{y}\}),$$

$$\{\xi^+ \leq x\} = \begin{cases} \varnothing, & x < 0, \\ \{\xi \leq x\}, & x \geq 0. \end{cases}$$

**Problem 1.4.7.** Let $\xi$ and $\eta$ be any two random variables with $\mathsf{D}\xi > 0$ and $\mathsf{D}\eta > 0$, and let $\rho = \rho(\xi, \eta)$ denote the correlation between $\xi$ and $\eta$. Prove that $|\rho| \leq 1$. In addition, prove that $|\rho| = 1$ implies that there are constants $a$ and $b$, for which one can write $\eta = a\xi + b$. Furthermore, if $\rho = 1$, then

$$\frac{\eta - \mathsf{E}\eta}{\sqrt{\mathsf{D}\eta}} = \frac{\xi - \mathsf{E}\xi}{\sqrt{\mathsf{D}\xi}},$$

(so that $\rho = 1$ implies that $a > 0$) and, if $\rho = -1$, then

$$\frac{\eta - \mathsf{E}\eta}{\sqrt{\mathsf{D}\eta}} = -\frac{\xi - \mathsf{E}\xi}{\sqrt{\mathsf{D}\xi}}$$

(so that $\rho = -1$ implies $a < 0$).

**Problem 1.4.8.** Suppose that $\xi$ and $\eta$ are two random variables with $\mathsf{E}\xi = \mathsf{E}\eta = 0$ and $\mathsf{D}\xi = \mathsf{D}\eta = 1$ and with correlation coefficient $\rho = \rho(\xi, \eta)$. Prove that

$$\mathsf{E}\max(\xi^2, \eta^2) \leq 1 + \sqrt{1 - \rho^2}.$$

*Hint.* Use the identity

$$\max(\xi^2, \eta^2) = \frac{1}{2}(\xi^2 + \eta^2 + |\xi^2 - \eta^2|)$$

and the Cauchy–Bunyakovski inequality.

**Problem 1.4.9.** By using the property $I_{\bigcup_{i=1}^{n} A_i} = 1 - \prod_{i=1}^{n}(1 - I_{A_i})$, associated with the indicators from Problem 1.4.1, verify the following "inclusion–exclusion" formula:

$$P(A_1 \cup \ldots \cup A_n) = \sum_{1 \leq i_1 \leq n} P(A_{i_1}) - \sum_{1 \leq i_1 < i_2 \leq n} P(A_{i_1} \cap A_{i_2}) + \ldots +$$

$$+ (-1)^{m+1} \sum_{1 \leq i_1 < \ldots < i_m \leq n} P(A_{i_1} \cap \ldots \cap A_{i_m}) + \ldots$$

$$+ (-1)^{n+1} P(A_1 \cap \ldots \cap A_n)$$

(comp. with Problem 1.1.12).

*Hint.* With the substitution $X_i = I_{A_i}$, prove first that the following "inclusion–exclusion" formula for indicators is in force:

$$1 - \prod_{i=1}^{n}(1 - X_i) = \sum_{1 \leq i \leq n} X_i - \sum_{1 \leq i_1 < i_2 \leq n} X_{i_1} X_{i_2} + \ldots +$$

$$+ (-1)^{m+1} \sum_{1 \leq i_1 < \ldots < i_m \leq n} X_{i_1} \ldots X_{i_m} + \ldots + (-1)^{n+1} X_1 \ldots X_n.$$

After that use the relation $P\left(\bigcup_{i=1}^{n} I_{A_i}\right) = \mathsf{E}I_{\bigcup_{i=1}^{n} A_i}$ (comp. with the hint in Problem 1.1.13).

**Problem 1.4.10.** Suppose that $\xi_1, \ldots, \xi_n$ are independent random variables and that $\varphi_1 = \varphi_1(\xi_1, \ldots, \xi_k)$ and $\varphi_2 = \varphi_2(\xi_{k+1}, \ldots, \xi_n)$ are any two random variables that can be written as functions, respectively, of $\xi_1, \ldots, \xi_k$ and $\xi_{k+1}, \ldots, \xi_n$. Prove that $\varphi_1$ and $\varphi_2$ are independent.

**Problem 1.4.11.** Prove that the random variables $\xi_1, \ldots, \xi_n$ are independent if and only if for every choice of the real numbers $x_1, \ldots, x_n$ one has

$$F_{\xi_1, \ldots, \xi_n}(x_1, \ldots, x_n) = F_{\xi_1}(x_1) \ldots F_{\xi_n}(x_n),$$

where $F_{\xi_1, \ldots, \xi_n}(x_1, \ldots, x_n) = P\{\xi_1 \leq x_1, \ldots, \xi_n \leq x_n\}$.

**Problem 1.4.12.** Prove that the random variable $\xi$ is independent from itself, i.e., one can claim that $\xi$ and $\xi$ are independent, if and only if $\xi(\omega) \equiv$ const, $\omega \in \Omega$.

**Problem 1.4.13.** Under what condition for the random variable $\xi$ can one claim that $\xi$ and $(\sin \xi)$ independent?

**Problem 1.4.14.** Suppose that $\xi$ and $\eta$ are independent random variables and that $\eta \neq 0$. Find expressions for $P\{\xi\eta \leq z\}$ and $P\left\{\frac{\xi}{\eta} \leq z\right\}$ in terms of the probabilities $P\{\xi \leq x\}$ and $P\{\eta \leq y\}$.

**Problem 1.4.15.** Suppose that the random variables $\xi$, $\eta$ and $\zeta$ are such that $|\xi| \leq 1$, $|\eta| \leq 1$, $|\zeta| \leq 1$. Prove *Bell's inequality*: $|E\xi\zeta - E\eta\zeta| \leq 1 - E\xi\eta$ (see [62], for example).
   *Hint.* Use the inequality $\xi(1 + \eta) \leq 1 + \eta$.

**Problem 1.4.16.** One throws, one-by-one and at random, $k$ balls in $n$ boxes (the probability that a given ball would fall in a given box is $1/n$). Find the expected number of the *non-empty* boxes.

**Problem 1.4.17.** Suppose that $\xi_1, \ldots, \xi_n$ are independent and identically distributed random variables with $P\{\xi_1 = 1\} = p$ and $P\{\xi_1 = 0\} = 1 - p$, for some $0 < p < 1$, and let $S_k = \xi_1 + \ldots + \xi_k$, $k \leq n$. Prove that, for $1 \leq m \leq n$, one has

$$P(S_m = k \mid S_n = l) = \frac{C_m^k\, C_{n-m}^{l-k}}{C_n^l}\,.$$

**Problem 1.4.18.** Suppose that $\xi_1, \ldots, \xi_n$ are independent random variables and let

$$\xi_{\min} = \min(\xi_1, \ldots, \xi_n) \quad \text{and} \quad \xi_{\max} = \max(\xi_1, \ldots, \xi_n)\,.$$

Prove that

$$P\{\xi_{\min} \geq x\} = \prod_{i=1}^{n} P\{\xi_i \geq x\} \quad \text{and} \quad P\{\xi_{\max} < x\} = \prod_{i=1}^{n} P\{\xi_i < x\}\,.$$

**Problem 1.4.19.** Let $S_{2n} = \xi_1 + \ldots + \xi_{2n}$ and set $M_{2n} = \max(S_1, \ldots, S_{2n})$. Prove that, for any $k \leq n$, one must have

$$P\{M_{2n} \geq k, S_{2n} = 0\} = P\{S_{2n} = 2k\}$$

and that, therefore,

$$P(M_{2n} \geq k \mid S_{2n} = 0) = \frac{P\{S_{2n} = 2k\}}{P\{S_{2n} = 0\}} = \frac{C_{2n}^{n+k}}{C_{2n}^n}\,.$$

Conclude from the last relation that

$$E(M_{2n} \mid S_{2n} = 0) = \frac{1}{2}\left[\frac{1}{P\{S_{2n} = 0\}} - 1\right].$$

**Problem 1.4.20.** Give an example of two random variables, $\xi$ and $\eta$, that share the same distribution function ($F_\xi = F_\eta$) and have the property $P\{\xi \neq \eta\} > 0$.

**Problem 1.4.21.** Suppose that $\xi$, $\eta$ and $\zeta$ are random variables, chosen so that the distribution functions of $\xi$ and $\eta$ coincide. Can one claim that the distribution functions of $\xi\zeta$ and $\eta\zeta$ also coincide?

**Problem 1.4.22.** Give an example of two independent random variables, $\xi$ and $\eta$, for which $\xi^2$ and $\eta^2$ are dependent.

**Problem 1.4.23.** Suppose that $\xi$ is some discrete random variable. Consider the following three definitions of the *median*, $\mu = \mu(\xi)$, of $\xi$ (see Problem 1.4.5):

(a) $\max(P\{\xi > \mu\}, P\{\xi < \mu\}) \leq 1/2$;
(b) $P\{\xi < \mu\} \leq 1/2 \leq P\{\xi \leq \mu\}$;
(c) $\mu = \inf\{x \in \mathbb{R} : P\{\xi \leq x\} \geq 1/2\}$.

Let $M_a$, $M_b$ and $M_c$ denote the sets of "medians" associated with definitions (a), (b) and (c), respectively. How do these three sets relate to each other?

**Problem 1.4.24.** A urn contains $N$ balls, of which $a$ are white, $b$ are black and $c$ are red, $a + b + c = N$. Suppose that $n$ balls are taken from the urn and suppose that among those $n$ balls there are $\xi$ white balls and $\eta$ red balls. Prove that: if the balls are sampled *with replacement*, one has

$$\mathrm{cov}(\xi, \eta) = -n\,p\,q\,,$$

where $p = a/N$ and $q = b/N$, and if the balls are sampled *without replacement*, one has

$$\mathrm{cov}(\xi, \eta) = -n\,p\,q\,\frac{N - n}{N - 1}.$$

Finally, prove that in both cases the correlation is given by

$$\rho(\xi, \eta) = -\sqrt{\frac{p\,q}{(1 - p)(1 - q)}}\,.$$

## 1.5   Bernoulli Scheme I: The Law of Large Numbers

**Problem 1.5.1.** Suppose that $\xi$ and $\eta$ are two random variables with correlation coefficient $\rho$. Verify the following *two-dimensional* analog of *Chebyshev's Inequality*:

$$P\{|\xi - E\xi| \geq \varepsilon\sqrt{D\xi} \text{ or } |\eta - E\eta| \geq \varepsilon\sqrt{D\eta}\} \leq \frac{1}{\varepsilon^2}(1 + \sqrt{1 - \rho^2})\,.$$

*Hint.* Without a loss of generality suppose that $E\xi = E\eta = 0$ and $D\xi = D\eta = 1$, in which case $P\{|\xi| \geq \varepsilon \text{ or } |\eta| \geq \varepsilon\} = P\{\max(\xi^2, \eta^2) \geq \varepsilon^2\}$. Then use the ("usual") Chebyshev inequality and the inequality established in Problem 1.4.8.

**Problem 1.5.2.** Suppose that $f = f(x)$ is some non-negative function which is even and is also non-decreasing for positive $x$. Given any random variable $\xi = \xi(\omega)$, with $|\xi(\omega)| \leq C, C > 0$, verify the following estimate:

$$P\{|\xi| \geq \varepsilon\} \geq \frac{Ef(\xi) - f(\varepsilon)}{f(C)}.$$

In particular, for $f(x) = x^2$ one must have

$$\frac{E\xi^2 - \varepsilon^2}{C^2} \leq P\{|\xi - E\xi| \geq \varepsilon\} \quad \left(\leq \frac{D\xi}{\varepsilon^2}\right).$$

*Hint.* Use the following relation

$$Ef(\xi) = Ef(|\xi|) \leq f(C)P\{|\xi| \geq \varepsilon\} + f(\varepsilon).$$

**Problem 1.5.3.** Let $\xi_1, \ldots, \xi_n$ be any sequence of independent random variables with $D\xi_i \leq C$. Prove that

$$P\left\{\left|\frac{\xi_1 + \ldots + \xi_n}{n} - \frac{E(\xi_1 + \ldots + \xi_n)}{n}\right| \geq \varepsilon\right\} \leq \frac{C}{n\varepsilon^2}.$$

(With the conventions adopted in [P §1.5, (8)], the above inequality gives a version of the law of large numbers, which is more general than the version obtained in the context of the Bernoulli scheme.)

**Problem 1.5.4.** Suppose that $\xi_1, \ldots, \xi_n$ are independent Bernoulli random variables with $P\{\xi_i = 1\} = p > 0$ and $P\{\xi_i = -1\} = 1 - p$. Verify *Bernstein's estimate*: there is some $a > 0$, for which

$$P\left\{\left|\frac{S_n}{n} - (2p - 1)\right| \geq \varepsilon\right\} \leq 2e^{-a\varepsilon^2 n},$$

where $S_n = \xi_1 + \ldots + \xi_n$ and $\varepsilon > 0$.
     *Hint.* See the proof of [P §1.6, (42)].

**Problem 1.5.5.** Let $\xi$ be any non-negative random variable and let $a > 0$. Find the maximal possible value for the probability $P\{\xi \geq a\}$ in each of the following three cases ($m$ and $\sigma$ are given real numbers):

   (i) $E\xi = m$;
   (ii) $E\xi = m, D\xi = \sigma^2$;
   (iii) $E\xi = m, D\xi = \sigma^2$ and $\xi$ is symmetric relative to its mean value $m$.

**Problem 1.5.6.** Let $S_0 = 0$ and $S_n = \xi_1 + \ldots + \xi_n$, $n \leq N$, where $\xi_1, \ldots, \xi_n$ is a Bernoulli sequence of independent random variables, with $P\{\xi_n = 1\} = p > 0$ and $P\{\xi_n = 0\} = q$, $n \leq N$, and let $P_n(k) = P\{S_n = k\}$. Prove that, for $n < N$ and $k \geq 1$, one has

$$P_{n+1}(k) = p\, P_n(k-1) + q\, P_n(k).$$

**Problem 1.5.7.** Suppose that $\xi_1, \ldots, \xi_N$ are independent Bernoulli random variables, with $P\{\xi_i = 1\} = P\{\xi_i = -1\} = 1/2$, $i = 1, \ldots, N$, and let $S_m = \xi_1 + \ldots + \xi_m$. Prove that for $2m \leq N$ one has

$$P\{S_1 \ldots S_{2m} \neq 0\} = 2^{-2m} C_{2m}^m.$$

**Problem 1.5.8.** Consider $M$ cells, labeled $1, \ldots, M$. Suppose that the cell with label $n$ contains one white ball and $n$ black balls. Consider a random sample of balls from the $M$ cells, let

$$\xi_n = \begin{cases} 1, & \text{if a white ball is drawn from the cell with label } n, \\ 0, & \text{if a black ball is drawn,} \end{cases}$$

and let $S_M = \xi_1 + \ldots + \xi_M$ denote the total number of white balls in the sample. Prove that for large $M$ the quantity $S_M$ "has order" $\ln M$, in the sense that, for any $\varepsilon > 0$, one has, as $M \to \infty$,

$$P\left\{ \left| \frac{S_M}{\ln M} - 1 \right| \geq \varepsilon \right\} \to 0$$

(with the convention adopted in formula [P §1.5, (8)]).

**Problem 1.5.9.** Suppose that $\xi_1, \ldots, \xi_n$ are some independent Bernoulli random variables, with $P\{\xi_k = 1\} = p_k$ and $P\{\xi_k = 0\} = 1 - p_k$, $1 \leq k \leq n$, and let $a = \frac{1}{n} \sum_{k=1}^{n} p_k$. Prove that, for any fixed $0 < a < 1$, the variance, $DS_n$, of the variable $S_n = \xi_1 + \ldots + \xi_n$ attains its maximal value when $p_1 = \ldots = p_n = a$.

**Problem 1.5.10.** Suppose that $\xi_1, \ldots, \xi_n$ are some independent Bernoulli random variables, with $P\{\xi_k = 1\} = p$ and $P\{\xi_k = 0\} = 1 - p$, $1 \leq k \leq n$. Find the conditional probability that the first 1 ("success") appears in the $m^{\text{th}}$ step, conditioned to the event that in all $n$ steps "success" occurs exactly once.

**Problem 1.5.11.** Let $(p_1, \ldots, p_r)$ and $(q_1, \ldots, q_r)$ be any two probability distributions. Prove the *Gibbs inequality*:

$$-\sum_{i=1}^{r} p_i \ln p_i \leq -\sum_{i=1}^{r} p_i \ln q_i.$$

In particular, the entropy $H = -\sum_{i=1}^{r} p_i \ln p_i$ must satisfy the relation $H \leq \ln r$— see [P §1.5, **4**].

**Problem 1.5.12.** In the context of Problem 1.5.10, prove the *Rényi inequality*:

$$P\left\{\left|\frac{S_n}{n} - p\right| \geq \varepsilon\right\} \leq \exp\left\{-\frac{n\varepsilon^2}{2pq(1 + \varepsilon/(2pq))^2}\right\}.$$

## 1.6   Bernoulli Scheme II: Limit Theorems (Local, Moivre–Laplace, Poisson)

**Problem 1.6.1.** Let $n = 100$ and consider the choices $p = 1/10, 2/10, 3/10, 4/10,$ $5/10$. By using the relevant tables for the binomial and the Poisson distributions (see [12], for example), or by using a computer, compare the exact values of the following probabilities:

$$P\{10 < S_{100} \leq 12\}, \quad P\{20 < S_{100} \leq 22\}, \tag{1.1}$$

$$P\{33 < S_{100} \leq 35\}, \quad P\{40 < S_{100} \leq 42\}, \tag{1.2}$$

$$P\{50 < S_{100} \leq 52\}, \tag{1.3}$$

with the respective values obtained by the normal and the the Poisson approximations.

**Problem 1.6.2.** Let $p = 1/2$ and let $Z_n = 2S_n - n$ (the aggregate *excess* of 1's vs. 0's in $n$ trials, the outcome from each trial being 0 or 1). Prove that

$$\sup_j \left|\sqrt{\pi n}\, P\{Z_{2n} = j\} - e^{-j^2/4^n}\right| \to 0, \quad \text{as } n \to \infty.$$

*Hint.* Setting $2n = m$ and $k = j/2 + n$, the proof comes down to showing that

$$\sup_k \left|\sqrt{\frac{\pi m}{2}}\, P\{S_m = k\} - e^{-\frac{(k-mp)^2}{2mpq}}\right| \quad \left(\equiv \sup_k \varepsilon(k, m)\right) \to 0, \quad \text{as } m \to \infty.$$

With this relation in mind, one must prove that

$$\sup_k \varepsilon(k, m) = \max(a_m, b_m),$$

where

$$a_m = \sup_{\{k:|k-mp|\leq (mpq)^s\}} \varepsilon(k, m), \quad b_m = \sup_{\{k:|k-mp|>(mpq)^s\}} \varepsilon(k, m),$$

for some $s \in (1/2, 2/3)$, and then verify that $a_m \to 0$ and $b_m \to 0$ as $m \to \infty$.

**Problem 1.6.3.** Prove that in the Poisson theorem (with $p = \lambda/n, \lambda > 0$) one has

$$\sup_k \left|P_n(k) - \frac{\lambda^k e^{-\lambda}}{k!}\right| \leq \frac{\lambda^2}{n}.$$

*Hint.* Let $\eta_1, \ldots, \eta_n$ and $\zeta_1, \ldots, \zeta_n$ be two different sets of independent random variables, distributed, respectively, according to Poisson's law with parameter $\lambda/n$ and Bernoulli's law with

$$P\{\zeta_i = 0\} = e^{\lambda/n}(1 - \lambda/n) \quad \text{and} \quad P\{\zeta_i = 1\} = 1 - e^{\lambda/n}(1 - \lambda/n).$$

Setting

$$\xi_i = \begin{cases} 0, & \text{if } \eta_i = 0, \zeta_i = 0, \\ 1 & \text{in all other cases}, \end{cases}$$

notice that $\xi_1, \ldots, \xi_n$ are independent Bernoulli random variables with

$$P\{\xi_i = 0\} = 1 - \frac{\lambda}{n}, \qquad P\{\xi_i = 1\} = \frac{\lambda}{n},$$

and that the distribution of $\xi = \xi_1 + \ldots + \xi_n$ is given by $P\{\xi = k\} = P_n(k)$. Then take into account that $\eta = \eta_1 + \ldots + \eta_n$ is distributed according to the Poisson law with parameter $\lambda$, and that, given any $k = 0, 1, 2, \ldots$, one has

$$|P\{\xi = k\} - P\{\eta = k\}| \le P\{\xi \ne \eta\} \le \frac{\lambda^2}{n}.$$

(Comp. with the results and the proofs in [P §3.12].)

**Problem 1.6.4.** Let $\xi_1, \ldots, \xi_n$ be independent and identically distributed random variables with $P\{\xi_k = 1\} = P\{\xi_k = -1\} = 1/2$ (this is a symmetric Bernoulli scheme), let $S_n = \xi_1 + \ldots + \xi_n$, and let $P_n(k) = P\{S_n = k\}$, for $k \in E_n = \{0, \pm 1, \ldots, \pm n\}$. By using the total probability formula (see [P §1.3, (3)]), verify the following recursive relation (a special case of the Kolmogorov–Chapman equation—see [P §1.12]):

$$P_{n+1}(k) = \frac{1}{2} P_n(k+1) + \frac{1}{2} P_n(k-1), \qquad k \in E_{n+1}, \tag{$*$}$$

which is equivalent to

$$P_{n+1}(k) - P_n(k) = \frac{1}{2} \Big[ P_n(k+1) - 2 P_n(k) + P_n(k-1) \Big]. \tag{$**$}$$

**Problem 1.6.5.** (*Continuation of Problem 1.6.4.*) The sequence of random variables $S_0 = 0$, $S_1 = \xi_1$, $S_2 = \xi_1 + \xi_2$, $\ldots$, $S_n = \xi_1 + \ldots + \xi_n$, may be identified with the *trajectory* of a random walk of a particle that starts from 0 and moves one unit up or down at integer times.

Suppose now that the up and down moves in the random walk occur only at times $\Delta, 2\Delta, \ldots, n\Delta$, for some $\Delta > 0$, and that the particle move up or down at distance $\Delta x$. Instead of the probabilities $P_n(k) = P\{S_n = k\}$, introduced in the previous problem, consider the probabilities

$$P_{n\Delta}(k\Delta x) = P\{S_{n\Delta} = k\Delta x\}.$$

Analogously to the recursive relation $(**)$, we find that

$$\frac{P_{(n+1)\Delta}(k\Delta x) - P_{n\Delta}(k\Delta x)}{\Delta} = \frac{1}{2}\Big[P_{n\Delta}((k+1)\Delta x)-$$

$$- 2P_{n\Delta}(k\Delta x) + P_{n\Delta}((k-1)\Delta x)\Big],$$

i.e., the (discrete) "first derivative" in the time-parameter coincides up to a factor of $\frac{1}{2}$ with the (discrete) "second derivative" in the space variable.

With $\Delta x = \sqrt{\Delta}$, $t > 0$, $x \in \mathbb{R}$, consider the special limiting procedure with $n \to \infty$ and $k \to \infty$, taken so that $n\Delta \to t$ and $k\sqrt{\Delta} \to x$, and prove that for this procedure one can claim that
  (a) the limit $P_t(x) = \lim P_{n\Delta}(k\sqrt{\Delta})$ exists, and,
  (b) as a function of $t$, satisfies the *heat equation*, namely,

$$\frac{\partial P_t(x)}{\partial t} = \frac{1}{2}\frac{\partial^2 P_t(x)}{\partial x^2}$$

(L. Bachelier, A. Einstein).

**Problem 1.6.6.** To generalize the result in the previous problem, suppose that the particle moves up at distance $\Delta x$ with probability $p(\Delta) = \frac{1}{2}+\Delta x$, and moves down at distance $\Delta x$ with probability $q(\Delta) = \frac{1}{2} - \Delta x$. Again set $\Delta x = \sqrt{\Delta}$ and suppose that $n\Delta \to t$ and $k\sqrt{\Delta} \to x$. Prove that, just as in the previous problem, one can claim that the limit $P_t(x) = \lim P_{n\Delta}(k\sqrt{\Delta})$ exists and satisfies the equation

$$\frac{\partial P_t(x)}{\partial t} = -\frac{\partial P_t(x)}{\partial x} + \frac{1}{2}\frac{\partial^2 P_t(x)}{\partial x^2}.$$

**Problem 1.6.7.** What should be changed in the limiting procedures in the last two problems, in order to claim that the function obtained in the limit satisfies the equation

$$\frac{\partial P_t(x)}{\partial t} = -\mu\frac{\partial P_t(x)}{\partial x} + \frac{1}{2}\sigma^2\frac{\partial^2 P_t(x)}{\partial x^2},$$

known as the *Fokker–Planck equation*, or *Kolmogorov forward equation*.

**Problem 1.6.8.** Suppose that $F_n = F_n(t)$, $t \in [0, 1]$, $n \geq 1$, is some sequence of nondecreasing functions, with the property $F_n(t) \to t$, for all rational $t \in \mathbb{Q} \cap [0, 1]$. Prove that this convergence must be *uniform*, i.e.,

$$\sup_{t \in [0,1]} |F_n(t) - t| \to 0 \quad \text{as } n \to \infty$$

(see also [P §3.1, (5)]).

**Problem 1.6.9.** Prove that, given any $x > 0$, one has

$$\frac{x}{1+x^2}\, \varphi(x) < 1 - \Phi(x) < \frac{\varphi(x)}{x} \,,$$

where $\varphi(x) = \frac{1}{\sqrt{2\pi}} e^{-x^2/2}$ and $\Phi(x) = \int_{-\infty}^{x} \varphi(y)\, dy$.

*Hint.* Take the derivatives $\varphi'(x)$ and $(x^{-1}\varphi(x))'$.

**Problem 1.6.10.** Prove that the Poisson distribution satisfies the following local theorem: given any $k = 0, 1, 2, \ldots$, as $\lambda \to \infty$ one has

$$\sqrt{\lambda}\, \left| \frac{\lambda^k}{k!}\, e^{-\lambda} - \frac{1}{\sqrt{2\pi\lambda}}\, \exp\left\{ -\frac{1}{2\lambda}(k - \lambda)^2 \right\} \right| \to 0\,.$$

*Hint.* Use Stirling's formula.

## 1.7   Estimate of the Probability for Success in Bernoulli Trials

**Problem 1.7.1.** A priori, it is known that the parameter $\theta$ takes values in the set $\Theta_0 \subseteq [0, 1]$. Explain when it might be possible to find an *unbiased* estimate for the parameter $\theta$, that takes values only in the set $\Theta_0$.

*Hint.* If $\Theta_0$ is a singleton ($\Theta_0 = \{\theta_0\}$), then the value $\theta_0$ must be the estimate itself. If $\Theta_0$ contains at least two points, then the following condition is necessary and sufficient for the existence of an unbiased estimate: $\{0\} \in \Theta_0$ and $\{1\} \in \Theta_0$. Verify this claim.

**Problem 1.7.2.** In the context of the previous problem, find an analog of the Rao–Cramér inequality and investigate the *efficiency* of the estimate.

**Problem 1.7.3.** In the context of the first problem, investigate the construction of *confidence intervals* for $\theta$.

**Problem 1.7.4.** As a continuation of Problem 1.2.21, investigate whether the estimate $\widehat{N}$ is unbiased and/or efficient, assuming that $N$ is sufficiently large, $N \gg M$, $N \gg n$. Analogously to the confidence intervals for the parameter $\theta$ (see [P1 § 1.7, (8) and (9)]), construct confidence intervals $[\widehat{N} - a(\widehat{N}), \widehat{N} + b(\widehat{N})]$ for $N$ with the property

$$\mathsf{P}_{N, M; n}\left\{ \widehat{N} - a(\widehat{N}) \le N \le \widehat{N} + b(\widehat{N}) \right\} \approx 1 - \varepsilon\,,$$

where $\varepsilon$ is some small positive number.

**Problem 1.7.5.** ($\chi^2$–*goodness-of-fit test*). Suppose that $\xi_1, \ldots, \xi_n$ are independent Bernoulli random variables with $\mathsf{P}\{\xi_i = 1\} = p$ and $\mathsf{P}\{\xi_i = 0\} = 1 - p$, $1 \le i \le n$. Unlike the main discussion in [P §1.7], which is concerned with estimates of the probability for "success," $p$, here we are concerned with the problem of *testing*, based on the observations $x = (x_1, \ldots, x_n)$, of the *hypothesis* $H_0$: $p = p_0$, i.e., the hypothesis that the true value of the parameter $p$ equals some given number $0 < p_0 < 1$. Let $S_n(\xi) = \xi_1 + \ldots + \xi_n$ and set

$$\chi_n^2(\xi) = \frac{(S_n(\xi) - np_0)^2}{np_0(1 - p_0)}.$$

Assuming that the hypothesis $H_0$ is true, prove that, for any $x \geq 0$, one must have

$$P\{\chi_n^2(\xi) \leq x\} \to \int_0^x \frac{1}{\sqrt{2\pi y}} e^{-y/2} \, dy, \quad \text{as } n \to \infty.$$

(According to [P §2.3, Table 3], $F(x) = \int_0^x \frac{1}{\sqrt{2\pi y}} e^{-y/2} \, dy$ is the cumulative distribution function of a $\chi^2$-random variable with one degree of freedom, i.e., the square of a standard $(0, 1)$-Gaussian random variable.)

The $\chi^2$-*goodness-of-fit* criterion for testing the hypothesis $H_0$: $p = p_0$ is based on the following argument. Choose the number $\varepsilon > 0$ so small that, in a single experiment, events that have probability $\varepsilon$ are extremely unlikely to occur. (If $\varepsilon$ is, say, 0.01, then by the law of large numbers—see the remark related to formula (8) in [P §1.5]—an event that occurs in each trial with probability 0.01 will occur "on average" only once in 100 independent trials.)

Next, consider $\varepsilon > 0$ as fixed and choose $\lambda(\varepsilon)$ so that $\int_{\lambda(\varepsilon)}^\infty \frac{1}{\sqrt{2\pi y}} e^{-y/2} \, dy = \varepsilon$. One can now test the hypothesis $H_0$: $p = p_0$ (by the $\chi^2$-goodness-of-fit test) in the following manner: if the value $\chi_n^2(x)$, calculated from the observations $x = (x_1, \ldots, x_n)$, exceeds the quantity $\lambda(\varepsilon)$, then $H_0$ is rejected and if $\chi_n^2(x) \leq \lambda(\varepsilon)$ then $H_0$ is accepted, i.e., one assumes that the observation $x = (x_1, \ldots, x_n)$ is in agreement with the property $p = p_0$.

(a) Based on the law of large numbers (see [P §1.5]), argue that, at least for very large values of $n$, using the $\chi^2$-goodness-of-fit criterion for testing $H_0$: $p = p_0$ is quite natural.

(b) By using the Berry–Esseen inequality [P §1.6, (24)], prove that, under the hypothesis $H_0$: $p = p_0$, one must have

$$\sup_x \left| P\{\chi_n^2(\xi) \leq x\} - \int_0^x \frac{1}{\sqrt{2\pi y}} e^{-y/2} \, dy \right| \leq \frac{2}{\sqrt{np_0(1 - p_0)}}.$$

(c) Suppose that $\lambda_n(\varepsilon)$ is chosen so that $P\{\chi_n^2(\xi) \geq \lambda_n(\varepsilon)\} \leq \varepsilon$. Find the rate of convergence of $\lambda_n(\varepsilon) \to \lambda(\varepsilon)$ and, this way, determine the error resulting from replacing the event "$\chi_n^2(\xi) \geq \lambda_n(\varepsilon)$" with the event "$\chi_n^2(\xi) \geq \lambda(\varepsilon)$," which is used in the $\chi^2$-goodness-of-fit test.

**Problem 1.7.6.** Let $\xi$ be any binomial random variable with distribution

$$P_\theta\{\xi = k\} = C_n^k \, \theta^k (1 - \theta)^{n-k}, \quad 0 \leq k \leq n,$$

where $n$ is some given number and $\theta$ is an "unknown parameter," which must be estimated by the (unique) observation over the random variable $\xi$.

A standard estimator for $\theta$ is given by the value $T(\xi) = \frac{\xi}{n}$. This estimator is unbiased: given any $\theta \in [0, 1]$, one has

$$\mathsf{E}_\theta T(\xi) = \theta .$$

Prove that, in the *class of unbiased estimators* $\widetilde{T} = \widetilde{T}(\xi)$, the estimator $T(\xi)$ is also *efficient*:

$$\mathsf{E}_\theta (T(\xi) - \theta)^2 = \inf_{\widetilde{T}} \mathsf{E}_\theta (\widetilde{T}(\xi) - \theta)^2 .$$

Argue that, for $n = 3$, if it is a priori known that $\theta \in (\frac{1}{4}, \frac{3}{4})$, then the estimator $\widehat{T}(\xi) \equiv \frac{1}{2}$, which is *unbiased* for every choice of $\theta \neq \frac{1}{2}$, is "better" than the unbiased estimator $T(\xi) = \frac{\xi}{3}$:

$$\mathsf{E}_\theta [\widehat{T}(\xi) - \theta]^2 < \mathsf{E}_\theta [T(\xi) - \theta]^2 ,$$

i.e.,

$$\mathsf{E}_\theta \left[ \frac{1}{2} - \theta \right]^2 < \mathsf{E}_\theta \left[ \frac{\xi}{3} - \theta \right]^2 .$$

Investigate the validity of this statement for arbitrary $n$.

**Problem 1.7.7.** Two correctors, A and B, are proof-reading a book. As a result, A detects $a$ misprints and B detects $b$ misprints, of which $c$ misprints are detected by both A and B. Assuming that the two correctors work independently from each other, give a "reasonable" estimate of the number of misprints that have remained undetected.

*Hint.* Based on a probabilistic argument, assuming that the number $n$ of all missprints in the book is quite large, one can suppose that $\frac{a}{n}$ and $\frac{b}{n}$ are reasonably close to the probabilities $p_a$ and $p_b$ for a misprint to be detected, respectively, by corrector A and corrector B.

## 1.8  Conditional Probabilities and Expectations with Respect to Partitions

**Problem 1.8.1.** Give an example of two random variables, $\xi$ and $\eta$, that are not independent and yet the relation

$$\mathsf{E}(\xi \mid \eta) = \mathsf{E}\xi$$

still holds (see [P §1.8, (22)]).

**Problem 1.8.2.** The *conditional variance* of the random variable $\xi$ with respect to the partition $\mathscr{D}$ is defined as the random variable

$$D(\xi \mid \mathscr{D}) = E[(\xi - E(\xi \mid \mathscr{D}))^2 \mid \mathscr{D}].$$

Prove that the variance of $\xi$ satisfies the relation:

$$D\xi = ED(\xi \mid \mathscr{D}) + DE(\xi \mid \mathscr{D}).$$

   *Hint.* Convince yourself that

$$ED(\xi \mid \mathscr{D}) = E\xi^2 - E[E(\xi \mid \mathscr{D})]^2 \quad \text{and} \quad DE(\xi \mid \mathscr{D}) = E[E(\xi \mid \mathscr{D})]^2 - (E\xi)^2.$$

**Problem 1.8.3.** Starting from the relation [$\underline{P}$ §1.8, (17)], prove that, given any function $f = f(\eta)$, the conditional expectation $E(\xi \mid \eta)$ has the property:

$$E[f(\eta)E(\xi \mid \eta)] = E[\xi f(\eta)].$$

**Problem 1.8.4.** Given two random variables, $\xi$ and $\eta$, prove that $\inf_f E(\eta - f(\xi))^2$ is achieved with the function $f^*(\xi) = E(\eta \mid \xi)$. (This way, the optimal mean-square-error-minimizing *estimator* of $\eta$ given $\xi$ can be identified with the conditional expectation $E(\eta \mid \xi)$.)
   *Hint.* Convince yourself that, for any function $f = f(x)$, one has

$$E(\eta - f(\xi))^2 = E(\eta - f^*(\xi))^2 + 2E[(\eta - f^*(\xi))(f^*(\xi) - f(\xi))] + E(f^*(\xi) - f(\xi))^2,$$

where the expected value of the variable in box brackets $E[\cdot]$ actually vanishes.

**Problem 1.8.5.** Let $\xi_1, \ldots, \xi_n$ and $\tau$ be *independent* random variables, such that $\xi_1, \ldots, \xi_n$ are identically distributed and $\tau$ takes its values in the set $1, \ldots, n$. Prove that *the sum of random number of random variables*, namely $S_\tau := \xi_1 + \ldots + \xi_\tau$, satisfies the relations

$$E(S_\tau \mid \tau) = \tau E\xi_1, \; D(S_\tau \mid \tau) = \tau D\xi_1$$

and

$$ES_\tau = E\tau \cdot E\xi_1, \; DS_\tau = E\tau \cdot D\xi_1 + D\tau \cdot (E\xi_1)^2.$$

   *Hint.* Use the relations

$$E(S_\tau \mid \tau) = \tau E\xi_1 \quad \text{and} \quad D(S_\tau \mid \tau) = \tau D\xi_1.$$

**Problem 1.8.6.** Suppose that the random variable $\xi$ is independent from the partion $\mathscr{D}$ (i.e., for any $D_i \in \mathscr{D}$, the random variables $\xi$ and $I_{D_i}$ are independent). Prove that

$$E(\xi \mid \mathscr{D}) = E\xi.$$

**Problem 1.8.7.** Let $\mathscr{E}$ be some experiment, with associated space of possible outcomes $\Omega = \{\omega_1, \ldots, \omega_k\}$, the respective probabilities (i.e., "weights" for the outcomes) being given by $p_i = p(\omega_i)$, $\sum_{i=1}^{k} p_i = 1$. It is established in [P §1.5, (14)] that the formula $H = -\sum_{i=1}^{k} p_i \ln p_i$ gives *the entropy* of the distribution $(p_1, \ldots, p_k)$, defined as a measure of the "uncertainty" in the experiment $\mathscr{E}$. In the same section it is also shown that the uncertainty is maximal in experiments where all $k$ outcomes are equally likely to occur, in which case one has $H = \ln k$.

The fact that, in the case where all outcomes are equally likely, the *logarithmic* function is a natural measure for the degree of uncertainty in the outcome of the experiment can be justified with the following argument, which is offered here as an exercise.

Suppose that the *degree of uncertainty* in an experiment $\mathscr{E}$, with $k$ outcomes, is given by some function $f(k)$, chosen so that $f(1) = 0$ and $f(k) > f(l)$ if $k > l$. In addition, suppose that $f(kl) = f(k) + f(l)$. (This reflects the requirement that, for independent experiments, $\mathscr{E}_1$ and $\mathscr{E}_2$, respectively, with $k$ outcomes and $l$ outcomes, the degree of uncertainty in the experiment $\mathscr{E}_1 \otimes \mathscr{E}_2$, which comes down to carrying out simultaneously $\mathscr{E}_1$ and $\mathscr{E}_2$, must be the sum of the degrees of uncertainty in the two experiments.)

Prove that under the above conditions $f(k)$ must be of the form: $f(k) = c \log_b k$, where $c > 0$ is some constant and the logarithm $\log_b k$ is taken with an arbitrary base $b > 0$.

*Remark.* As the transition from one logarithmic base to another is given by $\log_b k = \log_b a \cdot \log_a k$, it is clear that such a transition comes down to changing the unit in which the uncertainty is being measured. The most common choice is $b = 2$, which gives $\log_2 k = 1$ for $k = 2$ and therefore allows one to identify the selected *unit* of uncertainty with the uncertainty in an experiment with two equally likely outcomes. In communication theory (and, in particular, in coding theory) such an unit of uncertainty is called *bit of information*, or simply *bit*, which originates from the term *BInary digiT*. For example, in an experiment $\mathscr{E}$ with $k = 10$ equally likely outcomes, the degree of uncertainty equals $\log_2 10 \approx 3.3$ bits of information.

**Problem 1.8.8.** Let $(\Omega, \mathscr{F}, \mathsf{P})$ be any discrete probability space and suppose that $\xi = \xi(\omega)$, $\omega \in \Omega$, is any random variable that takes its values in the set $\{x_1, \ldots, x_k\}$, with respective probabilities $\mathsf{P}\{\xi = x_i\} = p_i$. The *entropy* of the random variable $\xi$ (or, equivalently, of the experiment $\mathscr{E}_\xi$, which comes down to observing the realization of $\xi$) is defined as

$$H(\xi) = -\sum_{i=1}^{k} p_i \log_2 p_i .$$

(Comp. with [P §1.5, (14)], where, instead of the binary logarithm $\log_2$, the natural logarithm $\ln$ is used—as explained above, this choice is inessential.)

Analogously, given a pair $(\xi, \eta)$ of random variables with $\mathsf{P}\{\xi = x_i, \eta = y_j\} = p_{ij}$, $i = 1, \ldots, k$, $j = 1, \ldots, l$, the *entropy* $H(\xi, \eta)$ is defined as

$$H(\xi, \eta) = -\sum_{i=1}^{k}\sum_{j=1}^{l} p_{ij} \log_2 p_{ij} \,.$$

Prove that if $\xi$ and $\eta$ are independent, then $H(\xi, \eta) = H(\xi) + H(\eta)$.

**Problem 1.8.9.** Consider a pair $(\xi, \eta)$ of random variables with values $(x_i, y_j)$, $i = 1, \ldots, k$, $j = 1, \ldots, l$. The *conditional entropy* of the random variable $\eta$, given the event $\{\xi = x_i\}$, is defined as

$$H_{x_i}(\eta) = -\sum_{j=1}^{l} \mathsf{P}\{\eta = y_j \mid \xi = x_i\} \log_2 \mathsf{P}\{\eta = y_j \mid \xi = x_i\} \,.$$

Then the mean conditional entropy of $\eta$ given $\xi$ is defined as

$$H_{\xi}(\eta) = \sum_{i=1}^{k} \mathsf{P}\{\xi = x_i\} H_{x_i}(\eta).$$

Prove that:
(a) $H(\xi, \eta) = H(\xi) + H_{\xi}(\eta)$;
(b) if $\xi$ and $\eta$ are independent, then

$$H(\xi, \eta) = H(\xi) + H_{\xi}(\eta);$$

(c) $0 \le H_{\xi}(\eta) \le H(\eta)$.

**Problem 1.8.10.** For a pair of random variables, $(\xi, \eta)$, the quantity

$$I_{\xi}(\eta) = H(\eta) - H_{\xi}(\eta)$$

gives the *amount of information* for the variable $\eta$ that is contained in the variable $\xi$. This terminology is justified by the fact that the difference $H(\eta) - H_{\xi}(\eta)$ represents the amount by which observations over $\xi$ decrease the uncertainty of $\eta$, i.e., decrease the quantity $H(\eta)$.
   Prove that:
(a) $I_{\xi}(\eta) = I_{\eta}(\xi) \ge 0$;
(b) $I_{\xi}(\eta) = H(\eta)$ if and only if $\eta$ happens to be a function of $\xi$;
(c) given any three random variables, $\xi$, $\eta$ and $\zeta$, one has

$$I_{(\xi,\zeta)}(\eta) = H(\eta) - H_{(\xi,\zeta)}(\eta) \ge I_{\xi}(\eta),$$

i.e., the information about $\eta$ contained in observations over $(\xi, \zeta)$ cannot be less than the information about $\eta$ contained in $\xi$ alone.

**Problem 1.8.11.** Let $\xi_1, \ldots, \xi_n$ be independent and identically distributed Bernoulli random variables, with $P\{\xi_i = 1\} = p$ and $P\{\xi_i = 0\} = 1 - p$, and let $S_n = \xi_1 + \ldots + \xi_n$. Prove that

(a) $\quad P(\xi_1 = x_1, \ldots, \xi_n = x_n \mid S_n = k) = \dfrac{I_{\{x\}}(k)}{C_n^x}$ ,

(b) $\quad P(S_n = x \mid S_{n+m} = k) = \dfrac{C_n^x \, C_m^{k-x}}{C_{n+m}^k}$ ,

where $x = x_1 + \ldots + x_n$, $x_i = 0, 1$, so that $x \leq k$ .

## 1.9    Random Walk I: Probability of Ruin and Time Until Ruin in Coin Tossing

**Problem 1.9.1.** Verify the following generalization of [ $\underline{P}$ §1.9, (33) and (34)]:

$$ES_{\tau_n^x}^x = x + (p - q)E\tau_n^x,$$

$$E[S_{\tau_n^x}^x - \tau_n^x E\xi_1]^2 = D\xi_1 \cdot E\tau_n^x + x^2.$$

**Problem 1.9.2.** Consider the quantities $\alpha(x)$, $\beta(x)$ and $m(x)$, which are defined in [$\underline{P}$ §1.9], and investigate the limiting behavior of these quantities as the level $A$ decreases to $-\infty$ ($A \searrow -\infty$).
   *Hint.* The answer is this:

$$\lim_{A \to -\infty} \alpha(x) = \begin{cases} 0, & p \geq q, \\ \dfrac{(q/p)^\beta - (q/p)^x}{(q/p)^\beta}, & p < q, \end{cases}$$

$$\lim_{A \to -\infty} m(x) = \begin{cases} \dfrac{\beta - x}{p - q}, & p > q, \\ \infty, & p \leq q. \end{cases}$$

**Problem 1.9.3.** Consider a Bernoulli scheme with $p = q = 1/2$ and prove that

$$E|S_n| \sim \sqrt{\frac{2}{\pi} n}, \quad \text{as } n \to \infty. \tag{$*$}$$

*Hint.* One can verify directly the following discrete version of *Tanaka's formula* (see Problem 7.9.8): for any $n \geq 1$, one has

$$|S_n| = \sum_{k=1}^n \text{sign}(S_{k-1}) \, \Delta S_k + N_n, \tag{$**$}$$

where $S_0 = 0$, $S_k = \xi_1 + \ldots + \xi_k$, $\Delta S_k = \xi_k$,

$$\text{sign}\, x = \begin{cases} 1, & x > 0, \\ 0, & x = 0, \\ -1, & x < 0, \end{cases}$$

and $N_n = \#\{0 \le k \le n - 1 : S_k = 0\}$ is the number of integers $k$, $0 \le k \le n - 1$, for which $S_k = 0$. Then prove that

$$\mathsf{E}|S_n| = \mathsf{E}N_n = \mathsf{E}\sum_{k=0}^{n-1} I(S_k = 0) = \sum_{k=0}^{n-1} \mathsf{P}\{S_k = 0\} \qquad (\ast\ast\ast)$$

and use the fact that $\mathsf{P}\{S_{2k} = 0\} = 2^{-2k} C_{2k}^k$ and that $\mathsf{P}\{S_k = 0\} = 0$ for odd $k$.

*Remark.* One can conclude from $(\ast\ast\ast)$ that

$$\mathsf{E}N_n \sim \sqrt{\frac{2}{\pi}n}, \quad \text{as } n \to \infty.$$

(See [P §7.9, Example 2]—in formula (15) in that example $2\pi$ must be changed to $2/\pi$.)

**Problem 1.9.4.** Two players are tossing symmetric coins (each player tosses his own coin). Prove that the probability that both players will have the same numbers of heads after $n$ tosses is given by $2^{-2n} \sum_{k=0}^{n} (C_n^k)^2$ and conclude that the following identity must hold: $\sum_{k=0}^{n}(C_n^k)^2 = C_{2n}^n$ (see Problem 1.2.2).

Let $\sigma_n$ be the first instant when the number of heads obtained by the two players in a total of $n$ trials coincide, with the understanding that $\sigma_n = n + 1$, if coincidence does not occur in the first $n$ trials. Calculate the probability $\mathsf{P}\{\sigma_n = k\}$, $1 \le k \le n + 1$, and the expected value $\mathsf{E}\min(\sigma_n, n)$.

*Hint.* Let $\xi_i^{(k)} = 1$ (or $-1$), if player $k$, $k = 1, 2$, obtains a head (or a tail) in the $i^{\text{th}}$ trial. Then

$$\mathsf{P}\left\{ \begin{array}{l} \text{the numbers of heads obtained by} \\ \text{the players after } n \text{ trials coincide} \end{array} \right\} = \mathsf{P}\left\{ \sum_{i=1}^{n} \xi_i^{(1)} = \sum_{i=1}^{n} \xi_i^{(2)} \right\}$$

$$= \sum_{j=0}^{n} \mathsf{P}\left\{ \sum_{i=1}^{n} \xi_i^{(1)} = 2j - n, \ \sum_{i=1}^{n} \xi_i^{(2)} = 2j - n \right\} = \sum_{j=0}^{n} 2^{-2n} (C_n^j)^2$$

and

$$\mathsf{P}\left\{ \sum_{i=1}^{n} \xi_i^{(1)} = \sum_{i=1}^{n} \xi_i^{(2)} \right\} = \mathsf{P}\left\{ \sum_{i=1}^{2n} \eta_i = 0 \right\} = 2^{-2n} C_{2n}^n,$$

where $\eta_1 = \xi_1^{(1)}$, $\eta_2 = -\xi_1^{(2)}$, $\eta_3 = \xi_2^{(1)}$, $\eta_4 = \xi_2^{(2)}$, $\ldots$ .

**Problem 1.9.5.** Suppose that $\xi_1, \ldots, \xi_N$ are independent Bernoulli random variables with $P\{\xi_i = 1\} = P\{\xi_i = -1\} = 1/2$ and let $S_n = \xi_1 + \ldots + \xi_n$, $1 \leq n \leq N$. Compute

$$P\left( \bigcup_{N_1 < n \leq N_2} \{S_n = 0\} \right),$$

i.e., compute the probability that at some moment $n \in (N_1 + 1, \ldots, N_2)$, $N_2 \leq N$, one has $S_n = 0$.

**Problem 1.9.6.** Suppose that $\xi_1, \ldots, \xi_N$ are independent Bernoulli random variables with $P\{\xi_i = 1\} = P\{\xi_i = -1\} = 1/2$, $1 \leq i \leq N$. Set $S_n = \xi_1 + \ldots + \xi_n$ and consider the *discrete telegraph signal* $X_n = \xi_0(-1)^{S_n}$, $1 \leq n \leq N$. Find the values and the variance of the random variables $X_n$, $1 \leq n \leq N$. Find also the conditional distribution $P\{X_n = 1 \mid \xi_0 = i\}$, $i = \pm 1$, $1 \leq n \leq N$.

**Problem 1.9.7.** Let $\xi_1, \ldots, \xi_N$ be independent Bernoully random variables, with $P\{\xi_i = 1\} = p$ and $P\{\xi_i = -1\} = 1 - p$, and let $S_i = \xi_1 + \ldots + \xi_i$, $1 \leq i \leq N$, $S_0 = 0$. Let $\mathscr{R}_N$ be the *span* (or the *breadth*) of—i.e., the total number of locations visited by—the random walk $\{S_0, S_1, \ldots, S_N\}$.

Calculate $E\mathscr{R}_N$ and $D\mathscr{R}_N$. Explain for what values of $p$ one can claim that the variables $\mathscr{R}_N$ satisfy the following version of the law of large numbers

$$P\left\{ \left| \frac{\mathscr{R}_N}{N} - c \right| > \varepsilon \right\} \to 0, \quad N \to \infty,$$

where $\varepsilon > 0$ and $c$ is some constant. (See Problem 2.6.87 and Problem 8.8.16.)

**Problem 1.9.8.** Let $\xi_1, \ldots, \xi_N$ be identically distributed random variables (not necessarily of Bernoulli type) and set $S_0 = 0$, $S_i = \xi_1 + \ldots + \xi_i$, $1 \leq i \leq N$.

Let

$$N_n = \sum_{k=1}^{n} I(S_k > 0)$$

be the total number of positive elements of the sequence $S_0, S_1, \ldots, S_n$. Prove the Sparre-Andersen identity:

$$P\{N_n = k\} = P\{N_k = k\}P\{N_{n-k} = 0\}, \quad 0 \leq k \leq n.$$

**Problem 1.9.9.** Let $\xi_1, \ldots, \xi_N$ be the Bernoulli random variables from Problem 1.9.7 and define the variables $X_1, \ldots, X_N$ by

$$X_1 = \xi_1, \quad X_n = \lambda X_{n-1} + \xi_n, \quad 2 \leq n \leq N, \lambda \in \mathbb{R}.$$

Calculate $EX_n$, $DX_n$ and $\text{cov}(X_n, X_{n+k})$.

## 1.10  Random Walk II: The Reflection Principle and the Arcsine Law

**Problem 1.10.1.** Define $\sigma_{2n} = \min\{1 \leq k \leq 2n : S_k = 0\}$, with the understanding that $\sigma_{2n} = \infty$ (or $\sigma_{2n} = 2n$), if $S_k \neq 0$ for all $1 \leq k \leq 2n$. What is the rate of convergence in $\mathsf{E}\min(\sigma_{2n}, 2n) \to \infty$ as $n \to \infty$?

*Hint.* Note that according to $[\,\underline{P}\ \S1.10,\ \mathbf{1}\,]$ one must have

$$\mathsf{E}\min(\sigma_{2n}, 2n) = \sum_{k=1}^{n} u_{2(k-1)} + 2n\, u_{2n}\,,$$

where $u_{2n} \sim 1/\sqrt{\pi n}$, and conclude that

$$\mathsf{E}\min(\sigma_n, 2n) \sim 4\sqrt{\frac{n}{\pi}}\,, \quad n \to \infty\,.$$

**Problem 1.10.2.** Let $\tau_n = \min\{1 \leq k \leq n : S_k = 1\}$, with the understanding that $\tau_n = \infty$ if $S_k < 1$, for all $1 \leq k \leq n$. What is the limit of $\mathsf{E}\min(\tau_n, n)$ as $n \to \infty$ in the case of a symmetric ($p = q = 1/2$) and non-symmetric ($p \neq q$) Bernoulli walk?

*Hint.* The answer here is this:

$$\mathsf{E}\min(\tau_n, n) \to \begin{cases} (p - q)^{-1}, & p > q, \\ \infty, & p \leq q. \end{cases}$$

**Problem 1.10.3.** Based on the concepts and the methods developed in $[\,\underline{P}\ \S1.10\,]$, prove that the *symmetric* ($p = q = 1/2$) Bernoulli random walk $\{S_k, k \leq n\}$, given by $S_0 = 0$ and $S_k = \xi_1 + \ldots + \xi_k$, $k \geq 1$, satisfies the following relations ($N$ is any positive integer):

$$\mathsf{P}\Big\{\max_{1 \leq k \leq n} S_k \geq N, S_n < N\Big\} = \mathsf{P}\{S_n > N\}\,,$$

$$\mathsf{P}\Big\{\max_{1 \leq k \leq n} S_k \geq N\Big\} = 2\mathsf{P}\{S_n \geq N\} - \mathsf{P}\{S_n = N\}\,,$$

$$\mathsf{P}\Big\{\max_{1 \leq k \leq n} S_k = N\Big\} = \mathsf{P}\{S_n = N\} + \mathsf{P}\{S_n = N + 1\} = 2^{-n} C_n^{\lfloor \frac{n+N+1}{2} \rfloor}\,,$$

$$\mathsf{P}\Big\{\max_{1 \leq k \leq n} S_k \leq 0\Big\} = \mathsf{P}\{S_n = 0\} + \mathsf{P}\{S_n = 1\} = 2^{-n} C_n^{\lfloor \frac{n}{2} \rfloor}\,,$$

$$\mathsf{P}\Big\{\max_{1 \leq k \leq n-1} S_k \leq 0, S_n > 0\Big\} = \mathsf{P}\{S_1 \neq 0, \ldots, S_n \neq 0, S_{n+1} = 0\}\,,$$

$$P\{S_1 > 0, \ldots, S_{2n-1} > 0, S_{2n} = 0\} = \frac{1}{n} 2^{-2n} C_{2n-2}^{n-1},$$

$$P\{S_1 \geq 0, \ldots, S_{2n-1} \geq 0, S_{2n} = 0\} = \frac{1}{n+1} 2^{-2n} C_{2n}^{n}.$$

In addition, prove that the relations

$$P\{S_{2n} = 2k\} = 2^{-2n} C_{2n}^{n-k}, \quad k = 0, \pm 1, \ldots, \pm n,$$

and

$$P\{S_{2n+1} = 2k + 1\} = 2^{-2n-1} C_{2n+1}^{n-k}, \quad k = -(n+1), 0, \pm 1, \ldots, \pm n,$$

can be re-written in the form:

$$P\{S_n = k\} = \begin{cases} 2^{-n} C_n^{(n-k)/2}, & \text{if } k \equiv n \pmod 2, \\ 0 & \text{in all other cases}, \end{cases}$$

where $k = 0, \pm 1, \ldots, \pm n$.

In addition to the above formula for $P\{\max_{1 \leq k \leq n} S_k = N\}$ for positive integers $N$, prove that

$$P\left\{ \max_{0 \leq k \leq n} S_k = r \right\} = C_n^{\lfloor (n-r)/2 \rfloor} \cdot 2^{-n},$$

for $r = 0, 1, \ldots, n$.

**Problem 1.10.4.** Let $\xi_1, \ldots, \xi_{2n}$ be independent Bernoulli random variables with $P\{\xi_k = 1\} = P\{\xi_k = -1\} = 1/2, k \leq 2n$. Let $S_0 = 0$ and $S_k = \xi_1 + \ldots + \xi_k$, for $k \geq 1$, and, finally, let

$$g_{2n} = \max\{0 < 2k \leq 2n : S_{2k} = 0\}$$

be the moment of the *last zero* in the sequence $(S_2, S_4, \ldots, S_{2n})$, where we set $g_{2n} = 0$ if no such moment exists.

Prove that

$$P\{g_{2n} = 2k\} = u_{2n} u_{2(n-k)}, \quad 1 \leq k \leq n,$$

where $u_{2k} = P\{S_{2k} = 0\} = 2^{-2k} C_{2k}^{k}$.

By comparing the distribution of $g_{2n}$ with the probability $P_{2k,2n}$ of the event that on the interval $[0, 2n]$ the random walk spends $2k$ units of time in the positive axis (see formula [P §1.10, (12)]), one finds that, just as in formula [P §1.10, (15)], the following property holds for $0 < x < 1$:

$$\sum_{\{k:0<\frac{k}{n}\leq x\}} P\{g_{2n} = 2k\} \to \frac{2}{\pi} \arcsin \sqrt{x}, \quad \text{as } n \to \infty;$$

i.e., the probability distribution of the last zero satisfies the asymptotic arcsine law.

**Problem 1.10.5.** In the context of the previous problem, let $\theta_{2n}$ denote the moment of the first maximum in the sequence $S_0, S_1, \ldots, S_{2n}$, i.e., $\theta_{2n} = k$, if $S_0 < S_k, \ldots, S_{k-1} < S_k$, while $S_{k+1} \leq S_k, \ldots, S_{2n} \leq S_k$, and $\theta_{2n} = 0$ if no such $k > 1$ exists. Prove that

$$P\{\theta_{2n} = 0\} = u_{2n}, \quad P\{\theta_{2n} = 2n\} = \frac{1}{2} u_{2n},$$

and that, for $0 < k < n$,

$$P\{\theta_{2n} = 2k \text{ or } 2k + 1\} = \frac{1}{2} u_{2k} u_{2n-2k}.$$

Then conclude that, just as in the previous problem, the law of the moment of the first maximum satisfies the arcsine law: given any $0 < x < 1$, one has

$$\sum_{\{k:0<\frac{k}{n}\leq x\}} P\{\theta_{2n} = 2k \text{ or } 2k + 1\} \to \frac{2}{\pi} \arcsin \sqrt{x}, \quad n \to \infty.$$

Consider also the case $x = 0$ and $x = 1$.

**Problem 1.10.6.** Let $S_k = \xi_1 + \ldots + \xi_k, k \leq 2n$, where $\xi_1, \ldots, \xi_{2n}$ are independent and identically distributed random variables with $P\{\xi_1 = 1\} = P\{\xi_1 = -1\} = 1/2$. Prove that:
  (a) For $r = \pm 1, \ldots, \pm n$, one has

$$P\{S_1 \neq 0, \ldots, S_{2n-1} \neq 0, S_{2n} = 2r\} = C_{2n}^{n+r} \frac{|r|}{n} 2^{-2n}.$$

  (b) For $r = 0, \pm 1, \ldots, \pm n$, one has

$$P\{S_{2n} = 2r\} = C_{2n}^{n-r} 2^{-2n}.$$

**Problem 1.10.7.** Let $\{S_k, k \leq n\}$, given by $S_0 = 0$ and $S_k = \xi_1 + \ldots + \xi_k$, for $k \geq 1$, be a symmetric Bernoulli random walk (with independent and identically distributed $\xi_1, \ldots, \xi_n$, with $P\{\xi_1 = 1\} = P\{\xi_1 = -1\} = 1/2$). Setting

$$M_n = \max_{0 \leq k \leq n} S_k, \quad m_n = \min_{0 \leq k \leq n} S_k,$$

prove that

$$(M_n - S_n, S_n - m_n, S_n) \overset{\text{law}}{=} (-m_n, M_n, S_n) \overset{\text{law}}{=} (M_n, -m_n, S_n),$$

where " $\overset{\text{law}}{=}$ " means that the respective triplets share the same joint distribution.

**Problem 1.10.8.** Let $S_0 = 0$ and $S_k = \xi_1 + \ldots + \xi_k$, $k \geq 1$, where $\xi_1, \xi_2, \ldots$ are independent random variables with $P\{\xi_k = 1\} = p$ and $P\{\xi_k = -1\} = q$, $p + q = 1$. Prove that

$$P\left\{ \max_{1 \leq k \leq n} S_k \geq N, S_n = m \right\} = C_n^u \, p^v q^{n-v},$$

where $u = N + (n-m)/2$ and $v = (n+m)/2$, and conclude that, for $p = q = 1/2$ and $m \leq N$, one has

$$P\left\{ \max_{1 \leq k \leq n} S_k = N, S_n = m \right\} = P\{S_n = 2N - m\} - P\{S_n = 2N - m + 2\}.$$

**Problem 1.10.9.** Let $\xi_1, \xi_2, \ldots$ be any *infinite* sequence of independent Bernoulli random variables with $P\{\xi_i = +1\} = P\{\xi_i = -1\} = 1/2$. Define $S_0 = 0$, $S_n = \xi_1 + \cdots + \xi_n$, $n \geq 1$, and, given any $x \in \mathbb{Z} = \{0, \pm 1, \pm 2, \ldots\}$, consider the moment (of the *first* visit of $x$ *after* time zero):

$$\sigma_1(x) = \inf\{n > 0 : S_n = x\},$$

with the understanding that $\sigma_1(x) = \infty$, if $\{\cdot\} = \varnothing$.

Prove that, for $x = 1, 2, \ldots$, one has

$$P\{\sigma_1(x) > n\} = P\left\{ \max_{0 \leq k \leq n} S_k < x \right\}, \quad P\{\sigma_1(1) = 2n + 1\} = \frac{2^{-2n-1}}{n+1} C_{2n}^n,$$

$$P\{\sigma_1(x) = n\} = \frac{x}{n} 2^{-n} C_n^{(n+x)/2}, \quad P\{\sigma_1(1) > n\} = 2^{-n} C_n^{[n/2]}.$$

*Remark.* With regard to the question of existence of an *infinite* sequences of independent random variables $\xi_1, \xi_2, \ldots$, see [P §1.5, 1].

**Problem 1.10.10.** Let everything be as in the previous problem. In addition to the moments $\sigma_1(x)$, define the moments

$$\sigma_k(x) = \inf\{n > \sigma_{k-1}(x) : S_n = x\}, \quad k = 2, 3, \ldots,$$

with the understanding that $\sigma_k(x) = \infty$ if $\{\cdot\} = \varnothing$. (The meaning of these moments should be clear: $\sigma_k(x)$ is the moment of the $k^{\text{th}}$ visit to $x$.)

Prove that, for $n = 1, 2, \ldots$, one has

$$P\{\sigma_1(0) = 2n\} = 2^{-2n+1} n^{-1} C_{2n-2}^{n-1}, \qquad P\{\sigma_1(0) < \infty\} = 1,$$

$$P\{\sigma_1(0) > 2n\} = 2^{-2n} C_{2n}^n = P\{S_{2n} = 0\}, \qquad E\sigma_1(0) = \infty.$$

Show also that $\sigma_1(0), \sigma_2(0) - \sigma_1(0), \sigma_3(0) - \sigma_2(0), \ldots$ is a squence of independent and identically distributed random variables. (This property is the basis for the method of "regenerating cycles," which is crucial in the study of random walk sequences—see Sect. A.7 in the Appendix for details.)

**Problem 1.10.11.** Let $\xi_1, \xi_2, \ldots$ be any infinite sequence of independent Bernoulli random variables and let $S_0 = 0$ and $S_n = \xi_1 + \cdots + \xi_n$, $n \geq 1$. Define

$$L_n(x) = \#\{k, 0 < k \leq n : S_k = x\}$$

and notice that $L_n(x)$ is nothing but the total number of moments $0 < k \leq n$, at which the random walk $(S_k)_{0 < k \leq n}$ happens to be in state $x$, $x = 0, \pm 1, \pm 2, \ldots$.—comp. this definition with the related quantity $N_n(x)$, introduced in Problem 7.9.8; see also Problem 1.9.3. The quantities $L_n(x)$ and $N_n(x)$ are commonly referred to as (discrete) local times in state $x$ on the time interval $\{k : 0 \leq k \leq n\}$.

Prove that, for $k = 0, 1, \ldots, n$, one has:

$$P\{L_{2n}(0) = k\} = P\{L_{2n+1}(0) = k\} = 2^{-2n+k} C_{2n-k}^n ,$$

$$P\{L_n(0) = k\} = 2^{-2\lfloor n/2 \rfloor + k} C_{2\lfloor n/2 \rfloor - k}^{\lfloor n/2 \rfloor} ,$$

$$P\{L_{2n}(0) < k\} = P\{\sigma_k(0) > 2n\} = 2^{-2n} \sum_{j=0}^{k-1} 2^j C_{2n-j}^n ,$$

$$P\{L_n(x) = 0\} = P\{\sigma_1(x) > n\} = \sum_{j=n+1}^{\infty} 2^{-j} \frac{x}{j} C_j^{(j+x)/2} ,$$

and, for $x = \pm 1, \pm 2, \ldots$, one has:

$$P\{L_{\sigma_1(0)}(x) = 0\} = \frac{2|x| - 1}{2|x|} , \qquad EL_{\sigma_1(0)}(x) = 1 .$$

(The quantities $\sigma_1(x)$ are defined in Problem 1.10.10.)

**Problem 1.10.12.** In the context of the previous problem, set

$$\mu(n) = \min \left\{ k, 0 \leq k \leq n : S_k = \max_{0 \leq j \leq k} S_j \right\},$$

and prove that

$$P\{\mu(2n) = k\} = \begin{cases} C_{2\lfloor k/2 \rfloor}^{\lfloor k/2 \rfloor} ; C_{2n-2\lfloor k/2 \rfloor}^{n-\lfloor k/2 \rfloor} ; 2^{-2n-1}, & k = 1, 2, 3, \ldots, 2n, \\ C_{2n}^n ; 2^{-2n}, & k = 0. \end{cases}$$

## 1.11    Martingales: Some Applications to the Random Walk

**Problem 1.11.1.** Let $\mathcal{D}_0 \preccurlyeq \mathcal{D}_1 \preccurlyeq \cdots \preccurlyeq \mathcal{D}_n$ be any nondecreasing sequence of partitions of $\Omega$, such that $\mathcal{D}_0 = \{\Omega\}$, and let $\eta_k$, $1 \leq k \leq n$, be some random variables on $\Omega$, chosen so that each $\eta_k$ is $\mathcal{D}_k$-measurable. Prove that the sequence $\xi = (\xi_k, \mathcal{D}_k)_{1 \leq k \leq n}$, given by

$$\xi_k = \sum_{l=1}^{k} [\eta_l - \mathsf{E}(\eta_l \mid \mathcal{D}_{l-1})],$$

is a martingale.

*Hint.* Prove that $\mathsf{E}(\xi_{k+1} - \xi_k \mid \mathcal{D}_k) = 0$.

**Problem 1.11.2.** Suppose that the random variables $\eta_1, \ldots, \eta_n$ are chosen so that $\mathsf{E}\eta_1 = 0$ and $\mathsf{E}(\eta_k \mid \eta_1, \ldots, \eta_{k-1}) = 0$, $1 \leq k \leq n$. Prove that the sequence $\xi = (\xi_k)_{1 \leq k \leq n}$, given by $\xi_1 = \eta_1$ and

$$\xi_{k+1} = \sum_{i=1}^{k} f_i(\eta_1, \ldots, \eta_i) \eta_{i+1}, \quad k < n,$$

for some choice of the functions $f_i(\eta_1, \ldots, \eta_i)$, represents a martingale.

**Problem 1.11.3.** Prove that any martingale $\xi = (\xi_k, \mathcal{D}_k)_{1 \leq k \leq n}$ has *independent increments*: if $a < b < c < d$, then

$$\mathsf{cov}(\xi_d - \xi_c, \xi_b - \xi_a) = 0.$$

(Recall that in the present chapter all random variables are assumed to take only finitely many values.)

**Problem 1.11.4.** Let $\xi = (\xi_1, \ldots, \xi_n)$ be any random sequence in which each $\xi_k$ is $\mathcal{D}_k$-measurable ($\mathcal{D}_1 \preccurlyeq \mathcal{D}_2 \preccurlyeq \ldots \preccurlyeq \mathcal{D}_n$). Prove that in order for this sequence to be a martingale (relative to the partitions $(\mathcal{D}_k)$), it is necessary and sufficient that, for any stopping time $\tau$ (relative to $(\mathcal{D}_k)$), one has $\mathsf{E}\xi_\tau = \mathsf{E}\xi_1$. (The phrase "for any stopping time" may be replaced by "for any stopping time that takes only two values".)

*Hint.* Let $\mathsf{E}\xi_\tau = \mathsf{E}\xi_1$, for any stopping time $\tau$ that takes only two values. For a fixed $k \in \{1, \ldots, n-1\}$ and $A \in \mathcal{D}_k$, consider the moment

$$\tau(\omega) = \begin{cases} k, & \text{if } \xi_k(\omega) \notin A, \\ k+1, & \text{if } \xi_k(\omega) \in A. \end{cases}$$

After showing that $\mathsf{E}\xi_\tau = \mathsf{E}\xi_k I_{\overline{A}} + \mathsf{E}\xi_k I_A$, conclude that $\mathsf{E}\xi_{k+1} I_A = \mathsf{E}\xi_k I_A$.

**Problem 1.11.5.** Prove that if $\xi = (\xi_k, \mathscr{D}_k)_{1 \le k \le n}$ is a martingale and $\tau$ is a stopping time, then for any $k \le n$ one has

$$\mathsf{E}[\xi_n I_{\{\tau=k\}}] = \mathsf{E}[\xi_k I_{\{\tau=k\}}].$$

**Problem 1.11.6.** Let $\xi = (\xi_k, \mathscr{D}_k)_{1 \le k \le n}$ and $\eta = (\eta_k, \mathscr{D}_k)_{1 \le k \le n}$ be any two martingales with $\xi_1 = \eta_1 = 0$. Prove that

$$\mathsf{E}\xi_n \eta_n = \sum_{k=2}^{n} \mathsf{E}(\xi_k - \xi_{k-1})(\eta_k - \eta_{k-1})$$

and that, in particular,

$$\mathsf{E}\xi_n^2 = \sum_{k=2}^{n} \mathsf{E}(\xi_k - \xi_{k-1})^2.$$

**Problem 1.11.7.** Let $\eta_1, \ldots, \eta_n$ be any sequence of independent and identically distributed random variables with $\mathsf{E}\eta_i = 0$. Prove that the sequence $\xi = (\xi_k)_{1 \le k \le n}$, given by

$$\xi_k = \left( \sum_{i=1}^{k} \eta_i \right)^2 - k \, \mathsf{E}\eta_1^2, \quad \text{or by} \quad \xi_k = \frac{\exp\{\lambda(\eta_1 + \ldots + \eta_k)\}}{(\mathsf{E}\exp\{\lambda\eta_1\})^k},$$

represents a martingale.

**Problem 1.11.8.** Let $\eta_1, \ldots, \eta_n$ be any sequence of independent and identically distributed random variables that take values only in the (finite) set $Y$. Let $f_0(y) = \mathsf{P}\{\eta_1 = y\} > 0$, $y \in Y$, and let $f_1(y)$ be any non-negative function with $\sum_{y \in Y} f_1(y) = 1$. Prove that the sequence $\xi = (\xi_k, \mathscr{D}_k^\eta)_{1 \le k \le n}$, with

$$\xi_k = \frac{f_1(\eta_1) \ldots f_1(\eta_k)}{f_0(\eta_1) \ldots f_0(\eta_k)}, \quad \mathscr{D}_k^\eta = \mathscr{D}_{\eta_1, \ldots, \eta_k},$$

forms a martingale. (The variables $\xi_k$ are known as *likelihood ratios* and play a fundamental role in statistics.)

**Problem 1.11.9.** We say that the sequence $\xi = (\xi_k, \mathscr{D}_k)_{0 \le k \le n}$ is a supermartingale (submartingale) if $\mathsf{P}$-a.s. one has

$$\mathsf{E}(\xi_{k+1} \mid \mathscr{D}_k) \le \xi_k \quad (\ge \xi_k), \ 0 \le k \le n.$$

Prove that every supermartingale (submartingale) can be represented (and in a unique way) in the form

$$\xi_k = m_k - a_k \quad (+ a_k),$$

where $m = (m_k, \mathscr{D}_k)_{0 \le k \le n}$ is a martingale and $a = (a_k, \mathscr{D}_k)_{0 \le k \le n}$ is a non-decreasing sequence such that $a_0 = 0$ and each $a_k$ is $\mathscr{D}_{k-1}$-measurable.

**Problem 1.11.10.** Let $\xi = (\xi_k, \mathscr{D}_k)_{0 \le k \le n}$ and $\eta = (\eta_k, \mathscr{D}_k)_{0 \le k \le n}$ be any two supermartingales and let $\tau$ be any stopping time, relative to the partition $(\mathscr{D}_k)_{0 \le k \le n}$, chosen so that $\mathsf{P}\{\xi_\tau \ge \eta_\tau\} = 1$. Prove that any sequence $\zeta = (\zeta_k, \mathscr{D}_k)_{0 \le k \le n}$ that switches from $\eta$ to $\xi$ at the random moment $\tau$, i.e., any $\zeta$ given either by

$$\zeta_k = \xi_k I(\tau > k) + \eta_k I(\tau \le k),$$

or by

$$\zeta_k = \xi_k I(\tau \ge k) + \eta_k I(\tau < k),$$

is also a supermartingale.

**Problem 1.11.11.** Let $\xi = (\xi_k, \mathscr{D}_k)_{0 \le k \le n}$ be any submartingale of the form

$$\xi_k = \sum_{m \le k} I_{A_m},$$

where $A_m \in \mathscr{D}_m$. Find the Doob decomposition for this submartingale.

**Problem 1.11.12.** Let $\xi = (\xi_k, \mathscr{D}_k)_{1 \le k \le n}$ be any submartingale. Verify the following "*maximal*" inequality:

$$\mathsf{E} \max_{l \le n} \xi_l^+ \le \frac{e}{1-e} [1 + \mathsf{E}(\xi_n^+ \ln^+ \xi_n^+)],$$

where $\ln^+ x = \max(\ln x, 0)$.

## 1.12   Markov Chains: The Ergodic Theorem: The Strong Markov Property

**Problem 1.12.1.** Let $\xi = (\xi_0, \xi_1, \dots, \xi_n)$ be a Markov chain with values in the space $X$ and let $f = f(x)$ $(x \in X)$ be some function on $X$. Does the sequence $(f(\xi_0), f(\xi_1), \dots, f(\xi_n))$ represent a Markov chain? Does the "reverse" sequence $(\xi_n, \xi_{n-1}, \dots, \xi_0)$ represent a Markov chain?

**Problem 1.12.2.** Let $\mathbb{P} = \|p_{ij}\|$, $1 \le i, j \le r$ be any stochastic matrix and let $\lambda$ be any eigenvalue of that matrix, i.e., $\lambda$ is a solution to the characteristic equation $\det[\mathbb{P} - \lambda I] = 0$. Prove that $\lambda_1 = 1$ is always an eigenvalue and that the absolute values of all remaining eigenvalues $\lambda_2, \dots, \lambda_r$ cannot exceed 1. Furthermore, if there is a number $n$, such that $\mathbb{P}^n > 0$ (in the sense that $p_{ij}^{(n)} > 0$), then $|\lambda_i| < 1$, $i = 2, \dots, r$. Show also that if all eigenvalues $\lambda_1, \dots, \lambda_r$ are different, then the transition probabilities $p_{ij}^{(k)}$ can be written as

$$p_{ij}^{(k)} = \pi_j + a_{ij}(2)\lambda_2^k + \dots + a_{ij}(r)\lambda_r^k,$$

where the quantities $\pi_j, a_{ij}(2), \dots, a_{ij}(r)$ can be expressed in terms of the entries of the matrix $\mathbb{P}$. (In particular, as a result of this *algebraic* approach to the study of the asymptotic properties of Markov chains, we find that, if $|\lambda_2| < 1, \dots, |\lambda_r| < 1$, then the limit $\lim_k p_{ij}^{(k)}$ exists for any $j$ and, in fact, does not depend on $i$.)

**Problem 1.12.3.** Let $\xi = (\xi_0, \xi_1, \dots, \xi_n)$ be any homogeneous Markov chain with (finite) state space $X$ and with transition probability matrix $\mathbb{P} = \|p_{xy}\|$. Denote by

$$T\varphi(x) = \mathsf{E}[\varphi(\xi_1) \mid \xi_0 = x] \quad \left(= \sum_y \varphi(y) p_{xy}\right)$$

the associated one-step *transition operator* and suppose that the function $\varphi = \varphi(x)$ satisfies the equation

$$T\varphi(x) = \varphi(x), \quad x \in X,$$

i.e., happens to be "harmonic." Prove that for any such choice of the function $\varphi$, the sequence

$$\zeta = (\zeta_k, \mathscr{D}_k^\xi)_{0 \le k \le n}, \quad \text{with} \quad \zeta_k = \varphi(\xi_k),$$

is a martingale, where $\mathscr{D}_k^\xi = \mathscr{D}_{\xi_0, \dots, \xi_k}$.

**Problem 1.12.4.** Let $\xi = (\xi_n, \Pi, \mathbb{P})$ and $\tilde{\xi} = (\tilde{\xi}_n, \tilde{\Pi}, \mathbb{P})$ be any two Markov chains that share the same transition matrix $\mathbb{P} = \|p_{ij}\|$, $1 \le i, j \le r$, but have two different initial distributions, resp. $\Pi = (p_1, \dots, p_r)$ and $\tilde{\Pi} = (\tilde{p}_1, \dots, \tilde{p}_r)$. Letting $\Pi^{(n)} = (p_1^{(n)}, \dots, p_r^{(n)})$ and $\tilde{\Pi}^{(n)} = (\tilde{p}_1^{(n)}, \dots, \tilde{p}_r^{(n)})$ denote the respective $n$-steps distributions, prove that if $\min_{i,j} p_{ij} \ge \varepsilon > 0$, then

$$\sum_{i=1}^r |\tilde{p}_i^{(n)} - p_i^{(n)}| \le 2(1 - r\varepsilon)^n.$$

*Hint.* Use induction in $n$.

**Problem 1.12.5.** Let $P$ and $Q$ be any two stochastic matrices. Prove that $PQ$ and $\alpha P + (1 - \alpha)Q$, for any choice of $0 \le \alpha \le 1$, are also stochastic matrices.

**Problem 1.12.6.** Consider any homogeneous Markov chain $(\xi_0, \xi_1, \dots, \xi_n)$ with state-space $X = \{0, 1\}$ and with transition probability matrix of the form

$$\begin{pmatrix} 1 - \alpha & \alpha \\ \beta & 1 - \beta \end{pmatrix},$$

for some $0 < \alpha < 1$ and $0 < \beta < 1$, and set $S_n = \xi_0 + \dots + \xi_n$. As a generalization of the Moivre–Laplace theorem (see [P §1.6]), prove that

$$\mathsf{P}\left\{ \frac{S_n - \frac{\alpha}{\alpha+\beta}n}{\sqrt{\frac{n\alpha\beta(2-\alpha-\beta)}{(\alpha+\beta)^3}}} \le x \right\} \to \Phi(x), \quad \text{as } n \to \infty.$$

Argue that if $\alpha + \beta = 1$ one can claim that the variables $\xi_0, \ldots, \xi_n$ are independent and that the last relation comes down to

$$\mathsf{P}\left\{ \frac{S_n - \alpha n}{\sqrt{n\alpha\beta}} \leq x \right\} \to \Phi(x), \quad \text{as } n \to \infty.$$

**Problem 1.12.7.** Let $\xi_0, \xi_1, \ldots, \xi_N$ be any Bernoulli sequence of independent random variables with $\mathsf{P}\{\xi_i = 1\} = \mathsf{P}\{\xi_i = -1\} = 1/2$. Consider the variables $\eta_0, \eta_1, \ldots, \eta_N$, defined by $\eta_0 = \xi_0$ and $\eta_n = \frac{\xi_{n-1} + \xi_n}{2}$, $1 \leq n \leq N$.
   (a) Is the sequence $\eta_0, \eta_1, \ldots, \eta_N$ Markovian?
   (b) Is the sequence $\zeta_0, \zeta_1, \ldots, \zeta_N$, given by $\zeta_0 = \xi_0$, and $\zeta_n = \xi_{n-1}\xi_n$, $1 \leq n \leq N$, Markovian?

**Problem 1.12.8.** Let $X_1, \ldots, X_n$ be any collection of independent and identically distributed random variables. With any such collection one can associate what is known as the *order statistics* and is defined as the sequence $X_1^{(n)}, \ldots, X_n^{(n)}$, obtained by arranging the values $X_1, \ldots, X_n$ in non-decreasing order. (So that $X_1^{(n)} = \min(X_1, \ldots, X_n), \ldots, X_n^{(n)} = \max(X_1, \ldots, X_n)$, with the understanding that when $X_{i_1} = \ldots = X_{i_k} = \min(X_1, \ldots, X_k)$ and $i_1 < \ldots < i_k$, then $X_1^{(n)}$ is the variable $X_{i_1}$. Similar convention is made in all analogous cases (see Problem 2.8.19, for example).

Note that, in general, the elements of the order statistics $X_1^{(n)}, \ldots, X_n^{(n)}$ (known as *rank statistics*) will not be independent even if the variables $X_1, \ldots, X_n$ are.

Prove that when each variable $X_i$ takes only two values, the rank statistics form a *Markov chain*. Prove by way of example that, in general, this claim cannot be made if each $X_i$ takes three values. (Note that if each $X_i$ is continuously distributed (see [$\underline{P}$ §2.3]), then the rank statistics always form a Markov chain.)

# Chapter 2
# Mathematical Foundations
# of Probability Theory

## 2.1 Probabilistic Models of Experiments with Infinitely Many Outcomes: Kolmogorov's Axioms

**Problem 2.1.1.** Let $\Omega = \{r : r \in [0, 1] \cap \mathbb{Q}\}$ denote the set of all rational numbers inside the interval $[0, 1]$, let $\mathscr{A}$ be the algebra of sets that can be expressed as finite unions of non-intersecting sets $A$ of the form $\{r : a < r < b\}$, $\{r : a \le r < b\}$, $\{r : a < r \le b\}$, or $\{r : a \le r \le b\}$, and let $\mathsf{P}(A) = b - a$. Prove that the set-function $\mathsf{P}(A)$, $A \in \mathscr{A}$, is finitely additive but not countably additive.

**Problem 2.1.2.** Let $\Omega$ be any countable set and let $\mathscr{F}$ denote the collection of all subsets of $\Omega$. Set $\mu(A) = 0$ if $A$ is finite and $\mu(A) = \infty$ if $A$ is infinite. Prove that the set-function $\mu$ is finitely additive but not countably additive.

**Problem 2.1.3.** Let $\mu$ be any countably additive measure on $(\Omega, \mathscr{F})$. Prove that
(a) If $A_n \uparrow A$, then $\mu(A_n) \uparrow \mu(A)$;
(b) If $A_n \downarrow A$ and $\mu(A_k) < \infty$ for some $k$, then $\mu(A_n) \downarrow \mu(A)$;
(c) If $\mu$ is finite ($\mu(\Omega) < \infty$) and $A = \lim A_n$, i.e., $A = \overline{\lim} A_n = \underline{\lim} A_n$, then $\mu(A) = \lim \mu(A_n)$.

(This problem continues in Problem 2.1.15.)
*Hint.* Use the relations

$$\underline{\lim} A_n = \bigcup_{n=1}^{\infty} \bigcap_{k=n}^{\infty} A_k \quad \text{and} \quad \overline{\lim} A_n = \bigcap_{n=1}^{\infty} \bigcup_{k=n}^{\infty} A_k.$$

**Problem 2.1.4.** Verify the following properties of the symmetric difference between sets:

$$(A \triangle B) \triangle C = A \triangle (B \triangle C), \quad (A \triangle B) \triangle (B \triangle C) = A \triangle C,$$

$$A \triangle B = C \iff A = B \triangle C.$$

A.N. Shiryaev, *Problems in Probability*, Problem Books in Mathematics, DOI 10.1007/978-1-4614-3688-1_2, © Springer Science+Business Media New York 2012

**Problem 2.1.5.** Prove that the "metrics" $\rho_1(A, B)$ and $\rho_2(A, B)$, defined by

$$\rho_1(A, B) = P(A \triangle B),$$

$$\rho_2(A, B) = \begin{cases} \frac{P(A \triangle B)}{P(A \cup B)}, & \text{if } P(A \cup B) \neq 0, \\ 0, & \text{if } P(A \cup B) = 0, \end{cases}$$

where $A \triangle B$ is the symmetric difference of $A$ and $B$, satisfy the "triangular inequality."

*Hint.* Use the relation $A \triangle C \subseteq (A \triangle B) \cup (B \triangle C)$.

**Problem 2.1.6.** Let $\mu$ be any finitely additive measure on some algebra $\mathscr{A}$. Show that if the sets $A_1, A_2, \cdots \in \mathscr{A}$ are non-intersecting and, in addition, one has $A = \sum_{i=1}^{\infty} A_i \in \mathscr{A}$, then one can claim that $\mu(A) \geq \sum_{i=1}^{\infty} \mu(A_i)$.

**Problem 2.1.7.** Prove that

$$\overline{\lim \sup A_n} = \lim \inf \overline{A_n}, \quad \overline{\lim \inf A_n} = \lim \sup \overline{A_n},$$

$$\lim \inf A_n \subseteq \lim \sup A_n, \quad \lim \sup(A_n \cup B_n) = \lim \sup A_n \cup \lim \sup B_n,$$

$$\lim \inf(A_n \cap B_n) = \lim \inf A_n \cap \lim \inf B_n,$$

$$\lim \sup A_n \cap \lim \inf B_n \subseteq \lim \sup(A_n \cap B_n) \subseteq \lim \sup A_n \cap \lim \sup B_n.$$

Prove also that if $A_n \uparrow A$, or if $A_n \downarrow A$, then

$$\lim \inf A_n = \lim \sup A_n.$$

**Problem 2.1.8.** Let $(x_n)$ be any sequence of real numbers and let $A_n = (-\infty, x_n)$. Prove that for $x = \lim \sup x_n$ and $A = \lim \sup A_n$ one has $(-\infty, x) \subseteq A \subseteq (-\infty, x]$. (In other words, $A$ must be either $(-\infty, x)$ or $(-\infty, x]$.)

**Problem 2.1.9.** Lt $A_1, A_2, \ldots$ be any sequence of subsets of the set $\Omega$. Prove that

$$\lim \sup (A_n \setminus A_{n+1}) = \lim \sup (A_{n+1} \setminus A_n) = (\lim \sup A_n) \setminus (\lim \inf A_n).$$

**Problem 2.1.10.** Give an example showing that, in general, a measure that can take the value $+\infty$ could be countably additive, but still not continuous at "the zero" $\varnothing$.

**Problem 2.1.11.** We say that the events $\{A_i \in \mathscr{F} : 1 \leq i \leq n\}$ are *exchangeable* (or *interchangeable*) if all probabilities $P(A_{i_1} \ldots A_{i_l})$ are identical $(= p_l)$ for every choice of the indices $1 \leq i_1 < \cdots < i_l \leq n$, and this property holds separately for every $1 \leq l \leq n$. Prove that for such events the following "inclusion–exclusion" formula is in place (see Problem 1.1.12):

$$P\left(\bigcup_{i=1}^{n} A_i\right) = np_1 - C_n^2 \, p_2 + C_n^3 \, p_3 - \cdots + (-1)^{n-1} p_n.$$

**Problem 2.1.12.** Let $(A_k)_{k\geq 1}$ be any *infinite* sequence of exchangeable events, i.e., for *every* $l \geq 1$ one can claim that the probability $P(A_{i_1} \ldots A_{i_l})$ $(= p_l)$ does not depend on choice of the indices $1 \leq i_1 < \cdots < i_l$. Prove that in any such situation one has

$$P\left(\varliminf_n A_n\right) = P\left(\bigcap_{k=1}^{\infty} A_k\right) = \lim_{l\to\infty} p_l,$$

$$P\left(\varlimsup_n A_n\right) = P\left(\bigcup_{k=1}^{\infty} A_k\right) = 1 - \lim_{l\to\infty} (-1)^l \Delta^l(p_0),$$

where $p_0 = 1$, $\Delta^1(p_n) = p_{n+1} - p_n$, $\Delta^l(p_n) = \Delta^1(\Delta^{l-1}(p_n))$, $l \geq 2$.

**Problem 2.1.13.** Let $(A_n)_{n\geq 1}$ be any sequence of sets and let $I(A_n)$, $n \geq 1$, be the associated sequence of indicator functions. Prove that

(a)   $I\left(\varliminf_n A_n\right) = \varliminf_n I(A_n), \quad I\left(\varlimsup_n A_n\right) = \varlimsup_n I(A_n),$

(b)   $\varlimsup_n I(A_n) - \varliminf_n I(A_n) = I\left(\varlimsup_n A_n \setminus \varliminf_n A_n\right),$

(c)   $I\left(\bigcup_{n=1}^{\infty} A_n\right) \leq \sum_{n=1}^{\infty} I(A_n).$

**Problem 2.1.14.** Prove that

$$I\left(\bigcup_{n=1}^{\infty} A_n\right) = \max_{n\geq 1} I(A_n), \quad I\left(\bigcap_{n=1}^{\infty} A_n\right) = \min_{n\geq 1} I(A_n).$$

**Problem 2.1.15.** (*Continuation of Problem 2.1.3.*) Let $\mu$ be any countably additive measure on $(\Omega, \mathscr{F})$. Prove that
(a) $\mu(\varliminf A_n) \leq \varliminf \mu(A_n)$.
(b) If, in addition, the measure $\mu$ happens to be finite ($\mu(\Omega) < \infty$), then

$$\mu(\varlimsup A_n) \geq \varlimsup \mu(A_n).$$

(c) In the special case of *probability* measures P, one has

$$P(\varliminf A_n) \leq \varliminf P(A_n) \leq \varlimsup P(A_n) \leq P(\varlimsup A_n)$$

("Fatou's lemma for sets").

Deduce from the above relations the following generalization of the "continuity" properties (2) and (3) of the probability P, mentioned in the Theorem of [ P §2.1, **2** ]: if $A = \lim_n A_n$ (i.e., $\varlimsup A_n = \varliminf A_n = A$), then $P(A) = \lim_n P(A_n)$.

**Problem 2.1.16.** Let $A^* = \overline{\lim} A_n$ and let $A_* = \underline{\lim} A_n$. Prove that $P(A_n - A_*) \to 0$ and $P(A^* - A_n) \to 0$.

**Problem 2.1.17.** Suppose that $A_n \to A$ (in the sense that $A = A^* = A_*$; see Problem 2.1.16). Prove that $P(A \triangle A_n) \to 0$.

**Problem 2.1.18.** Suppose that the sets $A_n$ converge to the set $A$, in the sense that $P(A \triangle \overline{\lim} A_n) = P(A \triangle \underline{\lim} A_n) = 0$. Prove that in that case one must have $P(A \triangle A_n) \to 0$.

**Problem 2.1.19.** Let $A_0, A_1, \ldots$ and $B_0, B_1, \ldots$ be any two sequences of subsets of $\Omega$. Verify the following properties of the symmetric difference:

$$A_0 \triangle B_0 = \overline{A_0} \triangle \overline{B_0},$$

$$A_0 \triangle \left( \bigcup_{n \geq 1} B_n \right) \subseteq \bigcup_{n \geq 1} (A_0 \triangle B_n),$$

$$A_0 \triangle \left( \bigcap_{n \geq 1} B_n \right) \supseteq \bigcap_{n \geq 1} (A_0 \triangle B_n),$$

$$\left( \bigcup_{n \geq 1} A_n \right) \triangle \left( \bigcup_{n \geq 1} B_n \right) \subseteq \bigcup_{n \geq 1} (A_n \triangle B_n),$$

$$\left( \bigcap_{n \geq 1} A_n \right) \triangle \left( \bigcap_{n \geq 1} B_n \right) \subseteq \bigcup_{n \geq 1} (A_n \triangle B_n).$$

**Problem 2.1.20.** Let $A, B, C$ be any three random events. Prove that

$$|P(A \cap B) - P(A \cap C)| \leq P(B \triangle C).$$

**Problem 2.1.21.** Prove that for any three events, $A$, $B$ and $C$, the probability that exactly one of these events will occur can be expressed as $P(A) + P(B) + P(C) - 2(P(AB) + P(AC) + P(CD)) + 3P(ABC)$. (Comp. with Problem 1.1.13.)

**Problem 2.1.22.** Let $(A_n)_{n \geq 1}$ be any sequence of events in $\mathscr{F}$, for which

$$\sum_n P(A_n \triangle A_{n+1}) < \infty.$$

Prove that

$$P\{(\overline{\lim} A_n) \triangle (\underline{\lim} A_n)\} = 0.$$

**Problem 2.1.23.** Prove that, for any two events $A$ and $B$, one has

$$\max(P(A), P(B)) \leq P(A \cup B) \leq 2 \max(P(A), P(B))$$

and

$$P(A \cup B)P(A \cap B) \le P(A)P(B).$$

When can one claim that the last relation is actually an identity?

Show also the *Boole inequality*: $P(A \cap B) \ge 1 - P(\overline{A}) - P(\overline{B})$.

**Problem 2.1.24.** Let $(A_n)_{n \ge 1}$ and $(B_n)_{n \ge 1}$ be any two sequences of events chosen so that $A_n \subseteq B_n$ for every $n \ge 1$. Prove that $\{A_n \text{ i. o. }\} \subseteq \{B_n \text{ i. o. }\}$ ("i.o." stands for "infinitely often," meaning that infinitely many events in the associated sequence occur).

**Problem 2.1.25.** Suppose again that $(A_n)_{n \ge 1}$ and $(B_n)_{n \ge 1}$ are two sequences of events such that

$$P\{A_n \text{ i. o. }\} = 1 \quad \text{and} \quad P\{\overline{B}_n \text{ i. o. }\} = 0.$$

Prove that $P\{A_n \cap B_n \text{ i. o. }\} = 1$.

**Problem 2.1.26.** Give an example of two *finite* measures, $\mu_1$ and $\mu_2$, defined on the same sample space $\Omega$ (i.e., two measures with $\mu_1(\Omega) < \infty$ and $\mu_2(\Omega) < \infty$), for which *the smallest measure*, $\nu$, with the property $\nu \ge \mu_1$ and $\nu \ge \mu_2$ is not, as one might think, $\max(\mu_1, \mu_2)$, but is actually $\mu_1 + \mu_2$.

**Problem 2.1.27.** Suppose that the measure space $(\Omega, \mathscr{F})$ is endowed with a sequence of probability measures, $P_1, P_2, \ldots$, and suppose that $P(A)$, $A \in \mathscr{F}$, is some set-function on $\mathscr{F}$, for which the following relation holds for every $A \in \mathscr{F}$:

$$P_n(A) \to P(A).$$

Prove the following properties, known as *the Vitali–Hahn–Saks theorem*:

(a) The set-function $P = P(\cdot)$ is a probability measure on $(\Omega, \mathscr{F})$.

(b) For any sequence $A_1, A_2, \ldots \in \mathscr{F}$ with $A_k \downarrow \varnothing$ as $k \to \infty$, one must have $\sup_n P_n(A_k) \downarrow 0$ as $k \to \infty$.

**Problem 2.1.28.** Consider the measure space $(\mathbb{R}, \mathscr{B}(\mathbb{R}))$ and give an example of a sequence of measures $\mu_n = \mu_n(A)$, $A \in \mathscr{B}(\mathbb{R})$, $n \ge 1$, such that for every $A \in \mathscr{B}(\mathbb{R})$ the sequence $(\mu_n(A))_{n \ge 1}$ is decreasing, but the limit $\nu(A) = \lim_n \mu_n(A)$, $A \in \mathscr{B}(\mathbb{R})$, does not represent a finitely-additive set function and, therefore, cannot be treated as a measure.

**Problem 2.1.29.** Let $(\Omega, \mathscr{F}, P)$ be any probability space and let $(A_n)_{n \ge 1}$ be any sequence of events inside $\mathscr{F}$. Suppose that $P(A_n) \ge c > 0$, $n \ge 1$, and let $A = \overline{\lim} A_n$. Prove that $P(A) \ge c$.

**Problem 2.1.30.** (*The Huygens problem.*) Two players, A and B, take turns tossing two fair dice. Player A wins if he gets a six before player B gets a seven (otherwise player A looses and player B wins). Assuming that player A tosses first, what is the probability that player A wins?

*Hint.* One must calculate the probability $P_A = \sum_{k=0}^{\infty} P(A_k)$, where $A_k$ is the event that player A wins after the $(k + 1)^{\text{st}}$ turn. (The answer is: $P_A = 30/61$.)

**Problem 2.1.31.** Suppose that the set $\Omega$ is at most countable and let $\mathscr{F}$ be any $\sigma$-algebra of subsets of $\Omega$. Prove that it is always possible to find a partition $\mathscr{D} = \{D_1, D_2, \ldots\}$ ($\bigcup_{i \in \mathbb{N}} D_i = \Omega$, $D_i \cap D_j = \varnothing$, $i \neq j$, $\mathbb{N} = \{1, 2, \ldots\}$) that generates $\mathscr{F}$, i.e.,

$$\mathscr{F} = \left\{ \bigcup_{i \in M} D_i : M \subseteq \mathbb{N} \right\}.$$

(Comp. with the analogous statement for the case of a finite set $\Omega$ formulated in [P §1.1, **3**])

*Hint.* Consider constructing $\mathscr{D}$ from the equivalence classes in the set $\Omega$ associated with the relation

$$\omega_1 \sim \omega_2 \Leftrightarrow (\omega_1 \in A \Leftrightarrow \omega_2 \in A \text{ for every } A \in \mathscr{F}).$$

## 2.2  Algebras and $\sigma$-algebras: Measurable Spaces

**Problem 2.2.1.** Let $\mathscr{B}_1$ and $\mathscr{B}_2$ be any two $\sigma$-algebras of subsets of the space $\Omega$. Can one claim that the following collections of sets form $\sigma$-algebras

$$\mathscr{B}_1 \cap \mathscr{B}_2 \equiv \{A : A \in \mathscr{B}_1 \text{ and } A \in \mathscr{B}_2\},$$

$$\mathscr{B}_1 \cup \mathscr{B}_2 \equiv \{A : A \in \mathscr{B}_1 \text{ or } A \in \mathscr{B}_2\}?$$

Let $\mathscr{B}_1 \vee \mathscr{B}_2$ be the smallest $\sigma$-algebra, $\sigma(\mathscr{B}_1, \mathscr{B}_2)$, that contains $\mathscr{B}_1$ and $\mathscr{B}_2$. Prove that $\mathscr{B}_1 \vee \mathscr{B}_2$ coincides with the smallest $\sigma$-algebra that contains all sets of the form $B_1 \cap B_2$, for all choices of $B_1 \in \mathscr{B}_1$ and $B_2 \in \mathscr{B}_2$.

*Hint.* Convince yourself that $\mathscr{B}_1 \cap \mathscr{B}_2$ is a $\sigma$-algebra and prove by way of example that, in general, $\mathscr{B}_1 \cup \mathscr{B}_2$ is not a $\sigma$-algebra. (Such an example can be constructed with a set $\Omega$ that has only three elements.)

**Problem 2.2.2.** Let $\mathscr{D} = \{D_1, D_2, \ldots\}$ be any countable partition of the set $\Omega$ and let $\mathscr{B} = \sigma(\mathscr{D})$. What is the cardinality of the $\sigma$-algebra $\mathscr{B}$?

*Hint.* With any sequence $x = (x_1, x_2, \ldots)$, that consists of 0's and 1's, one can associate the set $D^x = D_1^{x_1} \cup D_2^{x_2} \cup \ldots$, where $D_i^{x_i} = \varnothing$, if $x_i = 0$, and $D_i^{x_i} = D_i$, if $x_i = 1$.

**Problem 2.2.3.** Prove that

$$\mathscr{B}(\mathbb{R}^n) \otimes \mathscr{B}(\mathbb{R}) = \mathscr{B}(\mathbb{R}^{n+1}).$$

**Problem 2.2.4.** Prove that the sets (b)–(f) from [P §2.2, **4**] belong to $\mathscr{B}(\mathbb{R}^\infty)$.

*Hint.* In order to show, for example, (b), notice that

$$\left\{ x : \overline{\lim_n} x_n \leq a \right\} = \bigcap_{k=1}^{\infty} \bigcup_{m=1}^{\infty} \bigcap_{n=m}^{\infty} \left\{ x : x_n < a + \frac{1}{k} \right\}.$$

In other words, $\overline{\lim}_n x_n \leq a \ \Leftrightarrow \ \forall k \in N \ \exists m \in N : \forall n \geq m, \ x_n < a + \dfrac{1}{k}$. One can derive (c)–(f) with a similar argument.

**Problem 2.2.5.** Prove that the sets $A_2$ and $A_3$ from [ P §2.2, **5**] *do not* belong to $\mathscr{B}(\mathbb{R}^{[0,1]})$.

   *Hint.* As is the case with the set $A_1$, the desired property of $A_2$ and $A_3$ can be established by contradiction.

**Problem 2.2.6.** Verify that the function in [ P §2.2, (18)] is indeed a metric.

**Problem 2.2.7.** Prove that $\mathscr{B}_0(\mathbb{R}^n) = \mathscr{B}(\mathbb{R}^n), n \geq 1$, and $\mathscr{B}_0(\mathbb{R}^\infty) = \mathscr{B}(\mathbb{R}^\infty)$.

**Problem 2.2.8.** Let $C = C[0, \infty)$ be the space of continuous functions $x = (x_t)$, defined for $t \geq 0$. Prove that, relative to the metric

$$\rho(x, y) = \sum_{n=1}^{\infty} 2^{-n} \min[\ \sup_{0 \leq t \leq n} |x_t - y_t|, \ 1], \quad x, y \in C,$$

this space is a Polish space (just as $C = C[0, 1]$), i.e., *a complete, separable metric space*, and the $\sigma$-algebra $\mathscr{B}_0(C)$, generated by all open sets, coincides with the $\sigma$-algebra $\mathscr{B}(C)$, generated by all cylinder sets.

**Problem 2.2.9.** Show the equivalence between the group of conditions $\{(\lambda_a), (\lambda_b), (\lambda_c)\}$ and the group of conditions $\{(\lambda_a), (\lambda'_b), (\lambda'_c)\}$ in [ P §2.2, Definition 2]).

**Problem 2.2.10.** Prove [ P §2.2, Theorem 2] by using the statement in [ P §2.2, Theorem 1].

**Problem 2.2.11.** In the context of [ P §2.2, Theorem 3], prove that the system $\mathscr{L}$ is a $\lambda$-system.

**Problem 2.2.12.** A $\sigma$-algebra is said to be *countably generated*, or *separable*, if it is generated by some countable collection of sets.

   Prove that the $\sigma$-algebra $\mathscr{B}$, comprised of all Borel sets inside $\Omega = (0, 1]$, is countably generated.

   Prove by way of example that it is possible to find two $\sigma$-algebras, $\mathscr{F}_1$ and $\mathscr{F}_2$, such that $\mathscr{F}_2$ is countably generated, one has $\mathscr{F}_1 \subset \mathscr{F}_2$, and yet $\mathscr{F}_1$ is not countably generated.

**Problem 2.2.13.** Prove that, in order for the $\sigma$-algebra $\mathscr{G}$ to be countably generated, it is necessary and sufficient that $\mathscr{G} = \sigma(X)$, for some appropriate random variable $X$ (see [ P §2.2, **4**] for the definition of $\sigma(X)$).

**Problem 2.2.14.** Prove that $(X_1, X_2, \ldots)$ are independent random variables ([ P §2.2, **4**] and [ P §2.2, **5**]) whenever $\sigma(X_n)$ and $\sigma(X_1, \ldots, X_{n-1})$ are independent for every $n \geq 1$.

**Problem 2.2.15.** Let $(\Omega, \mathscr{F}, P)$ be any complete (see [ P §2.3, **1**] and Problem 2.2.34) probability space, let $\mathscr{G}$ be any sub-$\sigma$-algebra of $\mathscr{F}$ ($\mathscr{G} \subseteq \mathscr{F}$) and let $(\mathscr{E}_n)_{n \geq 1}$

be any non-increasing sequence of sub-$\sigma$-algebras of $\mathscr{F}$ ($\mathscr{E}_1 \supseteq \mathscr{E}_2 \supseteq \ldots$; $\mathscr{E}_n \subseteq \mathscr{F}$, $n \geq 1$). Suppose that all $\sigma$-algebras under consideration are completed with all P-negligible sets from $\mathscr{F}$. It may appear intuitive that, at least up to sets of measure zero, one must have

$$\bigcap_n \sigma(\mathscr{G}, \mathscr{E}_n) = \sigma\left(\mathscr{G}, \bigcap_n \mathscr{E}_n\right), \qquad (*)$$

or, in a different notation,

$$\bigcap_n (\mathscr{G} \vee \mathscr{E}_n) = \mathscr{G} \vee \bigcap_n \mathscr{E}_n, \qquad (**)$$

where $\mathscr{G} \vee \mathscr{E}_n \equiv \sigma(\mathscr{G}, \mathscr{E}_n)$ is the smallest $\sigma$-algebra generated by the sets from $\mathscr{G}$ and $\mathscr{E}$, and the identity in $(*)$ and $(**)$ is understood as "identity up to sets of measure zero" between two complete $\sigma$-algebras, say $\mathscr{H}_1 \subseteq \mathscr{F}$ and $\mathscr{H}_2 \subseteq \mathscr{F}$, in the sense that, for every $A \in \mathscr{H}_1$ one can find some $B \in \mathscr{H}_2$—and vice versa, for every $B \in \mathscr{H}_2$ one can find some $A \in \mathscr{H}_1$—so that $\mathsf{P}(A \triangle B) = 0$.

Nevertheless, the following example taken from [134] shows that, in general, the operations $\vee$ (the supremum) and $\cap$ (the intersection) between $\sigma$-algebras cannot be interchanged.

(a) Let $\xi_0, \xi_1, \xi_2, \ldots$ be any sequence of Bernoulli random variables with $\mathsf{P}\{\xi_i = 1\} = \mathsf{P}\{\xi_i = -1\} = 1/2$, and let $X_n = \xi_0 \xi_1 \ldots \xi_n$,

$$\mathscr{G} = \sigma(\xi_1, \xi_2, \ldots), \quad \text{and} \quad \mathscr{E}_n = \sigma(X_k, k > n).$$

Prove that

$$\bigcap_n \sigma(\mathscr{G}, \mathscr{E}_n) \neq \sigma\left(\mathscr{G}, \bigcap_n \mathscr{E}_n\right).$$

*Hint.* Prove that $\xi_0$ is measurable with respect to $\bigcap_n \sigma(\mathscr{G}, \mathscr{E}_n)$ $(= \sigma(\mathscr{G}, \xi_0))$, but is still independent from the events in $\sigma(\mathscr{G}, \bigcap_n \mathscr{E}_n)$ $(= \mathscr{G})$.

(This problem continues in Problem 7.4.25 below.)

(b) The fact that for $\sigma$-algebras the operations $\vee$ and $\cap$ do not commute follows from the (considerably simpler than (a)) claim (see [23]) that, if $\xi_1$ and $\xi_2$ are any two of the random variables described in (a) (i.e., any two independent and symmetric Bernoulli random variables) and if $\mathscr{E}_1 = \sigma(\xi_1)$, $\mathscr{E}_2 = \sigma(\xi_2)$ and $\mathscr{G} = \sigma(\xi_1 \xi_2)$, then

$$(\mathscr{G} \vee \mathscr{E}_1) \cap (\mathscr{G} \vee \mathscr{E}_2) \neq \mathscr{G} \vee (\mathscr{E}_1 \cap \mathscr{E}_2) \quad \text{and} \quad (\mathscr{G} \cap \mathscr{E}_1) \vee (\mathscr{G} \cap \mathscr{E}_2) \neq \mathscr{G} \cap (\mathscr{E}_1 \vee \mathscr{E}_2).$$

Prove this last statement.

**Problem 2.2.16.** Let $\mathscr{A}_1$ and $\mathscr{A}_2$ be any two independent collections of sets, every one of which represents a $\pi$-system. Prove that $\sigma(\mathscr{A}_1)$ and $\sigma(\mathscr{A}_2)$ are also independent. Give an example of two independent collections of sets, $\mathscr{A}_1$ and $\mathscr{A}_2$, neither of which is a $\pi$-systems, and $\sigma(\mathscr{A}_1)$ and $\sigma(\mathscr{A}_2)$ are not independent.

**Problem 2.2.17.** Assuming that $\mathscr{L}$ is a $\lambda$-system, prove that $(A, B \in \mathscr{L}, A \cap B = \varnothing) \implies (A \cup B \in \mathscr{L})$.

**Problem 2.2.18.** Let $\mathscr{F}_1$ and $\mathscr{F}_2$ be any two $\sigma$-algebras of subsets of the set $\Omega$ and let

$$d(\mathscr{F}_1, \mathscr{F}_2) = 4 \sup_{A_1 \in \mathscr{F}_1, A_2 \in \mathscr{F}_2} |P(A_1 A_2) - P(A_1)P(A_2)|.$$

Prove that the above quantity, which can be viewed as a measure of the dependence between $\mathscr{F}_1$ and $\mathscr{F}_2$, has the following properties:
(a) $0 \le d(\mathscr{F}_1, \mathscr{F}_2) \le 1$;
(b) $d(\mathscr{F}_1, \mathscr{F}_2) = 0$ if and only if $\mathscr{F}_1$ and $\mathscr{F}_2$ are independent;
(c) $d(\mathscr{F}_1, \mathscr{F}_2) = 1$ if and only if the intersection $\mathscr{F}_1$ and $\mathscr{F}_2$ contains an event that has probability equal to $1/2$.

**Problem 2.2.19.** Following the proof of [P §2.2, Lemma 1], prove the existence and the uniqueness of the classes $\lambda(\mathscr{E})$ and $\pi(\mathscr{E})$, which contain the system of sets $\mathscr{E}$.

**Problem 2.2.20.** Let $\mathscr{A}$ be any algebra of sets that has the following property: for any sequence, $(A_n)_{n\ge1}$, of non-intersecting sets $A_n \in \mathscr{A}$, one has $\bigcup_{n=1}^{\infty} A_n \in \mathscr{A}$. Prove that $\mathscr{A}$ is actually a $\sigma$-algebra.

**Problem 2.2.21.** Suppose that $(\mathscr{F}_n)_{n\ge1}$ is some increasing sequence of $\sigma$-algebras, i.e., $\mathscr{F}_n \subseteq \mathscr{F}_{n+1}$, $n \ge 1$. Prove that, generally, $\bigcup_{n=1}^{\infty} \mathscr{F}_n$ could only be claimed to be an algebra.

**Problem 2.2.22.** Let $\mathscr{F}$ be any algebra (resp., $\sigma$-algebra) and let $C$ be any set which does not belong to $\mathscr{F}$. Consider the smallest algebra (resp., $\sigma$-algebra), which contains the family $\mathscr{F} \cup \{C\}$. Prove that this algebra (resp., $\sigma$-algebra) is comprised of all sets of the form $(A \cap C) \cup (B \cap \overline{C})$, for all choices of $A, B \in \mathscr{F}$.

**Problem 2.2.23.** Let $\overline{\mathbb{R}} = \mathbb{R} \cup \{-\infty\} \cup \{\infty\}$ be the *extended* real line. The Borel $\sigma$-algebra $\mathscr{B}(\overline{\mathbb{R}})$ may be defined (comp. with [P §2.2, **2**]) as the $\sigma$-algebra generated by the sets $[-\infty, x]$, $x \in \mathbb{R}$, where $[-\infty, x] = \{-\infty\} \cup (-\infty, x]$. Prove that the $\sigma$-algebra $\mathscr{B}(\overline{\mathbb{R}})$ coincides with any of the $\sigma$-algebras that generated, respectively, by any of the following families of sets:
(a) $[-\infty, x)$, $x \in \mathbb{R}$;
(b) $(x, \infty]$, $x \in \mathbb{R}$ (where $(x, \infty] = (x, \infty) \cup \{\infty\}$);
(c) All finite intervals and $\{-\infty\}$ and $\{\infty\}$.

**Problem 2.2.24.** Consider the measurable space $(C, \mathscr{B}_0(C))$, in which $C = C[0, 1]$ is the space of all continuous functions $x = (x_t)_{0\le t\le1}$ and $\mathscr{B}_0(C)$ is the Borel $\sigma$-algebra for the metric $\rho(x, y) = \sup_{0\le t\le1} |x_t - y_t|$. Prove that:
(a) The space $C$ is complete (relative to the metric $\rho(\cdot, \cdot)$).
(b) The space $C$ is separable (relative to the metric $\rho(\cdot, \cdot)$).
*Hint.* Use the Bernstein polynomials—see [P §1.5].

(c) If treated as a subset of $C[0, 1]$, the subspace

$$C^d[0, 1] = \{x \in C[0, 1] : x \text{ is differentiable}\},$$

is *not* a Borel set.

**Problem 2.2.25.** Let $\mathscr{A}_0$ be any non-empty system of subsets of the sample space $\Omega$. Prove that $\alpha(\mathscr{A}_0)$, defined as the algebra generated by the system $\mathscr{A}_0$, can be constructed as follows: Set $\mathscr{A}_1 = \mathscr{A}_0 \cup \{\emptyset, \Omega\}$ and define, for $n \geq 1$,

$$\mathscr{A}_{n+1} = \{A \cup \overline{B} : A, B \in \mathscr{A}_n\}.$$

Then $\mathscr{A}_0 \subseteq \mathscr{A}_1 \subseteq \ldots \subseteq \mathscr{A}_n \subseteq \ldots$ and

$$\alpha(\mathscr{A}_0) = \bigcup_{n=1}^{\infty} \mathscr{A}_n.$$

**Problem 2.2.26.** For a given non-empty system of subsets of the sample space $\Omega$, denoted $\mathscr{A}_0$, in Problem 2.2.25 we gave a method for constructing the smallest *algebra*, $\alpha(\mathscr{A}_0)$, that contains the system $\mathscr{A}_0$. Analogously, we now define the systems:

$$\mathscr{A}_1 = \mathscr{A}_0 \cup \{\emptyset, \Omega\},$$

$$\mathscr{A}_2 = \mathscr{A}_1 \cup \left\{ \bigcup_{n=1}^{\infty} B_n : B_1, B_2, \ldots \in \mathscr{A}_1 \right\},$$

$$\mathscr{A}_3 = \mathscr{A}_2 \cup \{\overline{B} : B \in \mathscr{A}_2\},$$

$$\mathscr{A}_4 = \mathscr{A}_3 \cup \left\{ \bigcup_{n=1}^{\infty} B_n : B_1, B_2, \ldots \in \mathscr{A}_3 \right\},$$

$$\mathscr{A}_5 = \mathscr{A}_4 \cup \{\overline{B} : B \in \mathscr{A}_4\},$$

$$\mathscr{A}_6 = \mathscr{A}_5 \cup \left\{ \bigcup_{n=1}^{\infty} B_n : B_1, B_2, \ldots \in \mathscr{A}_5 \right\},$$

. . .

It may seem intuitive that the system $\mathscr{A}_{\infty} = \bigcup_{n=1}^{\infty} \mathscr{A}_n$ should give the smallest $\sigma$-algebra, $\sigma(\mathscr{A}_0)$, that contains the system $\mathscr{A}_0$, however, in general, this claim cannot be made: one always has $\mathscr{A}_{\infty} \subseteq \sigma(\mathscr{A}_0)$, while, in general, $\mathscr{A}_{\infty} \neq \sigma(\mathscr{A}_0)$, i.e., the above procedure may not give the entire $\sigma$-algebra $\sigma(A_0)$. Prove by way of example that $\sigma(\mathscr{A}_0)$ can be strictly larger than $\mathscr{A}_{\infty}$.

*Hint.* Consider the case where $\mathscr{A}_0$ is the system of all intervals on the real line $\mathbb{R}$.

Note that if one is to follow the above procedure, starting with $\mathscr{A}_\infty$ instead of $\mathscr{A}_0$, in general, one still cannot claim that $\sigma(\mathscr{A}_0)$ will be produced at the end. In order to produce $\sigma(\mathscr{A}_0)$, one must use transfinite induction ($\aleph_1$ "times"). We refer to [47, vol. 1, p. 235, vol. 2, p. 1068] for further explanation of G. Cantor's cardinality (or power) numbers and the related continuum hypothesis.

**Problem 2.2.27.** (*Suslin's counterexample.*) The construction and the conclusion given in the previous problem show that, in principle, a σ-algebra may have a rather complicated structure. In 1916 M. Suslin produced a counterexample, which proved that the following statement, due to H. Lebesgue is not true in general: *the projection of every Borel set B inside $\mathbb{R}^2$ onto one of the coordinate axes is a Borel set inside $\mathbb{R}^1$*. Just as M. Suslin did in 1916, construct a counterexample that disproves this statement.

Hint. M. Suslin's idea was to construct a concrete sequence, $A_1, A_2, \ldots$, of open sets in the plain $\mathbb{R}^2$ so that the projection of the intersection $\bigcap A_n$ on one of the coordinate axes is not a Borel set.

**Problem 2.2.28.** (*Sperner's lemma.*) Consider the set $A = \{1, \ldots, n\}$ and let $\{A_1, \ldots, A_k\}$ be any family of subsets of $A$, chosen so that no member of this family is included in some other member of the same family. Prove that the total number $K$ satisfies the estimate $K \le C_n^{[n/2]}$.

**Problem 2.2.29.** Let $\mathscr{E}$ be any system of subsets of $\Omega$ and let $\mathscr{F} = \sigma(\mathscr{E})$ be the smallest σ-algebra that includes the system $\mathscr{E}$ (i.e., the σ-algebra generated by $\mathscr{E}$). Suppose that $A \in \mathscr{F}$. By using the "suitable sets principle," prove that one can always find a countable family, $\mathscr{C} \subseteq \mathscr{E}$, for which one can claim that $A \in \sigma(\mathscr{C})$.

**Problem 2.2.30.** The *Borel σ-algebra* $\mathscr{E}$ associated with the metric space $(E, \rho)$ is defined as the σ-algebra generated by all *open sets* sets inside $E$ (relative to the metric $\rho$—see [P §3.1, **3**]). Prove that, for certain metric spaces, the σ-algebra $\mathscr{E}_0$, generated by all *open balls*, may be strictly smaller than $\mathscr{E}$ ($\mathscr{E}_0 \subset \mathscr{E}$).

**Problem 2.2.31.** Prove that there is no σ-algebra of cardinality $\aleph_0$ ("aleph-naught," the cardinality of the set of natural numbers), that has *countably infinitely many* elements. Plainly, the structure of *any* σ-algebra is always such that it has either *finitely many* elements (see Problem 1.1.10) or has *uncountably many* elements. For example, according to the next problem, as a set, the collection of all Borel subsets of $\mathbb{R}^n$ has power $\mathfrak{c} = 2^{\aleph_0}$ (i.e., the power of the continuum), which is the same as the power of the collection of all subsets of the set of natural numbers.

**Problem 2.2.32.** Just as in the previous problem, let $\mathfrak{c} = 2^{\aleph_0}$ denote the power of the continuum. Prove that, as a set, the collection of all Borell subsets of $\mathbb{R}^n$ has power $\mathfrak{c}$, while the σ-algebra of all Lebesgue subsets has power $2^{\mathfrak{c}}$.

**Problem 2.2.33.** Suppose that $B$ is some Borel subset of the real line $\mathbb{R}$ and let $\lambda$ denote the Lebesgue measure on $\mathbb{R}$. The *density* of the set $B$ is defined as the limit

$$D(B) = \lim_{T \to \infty} \frac{\lambda\{A \cap [-T, T]\}}{2T}, \qquad (*)$$

provided that the limit exists.

(a) Give an example of a set $B$ for which the density $D(B)$ does not exist, in that the limit $(*)$ does not exist.

(b) Prove that if $B_1$ and $B_2$ are two non-intersecting Borel subsets of the real line, then

$$D(B_1 + B_2) = D(B_1) + D(B_2),$$

in the sense that if either side of the above identity exists then so does also the other side and the identity holds.

(c) Construct a sequence, $B_1, B_2, \ldots$, of Borel sets inside $\mathbb{R}$, every one of which admits density $D(B_i)$, but, nevertheless, countrary to the intuition, one has

$$D\left(\sum_{i=1}^{\infty} B_i\right) \neq \sum_{i=1}^{\infty} D(B_i).$$

**Problem 2.2.34.** (*Completion of $\sigma$-algebras.*) Let $(\Omega, \mathscr{F}, \mathsf{P})$ be any probability space. We say that this probability space is *complete* (or, equivalently, P-complete, or, complete relative to the measure P), if $B \in \mathscr{F}$ and $\mathsf{P}(B) = 0$ implies that any set $A$ with $A \subseteq B$ must be an element of $\mathscr{F}$.

Let $\mathscr{N}$ denote the collection of all subsets $N \subseteq \Omega$ with the property that there is a set (possibly depending on $N$), $B_N \in \mathscr{F}$, with $\mathsf{P}(B_N) = 0$ and $N \subseteq B_N$. Let $\overline{\mathscr{F}}$ (sometimes written as $\mathscr{F}^{\mathsf{P}}$ or $\overline{\mathscr{F}}^{\mathsf{P}}$) denote the collection of all sets of the form $A \cup N$, for some choice of $A \in \mathscr{F}$ and $N \in \mathscr{N}$. Prove that:

(a) $\overline{\mathscr{F}}$ is a $\sigma$-algebra;

(b) If $B \subseteq \Omega$ and there are sets, $A_1$ and $A_2$, from $\mathscr{F}$, with $A_1 \subseteq B \subseteq A_2$ and $\mathsf{P}(A_2 \setminus A_1) = 0$, then $B \in \overline{\mathscr{F}}$;

(c) The probability space $(\Omega, \overline{\mathscr{F}}, \mathsf{P})$ is complete.

## 2.3 Methods for Constructing Probability Measures on Measurable Spaces

**Problem 2.3.1.** Suppose that $\mathsf{P}$ is a probability measure on $(\mathbb{R}, \mathscr{B}(\mathbb{R}))$ and let $F(x) = \mathsf{P}(-\infty, x]$, $x \in \mathbb{R}$. Prove that

$$P(a, b] = F(b) - F(a), \qquad P(a, b) = F(b-) - F(a),$$

$$P[a, b] = F(b) - F(a-), \qquad P[a, b) = F(b-) - F(a-),$$

$$P(\{x\}) = F(x) - F(x-),$$

where $F(x-) = \lim_{y \uparrow x} F(y)$.

**Problem 2.3.2.** Verify formula [P §2.3, (7)].

**Problem 2.3.3.** Give a complete proof of [ P §2.3, Theorem 2].

**Problem 2.3.4.** Prove that a (cumulative) distribution function $F = F(x)$, defined on the real line $\mathbb{R}$ can have at most countably many points of discontinuity. Does this statement have an analog for distribution functions defined on $\mathbb{R}^n$?

*Hint.* Consider using the relation $\{x : F(x) \neq F(x-)\} = \bigcup_{n=1}^{\infty}\{x : F(x) - F(x-) \geq \frac{1}{n}\}$. In general, for distribution functions defined in $\mathbb{R}^n$, one cannot claim that the points of discontinuity are at most countably many. To find a counter-example, consider the delta-measure

$$\delta_0(A) = \begin{cases} 1, & \text{if } 0 \in A, \\ 0, & \text{if } 0 \notin A, \end{cases} \qquad A \in \mathscr{B}(\mathbb{R}^n).$$

**Problem 2.3.5.** Prove that each of the functions

$$G(x, y) = \begin{cases} 1, & x + y \geq 0, \\ 0, & x + y < 0, \end{cases}$$

$$G(x, y) = \lfloor x + y \rfloor = \text{the integer part of } x + y,$$

is right continuous and increasing in each variable but, nevertheless, cannot be treated as a (generalized) distribution function in $\mathbb{R}^2$.

**Problem 2.3.6.** Let $\mu$ denote the Lebesgue–Stieltjes measure associated with some generalized distribution function and let $A$ be any at most countable set. Prove that $\mu(A) = 0$.

**Problem 2.3.7.** Prove that the Cantor set $\mathscr{N}$ is *uncountable*, *perfect* (meaning, a closed set in which every point is an accumulation point, or, equivalently, a closed set without isolated points), *nowhere dense* (meaning, a closed set without interior points) and has vanishing Lebesgue measure.

**Problem 2.3.8.** Let $(\Omega, \mathscr{F}, \mathsf{P})$ be any probability space and let $\mathscr{A}$ be any algebra of subsets of $\Omega$, such that $\sigma(\mathscr{A}) = \mathscr{F}$. Prove that, for every $\varepsilon > 0$ and for every set $B \in \mathscr{F}$, one can find a set $A_\varepsilon \in \mathscr{A}$, with the property

$$\mathsf{P}(A_\varepsilon \triangle B) \leq \varepsilon.$$

*Hint.* Consider the family $\mathscr{B} = \{B \in \mathscr{F} : \forall \varepsilon > 0 \, \exists A \in \mathscr{A} : \mathsf{P}(A \triangle B) \leq \varepsilon\}$ and prove that $\mathscr{B}$ is a $\sigma$-algebra, so that $\mathscr{F} \supseteq \mathscr{B} \supseteq \sigma(\mathscr{A}) = \mathscr{F}$.

**Problem 2.3.9.** Let $\mathsf{P}$ be any probability measure on $(\mathbb{R}^n, \mathscr{B}(\mathbb{R}^n))$. Prove that, given any $\varepsilon > 0$ and any $B \in \mathscr{B}(\mathbb{R}^n)$, one can find a *compact* set $A_1$ and an *open* set $A_2$ so that $A_1 \subseteq B \subseteq A_2$ and $\mathsf{P}(A_2 \setminus A_1) \leq \varepsilon$. (This result is used in the proof of [ P §2.3, Theorem 3].)

*Hint.* Consider the family

$$\mathscr{B} = \left\{ \begin{array}{l} B \in \mathscr{B}(\mathbb{R}^n) : \forall \varepsilon > 0 \text{ there is a compact set } A_1 \text{ and} \\ \qquad \text{an open set } A_2, \text{ so that the closure } \overline{A}_2 \text{ is compact} \\ \qquad \text{and one has } A_1 \subseteq B \subseteq A_2 \text{ and } \mathsf{P}(A_1 \setminus A_1) < \varepsilon \end{array} \right\}$$

and then prove that $\mathscr{B}$ constitutes a $\sigma$-algebra.

**Problem 2.3.10.** For a given probability measure $\mathsf{P}$, verify the compatibility of the measures $\{\mathsf{P}_\tau\}$, defined by $\mathsf{P}_\tau(B) = \mathsf{P}(\mathscr{I}_\tau(B))$ (see (21) and Theorem 4 in [$\underline{\mathrm{P}}$ §2.3, **5**]).

**Problem 2.3.11.** Verify that [$\underline{\mathrm{P}}$ §2.3, Tables 2 and 3] represent probability distributions, as claimed.

**Problem 2.3.12.** Prove that the system $\hat{\mathscr{A}}$, introduced in [$\underline{\mathrm{P}}$ §1.2, **3**], is a $\sigma$-algebra.

**Problem 2.3.13.** Prove that the set function $\mu(A)$, $A \in \hat{\mathscr{A}}$, introduced in Remark 2 in [$\underline{\mathrm{P}}$ §2.3, **1**] is a *measure*.

**Problem 2.3.14.** Prove by way of example that if the measure $\mu_0$, defined on the algebra $\mathscr{A}$, is *finitely additive* but is not countably additive, then $\mu_0$ cannot be extended to a countably additive measure on $\sigma(\mathscr{A})$.

**Problem 2.3.15.** Prove that any finitely additive probability measure, defined on some algebra, $\mathscr{A}$, of subsets of $\Omega$, can be extended to a finitely additive probability measure defined on *all* subsets of $\Omega$.

**Problem 2.3.16.** Let $\mathsf{P}$ be any probability measure defined on some $\sigma$-algebra $\mathscr{F}$ that consists of subsets of $\Omega$ and suppose that the set $C \subseteq \Omega$ is chosen so that $C \notin \mathscr{F}$. Prove that the measure $\mathsf{P}$ can be extended (countable additivity preserved) to a measure on the $\sigma$-algebra $\sigma(\mathscr{F} \cup \{C\})$.

**Problem 2.3.17.** Prove that the support, denoted by supp $F$, of any *continuous* distribution function $F$ must be a *perfect* set, i.e., a closed set without isolated points. (Recall that, for a given cumulative distribution function $F$ defined on $\mathbb{R}$, supp $F$ is the smallest closed set $G$ with the property $\mu(\mathbb{R} \setminus G) = 0$, where $\mu$ is the measure associated with $F$—see Sect. A.2.)

Give an example of a cumulative distribution function $F$, associated with some discrete probability measure on $\mathbb{R}$, for which one has supp $F = \mathbb{R}$, i.e., the support of $F$ is the entire real line $\mathbb{R}$.

**Problem 2.3.18.** Prove the following fundamental result (see the end of [$\underline{\mathrm{P}}$ §2.3, **1**]): every distribution function $F = F(x)$ can be expressed in the form

$$F = \alpha_1 F_{\mathrm{d}} + \alpha_2 F_{\mathrm{abc}} + \alpha_3 F_{\mathrm{sc}} ,$$

where $\alpha_i \geq 0$, $\alpha_1 + \alpha_2 + \alpha_3 = 1$, and

$F_d$   is a discrete distribution function (with jumps $p_k > 0$ at the points $x_k$):

$$F_d = \sum_{\{k : x_k \le x\}} p_k;$$

$F_{abc}$  is an absolutely continuous ditribution function:

$$F_{abc} = \int_{-\infty}^{x} f(t)\, dt$$

with density $f = f(t)$, which is non-negative, Borel-measurable and Lebesgue-integrable, i.e., $\int_{-\infty}^{\infty} f(t)\, dt = 1$;

$F_{sc}$  is a continuous and singular distribution function, i.e., continuous distribution function for which the points of increase form a set of Lebesgue measure 0.

What can be said about the uniqueness of the above decomposition of the distribution function $F = F(x)$?

**Problem 2.3.19.** (a) Prove that every real number $\omega \in [0, 1]$ admits a ternary (i.e., base 3) expansion of the form

$$\omega = \sum_{n=1}^{\infty} \frac{\omega_n}{3^n},$$

where $\omega_n \in \{0, 1, 2\}, n \ge 1$.

(b) Prove that if $\omega \in [0, 1]$ admits two ternary expansions, $\omega = \sum_{n=1}^{\infty} \frac{\omega_n}{3^n}$ and $\omega = \sum_{n=1}^{\infty} \frac{\omega_n'}{3^n}$, both of which are *non-terminating* (i.e., $\sum_{n=1}^{\infty} |\omega_n| = \infty$ and $\sum_{n=1}^{\infty} |\omega_n'| = \infty$), then one must have $\omega_n = \omega_n'$ for all $n \ge 1$ (uniqueness of the non-terminating expansions).

Notice that non-uniqueness of the ternary expansion is possible for reals $\omega \in [0, 1]$ that admit a terminating expansions of the form $\omega = \sum_{n=1}^{m} \frac{\omega_n}{3^n}, m < \infty$. In any such case, the following "canonical" expansion may be chosen:

1. If $\omega = \sum_{n=1}^{m-1} \frac{\omega_n}{3^n} + \frac{2}{3^m}$, set

$$\omega_n' = \begin{cases} \omega_n, & n \le m - 1, \\ 2, & n = m, \\ 0, & n \ge m + 1; \end{cases}$$

2. And if $\omega = \sum_{n=1}^{m-1} \frac{\omega_n}{3^n} + \frac{1}{3^m}$, set

$$\omega_n' = \begin{cases} \omega_n, & n \le m - 1, \\ 0, & n = m, \\ 2, & n \ge m + 1. \end{cases}$$

(c) Suppose that the set $\mathcal{N} \subseteq [0, 1]$ comprises all points of increase for the Cantor function on the interval $[0, 1]$ (recall that the Cantor function is a canonical example of a distribution function which is both continuous and singular—see [P §2.3, 1]). Prove that every $\omega \in \mathcal{N}$ admits an expansion of the form

$$\omega = \sum_{n=1}^{\infty} \frac{\omega_n}{3^n},$$

where $\omega_n \in \{0, 2\}, n \geq 1$.

*Remark.* It is interesting that (see [132]) if *decimal* expansions for the numbers $\omega \in \mathcal{N}$ are considered, then there will be precisely 14 numbers in the Cantor set $\mathcal{N}$ that admit terminating expansions. These numbers are

$$\frac{1}{4}, \frac{3}{4}, \frac{1}{10}, \frac{3}{10}, \frac{7}{10}, \frac{9}{10}, \frac{1}{40}, \frac{3}{40}, \frac{9}{40}, \frac{13}{40}, \frac{27}{40}, \frac{31}{40}, \frac{37}{40}, \frac{39}{40}.$$

**Problem 2.3.20.** Let $\mathcal{N}$ denote the Cantor set inside the interval $[0, 1]$.

(a) Prove that $\mathcal{N}$ has the same cardinality as the set $[0, 1]$.

(b) Describe the sets that can be identified with $\mathcal{N} \oplus \mathcal{N}$ and $\mathcal{N} \ominus \mathcal{N}$, i.e., the sets $\{\omega + \omega' : \omega \in \mathcal{N}, \omega' \in \mathcal{N}\}$ and $\{\omega - \omega' : \omega \in \mathcal{N}, \omega' \in \mathcal{N}\}$.

**Problem 2.3.21.** Let $C$ be any closed subset of the real line $\mathbb{R}$. Give an example of a distribution function $F$, for which one can claim that the support of $F$ is precisely the set $C$, i.e., supp $F = C$.

**Problem 2.3.22.** Give an example of a $\sigma$-finite measure $\mu$ which is defined on $(\mathbb{R}, \mathscr{B}(\mathbb{R}))$ and

(a) is not a *Lebesgue–Stieltjes measure*, in other words, one cannot find a non-decreasing and right-continuous function $G = G(x)$ (i.e., a generalized distribution function) with the property $\mu((a, b]) = G(b) - G(a), a < b$;

(b) is not a *locally finite* measure, in other words, every open neighborhood of every point $x \in \mathbb{R}$ has infinite measure.

**Problem 2.3.23.** Find a subset of the interval $[0, 1]$, which does not belong to the collection of all Lebesgue-measurable sets $\overline{\mathscr{B}}([0, 1])$—see [P §2.3, 1].

**Problem 2.3.24.** Give a probabilistic proof of *Euler's product formula for the Riemann zeta function*; namely, consider the Riemann zeta function $\zeta(\alpha) = \sum_{n=1}^{\infty} \frac{1}{n^{\alpha}}, 1 < \alpha < \infty$, and prove that the following representation, in which $p_1, p_2, \ldots$ is the sequence of all prime numbers greater than 1, is in force:

$$\zeta(\alpha)^{-1} = \prod_{n=1}^{\infty} \left(1 - \frac{1}{p_n^{\alpha}}\right).$$

*Hint.* Let $\mathbb{N} = \{1, 2, \ldots\}$ be the set of all natural numbers endowed with the $\sigma$-algebra $(2^{\mathbb{N}})$ comprised of all possible subsets of the (countable) set $\mathbb{N}$—see the notation for $N(\mathscr{A})$ at the end of [P §1.1, 3]. Then define a probability measure P on $(\mathbb{N}, 2^{\mathbb{N}})$ so that, given any $A \subseteq \mathbb{N}$, one has

$$P(A) = \zeta(\alpha)^{-1} \sum_{n \in A} n^{-\alpha}.$$

Let $A(p_i) = \{p_i, 2p_i, \ldots\}$ denote the collection of those numbers $n \in \mathbb{N}$ for which $p_i$ is a factor (i.e., a divisior) of $n$. Prove that
(a) $P(A(p_i)) = p_i^{-\alpha}$;
(b) The events $A(p_1), A(p_2), \ldots$ are independent;
(c) $\bigcap_{i=1}^{\infty} \overline{A}(p_i) = \{1\}$.
Furthermore, argue that, since (b) implies that the events $\overline{A}(p_1), \overline{A}(p_2), \ldots$ are also independent, then (c) and (a) imply that

$$P\left(\bigcap_{i=1}^{\infty} \overline{A}(p_i)\right) = P(\{1\}) = [\zeta(\alpha)]^{-1}$$

and, at the same time, that

$$P\left(\bigcap_{i=1}^{\infty} \overline{A}(p_i)\right) = \prod_{i=1}^{\infty} [1 - P(A(p_i))] = \prod_{i=1}^{\infty} \left(1 - \frac{1}{p_i^{\alpha}}\right),$$

which completes the proof.

**Problem 2.3.25.** Give a probabilistic proof of *Euler's product formula for Euler's totient function* $\varphi(n)$, which, for any $n \in \mathbb{N}$, gives the total number of positive integers $p$ that do not exceed $n$ and are also relatively prime to $n$ (i.e., the only common divisor of $n$ and $p$ is 1); namely, by using probabilistic reasoning, prove that

$$\frac{\varphi(n)}{n} = \prod_{p \mid n} \left(1 - \frac{1}{p}\right),$$

where the product is taken over all prime numbers $p$ that divide $n$ (i.e., all prime numbers $p$ that are factors of $n$).

*Hint.* Consider the usual uniform probability distribution $P(\{k\}) = 1/n$, $1 \le k \le n$ on the set $\{1, \ldots, n\}$. For a fixed $n \in \mathbb{N}$, let $A(p) = \{k \le n : p$ is a factor of $k\}$ and let $p_1, p_2, \ldots$ denote the (distinct) prime numbers that divide $n$. Prove that:
(a) $P(A(p_i)) = p_i^{-1}$.
(b) That the events $A(p_1), A(p_2), \ldots$ are independent.
(c) That $\prod_{p \mid n} \overline{A}(p)$ can be identified with the event that $k \le n$ is a prime number.

Then argue that

$$\frac{\varphi(n)}{n} = \mathsf{P}\left(\prod_{p \mid n} \overline{A}(p)\right) = \prod_{p \mid n} [1 - \mathsf{P}(A(p))] = \prod_{p \mid n} \left(1 - \frac{1}{p}\right).$$

**Problem 2.3.26.** Prove that the Lebesgue measure on $(\mathbb{R}^n, \mathscr{B}(\mathbb{R}^n))$ is invariant under translation when $n \geq 1$, and invariant under rotation when $n \geq 2$.

**Problem 2.3.27.** In the context of Carathéodory's theorem, prove by way of example that the requirement for the system of sets $\mathscr{A}$ to be an *algebra* is essential for both the *existence* and the *uniqueness* of the *extension* of the probability measure P, originally defined on $\mathscr{A}$, to a measure defined on the $\sigma$-algebra $\mathscr{F} = \sigma(\mathscr{A})$. Specifically:

(a) Construct a sample space $\Omega$, two systems of subsets of $\Omega$, $\mathscr{E}$ and $\mathscr{F}$, such that $\mathscr{E}$ is not an algebra and $\mathscr{F} = \sigma(\mathscr{E})$, and then construct a probability measure P, defined on $\mathscr{E}$, which cannot be extended to a probability measure on the $\sigma$-algebra $\mathscr{F}$.

(b) Construct a sample space $\Omega$, two systems of subsets of $\Omega$, $\mathscr{E}$ and $\mathscr{F}$ with $\mathscr{F} = \sigma(\mathscr{E})$, and also two distinct probability measures, P and Q, defined on the $\sigma$-algebra $\mathscr{F}$, such that their restrictions to $\mathscr{E}$, i.e., the measures $\mathsf{P}|\mathscr{E}$ and $\mathsf{Q}|\mathscr{E}$—see [P §2.3, 4]—coincide.

*Hint.* To prove (a), consider the sample space $\Omega = \{1, 2, 3\}$, set

$$\mathscr{E} = \{\varnothing, \{1\}, \{1, 2\}, \{1, 3\}, \Omega\},$$

define $\mathscr{F}$ to be the $\sigma$-algebra of all subsets of $\Omega$, and, finally, set $\mathsf{P}(\Omega) = \mathsf{P}(\{1, 2\}) = \mathsf{P}(\{1, 3\}) = 1, \mathsf{P}(\{1\}) = 1/2, \mathsf{P}(\varnothing) = 0$.

To prove (b), consider the sample space $\Omega = \{1, 2, 3, 4\}$, set

$$\mathscr{E} = \{\{1, 2\}, \{1, 3\}\},$$

define $\mathscr{F}$ to be the $\sigma$-algebra of all subsets of $\Omega$ and, finally, define $\mathsf{P}(\{2\}) = \mathsf{P}(\{4\}) = 1/2, \mathsf{Q}(\{2\}) = \mathsf{Q}(\{3\}) = 1/2$.

**Problem 2.3.28.** Let $F = F(x), x \in \mathbb{R}$, be any distribution function. Prove that for any $a \geq 0$ one has

$$\int_{\mathbb{R}} [F(x + a) - F(x)] \, dx = a.$$

**Problem 2.3.29.** The *density* of a given distribution function $F(x)$ is defined as any non-negative and Riemann integrable function $f(x), x \in \mathbb{R}$, for which one can write $F(x) = \int_{-\infty}^{x} f(t) \, dt$, for all $x \in \mathbb{R}$. Sometimes (say, when integrals are considered only in the sense of Riemann—not in the sense of Lebesgue) one does not suppose that the function $f(x)$ is Borel measurable.

Give an example of a function $f(x)$ which is *not Borel measurable*, but nevertheless represents a probability density (in the sense described above) and defines a probability measure on the Borel subsets $B$ of the real line $\mathbb{R}$, according to the formula $\mu(B) = \int_{-\infty}^{\infty} f(x) I_B(x)\, dx$.

*Hint.* The collection of all Borel subsets of the interval $[0, 1]$ has cardinality $\mathfrak{c}$, i.e., the cardinality of the continuum, while the cardinality of the Lebesgue subsets of $[0, 1]$ is $2^{\mathfrak{c}}$ (see Problem 2.2.32). By using this fact conclude that if $\mathcal{N}$ denotes the Cantor set inside the interval $[0, 1]$, then one can find a subset $D \subseteq [1/2, 1] \cap \mathcal{N}$, which is not a Borel-measurable set and has a Lebesgue measure 0. Then convince yourself that the function $f(x) = 2I_{[1/2,1] \setminus D}(x)$, not being Borel measurable, is actually Riemann integrable and the integral $\int f(x) I_B(x)\, dx$ is well defined and gives rise to a probability measure on the Borel sets $B \subseteq [0, 1]$.

**Problem 2.3.30.** Find two sets, $A$ and $B$, inside the real line $\mathbb{R}$, which have Lebesgue measure equal to 0, and yet have the property $A \oplus B = \mathbb{R}$.

**Problem 2.3.31.** Let $\mathscr{A}$ be any $\sigma$-algebra of subsets of the set $\Omega$ and let $\mathscr{F} = \sigma(\mathscr{A})$. Let $\mu$ be any $\sigma$-finite measure on $\mathscr{F}$. Prove that:

(a) The measure $\mu$ may not be $\sigma$-finite on $\mathscr{A}$.

(b) If the measure $\mu$ is $\sigma$-finite on $\mathscr{A}$ then the analog of the property stated in Problem 2.3.8 still holds, i.e., for every $\varepsilon > 0$ and every $B \in \mathscr{F}$ with $\mu(B) < \infty$ one can find a subset $A_\varepsilon \in \mathscr{A}$ such that $\mu(A_\varepsilon \triangle B) < \varepsilon$.

(c) If the measure $\mu$ is not $\sigma$-finite on $\mathscr{A}$, then the claim made in b) may be false.

**Problem 2.3.32.** Prove that a probability measure $\mu$ defined on $(\mathbb{R}^d, \mathscr{B}(\mathbb{R}^d))$ is always *regular*, in the sense that for any Borel set $B \in \mathscr{B}(\mathbb{R}^d))$ one has

$$\mu(B) = \inf_U \{ \mu(U) : U \supseteq B,\ U \text{ is an open set}\},$$

and $\quad \mu(B) = \sup_F \{ \mu(F) : F \subseteq B,\ F \text{ is a closed set}\}.$

In addition, prove that the following relation holds for any Borel set $B \in \mathscr{B}(\mathbb{R}^d))$:

$$\mu(B) = \sup_K \{ \mu(K) : K \subseteq B,\ K \text{ is a compact set}\}.$$

**Problem 2.3.33.** (*Bertrand's Paradox.*) The well known Bertrand's Paradox is a good illustration of the fact that in many probabilistic models (in particular, models involving geometric probabilities, which the paradox is concerned with) one must be careful in the formulation of the model and in giving meaning to phrases like "a randomly chosen point," "a randomly chosen figure," etc. (This was already discussed in Problem 1.1.12.)

The problem, found in Bertrand's book [7], and its contradicting answers (whence the term"paradox"), found by using different calculation methods, were understood to imply that in random experiments with infinitely many outcomes there are events to which it is impossible to assign probabilities in a meaningful way.

Bertrand's problem may be stated as follows: Suppose that a chord $AB$, with end-points $A$ and $B$, is chosen *at random* on a circle of radius $r$. What is the probability that the length $|AB|$ of the (random) chord $AB$ is smaller than the radius $r$?

Consider the following three possible formulations of this problem:

(a) The phrase "the chord $AB$ is chosen at random" is understood to mean that the points $A$ and $B$ are sampled *independently* from the uniform distribution on the circle.

Prove that in this case $P_a\{|AB| < r\} = 1/3$. (Fix the point $A$ and consider the regular hexagon inscribed in the circle so that one of its vertices coinsides with $A$.)

(b) Every chord $AB$ is uniquely determined by the point $M \in AB$, chosen so that $OM \perp AB$, $O$ being the center of the circle. The phrase "the chord $AB$ is chosen at random" is understood to mean that the point $M$ is sampled from the uniform distribution on the disc (surrounded by the circle).

Prove that in this case $P_b\{|AB| < r\} = 1/4$. (Convince yourself that the event $\{|AB| < r\}$ is the same as the event that $M$ belongs to the ring surrounded by the circle of radius $r$ and radius $r\sqrt{3/4}$.)

(c) As the length of the chord $AB$ is determined by its distance to the center of the circle and not by its position on the circle, one may suppose that $AB$ is parellel to the horizontal diameter $CD$, while the random point $M \in AB$, defined as the intersection between $AB$ and the vertical diameter $EF$ (which is perpendicular to $CD$) is uniformly distributed on $EF$.

Prove that in this case

$$P_c\{|AB| < r\} = 1 - \frac{\sqrt{3}}{2} \quad (\approx 0.13).$$

(One must prove that the event $\{|AB| < r\}$ coincides with the event $\{|OM| > \sqrt{3}r/2\}$, where $|OM|$ is the distance between $M$ and the center, $O$, of the circle.)

**Problem 2.3.34.** (*Continuation of Problem 2.3.33.*) Argue that the situation described in Problem 2.3.33 can actually be connected to three different problems. More specifically, let $\rho = |OM|$, where $O$ is the center of the circle and the point $M$ is defined as in part (b) in the previous problem, and let $\theta$ denote the angle between the chord $AB$ and some fixed direction, so that, assuming that $r = 1$, for a chord with $|AB| > 0$ one must have $0 < \rho \leq 1, 0 \leq \theta < 2\pi$.

Prove that in parts (a), (b) and (c) in the previous problem the joint distribution of $(\rho, \theta)$ is given, respectively, by the densities

$$p_a(\rho, \theta) = \frac{1}{2\pi^2 \sqrt{1 - \rho^2}}, \qquad p_b(\rho, \theta) = \frac{\rho}{\pi}, \qquad p_c(\rho, \theta) = \frac{1}{2\pi}.$$

Consequently, there is no "paradox", as the phrase "the chord $AB$ is chosen at random" is given a completely different probabilistic meaning in parts (a), (b) and (c).

**Problem 2.3.35.** Let $(X, \mathscr{X}, \mu)$ be the space associated with some measurable structure $(X, \mathscr{X})$ and some countably additive measure $\mu$ (see [P §2.3, Definitions 5 and 6]).

A measurable set $A$ is said to be an *atom* relative to the measure $\mu$, or, equivalently, a $\mu$-atom, if $\mu(A) > 0$ and for every measurable set $B$ one has either $\mu(A \cap B) = 0$ or $\mu(A \setminus B) = 0$. The measure $\mu$ is said to be *atomic*, if every measurable set with a positive $\mu$-measure contains an atom.

The measure $\mu$ is said to be *non-atomic* if no $\mu$-atoms exist.

The measure $\mu$ is said to be a *diffusion measure* if every one-point set is a measurable $\mu$-null set.

Give examples of atomic, non-atomic and diffusion measures and also an example of a measure which is simultaneously an atomic measure and a diffusion measure.

Prove that the sum of an atomic and a non-atomic measure may be an atomic measure.

**Problem 2.3.36.** Let $\mathsf{P}$ and $\widetilde{\mathsf{P}}$ be any two probability measures on $(\Omega, \mathscr{F})$ such that $\mathsf{P}(A) = \widetilde{\mathsf{P}}(A)$, for any $A \in \mathscr{F}$ with $\mathsf{P}(A) \leq 1/2$. Prove that when this condition is satisfied then $\mathsf{P}(A) = \widetilde{\mathsf{P}}(A)$ for *every* set $A \in \mathscr{F}$.

## 2.4   Random Variables I

**Problem 2.4.1.** Prove that the random variable $\xi$ has a continuous distribution, or, "$\xi$ is continuous" for short, if and only if $\mathsf{P}\{\xi = x\} = 0$ for any $x \in \mathbb{R}$.

**Problem 2.4.2.** Can one claim that if $|\xi|$ is $\mathscr{F}$-measurable then $\xi$ also must be $\mathscr{F}$-measurable?

**Problem 2.4.3.** Prove that $x^n$, $x^+ = \max(x, 0)$, $x^- = -\min(x, 0)$ and $|x| = x^+ + x^-$ are all Borel functions of $x$. Prove that the following more general statement: every continuous function $f = f(x)$, $x \in \mathbb{R}$ is Borel measurable.

*Hint.* Given any $\alpha \in \mathbb{R}$, consider the open set $\{\omega \in \mathbb{R} : f(\omega) < \alpha\}$ and use the result established in Problem 2.2.7.

**Problem 2.4.4.** Prove that if $\xi$ and $\eta$ are $\mathscr{F}$-measurable then

$$\{\omega : \xi(\omega) = \eta(\omega)\} \in \mathscr{F}.$$

**Problem 2.4.5.** Let $\xi$ and $\eta$ be any two random variables on $(\Omega, \mathscr{F})$ and let $A \in \mathscr{F}$. Then the function

$$\zeta(\omega) = \xi(\omega)I_A + \eta(\omega)I_{\overline{A}}$$

also must be a random variable.

**Problem 2.4.6.** Let $\xi_1, \ldots, \xi_n$ be any $n \geq 1$ random variables and let $\varphi(x_1, \ldots, x_n)$ be any Borel-measurable function. Prove that $\varphi(\xi_1(\omega), \ldots, \xi_n(\omega))$ is a random variable.

*Hint.* Show first that the map

$$\omega \rightsquigarrow (\xi_1(\omega), \ldots, \xi_n(\omega)) \in \mathbb{R}^n$$

is $\mathscr{F}/\mathscr{B}(\mathbb{R}^n)$-measurable. Then use the fact that the map $\omega \rightsquigarrow \varphi(\xi_1(\omega), \ldots, \xi_n(\omega))$ is a composition of measurable maps.

**Problem 2.4.7.** Let $\xi$ and $\eta$ be any two random variables with values in the set $\{1, \ldots, N\}$ and suppose that $\mathscr{F}_\xi = \mathscr{F}_\eta$. Prove that there is a permutation $(i_1, \ldots, i_N)$ of the set $(1, \ldots, N)$ for which one can claim that for any $j = 1, \ldots, N$ the sets $\{\omega \colon \xi = j\}$ and $\{\omega \colon \eta = i_j\}$ coincide.

*Hint.* Consider using [P §2.4, Theorem 3], according to which there are functions $\varphi$ and $\psi$ such that $\xi = \varphi(\eta)$ and $\eta = \psi(\xi)$. Then argue that $i_j = \psi(j)$ gives the desired permutation.

**Problem 2.4.8.** Give an example of a random variable $\xi$ that admits a probability density $f(x)$ such that $\lim_{x \to \infty} f(x)$ does not exist and, therefore, the function $f(x)$ does not vanish at infinity.

**Problem 2.4.9.** Let $\xi$ and $\eta$ be any two bounded random variables with $|\xi| \leq c_1$, $|\eta| \leq c_2$. Prove that if

$$\mathsf{E}\xi^m \eta^n = \mathsf{E}\xi^m \cdot \mathsf{E}\eta^n,$$

for any $m, n \geq 1$, then $\xi$ and $\eta$ must be independent.

**Problem 2.4.10.** Let $\xi$ and $\eta$ be any two random variables whose distribution functions, $F_\xi$ and $F_\eta$ coincide. Prove that if $x \in \mathbb{R}$ and $\{\omega \colon \xi(\omega) = x\} \neq \varnothing$, then there is a real number $y \in \mathbb{R}$ such that $\{\omega \colon \xi(\omega) = x\} = \{\omega \colon \eta(\omega) = y\}$.

**Problem 2.4.11.** Let $E$ be any at most countable subset of $\mathbb{R}$ and consider the map $\xi \colon \Omega \mapsto E$. Prove that $\xi$ is a random variable on $(\Omega, \mathscr{F})$ if and only if $\{\omega \colon \xi(\omega) = x\} \in \mathscr{F}$ for any $x \in E$.

**Problem 2.4.12.** Let $\xi$ be any random variable with the property $\mathsf{P}\{\xi \neq 0\} > 0$. Suppose that for some $a$ and $b$ the random variables $a\xi$ and $b\xi$ have one and the same distribution, i.e., $F_{a\xi}(x) = F_{b\xi}(x)$, $x \in \mathbb{R}$. Can one claim that this is possible only if $a = b$? Does the assumption $a \geq 0$ and $b \geq 0$ change the answer to the last question?

**Problem 2.4.13.** Let $(\Omega, \mathscr{F}, \mathsf{P})$ be any probability space and let $(\Omega, \overline{\mathscr{F}}^{\mathsf{P}}, \mathsf{P})$ be its completion relative to the measure $\mathsf{P}$ (see Problem 2.2.34 and Remark 1 in [P §2.3, 1]). Prove that, given any random variable $\overline{\xi} = \overline{\xi}(\omega)$ on $(\Omega, \overline{\mathscr{F}}^{\mathsf{P}}, \mathsf{P})$, it is always possible to find a random variable $\xi = \xi(\omega)$, defined on $(\Omega, \mathscr{F}, \mathsf{P})$, for which one can claim that $\mathsf{P}\{\xi \neq \overline{\xi}\} = 0$, i.e., $\overline{\xi}$ and $\xi$ differ only on a set with probability 0.

**Problem 2.4.14.** Let $\xi$ be any random variable and $B$ be any Borel set in $\mathbb{R}$. Prove that

$$\sigma(\xi I(\xi \in B)) = \xi^{-1}(B) \cap \sigma(\xi).$$

**Problem 2.4.15.** Let $\xi_1, \xi_2, \ldots$ be any sequence of independent random variables every one of which is uniformly distributed in the interval $[0, 1]$. Given any $\omega \in \Omega$, consider the set $A(\omega) \subseteq [0, 1]$ which consists of all values $\xi_1(\omega), \xi_2(\omega), \ldots$. Prove that for almost every $\omega \in \Omega$ one can claim that the set $A(\omega)$ is *everywhere dense* in $[0, 1]$.

**Problem 2.4.16.** Let $\xi_1, \xi_2, \ldots$ be any sequence of Bernoulli random variables, such that $P\{\xi_k = 1\} = P\{\xi_k = -1\} = 1/2$, $k \geq 1$. Consider the random walk $S = (S_n)_{n \geq 0}$, defined by $S_0 = 0$ and $S_n = \xi_1 + \ldots + \xi_n$, for $n \geq 1$.

Let $\sigma_0 = \inf\{n > 0 : S_n = 0\}$ be the first moment (after $n = 0$) at which the random walk returns to $0$, with the understanding that $\sigma_0 = \infty$ if $\{n > 0 : S_n = 0\} = \varnothing$.

Prove that

$$P\{\sigma_0 > 2n\} = C_{2n}^n \left(\frac{1}{2}\right)^{2n} \quad \text{and} \quad P\{\sigma_0 = 2n\} = \frac{1}{2n-1} C_{2n}^n \left(\frac{1}{2}\right)^{2n}.$$

By using Stirling's formula, argue that for large $n$ one has

$$P\{\sigma_0 > 2n\} \sim \frac{1}{\sqrt{\pi n}} \quad \text{and} \quad P\{\sigma_0 = 2n\} \sim \frac{1}{2\sqrt{\pi}\, n^{3/2}}$$

(comp. with the formulas for $u_{2k}$ and $f_{2k}$ given in [$\underline{P}$ §1.10]). For example, the above formulas imply that $P\{\sigma_0 < \infty\} = 1$ and $E\sigma_0^\alpha < \infty$ if and only if $\alpha < 1/2$—see [$\underline{P}$ §1.9] for related results.

**Problem 2.4.17.** In the context of the previous problem, let $\sigma_k = \inf\{n \geq 1 : S_n = k\}$, $k = 1, 2, \ldots$ Prove that

$$P\{\sigma_k = n\} = \frac{k}{n} P\{S_n = k\}$$

and conclude that

$$P\{\sigma_k = n\} = \frac{k}{n} C_n^{\frac{n+k}{2}} \left(\frac{1}{2}\right)^n.$$

**Problem 2.4.18.** Let $\xi = \xi(\omega)$ be any non-degenerate random variable, such that, with some constants $a > 0$ and $b$, the distribution of $a\xi + b$ coincides with the distribution of $\xi$. Prove that this is only possible if $a = 1$ and $b = 0$.

**Problem 2.4.19.** Let $\xi_1$ and $\xi_2$ be any two exchangeable random variables, i.e., $\xi_1$ and $\xi_2$ are such that the distribution law of $(\xi_1, \xi_2)$ coincides with the distribution law of $(\xi_2, \xi_1)$. Prove that if $f = f(x)$ and $g = g(x)$ are any two non-negative and non-decreasing functions, then

$$E f(\xi_1) g(\xi_1) \geq E f(\xi_1) g(\xi_2).$$

**Problem 2.4.20.** (*On [ P §2.4, Theorem 2].*) Let $\xi_1, \xi_2, \ldots$ be any sequence of real-valued random variables. Prove that

$$B := \{\omega : \lim \xi_n(\omega) \text{ exists and is finite}\} \in \mathscr{F}.$$

*Hint.* Use the fact that $B$ may be expressed as:

$$B = \{\underline{\lim} \, \xi_n > -\infty\} \cap \{\overline{\lim} \, \xi_n < \infty\} \cap \{\overline{\lim} \, \xi_n - \underline{\lim} \, \xi_n = 0\}.$$

**Problem 2.4.21.** Let $\xi_1, \xi_2, \ldots$ be any sequence of independent and identically distributed random variables that share the same *continuous* distribution function. Let $A_1, A_2, \ldots$ be any sequence of events, such that $A_1 = \Omega$ and

$$A_n = \{\xi_n > \xi_m \text{ for all } m < n\}, \quad n \geq 2,$$

i.e., $A_n$ is the event that a "record" occurs at time $n$. Prove that the events $A_1, A_2, \ldots$ are independent and that $P(A_n) = 1/n, n \geq 1$.

**Problem 2.4.22.** Let $\xi$ and $\eta$ be any two random variables, such that $\mathrm{Law}(\eta)$, i.e., the distribution law of $\eta$, is absolutely continuous (in the sense that the associated distribution function $F_\eta$ is absolutely continuous). Prove that:

(a) If $\xi$ and $\eta$ are *independent*, then $\mathrm{Law}(\xi + \eta)$ is also absolutely continuous.

(b) If $\xi$ and $\eta$ are *not independent*, then $\mathrm{Law}(\xi + \eta)$ may not be absolutely continuous.

**Problem 2.4.23.** Let $\xi$ and $\eta$ be any two random variables, such that $\xi$ is discrete and $\eta$ is singular, i.e., $F_\xi$ is a discrete distribution function and $F_\eta$ is a singular distribution function. Prove that the distribution function $F_{\xi+\eta}$, associated with the random variable $\xi + \eta$, is singular.

**Problem 2.4.24.** Let $(\Omega, \mathscr{F})$ be any measurable space, such that the $\sigma$-algebra $\mathscr{F}$ is (countably) generated by some partition $\mathscr{D} = \{D_1, D_2, \ldots\}$ (see [ P §2.2, **1** ]). Prove that the $\sigma$-algebra $\mathscr{F}$ can be identified with the $\sigma$-algebra $\mathscr{F}_X$, generated by the random variable

$$X(\omega) = \sum_{n=1}^{\infty} \frac{\varphi(I_{D_n}(\omega))}{10^n},$$

where $\varphi(0) = 3$ and $\varphi(1) = 5$.

**Problem 2.4.25.** (a) Suppose that the random variable $X$ has a symmetric distribution, i.e., $\mathrm{Law}(X) = \mathrm{Law}(-X)$. Prove that $\mathrm{Law}(X) = \mathrm{Law}(\xi Y)$, where $\xi$ and $Y$ are independent random variables, such that $P\{\xi = 1\} = P\{\xi = -1\} = 1/2$ and $\mathrm{Law}(Y) = \mathrm{Law}(|X|)$.

(b) Suppose that $\xi$ and $Y$ are two independent random variables and that $P\{\xi = 1\} = P\{\xi = -1\} = 1/2$. Prove that $\xi$ and $\xi Y$ are independent if and only if $Y$ has a symmetric distribution, i.e., $\mathrm{Law}(Y) = \mathrm{Law}(-Y)$.

**Problem 2.4.26.** Suppose that the random variable $X$ takes only two values, $x_1$ and $x_2$, $x_1 \neq x_2$, and that the random variable $Y$ also takes only two values $y_1$ and $y_2$, $y_1 \neq y_2$. Prove that if $\text{cov}(X, Y) = 0$ then $X$ and $Y$ must be independent.

**Problem 2.4.27.** Suppose that $\xi_1, \xi_2, \ldots$ are independent and identically distributed random variables, all being uniformly distributed in the interval $[0, 1]$. Given any $0 < x < 1$, set

$$\tau(x) = \min\{n \geq 1 : \xi_1 + \cdots + \xi_n > x\}.$$

Prove that $P\{\tau(x) > n\} = x^n / n!, n \geq 1$.

**Problem 2.4.28.** Suppose that $X_1$, $X_2$ and $X_3$ are independent and identically distributed random variables with exponential density $f(x) = e^{-x} I(x > 0)$. Define the random variables

$$Y_1 = \frac{X_1}{X_1 + X_2}, \quad Y_2 = \frac{X_1 + X_2}{X_1 + X_2 + X_3} \quad \text{and} \quad Y_3 = X_1 + X_2 + X_3.$$

Prove that the above random variables, $Y_1$, $Y_2$ and $Y_3$ are independent.

**Problem 2.4.29.** Suppose that $X_1$ and $X_2$ are independent random variables, both having a $\chi^2$-distribution, respectively, with $r_1$ and $r_2$ degrees of freedom (see formula [$\underline{P}$ §2.8, (34)], or [$\underline{P}$ §2.3, Table 3]). Prove that the random variables $Y_1 = X_1 / X_2$ and $Y_2 = X_1 + X_2$ are independent (comp. with the statements of Problems 2.13.34 and 2.13.39).

## 2.5   Random Elements

**Problem 2.5.1.** Let $\xi_1, \ldots, \xi_n$ be any family of $n$ discrete random variables. Prove that these random variables are independent if and only if for every choice of the real numbers $x_1, \ldots, x_n$ one has

$$P\{\xi_1 = x_1, \ldots, \xi_n = x_n\} = \prod_{i=1}^{n} P\{\xi_i = x_i\}.$$

**Problem 2.5.2.** Give a complete proof of the fact that every random function $X(\omega) = (\xi_t(\omega))_{t \in T}$ is a random process in the sense of [$\underline{P}$ §2.5, Definition 3] and vice versa.

   *Hint.* If $X = X(\omega)$ is a $\mathscr{F}/\mathscr{B}(\mathbb{R}^T)$-measurable function, then for every $t \in T$ and $B \in \mathscr{B}(\mathbb{R})$ one has

$$\{\omega : \xi_t(\omega) \in B\} = \{\omega : X(\omega) \in C\} \in \mathscr{F}, \quad \text{where } C = \{x \in \mathbb{R}^T : x_t \in B\}.$$

Conversely, it is enough to consider sets $C \in \mathscr{B}(\mathbb{R}^T)$ of the form $\{x : x_{t_1} \in B_1, \ldots, x_{t_n} \in B_n\}$, $B_1, \ldots, B_n \in \mathscr{B}(\mathbb{R})$, which, obviously, belong to $\mathscr{F}$.

**Problem 2.5.3.** Let $X_1, \ldots, X_n$ be random elements with values, respectively, in $(E_1, \mathscr{E}_1), \ldots, (E_n, \mathscr{E}_n)$. Furthermore, suppose that $(E'_1, \mathscr{E}'_1), \ldots, (E'_n, \mathscr{E}'_n)$ are measurable spaces and that $g_1, \ldots, g_n$ are, respectively, $\mathscr{E}_1/\mathscr{E}'_1, \ldots, \mathscr{E}_n/\mathscr{E}'_n$-measurable functions. Prove that if $X_1, \ldots, X_n$ are independent, then the random elements $g_1 \circ X_1, \ldots, g_n \circ X_n$ also must be independent, where $g_i \circ X_i = g_i(X_i), i = 1, \ldots, n$.

*Hint.* It is enough to notice that for any $B_i \in \mathscr{E}_i, i = 1, \ldots, n$, one has

$$P\{g_1(X_1) \in B_1, \ldots, g_n(X_n) \in B_n\} = P\{X_1 \in g_1^{-1}(B_1), \ldots, X_n \in g_n^{-1}(B_n)\}.$$

**Problem 2.5.4.** Let $X_1, X_2, \ldots$ be any *infinite* sequence of *exchangeable* random variables, i.e., the joint distribution of any $k$ elements of the sequence with distinct indices, say, $X_{i_1}, \ldots, X_{i_k}$, depends on $k$ but not on the choice or the order of the indices $i_1, \ldots, i_k$—comp. with the definition in Problem 2.1.11. Prove that if $\mathsf{E}X_n^2 < \infty, n \geq 1$, then the covariance of $X_1$ and $X_2$ satisfies $\mathrm{cov}(X_1, X_2) \geq 0$.

*Hint.* Using the exchangeability, write the variance $\mathsf{D}\left(\sum_{i=1}^n X_i\right)$ in terms of the first two moments and the covariances and then take the limit as $n \to \infty$.

**Problem 2.5.5.** Let $\xi_1, \ldots, \xi_m$ and $\eta_1, \ldots, \eta_n$ be any two (arbitrarily chosen) sets of random variables. Define the vectors $X = (\xi_1, \ldots, \xi_m)$ and $Y = (\eta_1, \ldots, \eta_n)$ and suppose that the following conditions are satisfied:

  (i) the random variables $\xi_1, \ldots, \xi_m$ are independent;
  (ii) the random variables $\eta_1, \ldots, \eta_n$ are independent;
  (iii) the random vectors $X$ and $Y$, treated as (random) elements of, respectively, $\mathbb{R}^m$ and $\mathbb{R}^n$ are independent.

Prove that the random variables $\xi_1, \ldots, \xi_m, \eta_1, \ldots, \eta_n$ are independent.

**Problem 2.5.6.** Consider the random vectors $X = (\xi_1, \ldots, \xi_m)$ and $Y = (\eta_1, \ldots, \eta_n)$ and suppose that their components $\xi_1, \ldots, \xi_m, \eta_1, \ldots, \eta_n$ are independent.

  (a) Prove that the random vectors $X$ and $Y$, treated as random elements, are independent (comp. with Problem 2.5.5).

  (b) Let $f: \mathbb{R}^m \to \mathbb{R}$ be $g: \mathbb{R}^n \to \mathbb{R}$ be two Borel functions. Prove that the random variables $f(\xi_1, \ldots, \xi_m)$ and $g(\eta_1, \ldots, \eta_n)$ are independent.

**Problem 2.5.7.** Suppose that $(\Omega, \mathscr{F})$ is a measurable space and let $(E, \mathscr{E}, \rho)$ be a metric space endowed with metric $\rho$ and a Borel $\sigma$-algebra $\mathscr{E}$, associated with the metric $\rho$—see [P §2.2]. Let $X_1(\omega), X_2(\omega), \ldots$ be some sequence of $\mathscr{F}/\mathscr{E}$-measurable functions (i.e., random elements), such that for any $\omega \in \Omega$ the limit

$$X(\omega) = \lim_{n \to \infty} X_n(\omega)$$

exists. Prove that the limit $X(\omega)$, treated as a function of $\omega \in \Omega$, must be $\mathscr{F}/\mathscr{E}$-measurable.

**Problem 2.5.8.** Let $\xi_1, \xi_2, \ldots$ be any sequence of independent and identically distributed random variables, let $\mathscr{F}_n = \sigma(\xi_1, \xi_2, \ldots)$, $n \geq 1$, and let $\tau$ be any stopping time (relative to $(\mathscr{F}_n)_{n \geq 1}$). Set

$$\eta_n(\omega) = \xi_{n+\tau(\omega)}(\omega).$$

Prove that the sequence $(\eta_1, \eta_2, \ldots)$ has the same distribution as the sequence $(\xi_1, \xi_2, \ldots)$.

## 2.6 The Lebesgue Integral: Expectation

**Problem 2.6.1.** Prove that the representation in [P §2.6, (6)] is indeed in force.

  *Hint.* Let $S$ denote the space of simple functions $s$. If $s \in \{s \in S : s \leq \xi\}$ and if $(\xi_n)_{n \geq 1}$ is some sequence of simple random variables such that $\xi_n \uparrow \xi$, then $\max(\xi_n, s) \uparrow \xi$ and $Es \leq E\max(\xi_n, s)$. From the last inequality one can conclude that $Es \leq E\xi$ and that $\sup_{\{s \in S : s \leq \xi\}} Es \leq E\xi$. The opposite inequality follows directly from the construction of $E\xi$.

**Problem 2.6.2.** Verify the following generalization of property **E**, described in [P §2.6, **3**]. Suppose that $\xi$ and $\eta$ are two random variables for which the expectations $E\xi$ and $E\eta$ are well defined and the expression $E\xi + E\eta$ is meaningful, in the sense that it does not have the form $\infty - \infty$ or the form $-\infty + \infty$. Then one can write

$$E(\xi + \eta) = E\xi + E\eta.$$

  *Hint.* Just as in the proof of property **E**, one must consider the infinities arrising from the representations $\xi = \xi^+ - \xi^-$ and $\eta = \eta^+ - \eta^-$. If, for example, $E\xi^+ = \infty$, then, by using the assumptions in the problem, one can prove by contradiction that $E(\xi + \eta)^+ = \infty$.

**Problem 2.6.3.** Generalize property **G** in [P §2.6, **3**] by showing that if $\xi = \eta$ (a. e.) and $E\xi$ is well defined, then $E\eta$ also well defined and $E\eta = E\xi$.

**Problem 2.6.4.** Let $\xi$ be any extended random variable and let $\mu$ be any $\sigma$-finite measure with the property $\int_\Omega |\xi| \, d\mu < \infty$. Prove that $|\xi| < \infty$ ($\mu$-a. e.). (Comp. with Property **J**.)

**Problem 2.6.5.** Suppose that $\mu$ is some $\sigma$-finite measure and that $\xi$ and $\eta$ are extended random variables for which $\int \xi \, d\mu$ and $\int \eta \, d\mu$ are well defined. Prove that if one can claim that $\int_A \xi \, d\mu \leq \int_A \eta \, d\mu$ for any set $A \in \mathscr{F}$, then one can also claim that $\xi \leq \eta$ ($\mu$-a. e.). (Comp. with property **I**.)

**Problem 2.6.6.** Assuming that $\xi$ and $\eta$ are two independent random variables, prove that $E\xi\eta = E\xi \cdot E\eta$.

*Hint.* Instead of $\xi$ and $\eta$ consider the simple random variables $\xi_n$ and $\eta_n$, chosen so that $\xi_n \uparrow \xi$ and $\eta_n \uparrow \eta$. According to [P §2.6, Theorem 6] one must have $E\xi_n\eta_n = E\xi_n E\eta_n$. The proof can be completed by using the monotone convergence theorem.

**Problem 2.6.7.** By using Fatou's lemma prove that

$$P(\varliminf A_n) \leq \varliminf P(A_n), \quad P(\varlimsup A_n) \leq \varlimsup P(A_n).$$

**Problem 2.6.8.** Construct an example that proves that, in general, in the dominated convergence theorem one cannot relax the condition "$|\xi_n| \leq \eta$, $E\eta < \infty$".

*Hint.* Let $\Omega = [0, 1]$, let $\mathscr{F} = \mathscr{B}([0, 1])$, suppose that $P$ is the Lebesgue measure on $[0, 1]$, and then consider the random variables $\xi_n(\omega) = -nI(\omega \leq 1/n)$, $n \geq 1$.

**Problem 2.6.9.** By way of example, prove that, in general, in the dominated convergence theorem one cannot remove the condition "$\xi_n \leq \eta$, $E\eta > -\infty$".

**Problem 2.6.10.** Prove the following variant of Fatou's lemma: if the family of random variables $\{\xi_n^+, n \geq 1\}$ is uniformly integrable, then

$$\varlimsup E\xi_n \leq E \varlimsup \xi_n.$$

*Hint.* Use the fact that for any $\varepsilon > 0$ one can find some $c > 0$ such that $E\xi_n I(\xi_n > c) < \varepsilon$, for all $n \geq 1$.

**Problem 2.6.11.** The Dirichlet function is given by

$$d(x) = \begin{cases} 1, & x \text{ is rational} \\ 0, & x \text{ is irrational} \end{cases}, \quad x \in [0, 1].$$

This function is Lebesgue-integrable (on $[0, 1]$), but is not Riemann-integrable. Why?

**Problem 2.6.12.** Give an example of a sequence of Riemann-integrable functions $(f_n)_{n \geq 1}$, which are defined on $[0, 1]$ and are such that $|f_n| \leq 1$, $f_n \to f$ Lebesgue-almost everywhere, and yet the limit $f$ is not Riemann-integrable.

*Hint.* Consider the function $f_n(x) = \sum_{i=1}^n I_{\{q_i\}}(x)$, where $\{q_1, q_2, \dots\}$ is the set of all rational numbers in $[0, 1]$.

**Problem 2.6.13.** Let $\{a_{ij}; i, j \geq 1\}$ be any sequence of real numbers with $\sum_{i,j} |a_{ij}| < \infty$. By using Fubini's theorem, prove that

$$\sum_{(i,j)} a_{ij} = \sum_i \left( \sum_j a_{ij} \right) = \sum_j \left( \sum_i a_{ij} \right). \tag{$*$}$$

*Hint.* Consider an arbitrary sequence of positive numbers $p_1, p_2, \ldots$ with $\sum_{i=1}^{\infty} p_i = 1$ and define the probability measure $\mathsf{P}$ on $\Omega = N = \{1, 2, \ldots\}$ according to the formula $\mathsf{P}(A) = \sum_{i \in A} p_i$. Then define the function $f(i, j) = \frac{a_{ij}}{p_i p_j}$, observe that

$$\int_{\Omega \times \Omega} |f(\omega_1, \omega_2)| \, d(\mathsf{P} \times \mathsf{P}) = \sum_{i,j} |f(i, j)| p_i p_j = \sum_{i,j} |a_{ij}| < \infty,$$

and use Fubini's theorem.

**Problem 2.6.14.** Give an example of a sequence $(a_{ij}; i, j \geq 1)$ for which $\sum_{i,j} |a_{ij}| = \infty$, but the second identity in $(*)$ (Problem 2.6.13) does not hold.

*Hint.* Consider the sequence

$$a_{ij} = \begin{cases} 0, & i = j \\ (i - j)^{-3}, & i \neq j \end{cases}.$$

**Problem 2.6.15.** Starting with simple functions and using the results concerning the passage to the limit under the Lebesgue integral, prove the following version of the *change of variables theorem.*

Let $h = h(y)$ be any non-decreasing and continuously differentiable function defined on the interval $[a, b]$ and let $f(x)$ be any integrable (relative to the standard Lebesgue measure $dx$) function on the interval $[h(a), h(b)]$. Then the function $f(h(y))h'(y)$ is Lebesgue-integrable on the interval $[a, b]$ and

$$\int_{h(a)}^{h(b)} f(x) \, dx = \int_a^b f(h(y))h'(y) \, dy.$$

*Hint.* First prove the result for functions $f$ that can be written as finite linear combinations of indicators of Borel sets. By using the monotone convergence theorem then extend the result for all non-negative functions $f$ and, finally, prove the result for arbitrary functions $f$ by using the usual representation $f = f^+ - f^-$.

**Problem 2.6.16.** Verify formula [$\underline{P}$ §2.6, (70)].

*Hint.* Consider the random variable $\widetilde{\xi} = -\xi$, which has a distribution function $\widetilde{F}(x) = 1 - F((-x)-)$, and notice that $\int_{-\infty}^{0} |x|^n \, dF(x) = \int_0^{\infty} x^n \, d\widetilde{F}(x)$. Use formula [$\underline{P}$ §2.6, (69)].

**Problem 2.6.17.** Let $\xi, \xi_1, \xi_2, \ldots$ be any sequence of *non-negative* random variables that converges in probability $\mathsf{P}$ to the random variable $\xi$, i.e., $\mathsf{P}(|\xi_n - \xi| > \varepsilon) \to 0, n \to \infty$ (notation: $\xi_n \overset{\mathsf{P}}{\to} \xi$—see [$\underline{P}$ §2.10]).

(a) Generalize [$\underline{P}$ §2.6, Theorem 5] by showing that if $\mathsf{E}\xi_n < \infty, n \geq 1$, then the following claim can be made: $\mathsf{E}\xi_n \to \mathsf{E}\xi < \infty$ if and only if the family $\{\xi_n, n \geq 1\}$ is uniformly integrable; in other words, the statement of Theorem 5 remains valid if the convergence with probability 1 is replaced by *convergence in probability.*

(b) Prove that if all random variables $\xi, \xi_1, \xi_2, \ldots$ are integrable, i.e., $E\xi < \infty$ and $E\xi_n < \infty, n \geq 1$, then

$$E\xi_n \to E\xi \quad \Longrightarrow \quad E|\xi_n - \xi| \to 0.$$

*Hint.* (a) The sufficiency follows from [P §2.6, Theorem 4] and Problem 2.10.1. The necessity can be established, as in [P §2.6, Theorem 5], by replacing the almost everywhere convergence with convergence in probability (one must again use Problem 2.10.1).

(b) Given any $c > 0$ one has

$$E|\xi - \xi_n| \leq E|\xi - (\xi \wedge c)| + E|(\xi \wedge c) - (\xi_n \wedge c)| + E|(\xi_n \wedge c) - \xi_n|.$$

By keeping $\varepsilon > 0$ fixed and by choosing $c > 0$ so that $E|\xi - (\xi \wedge c)| < \varepsilon$, one can claim (due to the assumptions) that $E|(\xi \wedge c) - (\xi_n \wedge c)| \leq \varepsilon$ and $E|(\xi_n \wedge c) - \xi_n| \leq 3\varepsilon$, for all sufficiently large $n$. Consequently, $E|\xi - \xi_n| \leq 5\varepsilon$, for all sufficiently large $n$.

**Problem 2.6.18.** Let $\xi$ be any integrable random variable, i.e., $E|\xi| < \infty$.

(a) Prove that for any $\varepsilon > 0$ one can find some $\delta > 0$ with the property that for any $A \in \mathscr{F}$ with $P(A) < \delta$ one has $EI_A|\xi| < \varepsilon$ (*absolute continuity property of the Lebesgue integral*).

(b) Conclude from (a) that if $(A_n)_{n \geq 1}$ is some sequence of events for which $\lim_n P(A_n) = 0$, then $E(\xi I(A_n)) \to 0$, as $n \to \infty$. *Hint.* Use Lemma 2 in [P §2.6, 5].

*Remark.* Comp. with (b) from [P §2.6, Theorem 3].

**Problem 2.6.19.** Suppose that the random variables $\xi, \eta, \zeta$ and $\xi_n, \eta_n, \zeta_n, n \geq 1$, are such that (see the definition of convergence in probability $\overset{P}{\to}$ in Problem 2.6.17)

$$\xi_n \overset{P}{\to} \xi, \; \eta_n \overset{P}{\to} \eta, \; \zeta_n \overset{P}{\to} \zeta, \quad \eta_n \leq \xi_n \leq \zeta_n, \; n \geq 1,$$
$$E\zeta_n \to E\zeta, \quad E\eta_n \to E\eta,$$

and the expectations $E\xi, E\eta, E\zeta$ are all finite. Prove the following result known as *Pratt's lemma*:

(a) $E\xi_n \to E\xi$.

(b) If, in addition, $\eta_n \leq 0 \leq \zeta_n$, then $E|\xi_n - \xi| \to 0$.

Conclude that if $\xi_n \overset{P}{\to} \xi, E|\xi_n| \to E|\xi|$ and $E|\xi| < \infty$, then $E|\xi_n - \xi| \to 0$.

Give an example showing that if condition (b) is removed then it is possible that $E|\xi_n - \xi| \not\to 0$.

*Hint.* For the random variables $\widetilde{\eta}_n = 0$, $\widetilde{\xi}_n = \xi_n - \eta_n$, $\widetilde{\zeta}_n = \zeta_n - \eta_n$ and $\widetilde{\eta} = 0$, $\widetilde{\xi} = \xi - \eta$, $\widetilde{\zeta} = \zeta - \eta$, one has $0 \le \widetilde{\zeta}_n \overset{\text{P}}{\to} \widetilde{\zeta}$ and $\mathsf{E}\zeta_n \to \mathsf{E}\zeta$. According to part (a) in Problem 2.6.17 the family $\{\zeta_n, n \ge 1\}$ is uniformly integrable and, since $0 \le \widetilde{\xi}_n \le \widetilde{\zeta}_n$, the family $\{\widetilde{\xi}_n, n \ge 1\}$ also must be uniformly integrable. Consequently, one can claim that $\mathsf{E}\widetilde{\xi}_n \to \mathsf{E}\widetilde{\xi}$ (and even that $\mathsf{E}|\widetilde{\xi}_n - \widetilde{\xi}| \to 0$). Because of the assumption $\mathsf{E}\eta_n \to \mathsf{E}\eta$, it follows that $\mathsf{E}\xi_n \to \mathsf{E}\xi$.

**Problem 2.6.20.** Prove that $L_* f \le L^* f$ and, if the function $f$ is bounded and the measure $\mu$ is finite, then $L_* f = L^* f$ (see Remark 2 in [P §2.6, **11**]).

**Problem 2.6.21.** Prove that for any *bounded* function $f$ one has $\mathsf{E}f = L_* f$ (see Remark 2 in [P §2.6, **11**]).

**Problem 2.6.22.** Prove the final statement in Remark 2 in [P §2.6, **11**].

**Problem 2.6.23.** Let $F = F(x)$ be the distribution function of the random variable $X$. Prove that:

(a)   $\mathsf{E}|X| < \infty \iff \displaystyle\int_{-\infty}^{0} F(x)\,dx < \infty$ and $\displaystyle\int_{0}^{\infty}(1 - F(x))\,dx < \infty$;

(b)   $\mathsf{E}X^{+} < \infty \iff \displaystyle\int_{a}^{\infty} \ln\frac{1}{F(x)}\,dx < \infty$ for some $a$.

*Hint.* (b) Verify the following inequality

$$\mathsf{E}[XI(X > a)] \le \int_{a}^{\infty} \ln\frac{1}{F(x)}\,dx \le \frac{1}{F(a)}\mathsf{E}[XI(x > a)], \qquad a > 0.$$

**Problem 2.6.24.** Prove that if $p > 0$ and $\lim_{x\to\infty} x^p \mathsf{P}\{|\xi| > x\} = 0$, then $\mathsf{E}|\xi|^r < \infty$ for all $r < p$. Give an example showing that if $r = p$ then one can have $\mathsf{E}|\xi|^r = \infty$.

**Problem 2.6.25.** Give an example of a probability density $f(x)$, which is not an even function, but nevertheless all odd moments vanish, i.e., $\int_{-\infty}^{\infty} x^k f(x)\,dx = 0$, $k = 1, 3, \ldots$.

**Problem 2.6.26.** Give an example of a sequence of random variables $\xi_n$, $n \ge 1$, that has the following property:

$$\mathsf{E}\sum_{n=1}^{\infty}\xi_n \ne \sum_{n=1}^{\infty}\mathsf{E}\xi_n.$$

**Problem 2.6.27.** Suppose that the random variable $X$ is such that for any $\alpha > 1$ one has

$$\frac{\mathsf{P}\{|X| > \alpha n\}}{\mathsf{P}\{|X| > n\}} \to 0 \quad \text{as } n \to \infty.$$

Prove that then $X$ admits finite moments of all orders.

*Hint.* Use the formula

$$E|X|^N = N \int_0^\infty x^{N-1} P(|X| > x)\, dx, \qquad N \geq 1.$$

**Problem 2.6.28.** Let $X$ be any random variable that takes the values $k = 0, 1,$ $2, \ldots$ with probabilities $p_k$. The function $G(s) = \mathsf{E}s^X$ $(= \sum_{k=0}^\infty p_k s^k)$, $|s| \leq 1$, is known as the *generating function* of the random variable $X$ (see Sect. A.3). Verify the following formulas:

(a) If $X$ is a Poisson random variable, i.e., $p_k = e^{-\lambda} \lambda^k / k!$, $k = 0, 1, 2, \ldots,$, for some $\lambda > 0$, then

$$G(s) = \mathsf{E}e^X = e^{-\lambda(1-s)}, \quad |s| \leq 1.$$

(b) If the random variable $X$ has a geometric distribution, i.e., if $p_k = pq^k$, $k = 0, 1, 2, \ldots,$ for some $0 < p < 1$ and $q = 1 - p$, then

$$G(s) = \frac{p}{1 - sq}, \quad |s| \leq 1.$$

(c) If $X_1, \ldots, X_n$ are independent and identically distributed random variables with $P\{X_1 = 1\} = p$, $P\{X_1 = 0\} = p$ $(q = 1 - p)$, then

$$G(s) = (ps + q)^n \quad \left( = \sum_{k=0}^n [C_n^k\, p^k q^{n-k}] s^k \right)$$

and, consequently, $P\{X_1 + \ldots + X_n = k\} = C_n^k\, p^k q^{n-k}$.

**Problem 2.6.29.** Let $X$ be any random variable that takes values in the set $\{0, 1, 2, \ldots\}$ and let $G(s) = \sum_{n=0}^\infty p_k s^k$, where $p_k = P\{X = k\}$, $k \geq 0$. Assuming that $r \geq 1$, prove that:

(a) If $\mathsf{E}X^r < \infty$, then the factorial moment $\mathsf{E}(X)_r \equiv \mathsf{E}X(X-1)\ldots(X-r+1)$ is finite and $\mathsf{E}(X)_r = \lim_{s \to 1} G^{(r)}(s)$ $(= G^{(r)}(1))$, where $G^{(r)}(s)$ is the $r$-th derivative of $G(s)$.

(b) If $\mathsf{E}X^r = \infty$, then $\mathsf{E}(X)_r = \infty$ and $\lim_{s \to 1} G^{(r)} = \infty$.

**Problem 2.6.30.** Let $X$ be any random variable which is uniformly distributed in the set $\{0, 1, \ldots, n\}$, i.e., $P\{X = k\} = \frac{1}{n+1}$, where $k = 0, 1, \ldots, n$. Prove that $G(s) = \frac{1}{n+1} \frac{1-s^{n+1}}{1-s}$ and, after computing $\mathsf{E}X$ and $\mathsf{E}X^2$, establish the following relations:

$$\sum_{k=1}^n k = \frac{n(n+1)}{2}, \qquad \sum_{k=1}^n k^2 = \frac{n(n+1)(n+2)}{6}.$$

**Problem 2.6.31.** (*Continuation of Problem 1.1.13.*) Consider the (not-necessarily independent) events $A_1, \ldots, A_n$, let $X_i = I_{A_i}$, $i = 1, \ldots, n$, and let $\Sigma_n = I_{A_1} + \ldots + I_{A_n}$. Prove that the generating function $G_{\Sigma_n}(s) = \mathsf{E} s^{\Sigma_n}$ is given by the formula:

$$G_{\Sigma_n}(s) = \sum_{m=0}^{n} S_m (s-1)^m,$$

where

$$S_m = \sum_{1 \le i_1 < \ldots < i_m \le n} \mathsf{P}(A_{i_1} + \ldots + A_{i_m}) \left( = \sum_{1 \le i_1 < \ldots < i_m \le n} \mathsf{P}\{X_{i_1} = 1, \ldots, X_{i_m} = 1\} \right)$$

(see Problem 1.1.12). Conclude that the probabilities of the events $B_m = \{\Sigma_n = m\}$ are given by the formula

$$\mathsf{P}(B_m) = \sum_{k=m}^{n} (-1)^{k-m} C_k^m \, S_k.$$

*Hint.* Use the relations $G_{\Sigma_n}(s) = \mathsf{E} \prod_{i=1}^{n} (1 + X_i(s-1))$ and

$$\prod_{i=1}^{n} (1 + X_i(s-1)) = 1 + \sum_{i=1}^{m} X_i(s-1) + \sum_{1 \le i_1 < i_2 \le n} X_{i_1} X_{i_2}(s-1)^2 + \ldots + \prod_{i=1}^{n} X_i(s-1)^n.$$

**Problem 2.6.32.** In addition to the generating functions $G(s)$, it is often useful to work with *moment generating functions*, which are defined as $M(s) = \mathsf{E} e^{sX}$, assuming that $s$ is chosen so that $\mathsf{E} e^{sX} < \infty$. Note that if the random variable $X$ is non-negative and $s = -\lambda$, where $\lambda \ge 0$, then the function $\widehat{F}(\lambda) = M(-\lambda)$ ($= \mathsf{E} e^{-\lambda X}$) is nothing but the *Laplace transform* of the random variable $X$ with c.d.f. $F = F(x)$.

(a) Prove that if the moment generating function $M(s)$ is defined for all $s$ in some neighborhood of the origin ($s \in [-a, a]$, $a > 0$), then all derivatives $M^{(k)}(s)$, $k = 1, 2, \ldots$, exist at $s = 0$ and

$$M^{(k)}(0) = \mathsf{E} X^k.$$

This observation justifies the term "moment generating function" in reference to the function $M(s)$.

(b) Give an example of a random variable for which $M(s) = \infty$ for every $s > 0$.

(c) Prove that if $X$ has a Poisson distribution with $\lambda > 0$ then $M(s) = e^{-\lambda(1-e^s)}$ for all $s \in \mathbb{R}$.

(d) Give an example of two random variables, $X$ and $Y$, which are *not* independent and, at the same time, $M_{X+Y}(s) = Ee^{s(X+Y)}$ is the product of the moment generating functions $M_X(s) = Ee^{sX}$ and $M_Y(s) = Ee^{sY}$.

**Problem 2.6.33.** Prove that if $0 < r < \infty$, $X_n \in L^r$ and $X_n \overset{P}{\to} X$, then the following conditions are equivalent:

(i) The family $\{|X_n|^r, n \geq 1\}$ is uniformly integrable.
(ii) $X_n \to X$ in $L^r$.
(iii) $E|X_n|^r \to E|X|^r < \infty$.

**Problem 2.6.34.** (*Spitzer identity.*) Let $X_1, X_2, \ldots$ be independent and identically distributed random variables with $E|X_1| < \infty$, and let $S_k = X_1 + \cdots + X_k$, $M_k = \max(0, S_1, \ldots, S_k)$, $k \geq 1$. Prove that, for any $n \geq 1$,

$$EM_n = \sum_{k=1}^{n} \frac{1}{k} ES_k^+, \tag{$*$}$$

where $S_k^+ = \max(0, S_k)$.

*Hint.* By using the relations

$$M_n = I(S_n > 0)M_n + I(S_n \leq 0)M_n,$$

$$E[I(S_n > 0)M_n] = E[I(S_n > 0)X_1] + E[I(S_n > 0)M_{n-1}]$$

and $E[I(S_n > 0)X_1] = n^{-1}ES_n^+$, one can prove by induction that

$$EM_n = \frac{1}{n}ES_n^+ + EM_{n-1} = \frac{1}{n}ES_n^+ + \left[\frac{1}{n-1}ES_{n-1}^+ + EM_{n-2}\right] = \cdots =$$

$$= \sum_{k=2}^{n} \frac{1}{k}ES_k^+ + EM_1 = \sum_{k=1}^{n} \frac{1}{k}ES_k^+.$$

*Remark.* One can derive $(*)$ by differentiating in $t$ the more general Spitzer identity, according to which, for any $0 < u < 1$, one has

$$\sum_{k=0}^{\infty} u^k Ee^{itM_k} = \exp\left\{\sum_{k=1}^{\infty} \frac{u^k}{k} Ee^{itS_k^+}\right\}.$$

The proof of the above relation is somewhat more involved than the proof of $(*)$.

**Problem 2.6.35.** Let $S_0 = 0$, $S_n = X_1 + \cdots + X_n$, $n \geq 1$, be a simple random walk (see [P §8.8]) and let $\sigma = \min\{n > 0: S_n \geq 0\}$. Prove that

$$E\min(\sigma, 2m) = 2E|S_{2m}| = 4mP\{S_{2m} = 0\}, \quad m \geq 0.$$

**Problem 2.6.36.** (a) Let $\xi$ be any standard Gaussian random variable ($\xi \sim \mathcal{N}(0, 1)$). By using integration by parts, prove that $\mathsf{E}\xi^k = (k - 1)\mathsf{E}\xi^{k-2}$, $k \geq 2$, and conclude that, for $k \geq 1$,

$$\mathsf{E}\xi^{2k-1} = 0 \quad \text{and} \quad \mathsf{E}\xi^{2k} = 1 \cdot 3 \cdot \cdots \cdot (2k - 3) \cdot (2k - 1) \ (= (2k - 1)!!).$$

(b) Prove that for any random variable $X$ that has Gamma distribution (see [$\underline{P}$ §2.3, Table 3]) with $\beta = 1$ one has

$$\mathsf{E}X^k = \frac{\Gamma(k + \alpha)}{\Gamma(\alpha)}, \quad k \geq 1.$$

In particular, $\mathsf{E}X = \alpha$, $\mathsf{E}X^2 = \alpha(\alpha + 1)$ and, therefore, $\mathsf{D}X = \alpha$. Find an analog of the above formula when $\beta \neq 1$.

(c) Prove that for any Beta-distributed random variable $X$ (see Table 3 in § 3) one must have

$$\mathsf{E}X^k = \frac{B(r + k, s)}{B(r, s)}, \quad k \geq 1.$$

**Problem 2.6.37.** Prove that the function

$$\xi(\omega_1, \omega_2) = e^{-\omega_1 \omega_2} - 2e^{-2\omega_1 \omega_2}, \quad \omega_1 \in \Omega_1 = [1, \infty), \ \omega_2 \in \Omega_2 = (0, 1],$$

has the following properties:
(a) for any fixed $\omega_2$, $\xi$ is Lebesgue-integrable in the variable $\omega_1 \in \Omega_1$;
(b) for any fixed $\omega_1$, $\xi$ is Lebesgue-integrable in the variable $\omega_2 \in \Omega_2$,

and yet Fubini's theorem does not hold.

**Problem 2.6.38.** Prove *Beppo Levi's theorem*, which claims the following: if the random variables $\xi_1, \xi_2, \ldots$ are integrable (i.e., $\mathsf{E}|\xi_n| < \infty$ for all $n \geq 1$), if $\sup_n \mathsf{E}\xi_n < \infty$, and if $\xi_n \uparrow \xi$ for some random variable $\xi$, then $\xi$ is integrable and one has $\mathsf{E}\xi_n \uparrow \mathsf{E}\xi$ (comp. with [$\underline{P}$ §2.6, Theorem 1a]).

**Problem 2.6.39.** Prove the following variation of *Fatou's lemma*: if $0 \leq \xi_n \to \xi$ (P-a.s.) and $\mathsf{E}\xi_n \leq A < \infty$, $n \geq 1$, then $\xi$ must be integrable and $\mathsf{E}\xi \leq A$.

**Problem 2.6.40.** (*On the connection between the Riemann and the Lebesgue integrals.*) Suppose that the Borel function $f = f(x)$, $x \in \mathbb{R}$, is integrable with respect to the Lebesgue measure $\lambda$, i.e., $\int_{\mathbb{R}} |f(x)| \lambda(dx) < \infty$. Prove that for any $\varepsilon > 0$ one can find:

(a) a *step function* of the form $f_\varepsilon(x) = \sum_{i=1}^n f_i I_{A_i}(x)$, the sets $A_i$ being bounded intervals, such that $\int_{\mathbb{R}} |f(x) - f_\varepsilon(x)| \lambda(dx) < \varepsilon$;

(b) an integrable *continuous function* $g_\varepsilon(x)$ that has bounded support and is such that $\int_{\mathbb{R}} |f(x) - g_\varepsilon(x)| \lambda(dx) < \varepsilon$.

**Problem 2.6.41.** Prove that if $\xi$ is any integrable random variable then

$$E\xi = \int_0^\infty P\{\xi > x\}\,dx - \int_{-\infty}^0 P\{\xi < x\}\,dx.$$

Show also that for any $a > 0$ one must have

$$E[\xi\, I(\xi > a)] = \int_a^\infty P\{\xi > x\}\,dx + aP\{\xi > a\}$$

and that if $\xi \geq 0$ then

$$E[\xi I(\xi \leq a)] = \int_0^a P\{x < \xi \leq a\}\,dx.$$

**Problem 2.6.42.** Let $\xi$ and $\eta$ be any two integrable random variables. Prove that

$$E\xi - E\eta = \int_{-\infty}^\infty [P\{\eta < x \leq \xi\} - P\{\xi < x \leq \eta\}]\,dx.$$

**Problem 2.6.43.** Let $\xi$ be any non-negative random variable ($\xi \geq 0$) with Laplace transform $\widehat{F}(\lambda) = Ee^{-\lambda\xi}$, $\lambda \geq 0$.
  (a) Prove that for any $0 < r < 1$ one has

$$E\xi^r = \frac{r}{\Gamma(1-r)} \int_0^\infty \frac{1 - \widehat{F}(\lambda)}{\lambda^{r+1}}\,d\lambda.$$

*Hint*: use the fact that

$$\frac{1}{r}\,\Gamma(1-r)s^r = \int_0^\infty \frac{1 - e^{-s\lambda}}{\lambda^{r+1}}\,d\lambda,$$

for any $s \geq 0$ and any $0 < r < 1$.
  (b) Prove that for any $r > 0$ one has

$$E\xi^{-r} = \frac{1}{r\,\Gamma(r)} \int_0^\infty \widehat{F}(\lambda^{1/r})\,d\lambda.$$

*Hint*: use the fact that

$$s = \frac{r}{r\,\Gamma(1/r)} \int_0^\infty \exp\{-(\lambda/s)^r\}\,d\lambda,$$

for any $s \geq 0$ and any $r > 0$.

**Problem 2.6.44.** (a) Prove that in Hölder's inequality [$\underline{P}$ §2.6, (29)] the *identity* is attained if and only if $|\xi|^p$ and $|\eta|^q$ are linearly dependent P-a.e., i.e., one can find constants $a$ and $b$, that are not simultaneously null, for which one has P-a.e. $a|\xi|^p = b|\eta|^q$.

(b) Prove that in Minkowski's inequality [$\underline{P}$ §2.6, (31)] (with $1 < p < \infty$) the *identity* is attained if and only if one can find two constants, $a$ and $b$, that are not simultaneously null, for which one has $a\xi = b\eta$, P-a.e..

(c) Prove that in Cauchy–Bunyakovsky's inequality [$\underline{P}$ §2.6, (24)] the *identity* is attained if and only if $\xi$ and $\eta$ are linearly dependent P-a.e., i.e., $a\xi = b\eta$, P-a.e., for some constants $a$ and $b$ that are not simultaneously null.

**Problem 2.6.45.** Suppose that $X$ is a random variable with $P\{a \le X \le b\} = 1$, for some choice of $a, b \in \mathbb{R}$, $a < b$. Setting $m = EX$ and $\sigma^2 = DX$, prove that $\sigma^2 \le (m - a)(b - m)$, where equality is reached if and only if $P\{X = a\} + P\{X = b\} = 1$.

**Problem 2.6.46.** Assuming that $X$ is a random variable with $E|X| < \infty$, prove that:

(a) If $X > 0$ (P-a.s.) then

$$E\frac{1}{X} \ge \frac{1}{EX}, \quad E \ln X \le \ln EX, \quad E(X \ln X) \ge EX \cdot \ln EX,$$

where we suppose that $0 \cdot \ln 0 = 0$.

(b) If $X$ takes values only in the interval $[a, b]$, $0 < a < b < \infty$, then

$$1 \le EX \cdot E\frac{1}{X} \le \frac{(a + b)^2}{4ab}$$

(when do the equalities in the last relation hold?).

(c) If the random variable $X$ is positive and if $EX^2 < \infty$, then the following lower bound estimate, known as the *Paley–Zygmund inequality*, holds: for any $0 < \lambda < 1$

$$P\{X > \lambda EX\} \ge (1 - \lambda)^2 \frac{(EX)^2}{EX^2}.$$

(d) By using the above inequality, prove that if $P\{X \le u\} \le c$ for some $u > 0$ and $c \ge 0$, then for every $r > 0$ one has

$$EX^r \le \frac{u^r}{1 - (cEX^{2r})^{1/2}/EX^r}.$$

provided that the expression in the denominator is well defined and strictly positive.

(e) Prove that if $X$ is a non-negative integer-valued random variable then

$$P\{X > 0\} \ge \frac{(EX)^2}{EX^2}.$$

**Problem 2.6.47.** Suppose that $\xi$ is a random variable with $E\xi = m$ and $E(\xi - m)^2 = \sigma^2$. Prove *Cantelli's inequalities*:

$$\max \left( P\{\xi - m > \varepsilon\}, P\{\xi - m < \varepsilon\} \right) \leq \frac{\sigma^2}{\sigma^2 + \varepsilon^2}, \qquad \varepsilon > 0;$$

$$P\{|\xi - m| \geq \varepsilon\} \leq \frac{2\sigma^2}{\sigma^2 + \varepsilon^2}, \qquad \varepsilon > 0.$$

**Problem 2.6.48.** Suppose that $\xi$ is some random variable with $E|\xi| < \infty$ and let $g = g(x)$ be any strictly convex function defined on the real line $\mathbb{R}$. Prove that $Eg(\xi) = g(E\xi)$ if and only if $\xi = E\xi$ (P-a.s.).

**Problem 2.6.49.** Let $\xi$ be any integrable random variable, i.e., $E|\xi| < \infty$. Prove that for any $\varepsilon$ one can find a simple random variables $\xi_\varepsilon > 0$ so that $E|\xi - \xi_\varepsilon| \leq \varepsilon$.

**Problem 2.6.50.** Consider the equation

$$Z_t = B(t) + \int_0^t Z_{s-} dA(s)$$

(comp. with equation [P §2.6, (74)]), where $A(t)$ and $B(t)$ are functions with locally bounded variations, which are right-continuous (for $t \geq 0$), admit left limits (for $t > 0$) and are such that $A(0) = B(0) = 0$, $\Delta A(t) > -1$, where $\Delta A(t) = A(t) - A(t-), t > 0, \Delta A(0) = 0)$.

Prove that, in the class of all locally bounded functions, the above equation admits a unique solution $\mathscr{E}_t(A, B)$, which, for any $t > 0$, is given by the formula:

$$\mathscr{E}_t(A, B) = \mathscr{E}_{t-}(A) \int_0^t \frac{1}{\mathscr{E}_{s-}(A)} dB(s).$$

**Problem 2.6.51.** Let $V(t)$ be any function with locally bounded variation, which is right-continuous (for $t \geq 0$), admits left limits (for $t > 0$), and satisfies the relation

$$V(t) \leq K + \int_0^t V(s-) dA(s),$$

with some constant $K \geq 0$ and some non-decreasing and right continuous function $A(t)$, which admits left limits and has the property $A(0) = 0$.

Prove that

$$V(t) \leq K\mathscr{E}_t(A), \quad t \geq 0;$$

in particular, if $A(t) = \int_0^t a(s) ds, a(s) \geq 0$, then the function $V(t)$ satisfies *the Gronwall–Bellman inequality*:

$$V(t) \leq K \exp \left\{ \int_0^t a(s) ds \right\}, \quad t \geq 0.$$

**Problem 2.6.52.** The derivation of Hölder's inequality [P §2.6, (29)] uses the inequality

$$ab \le \frac{a^p}{p} + \frac{b^q}{q},$$

in which $a > 0$, $b > 0$ and $p > 1$ and $q > 1$ are such that $\frac{1}{p} + \frac{1}{q} = 1$. Prove that the above inequality is a special case (with $h(x) = x^{p-1}$) of *Young's inequality*:

$$ab \le H(a) + \widetilde{H}(b), \qquad a > 0, \quad b > 0,$$

where

$$H(x) = \int_0^x h(y)\, dy, \qquad \widetilde{H}(x) = \int_0^x \widetilde{h}(y)\, dy,$$

and $h = h(y)$, $y \in \mathbb{R}_+$, is some continuous and strictly increasing function with $h(0) = 0$, $\lim_{y \to \infty} h(y) = \infty$, while $\widetilde{h} = \widetilde{h}(y)$, $y \in \mathbb{R}_+$, is the inverse of the function $h = h(y)$, i.e.,

$$\widetilde{h}(y) = \inf\{t : h(t) > y\}.$$

Note that since $h = h(y)$ is continuous and strictly increasing then one has $\widetilde{h}(y) = h^{-1}(y)$.

**Problem 2.6.53.** Let $X$ be any random variable. Prove that the following implications are in force for any $a > 0$:

$$\mathsf{E}|X|^a < \infty \iff \sum_{n=1}^{\infty} n^{a-1} \mathsf{P}\{|X| \ge n\} < \infty.$$

**Problem 2.6.54.** Let $\xi$ be any non-negative random variable. Prove that for any $r > 1$ one has

$$\int_0^\infty \frac{\mathsf{E}(\xi \wedge x^r)}{x^r}\, dx = \frac{r}{r-1} \mathsf{E}\xi^{1/r}.$$

In particular,

$$\int_0^\infty \frac{\mathsf{E}(\xi \wedge x^2)}{x^2}\, dx = 2\mathsf{E}\sqrt{\xi}.$$

**Problem 2.6.55.** Let $\xi$ be any random variable with $\mathsf{E}\xi \ge 0$, $0 < \mathsf{E}\xi^2 < \infty$ and let $\varepsilon \in [0, 1]$. Verify the following "inverse" of the Chebyshev inequality

$$\mathsf{P}\{\xi > \varepsilon \mathsf{E}\xi\} \ge (1 - \varepsilon) \frac{2(\mathsf{E}\xi)^2}{\mathsf{E}\xi^2}.$$

**Problem 2.6.56.** Let $(\Omega, \mathscr{F})$ be any measurable space and define the set function $\mu = \mu(B)$, $B \in \mathscr{F}$, so that

$$\mu(B) = \begin{cases} |B|, & \text{if } B \text{ is finite}, \\ \infty & \text{if } B \text{ is not finite}, \end{cases}$$

where $|B|$ denotes the cardinality of the set $B$. Prove that the set function $\mu$ defined above is a *measure* (in the sense of [P §2.1, Definition 6]). This measure is known as *counting measure*. It is $\sigma$-finite if and only if the set $\Omega$ is at most countable.

**Problem 2.6.57.** (*On the Radon–Nikodym Theorem I.*) Let $\lambda$, $\mu$ and $\nu$ be $\sigma$-finite measures defined on the measurable space $(\Omega, \mathscr{F})$ and suppose that the Radon–Nikodym derivatives $\frac{d\nu}{d\mu}$ and $\frac{d\mu}{d\lambda}$ exist. Then show that the Radon–Nikodym derivative $\frac{d\nu}{d\lambda}$ also exists and

$$\frac{d\nu}{d\lambda} = \frac{d\nu}{d\mu}\frac{d\mu}{d\lambda} \quad (\lambda\text{-a.e.}).$$

**Problem 2.6.58.** (*On the Radon–Nikodym Theorem II.*) Consider the measure space $(\Omega, \mathscr{F}) = ([0, 1], \mathscr{B}([0, 1]))$, let $\lambda$ be the Lebesgue measure and let $\mu$ be any counting measure (as in Problem 2.6.56) on $\mathscr{F}$. Prove that $\mu \ll \lambda$, but, at the same time, the Radon–Nikodym theorem, which guarantees the existence of the density $\frac{d\mu}{d\lambda}$, is not valid.

**Problem 2.6.59.** Let $\lambda$ and $\mu$ be any two $\sigma$-finite measures on $(\Omega, \mathscr{F})$ and let $f = \frac{d\lambda}{d\mu}$. Prove that if $\mu\{\omega : f = 0\} = 0$, then the density $\frac{d\mu}{d\lambda}$ exists and can be represented by the function

$$\varphi = \begin{cases} \frac{1}{f} & \text{on the set } \{f \neq 0\}, \\ c & \text{on the set } \{f = 0\}, \end{cases}$$

where $c$ is some arbitrary constant.

**Problem 2.6.60.** Prove that the following function on the interval $[0, \infty)$

$$f(x) = \begin{cases} 1, & x = 0, \\ \frac{\sin x}{x}, & x > 0, \end{cases}$$

is Riemann-integrable (in fact, (R) $\int_0^\infty f(x)\,dx = \frac{\pi}{2}$), but is not Lebesgue integrable.

**Problem 2.6.61.** Give an example of a function $f = f(x)$ defined on $[0, 1]$, which is bounded and Lebesgue integrable, and yet one cannot find a Riemann integrable function $g = g(x)$ which coincides with $f = f(x)$ Lebesgue-almost everywhere in $[0, 1]$.

**Problem 2.6.62.** Give an example of a bounded Borel function $f = f(x, y)$ defined on $\mathbb{R}^2$, which is not Lebesgue-integrable on $(\mathbb{R}^2, \mathscr{B}(\mathbb{R}^2))$, but is such that, for every $y \in \mathbb{R}$ and every $x \in \mathbb{R}$, one has, respectively,

$$\int_{\mathbb{R}} f(x, y)\,\lambda(dx) = 0 \quad \text{and} \quad \int_{\mathbb{R}} f(x, y)\,\lambda(dy) = 0,$$

where both integrals are understood to exist in the sense of Lebesgue.

**Problem 2.6.63.** (*On Fubini's Theorem I.*) Let $\lambda = \lambda(dx)$ denote the Lebesgue measure on $[0, 1]$ and let $\mu = \mu(dy)$ be any counting measure on $[0, 1]$. Let $D$ denote the diagonal of the unit square $[0, 1]^2$. Prove that

$$\int_{[0,1]} \left[ \int_{[0,1]} I_D(x, y) \, \lambda(dx) \right] \mu(dy) = 0$$

and

$$\int_{[0,1]} \left[ \int_{[0,1]} I_D(x, y) \, \mu(dy) \right] \lambda(dx) = 1.$$

The above relations show that the property [**P** §2.6, (49)], in the conclusion of the Fubini theorem ([**P** §2.6, Theorem 3]), cannot hold without the *finiteness* assumption for the measure.

**Problem 2.6.64.** (*On Fubini's Theorem II.*) Prove that Fubini's theorem remains valid even if the requirement for the two participating measures to be finite is replaced with the requirement that these measures are $\sigma$-finite. Prove that, in general, the assumption for $\sigma$-finiteness of the participating measures cannot be relaxed further (see Problem 2.6.63).

**Problem 2.6.65.** (*Part a*) in [**P** §2.6, Theorem 10].) Give an example of a bounded *non-Borel* function which is Riemann integrable (a reformulation of Problem 2.3.29).

**Problem 2.6.66.** Let $f = f(x)$ be any Borel-measurable function defined on the measurable structure $(\mathbb{R}^n, \mathscr{B}(\mathbb{R}^n))$, which is endowed with the Lebesgue measure $\lambda = \lambda(dx)$. Assuming that $\int_{\mathbb{R}^n} |f(x)| \, \lambda(dx) < \infty$, prove that:

$$\lim_{h \to 0} \int_{\mathbb{R}^n} |f(x + h) - f(x)| \, \lambda(dx) = 0.$$

*Hint.* Use part (b) in Problem 2.6.40.

**Problem 2.6.67.** For any *finite* number of independent and integrable random variables $\xi_1, \ldots, \xi_n$ one has

$$\mathsf{E} \prod_{k=1}^{n} \xi_k = \prod_{k=1}^{n} \mathsf{E} \xi_k$$

(see [**P** §2.6, Theorem 6]). Prove that if $\xi_1, \xi_2, \ldots$ is any sequence of independent and integrable random variables, then, in general, one has

$$\mathsf{E} \prod_{k=1}^{\infty} \xi_k \neq \prod_{k=1}^{\infty} \mathsf{E} \xi_k.$$

**Problem 2.6.68.** Suppose that the random variable $\xi$ is such that $E\xi < 0$ and $Ee^{\theta\xi} = 1$ for some $\theta \neq 0$. Prove that this is only possible if $\theta > 0$.

**Problem 2.6.69.** Let $h = h(t, x)$ be any function defined on the set $[a, b] \times \mathbb{R}$, where $a, b \in \mathbb{R}$ and $a < b$.

(a) Suppose that

1. For any fixed $x^\circ \in \mathbb{R}$ the function $h(t, x^\circ)$, $t \in [a, b]$, is continuous;
2. For any fixed $t^\circ \in [a, b]$ the function $h(t^\circ, x)$, $x \in \mathbb{R}$, is $\mathcal{B}(\mathbb{R})$-(i.e., Borel)-measurable,

and prove that when the above conditions are satisfied one can claim that the function $h = h(t, x)$, $t \in [a, b]$, $x \in \mathbb{R}$, is $\mathcal{B}([a, b]) \times \mathcal{B}(\mathbb{R})$-measurable.

(b) Assume that $\xi$ is a random variable, defined on some probability space, and that when conditions 1 and 2 above are satisfied, together with the condition

3. The family of random variables $\{h(t, \xi), t \in [a, b]\}$ is uniformly integrable. Show that:
   (i) The expected value $Eh(t, \xi)$ is a continuous function of the variable $t \in [a, b]$.
   (ii) If $H(t, x) = \int_a^t h(s, x)\, ds$, then the derivative $\frac{d}{dt}EH(t, \xi)$ exists for all $t \in (a, b)$ and equals $Eh(t, \xi)$, i.e.,

$$\frac{d}{dt}E \int_0^t h(s, \xi)\, ds = Eh(t, \xi).$$

**Problem 2.6.70.** (*On [ P §2.6, Lemma 2].*) (a) Let $\xi$ be any random variable with $E|\xi| < \infty$. Prove that for any $\varepsilon > 0$ one can find a $\delta > 0$, so that $P(A) < \delta$, $A \in \mathscr{F}$, implies $E(|\xi|I_A) < \varepsilon$. Conclude that, given any random variable $\xi$ with $E|\xi| < \infty$ and any $\varepsilon > 0$, one can find a constant $K = K(\varepsilon)$ so that

$$E(|\xi|; |\xi| > K) \equiv E|\xi|I(|\xi| > K) < \varepsilon.$$

(b) Prove that if $\{\xi_n, n \geq 1\}$ is a uniformly integrable family of random variables, then the family $\left\{\frac{1}{n}\sum_{k=1}^n \xi_k, n \geq 1\right\}$ also must be uniformly integrable.

**Problem 2.6.71.** Prove that Jensen's inequality [P §2.6, (25)] remains valid even when the function $g = g(x)$, assumed to be convex, is defined not on the entire real line $\mathbb{R}$, but only on some *open* set $G \subseteq \mathbb{R}$, and the random variable $\xi$ is such that $P\{\xi \in G\} = 1$ and $E|\xi| < \infty$. Prove that a function $g = g(x)$, which is defined on an open set $G$ and is convex, must be continuous. Prove that any such function admits the representation:

$$g(x) = \sup_n(a_n x + b_n), \quad x \in G,$$

where $a_n$ and $b_n$ are some appropriate constants.

**Problem 2.6.72.** Prove that for any $a, b \in \mathbb{R}$ and any $r \geq 0$ one has

$$|a + b|^r \leq c_r(|a|^r + |b|^r),$$

where $c_r = 1$ when $r \leq 1$ and $c_r = 2^{r-1}$ when $r \geq 1$. The above relation is known as the $c_r$-*inequality*.

**Problem 2.6.73.** Assuming that $\xi$ and $\eta$ are two non-negative random variables with the property

$$P\{\xi \geq x\} \leq x^{-1} E[\eta I(\xi \geq x)], \quad \text{for every } x > 0,$$

prove that

$$E\xi^p \leq \left(\frac{p}{p-1}\right)^p E\eta^p, \quad \text{for every } p > 1.$$

*Hint.* Consider first the case where $\xi$ is bounded, i.e., replace $\xi$ by $\xi_c = \xi \wedge c$, $c > 0$, in which case, according to [P §2.6, (69)], one must have

$$E\xi_c^p = p \int_0^c x^{p-1} P\{\xi > x\} \, dx.$$

Then prove the required property for $E\xi_c^p$ and pass to the limit as $c \uparrow \infty$.

**Problem 2.6.74.** Prove the following analog of the *integration-by-substitution rule* (see Problem 2.6.15 and [P §2.6, Theorem 7], regarding the change of variables in the Lebesgue integral).

Let $I$ be any open subset of $\mathbb{R}^n$ and let $y = \varphi(x)$ be any function which is defined on $I$ and takes values in $\mathbb{R}^n$ (if $x = (x_1, \ldots, x_n) \in I$, then $y = (y_1, \ldots, y_n)$ with $y_i = \varphi_i(x_1, \ldots, x_n)$, $i = 1, \ldots, n$). Suppose that all derivatives $\frac{\partial \varphi_i}{\partial x_j}$ are well defined and continuous and that $|J_\varphi(x)| > 0$, $x \in I$, where $J_\varphi(x)$ stands for the determinant of the Jacobian of the function $\varphi$, i.e.,

$$J_\varphi(x) = \det \left\| \frac{\partial \varphi_i}{\partial x_j}, \quad 1 \leq i, j \leq n \right\|.$$

As a consequence of the above assumptions, the set $\varphi(I) \subseteq \mathbb{R}^n$ is open, the function $\varphi$ admits an inverse $h = \varphi^{-1}$, and the Jacobian $J_h(y)$ exists and is continuous on $\varphi(I)$, with $|J_h(y)| > 0$, $y \in \varphi(I)$.

Prove that for every non-negative or integrable function $g = g(x)$, $x \in I$, one has

$$\int_I g(x) \, dx = \int_{\varphi(I)} g(h(y)) |J_h(y)| \, dy,$$

which can be written also as

$$\int_I g(x) \, dx = \int_{\varphi(I)} g(\varphi^{-1}(y)) |J_{\varphi^{-1}}(y)| \, dy,$$

where all integrals are understood as Lebesgue integrals in $\mathbb{R}^n$.

**Problem 2.6.75.** Let $F = F(x)$ be a cummulative distribution function with $F(0) = 0$ and $F(1) = 1$, which is Lipschitz continuous, in that $|F(x) - F(y)| \leq L|x - y|$, $x, y \in [0, 1]$. Let $m$ be the measure on $[0, 1]$ given by $m(B) = \int_B dF(x)$, $B \in \mathscr{B}([0, 1])$, and let $\lambda$ be the Lebesgue measure on $[0, 1]$.

Prove that $m \ll \lambda$ and

$$\frac{dm}{d\lambda} \leq L \quad (\lambda\text{-a.e.}).$$

**Problem 2.6.76.** Suppose that $g = g(x)$ is some function which is defined on the interval $[a, b] \subseteq \mathbb{R}$ and is convex, i.e., $g((1 - \lambda)x + \lambda y) \leq (1 - \lambda)g(x) + \lambda g(y)$ for any $x, y \in [a, b]$ and any $0 \leq \lambda \leq 1$). Prove that this function must be continuous on the interval $(a, b)$ and conclude that it is a Borel function.

*Hint.* Argue that convexity implies that for every choice of $x, y, z \in [a, b]$, $x < y < z$, one has

$$\frac{g(y) - g(x)}{y - x} \leq \frac{g(z) - g(y)}{z - y}.$$

By using the above relation conclude that $g = g(x)$ must be continuous on the interval $(a, b)$.

**Problem 2.6.77.** Consider the generating function $G(s) = \sum_{k=0}^{\infty} p_k s^k$, associated with the discrete random variable $X$ with $P\{X = k\} = p_k$, $k = 0, 1, 2, \ldots$, $\sum_{k=0}^{\infty} p_k = 1$ (see Problem 2.6.28), and let

$$q_k = P\{X > k\}, \qquad r_k = P\{X \leq k\}, \qquad k = 0, 1, 2, \ldots .$$

Prove that the generating functions for the sequences $q = (q_k)_{k \geq 0}$ and $r = (r_k)_{k \geq 0}$ are given, respectively, by

$$G_q(s) = \frac{1 - G(s)}{1 - s}, \quad |s| < 1, \qquad G_r(s) = \frac{G(s)}{1 - s}, \quad |s| < 1.$$

**Problem 2.6.78.** (*On the "probability for ruin"—see [P §1.9].*) Let $S_0 = x$ and let $S_n = x + \xi_1 + \ldots + \xi_n$, $n \geq 1$, where $(\xi_k)_{k \geq 1}$ is some sequence of independent and identically distributed random variables with $P\{\xi_k = 1\} = p$, $P\{\xi_k = -1\} = q$, $p + q = 1$, and $x$ is some integer number with $0 \leq x \leq A$. Consider the stopping time for the random walk (or for the "game" between two players—see [P §1.9]), which is given by

$$\tau = \inf\{n \geq 0 : S_n = 0 \text{ or } S_n = A\}.$$

Consider also the probability $p_x(n) = P\{\tau = n, S_n = 0\}$ for the stopping to occur with "ruin" (i.e., $\{S_n = 0\}$) in the $n^{\text{th}}$-period.

Prove that the generating function $G_x(s) = \sum_{n=0}^{\infty} p_x(n)s^n$ satisfies the following recursive relation:

$$G_x(s) = ps\,G_{x+1}(s) + qs\,G_{x-1}(s),$$

with $G_0(s) = 1$ and $G_A(s) = 0$. By using this relation prove the formula

$$G_x(s) = \left(\frac{q}{p}\right)^x \frac{\lambda_1^{A-x}(s) - \lambda_2^{A-x}(s)}{\lambda_1^A(s) - \lambda_2^A(s)},$$

where

$$\lambda_1(s) = \frac{1}{2ps}\{1 + \sqrt{1 - 4pqs^2}\} \quad \text{and} \quad \lambda_2(s) = \frac{1}{2ps}\{1 - \sqrt{1 - 4pqs^2}\}.$$

**Problem 2.6.79.** Consider a lottery with tickets numbered 000000, ..., 999999 and suppose that one of these tickets is chosen at random. Find the probability, $P_{21}$, that the sum of the six digits on this ticket equals 21.

*Hint.* Use the methodology based on generating functions, developed in Sect. A.3 (pp. 372–373). The answer is $P_{21} = 0.04$.

**Problem 2.6.80.** Suppose that $\xi$ is a random variable with unimodal probability density $f(x)$ that has maximum at the point $x_0$ (referred to as the mode, or the peak, of the respective distribution), so that $f(x)$ is non-decreasing for $x < x_0$ and is non-increasing for $x > x_0$.

Prove *Gauss inequality*:

$$P\{|\xi - x_0| \geq \varepsilon E|\xi - x_0|^2\} \leq \frac{4}{9\varepsilon^2}, \quad \text{for every } \varepsilon > 0.$$

*Hint.* If the function $g(y)$ does not increase for $y > 0$, then

$$\varepsilon^2 \int_\varepsilon^\infty g(y)\,dy \leq \frac{4}{9} \int_0^\infty y^2 g(y)\,dy, \quad \text{for any } \varepsilon > 0.$$

One can conclude from the above inequality that, given any $\varepsilon > 0$ and $d^2 = E|\xi - x_0|^2$, one has

$$P\{|\xi - x_0| \geq \varepsilon d\} \leq \frac{4}{9} \frac{E[(\xi - x_0)/d]^2}{\varepsilon^2} = \frac{4}{9\varepsilon^2}.$$

**Problem 2.6.81.** Suppose that the random variables $\xi_1, \ldots, \xi_n$ are independent and identically distributed, with $P\{\xi_i > 0\} = 1$ and $D \ln \xi_1 = \sigma^2$. Given any $\varepsilon > 0$, prove that

$$P\{\xi_1 \ldots \xi_n \leq (E \ln \xi_1)^n e^{n\varepsilon}\} \geq 1 - \frac{\sigma^2}{n\varepsilon^2}.$$

*Hint.* Use Chebyshev's inequality

$$P\{|Y_n - EY_n| \leq n\varepsilon\} \geq 1 - \frac{DY_n}{n^2\varepsilon^2},$$

with $Y_n = \sum_{i=1}^n \ln \xi_i$.

**Problem 2.6.82.** Let $P$ and $\widetilde{P}$ be probability measures on $(\Omega, \mathcal{F})$, such that $\widetilde{P}$ is absolutely continuous with respect to $P$ ($\widetilde{P} \ll P$), with density that is bounded by some constant $c \geq 1$:

$$\frac{d\widetilde{P}}{dP} \leq c \quad \text{(P-a.s.)}.$$

Prove that there is a number $\alpha \in (0, 1]$ and a *probability* measure $Q$ for which one can write

$$P = \alpha\widetilde{P} + (1 - \alpha)Q.$$

*Hint.* Choose an arbitrary constant $C > c$ and set $\alpha = 1/C$ and

$$Q(A) = \frac{1}{1 - \alpha} \int_A \left(1 - \alpha\frac{d\widetilde{P}}{dP}\right) dP.$$

**Problem 2.6.83.** Let $\xi$ and $\eta$ be any two independent random variables with $E\xi = 0$. Prove that $E|\xi - \eta| \geq E|\eta|$.

**Problem 2.6.84.** Let $\xi_1, \xi_2, \ldots$ be any sequence of independent and identically distributed random variables taking values in $\mathbb{R} = (-\infty, \infty)$ and set $S_0 = 0$, $S_n = \xi_1 + \ldots + \xi_n$. The so-called "ladder indexes" (a.k.a. "ladder moments") are defined by the following recursive rule:

$$T_0 = 0, \qquad T_k = \inf\{n > T_{k-1} : S_n - S_{T_{k-1}} > 0\}, \quad k \geq 1,$$

where, as usual, we set $\inf \varnothing = \infty$. It is clear that

$$P\{T_1 = n\} = P\{S_1 \leq 0, \ldots, S_{n-1} \leq 0, S_n > 0\}, \quad \text{for all } n \geq 1.$$

Prove that the generating function $G(s) = \sum_{n=1}^{\infty} f_n s^n$ for the random variable $T_1$ (with $f_n = P\{T_1 = n\}$) is given by the formula

$$G(s) = \exp\left\{-\sum_{n=1}^{\infty} \frac{s^n}{n} P\{S_n > 0\}\right\}, \qquad |s| < 1.$$

**Problem 2.6.85.** With the notation adopted in the previous problem, setting

$$A = \sum_{n=1}^{\infty} \frac{1}{n} P\{S_n \leq 0\}, \qquad B = \sum_{n=1}^{\infty} \frac{1}{n} P\{S_n > 0\},$$

prove that

$$P\{T_1 < \infty\} = \begin{cases} 1, & \text{if } B = \infty, \\ 1 - e^{-B}, & \text{if } B < \infty. \end{cases}$$

Show also that if $B = \infty$, then

$$ET_1 = \begin{cases} e^A, & \text{if } A < \infty, \\ \infty, & \text{if } A = \infty. \end{cases}$$

**Problem 2.6.86.** Just as in Problem 2.6.84, let $\xi_1, \xi_2, \ldots$ be any sequence of independent and identically distributed random variables with $E\xi_1 > 0$, set $S_0 = 0$, $S_n = \xi + \cdots + \xi_n$, and let

$$\tau = \inf\{n \geq: S_n > 0\}.$$

Prove that $E\tau < \infty$.

**Problem 2.6.87.** Let everything be as in Problem 2.6.84, let $\mathscr{R}_N$ denote the *breadth* (*span*) of the sequence $S_0, S_1, \ldots, S_N$, i.e., the total number of different values that can be found in that sequence, and let

$$\sigma(0) = \inf\{n > 0 : S_n = 0\}$$

be the moment of the first return to 0.
    Prove that for $N \to \infty$ one has

$$E\frac{\mathscr{R}_N}{N} \to P\{\sigma(0) = \infty\}.$$

Note that $P\{\sigma(0) = \infty\}$ is the probability for *no-return* to 0—comp. this result with Problem 1.9.7.

    *Remark.* According to Problem 8.8.16, in the case of a simple random walk with $P\{\xi_1 = 1\} = p$ and $P\{\xi_1 = -1\} = q$, one must have

$$E\frac{\mathscr{R}_N}{N} \to |p - q| \quad \text{as } N \to \infty \, ;$$

in other words,

$$E\frac{\mathscr{R}_N}{N} \to \begin{cases} p - q, & \text{if } p > 1/2, \\ 0, & \text{if } p = 1/2, \\ q - p, & \text{if } p < 1/2. \end{cases}$$

**Problem 2.6.88.** Let $\xi_1, \xi_2, \ldots$ be any sequence of independent random variables that are uniformly distributed in the interval $(0, 1)$ and let

$$\nu = \min\{n \geq 2 : \xi_n > \xi_{n-1}\}, \qquad \mu(x) = \min\{n \geq 1 : \xi_1 + \cdots + \xi_n > x\},$$

where $0 < x \leq 1$. Prove that:

(a) $P\{\mu(x) > n\} = x^n/n!, n \geq 1$;

(b) $\text{Law}(\nu) = \text{Law}(\mu(1))$;

(c) $E\nu = E\mu(1) = e$.

*Hint.* (a) Consider proof by induction.

(b) One must show that $P\{\nu > n\} = P\{\xi_1 > \xi_2 > \cdots > \xi_n\} = 1/n!$.

## 2.7   Conditional Probabilities and Conditional Expectations with Respect to $\sigma$-algebras

**Problem 2.7.1.** Suppose that $\xi$ and $\eta$ are two independent and identically distributed random variables, such that $E\xi$ exists. Prove that

$$E(\xi \mid \xi + \eta) = E(\eta \mid \xi + \eta) = \frac{\xi + \eta}{2} \quad \text{(a. e.)}.$$

*Hint.* Observe that for any $A \in \sigma(\xi + \eta)$ one has $E\xi I_A = E\eta I_A$.

**Problem 2.7.2.** Suppose that $\xi_1, \xi_2, \ldots$ are independent and identically distributed random variables, such that $E|\xi_i| < \infty$. Prove that

$$E(\xi_1 \mid S_n, S_{n+1}, \ldots) = \frac{S_n}{n} \quad \text{(a. e.)},$$

where $S_n = \xi_1 + \cdots + \xi_n$.

*Hint.* Use the fact that $E\xi_i I_A = E\xi_j I_A$ for any $A \in \sigma(S_n, S_{n+1}, \ldots)$.

**Problem 2.7.3.** Suppose that the random elements $(X, Y)$ are such that there is a regular distribution of the form $P_x(B) = P(Y \in B \mid X = x)$. Prove that if $E|g(X, Y)| < \infty$, for some appropriate function $g$, then $P_x$-a. e. one has

$$E[g(X, Y) \mid X = x] = \int g(x, y) P_x(dy).$$

*Hint.* By using the definition of a regular conditional distribution and the notion of "$\pi$-$\lambda$-system" (see [P §2.2]), prove that for any function $g(x, y)$ of the form $\sum_{i=1}^{n} \lambda_i I_{A_i}$, where $A_i \in \mathscr{B}(\mathbb{R}^2)$, the map

$$x \rightsquigarrow \int_{\mathbb{R}} g(x, y) P_x(dy)$$

must be $\mathscr{B}(\mathbb{R})/\mathscr{B}(\mathbb{R})$-measurable and

$$\mathsf{E}g(X, Y)I_B = \int_B \left( \int_{\mathbb{R}} g(x, y)\, P_x(dy) \right) Q(dx), \quad \text{for every } B \in \mathscr{B}(\mathbb{R}),$$

where $Q$ stands for the distribution of the random variable $X$. Prove that these properties hold for all bounded $\mathscr{B}(\mathbb{R}^2)$-measurable functions, and then conclude that they must hold for all $g(x, y)$ with $\mathsf{E}|g(X, Y)| < \infty$.

**Problem 2.7.4.** Let $\xi$ be a random variable with (cummulative) distribution function $F_\xi(x)$. Prove that

$$\mathsf{E}(\xi \mid a < \xi \le b) = \frac{\int_a^b x\, dF_\xi(x)}{F_\xi(b) - F_\xi(a)},$$

where we suppose that $F_\xi(b) - F_\xi(a) > 0$.

Hint. Use the fact that, by the very definition of conditional expectation, if $\mathsf{E}[I(a < \xi \le b)] > 0$ one can write

$$\mathsf{E}(\xi \mid a < \xi \le b) = \frac{\mathsf{E}[\xi I(a < \xi \le b)]}{\mathsf{E}[I(a < \xi \le b)]}.$$

**Problem 2.7.5.** Let $g = g(x)$ be any function defined on $\mathbb{R}$ which is convex and is such that $\mathsf{E}|g(\xi)| < \infty$. Prove that *Jensen's inequality* holds P-a. e. for *conditional expected values*, namely,

$$g(\mathsf{E}(\xi \mid \mathscr{G})) \le \mathsf{E}(g(\xi) \mid \mathscr{G}) \quad \text{(a.e.)}.$$

Hint. First use the fact that for the regular conditional distribution $Q(x; B)$, associated with the random variable $\xi$, relative to the $\sigma$-algebra $\mathscr{G}$ one can write

$$\mathsf{E}(g(\xi) \mid \mathscr{G})(\omega) = \int_{\mathbb{R}} g(x)\, Q(\omega; dx)$$

(see [P §2.7, Theorem 3]), and then use Jensen's inequality for standard expected values.

**Problem 2.7.6.** Prove that the random variable $\xi$ and the $\sigma$-algebra $\mathscr{G}$ are independent (i. e., the random variables $\xi$ and $I_B(\omega)$ are independent for every choice of $B \in \mathscr{G}$) if and only if $\mathsf{E}(g(\xi) \mid \mathscr{G}) = \mathsf{E}g(\xi)$ for any Borel function $g(x)$, such that $\mathsf{E}|g(\xi)| < \infty$.

Hint. If $A \in \mathscr{G}$ and $B \in \mathscr{B}(\mathbb{R})$, then, due to the independence between $\xi$ and $\mathscr{G}$, we have $\mathsf{P}(A \cap \{g(\xi) \in B\}) = \mathsf{P}(A)\mathsf{P}\{g(\xi) \in B\}$, and, therefore, $\mathsf{E}(g(\xi) \mid \mathscr{G}) = \mathsf{E}g(\xi)$. Conversely, when the last relation holds, setting $g(\xi) = I(\xi \in B)$, one finds that

$$\mathsf{P}(A \cap \{\xi \in B\}) = \mathsf{P}(A)\mathsf{P}\{\xi \in B\}.$$

The independence between $\xi$ and $\mathcal{G}$ follows from the fact that in the last relation $A \in \mathcal{G}$ and $B \in \mathcal{B}(\mathbb{R})$ are chosen arbitrarily.

**Problem 2.7.7.** Suppose that $\xi$ is a non-negative random variable and let $\mathcal{G}$ be some $\sigma$-algebra, $\mathcal{G} \subseteq \mathcal{F}$. Prove that $E(\xi \mid \mathcal{G}) < \infty$ (a. e.) if and only if the measure $Q$, defined on the sets $A \in \mathcal{G}$ by $Q(A) = \int_A \xi \, dP$, is $\sigma$-finite.

Hint. To prove the *sufficiency* part, set $A_\infty = \{E(\xi \mid \mathcal{G}) = \infty\}$, $A_n = \{E(\xi \mid \mathcal{G}) \le n\}$, and check that $Q(A_n) \le n$, which implies that $Q$ is $\sigma$-finite.

To prove the *necessity* part, one must conclude from the existence of the sets $A_1, A_2, \ldots \in \mathcal{G}$, with $\bigcup_{i=1}^{\infty} A_i = \Omega$ and $Q(A_i) < \infty$, $i = 1, \ldots, n$, that $E(\xi \mid \mathcal{G}) < \infty$ (P-a.s.).

**Problem 2.7.8.** Prove that the conditional probability $P(A \mid B)$ can be claimed to be "continuous", in the sense that if $\lim_n A_n = A$ and $\lim_n B_n = B$, with $P(B_n) > 0$ and $P(B) > 0$, then $\lim_n P(A_n \mid B_n) = P(A \mid B)$.

**Problem 2.7.9.** Let $\Omega = (0, 1)$, $\mathcal{F} = \mathcal{B}((0, 1))$, and let $P$ denote the Lebesgue measure. Suppose that $X(\omega)$ and $Y(\omega)$, $\omega \in \Omega$, are two independent random variables that are uniformly distributed on $(0, 1)$. Consider a third random variable, $Z(\omega) = |X(\omega) - Y(\omega)|$, which represents the distance between the random points $X(\omega)$ and $Y(\omega)$. Prove that the distribution of $Z(\omega)$ admits density $f_Z(z) = 2(1-z)$, $0 \le z \le 1$, and conclude that $EZ = 1/3$.

**Problem 2.7.10.** Two points, $A_1$ and $A_2$, are chosen at random on the circle $\{(x, y): x^2 + y^2 \le R^2\}$; more specifically, $A_1$ and $A_2$ are sampled independently and in such a way that (in polar coordinates, $A_i = (\rho_i, \theta_i)$, $i = 1, 2$)

$$P(\rho_i \in dr, \theta_i \in d\theta) = \frac{r \, dr \, d\theta}{\pi R^2}, \quad i = 1, 2.$$

Prove that the random variable $\rho$, which represents the distance between $A_1$ and $A_2$, admits density $f_\rho(r)$, given by

$$f_\rho(r) = \frac{2r}{\pi R^2} \left[ 2 \arccos\left(\frac{r}{2R}\right) - \frac{r}{R} \sqrt{1 - \left(\frac{r}{2R}\right)^2} \right],$$

where $0 < r < 2R$.

**Problem 2.7.11.** The point $P = (x, y)$ is sampled *randomly* (clarify!) from the unit square, i.e., from the square with vertices $(0, 0)$, $(0, 1)$, $(1, 1)$, $(1, 0)$. Find the probability that the point $P$ is closer to the point $(1, 1)$, than to the point $(1/2, 1/2)$.

**Problem 2.7.12.** (The "random meeting" problem.) Person $A$ and person $B$ have agreed to meet between 7:00 p.m. and 8:00 p.m. at a particular location. They have both forgotten the exact meeting time and choose their respective arrival times randomly and independently from each other between 7:00 p.m. and 8:00 p.m., according to the uniform distribution on the interval [7:00, 8:00]. They both have patience to wait no longer than 10 min. Prove that the probability that $A$ and $B$ will actually meet equals $11/36$.

**Problem 2.7.13.** Let $X_1, X_2, \ldots$ be a sequence of independent random variables and let $S_n = \sum_{i=1}^{n} X_i$. Prove that $S_1$ and $S_3$ are conditionally independent relative to the $\sigma$-algebra $\sigma(S_2)$, generated by the random variable $S_2$.

**Problem 2.7.14.** We say that the $\sigma$-algebras $\mathcal{G}_1$ and $\mathcal{G}_2$ are conditionally independent relative to the $\sigma$-algebra $\mathcal{G}_3$ if

$$P(A_1 A_2 \mid \mathcal{G}_3) = P(A_1 \mid \mathcal{G}_3) P(A_2 \mid \mathcal{G}_3), \quad \text{for all } A_i \in \mathcal{G}_i, i = 1, 2.$$

Prove that the conditional independence of $\mathcal{G}_1$ and $\mathcal{G}_2$ from $\mathcal{G}_3$ is equivalent to the claim that any of the following conditions holds P-a. e.:
(a) $P(A_1 \mid \sigma(\mathcal{G}_2 \cup \mathcal{G}_3)) = P(A_1 \mid \mathcal{G}_3)$, for all $A_1 \in \mathcal{G}_1$;
(b) $P(B \mid \sigma(\mathcal{G}_2 \cup \mathcal{G}_3)) = P(B \mid \mathcal{G}_3)$ for all $B \in \mathcal{P}_1$, where $\mathcal{P}_1$ is any $\pi$-system of subsets of $\mathcal{G}_1$, such that $\mathcal{G}_1 = \sigma(\mathcal{P}_1)$;
(c) $P(B_1 B_2 \mid \sigma(\mathcal{G}_2 \cup \mathcal{G}_3)) = P(B_1 \mid \mathcal{G}_3) P(B_2 \mid \mathcal{G}_3)$ for all sets $B_1 \in \mathcal{P}_1$ and $B_2 \in \mathcal{P}_2$, where $\mathcal{P}_1$ and $\mathcal{P}_2$ are any two $\pi$-systems of subsets of, respectively, $\mathcal{G}_1$ and $\mathcal{G}_2$, chosen so that $\mathcal{G}_1 = \sigma(\mathcal{P}_1)$ and $\mathcal{G}_2 = \sigma(\mathcal{P}_2)$;
(d) $E(X \mid \sigma(\mathcal{G}_2 \cup \mathcal{G}_3)) = E(X \mid \mathcal{G}_3)$ for any $\sigma(\mathcal{G}_2 \cup \mathcal{G}_3)$-measurable random variable $X$, for which the expectation $EX$ exists (see Definition 2 in § 6).

**Problem 2.7.15.** Prove the following generalization of *Fatou's lemma* for *conditional expectations* (comp. with (d) from [ P §2.7, Theorem 2]).
Let $(\Omega, \mathcal{F}, P)$ be any probability space and let $(\xi_n)_{n \geq 1}$ be any sequence of random variables, chosen so that the expectations $E\xi_n$, $n \geq 1$, are *well defined* and the limit $E \varliminf \xi_n$ exists (and may equal $\pm \infty$—see [ P §2.6, Definition 2]).
Suppose that $\mathcal{G}$ is some $\sigma$-algebra of events inside $\mathcal{F}$ chosen so that

$$\sup_{n \geq 1} E(\xi_n^- I(\xi_n \geq a) \mid \mathcal{G}) \to 0 \quad \text{as } a \to \infty \quad \text{(P-a. e.).}$$

Then

$$E(\varliminf \xi_n \mid \mathcal{G}) \leq \varliminf E(\xi_n \mid \mathcal{G}) \quad \text{(P-a. e.).}$$

**Problem 2.7.16.** Just as in the previous problem, let $(\xi_n)_{n \geq 1}$ be any sequence of random variables, chosen so that all expectations $E\xi_n$, $n \geq 1$, exist, and suppose that $\mathcal{G}$ is some $\sigma$-algebra of events inside $\mathcal{F}$ chosen so that

$$\sup_n \lim_{k \to \infty} E(|\xi_n| I(|\xi_n| \geq k) \mid \mathcal{G}) = 0 \quad \text{(P-a. e.).} \tag{$*$}$$

Prove that if $\xi_n \to \xi$ (P-a. e.) and the expected value $E\xi$ exists, then

$$E(\xi_n \mid \mathcal{G}) \to E(\xi \mid \mathcal{G}) \quad \text{(P-a. e.).} \tag{$**$}$$

**Problem 2.7.17.** Let everything be as in the previous problem, but suppose that $(*)$ is replaced with the condition $\sup_n E(|\xi_n|^\alpha \mid \mathcal{G}) < \infty$ (P-a. e.), for some $\alpha > 1$. Prove that the convergence in $(**)$ still holds.

**Problem 2.7.18.** Prove that if $\xi_n \xrightarrow{L^p} \xi$, for some $p \geq 1$, then

$$\mathsf{E}(\xi_n \mid \mathscr{G}) \xrightarrow{L^p} \mathsf{E}(\xi \mid \mathscr{G}),$$

for any sub-$\sigma$-algebra $\mathscr{G} \subseteq \mathscr{F}$.

**Problem 2.7.19.** Let $X$ and $Y$ be any two random variables with $\mathsf{E}X^2 < \infty$ and $\mathsf{E}|Y| < \infty$.
   (a) Setting $\mathsf{D}(X \mid Y) \equiv \mathsf{E}[(X - \mathsf{E}(X \mid Y))^2 \mid Y]$, prove that $\mathsf{D}X = \mathsf{E}\mathsf{D}(X \mid Y) + \mathsf{D}\mathsf{E}(X \mid Y)$ (see Problem 1.8.2).
   (b) Prove that $\mathrm{cov}(X, Y) = \mathrm{cov}(X, \mathsf{E}(Y \mid X))$.

**Problem 2.7.20.** Is the sufficient statistics $T(\omega) = s(X_1(\omega)) + \cdots + s(X_n(\omega))$ in [ P §2.7, Example 5] minimal?

**Problem 2.7.21.** Prove the factorization identity [ P §2.7, (57)].

**Problem 2.7.22.** In the context of [ P §2.7, Example 2], prove that $\mathsf{E}_\theta(X_i \mid T) = \frac{n+1}{2n}T$, where $X_i(\omega) = x_i$ for $\omega = (x_1, \ldots, x_n)$, $i = 1, \ldots, n$.

**Problem 2.7.23.** Let $(\Omega, \mathscr{F}, \mathsf{P})$ be a probability space, let $A$, $B$ and $C_1, \ldots, C_n$ be events chosen from the $\sigma$-algebra $\mathscr{F}$, and suppose that for any $i = 1, \ldots, n$ one has

$$\mathsf{P}(C_i) > 0, \quad \mathsf{P}(A \mid C_i) \geq \mathsf{P}(B \mid C_i),$$

and $\bigcup_{i=1}^{n} C_i = \Omega$. Can one claim that $\mathsf{P}(A) \geq \mathsf{P}(B)$?

**Problem 2.7.24.** Let $X$ and $Y$ be any two random variables with $\mathsf{E}|X| < \infty$, $\mathsf{E}|Y| < \infty$, such that $\mathsf{E}(X \mid Y) \geq Y$ and $\mathsf{E}(Y \mid X) \geq X$ (P-a. e.). Prove that $X = Y$ (P-a. e.).
   *Hint.* Prove that it is enough to consider the case where the inequalities $\geq$ are replaced by equalities.
   *Method I.* Consider the function $g(u) = \arctan u$. Then $(X - Y)(g(X) - g(Y)) \geq 0$ (P-a. e.), and, at the same time, one can show that $\mathsf{E}[(X - Y)(g(X) - g(Y))] = 0$, from where one can conclude that $X = Y$ (P-a. e.).
   *Method II.* Argue that $\mathsf{E}\frac{X^+ + 1}{Y^+ + 1} = 1$, $\mathsf{E}\frac{Y^+ + 1}{X^+ + 1} = 1$. Then conclude setting $Z = \frac{X^+ + 1}{Y^+ + 1}$, show that $\mathsf{E}(\sqrt{Z} - \frac{1}{\sqrt{Z}})^2 = 1$. Then conclude that $\mathsf{P}\{X^+ = Y^+\} = 1$. One can show, analogously, that $\mathsf{P}\{X^- = Y^-\} = 1$.
   *Method III.* Prove that if $\mathsf{P}\{X < Y\} > 0$ then there is a constant $c$ with the property $\mathsf{P}\{X \leq c < Y\} > 0$. Consider the sets $A = \{X \leq c\}$ and $B = \{Y > c\}$ and argue that

$$\int_{\Omega} (Y - X) \, d\mathsf{P} = \int_{A} (Y - X) \, d\mathsf{P} + \int_{B \setminus A} (Y - X) \, d\mathsf{P} + \int_{A \cap B} (Y - X) \, d\mathsf{P} < 0,$$

which contradicts to $\int_{\Omega} (Y - X) \, d\mathsf{P} = 0$.

*Remark.* It is not possible to find two random variables $X$ and $Y$ with $E|X| < \infty$ and $E|Y| < \infty$, such that the *strict* inequalities $E(X \mid Y) > Y$ and $E(Y \mid X) > X$ both hold with probability 1. Indeed, assuming that such random variables exist would lead to a contradiction: $EX = EE(X \mid Y) > EY = EE(Y \mid X) > EX$.

**Problem 2.7.25.** Assuming that $X$ is some geometrically distributed random variable with

$$P\{X = k\} = p\,q^{k-1}, \quad k = 1, 2, \ldots, \quad 0 \le p \le 1, \quad q = 1 - p, \qquad (*)$$

prove that

$$P\{X > n + m \mid X > n\} = P\{X > m\}, \quad \text{for } m, n \in \{1, 2, \ldots\}. \qquad (**)$$

What is the interpretation of the above property?

In addition, prove the converse statement: if a discrete random variable takes values in the set $\{1, 2, \ldots\}$ and has the property $(**)$, then it must also have the property $(*)$.

(Comp. with Problem 2.7.45.)

**Problem 2.7.26.** Prove that the random vectors $(X, Y)$ and $(\widetilde{X}, Y)$ share the same distribution $((X, Y) \overset{d}{=} (\widetilde{X}, Y))$, if and only if $P\{X \in A \mid Y\} = P\{\widetilde{X} \in A \mid Y\}$ (P-a. e.), for any event $A$.

**Problem 2.7.27.** Let $X$ and $Y$ be any two independent Poisson random variables with parameters, respectively, $\lambda > 0$ and $\mu > 0$. Prove that the conditional distribution $\mathrm{Law}(X \mid X + Y)$ is binomial, i.e.,

$$P(X = k \mid X + Y = n) = C_n^k \left(\frac{\lambda}{\lambda + \mu}\right)^k \left(\frac{\mu}{\lambda + \mu}\right)^{n-k}, \quad \text{for} \quad 0 \le k \le n.$$

**Problem 2.7.28.** Suppose that $\xi$ is a random variable that is uniformly distributed in the interval $[-a, b]$, with $a > 0$, $b > 0$, and, setting $\mathscr{G}_1 = \sigma(|\xi|)$ and $\mathscr{G}_2 = \sigma(\mathrm{sign}\,\xi)$, calculate the conditional probabilities $P(A \mid \mathscr{G}_i)$, $i = 1, 2$, for the events $A = \{\xi > 0\}$ and $A = \{\xi \le \alpha\}$, where $\alpha \in [-a, b]$.

**Problem 2.7.29.** Prove by way of example that the relation $E(\xi + \eta \mid \mathscr{G}) = E(\xi \mid \mathscr{G}) + E(\eta \mid \mathscr{G})$ (P-a. e.) does not always hold (comp. with property $\mathbf{D}^*$ in [$\underline{P}$ §2.7, **4**]).

*Hint.* It may happen that the expected value $E(\xi + \eta \mid \mathscr{G})$ is well defined and equals zero (P-a. e.), while, at the same time, the sum $E(\xi \mid \mathscr{G}) + E(\eta \mid \mathscr{G})$ is not defined.

**Problem 2.7.30.** In the definition of the conditional probability $P(B \mid \mathscr{G})(\omega)$ (of the event $B \in \mathscr{F}$ relative to the $\sigma$-algebra $\mathscr{G} \subseteq \mathscr{F}$—see Definition 2 in [$\underline{P}$ §2.7, **2**]), the map $P(\cdot \mid \mathscr{G})(\omega)$ is *not* required to be a *measure* on $(\Omega, \mathscr{F})$ for P-a.e. $\omega \in \Omega$. Prove that such a requirement cannot be imposed; namely, construct an example where the set of all $\omega \in \Omega$ for which $P(\cdot \mid \mathscr{G})(\omega)$ *fails* to be a measure is not P-negligible.

**Problem 2.7.31.** Give an example of two independent random variables, $X$ and $Y$, and a $\sigma$-algebra $\mathscr{G}$, chosen so that for some choice of the events $A$ and $B$ one has

$$P(X \in A, Y \in B \,|\, \mathscr{G})(\omega) \neq P(X \in A \,|\, \mathscr{G})(\omega)\, P(Y \in B \,|\, \mathscr{G})(\omega),$$

for all $\omega$ inside some set of positive P-measure. In other words, show that independence does not imply conditional independence.

**Problem 2.7.32.** If the family of random variables $\{\xi_n\}_{n\geq 1}$ is uniformly integrable and $\xi_n \to \xi$ (P-a. e.), then $\mathsf{E}\xi_n \to \mathsf{E}\xi$ (see b) in Theorem 4 from [P §2.6]). At the same time the P-a. e. convergence of the conditional expectations $\mathsf{E}(\xi_n \,|\, \mathscr{G}) \to \mathsf{E}(\xi \,|\, \mathscr{G})$ can be established (see a) in Theorem 2 from [P §2.6]) under the assumption that $|\xi_n| \leq \eta$, $\mathsf{E}\eta < \infty$, $n \geq 1$, and $\xi_n \to \xi$ (P-a. e.).

Give an example showing that if the condition "$|\xi_n| \leq \eta$, $\mathsf{E}\eta < \infty$, $n \geq 1$" is replaced by the condition "the family $\{\xi_n\}_{n\geq 1}$ is uniformly integrable," then the convergence $\mathsf{E}(\xi_n \,|\, \mathscr{G}) \to \mathsf{E}(\xi \,|\, \mathscr{G})$ (P-a. e.) may not hold. Analogous claim can be made about condition a) in [P §2.6, Theorem 4] (i.e., Fatou's Lemma for uniformly integrable random variables) in the case of conditional expected values (see, however, Problems 2.7.15–2.7.17 above).

**Problem 2.7.33.** Suppose that $(\Omega, \mathscr{F}, \mathsf{P})$ is identified with the probability space $([0, 1], \mathscr{F}, \lambda)$, where $\lambda$ is the Lebesgue measure and $\mathscr{F}$ is the Borel $\sigma$-algebra on $[0, 1]$. Give an example of a sub-$\sigma$-algebra $\mathscr{G} \subseteq \mathscr{F}$, chosen so that the Dirichlet function

$$d(\omega) = \begin{cases} 1, & \text{if } \omega \text{ is irrational,} \\ 0, & \text{if } \omega \text{ is rational,} \end{cases}$$

is a *version* of the conditional expectation $\mathsf{E}(1 \,|\, \mathscr{G})(\omega)$. In particular, the conditional expectation $\mathsf{E}(\xi \,|\, \mathscr{G})(\omega)$ of some "smooth" function $\xi = \xi(\omega)$ (for example, $\xi(\omega) \equiv 1$) may have a version, which, as a function of $\omega$, may be "extremely non-smooth".

**Problem 2.7.34.** If, given a random variable $\xi$, the expected value $\mathsf{E}\xi$ exists, then, by property $\mathsf{G}^*$ in [P §2.7, 4], one can write $\mathsf{E}(\mathsf{E}(\xi \,|\, \eta)) = \mathsf{E}\xi$ for any random variable $\eta$. Give an example of two random variables $\xi$ and $\eta$, for which $\mathsf{E}(\mathsf{E}(\xi \,|\, \eta))$ is well defined, while $\mathsf{E}\xi$ is not.

**Problem 2.7.35.** Consider the sample space $\Omega = \{0, 1, 2, \ldots\}$ and suppose that this space is endowed with a family of Poisson distribution laws $\mathsf{P}_\theta\{k\} = \frac{e^{-\theta}\theta^k}{k!}$, $k \in \Omega$, parameterized by $\theta > 0$. Prove that it is *not* possible to construct an unbiased estimator $T = T(\omega)$ for the parameter $\frac{1}{\theta}$, i.e., one cannot construct a random variable $T = T(\omega)$, $\omega \in \Omega$ with the property $\mathsf{E}_\theta|T| < \infty$, for all $\theta > 0$, and $\mathsf{E}_\theta T = \frac{1}{\theta}$, for all $\theta > 0$.

**Problem 2.7.36.** Consider the statistical model $(\Omega, \mathscr{F}, \mathscr{P})$, where $\mathscr{P} = \{\mathsf{P}\}$ is some *dominated* family of probability measures $\mathsf{P}$. Prove that if $\mathscr{G} \subseteq \mathscr{F}$ is some *sufficient* $\sigma$-algebra then any $\sigma$-algebra $\widehat{\mathscr{G}}$ with $\mathscr{G} \subseteq \widehat{\mathscr{G}} \subseteq \mathscr{F}$ is also sufficient. (Burkholder's example [18] shows that if the family $\mathscr{P}$ is *not dominated*, then, in general, this claim cannot be made.)

**Problem 2.7.37.** Prove that each of the following structures represents a Borel space:

(a) $(\mathbb{R}^n, \mathcal{B}(\mathbb{R}^n))$;

(b) $(\mathbb{R}^\infty, \mathcal{B}(\mathbb{R}^\infty))$;

(c) any complete separable metric space, i.e., any Polish space.

**Problem 2.7.38.** Assuming that $(E, \mathcal{E})$ is a Borel space, prove that there is a *countably-generated* algebra $\mathcal{A}$, for which one can claim that $\sigma(\mathcal{A}) = \mathcal{E}$.

**Problem 2.7.39.** (*On the property* **K\***.) Let $\eta$ be any $\mathcal{G}$-measurable random variable, $\xi$ be any $\mathcal{F}$-measurable random variable and suppose that $\mathsf{E}|\eta|^q < \infty$ and $\mathsf{E}|\xi|^p < \infty$, where $p > 1$ and $\frac{1}{p} + \frac{1}{q} = 1$. Prove that $\mathsf{E}(\xi\eta \mid \mathcal{G}) = \eta\mathsf{E}(\xi \mid \mathcal{G})$.

**Problem 2.7.40.** Given some symmetrically distributed random variable $X$ (i.e., $\text{Law}(X) = \text{Law}(-X)$), calculate the conditional distribution

$$\mathsf{P}(X \leq x \mid \sigma(|X|))(\omega), \quad x \in \mathbb{R},$$

in terms of the (cumulative) distribution function $F(x)$, where $\sigma(|X|)$ stands for the $\sigma$-algebra generated by $|X|$.

**Problem 2.7.41.** Assuming that $A$ and $B$ are two events with $\mathsf{P}(A) = \alpha$ and $\mathsf{P}(B) = 1 - \beta$, where $0 \leq \beta < 1$ and $\beta \leq \alpha$, prove that

$$\frac{\alpha - \beta}{1 - \beta} \leq \mathsf{P}(A \mid B).$$

**Problem 2.7.42.** Let $p_k$ denote the probability that a given family has $k$ children, and suppose that

$$p_0 = p_1 = a \ (< 1/2) \quad \text{and} \quad p_k = (1 - 2a)2^{-(k-1)}, \quad k \geq 2.$$

It is assumed that in any given birth the probability for a boy and the probability for a girl both equal to 1/2.

Assuming that a particular family already has two boys, what is the probability that:

(a) The family has only two children;

(b) The family also has two girls.

*Hint.* The solution is a straight-forward application of Bayes' formula. The two probabilities are respectively 27/64 and 81/512.

**Problem 2.7.43.** Suppose that $X$ is some symmetrically distributed random variable $\left(\text{i.e., } X \overset{d}{=} -X\right)$ and the function $\varphi = \varphi(x)$, $x \in \mathbb{R}$, is chosen so that $\mathsf{E}|\varphi(X)| < \infty$. Prove that

$$\mathsf{E}[\varphi(X) \mid |X|] = \frac{1}{2}[\varphi(|X|) + \varphi(-|X|)] \quad (\text{P-a. e.}).$$

**Problem 2.7.44.** Assuming that $X$ is some non-negative random variable, calculate the conditional probabilities

$$P(X \leq x \,|\, \lfloor X \rfloor) \quad \text{and} \quad P(X \leq x \,|\, \lceil X \rceil),$$

where $\lfloor X \rfloor$ stands for the largest integer which does not exceed $X$ and $\lceil X \rceil$ stands for the smallest integer which is greater than or equal to $X$.

**Problem 2.7.45.** Assuming that $X$ is some exponentially distributed random variable with parameter $\lambda > 0$, i.e., $P\{X > x\} = e^{-\lambda x}$, $x \geq 0$, prove that for any two non-negative real numbers, $x$ and $y$, one has

$$P(X > x + y \,|\, X > x) = P\{X > y\}.$$

The last relation is often interpreted as the "*lack of memory*" in the values of $X$. Prove that if some extended (i.e., with values in $[0, \infty]$) random variable $X$ lacks memory, i.e., has the above property, then only one of the following three cases is possible: $P\{X = 0\} = 1$, or $P\{X = \infty\} = 1$, or $X$ is exponentially distributed with some parameter $0 < \lambda < \infty$.

  *Hint.* Setting $f(x) = P\{X > x\}$, the lack of memory property can be expressed as $f(x + y) = f(x) f(y)$. Consequently, the proof comes down to showing that, in the class of all right-continuous functions $f(\cdot)$ with values in the interval $[0, 1]$, the solution to the equation $f(x + y) = f(x) f(y)$ can be either of the form $f(x) \equiv 0$, or of the form $f(x) \equiv 1$, or of the form $f(x) \equiv e^{-\lambda x}$, for some parameter $0 < \lambda < \infty$.

**Problem 2.7.46.** Assuming that the random variables $X$ and $Y$ have finite second moments, prove that:
  (a) $\operatorname{cov}(X, Y) = \operatorname{cov}(X, \mathsf{E}(Y \,|\, X))$;
  (b) If $\mathsf{E}(Y \,|\, X) = 1$ then $\mathsf{D}X \leq \mathsf{D}XY$.

## 2.8  Random Variables II

**Problem 2.8.1.** Establish the validity of formulas (9), (10), (24), (27), (28) and (34)–(38) in [$\underline{P}$ §2.8].

**Problem 2.8.2.** Suppose that $\xi_1, \ldots, \xi_n$, $n \geq 2$, are independent and identically distributed random variables with (cumulative) distribution function $F(x)$ and, if it exists, density $f(x)$. Let $\overline{\xi} = \max(\xi_1, \ldots, \xi_n)$, $\underline{\xi} = \min(\xi_1, \ldots, \xi_n)$ and $\rho = \overline{\xi} - \underline{\xi}$. Prove that:

$$F_{\overline{\xi},\underline{\xi}}(y, x) = \begin{cases} (F(y))^n - (F(y) - F(x))^n, & y > x, \\ (F(y))^n, & y \leq x; \end{cases}$$

$$f_{\underline{\xi},\underline{\xi}}(y,x) = \begin{cases} n(n-1)[F(y)-F(x)]^{n-2} f(x)f(y), & y > x, \\ 0, & y \le x; \end{cases}$$

$$F_\rho(x) = \begin{cases} n \int_{-\infty}^{\infty} [F(y)-F(y-x)]^{n-1} f(y)\,dy, & x \ge 0, \\ 0, & x < 0; \end{cases}$$

$$f_\rho(x) = \begin{cases} n(n-1) \int_{-\infty}^{\infty} [F(y)-F(y-x)]^{n-2} f(y-x)f(y)\,dy, & x > 0, \\ 0, & x < 0. \end{cases}$$

**Problem 2.8.3.** Assuming that $\xi_1$ and $\xi_2$ are two independent Poisson random variables with parameters, respectively, $\lambda_1 > 0$ and $\lambda_2 > 0$, prove that:
(a) $\xi_1 + \xi_2$ has Poisson distribution with parameter $\lambda_1 + \lambda_2$.
(b) The distribution of $\xi_1 - \xi_2$ is given by

$$\mathsf{P}\{\xi_1 - \xi_2 = k\} = e^{-(\lambda_1+\lambda_2)} \left(\frac{\lambda_1}{\lambda_2}\right)^{k/2} I_k(2\sqrt{\lambda_1\lambda_2}), \quad k = 0, \pm1, \pm2, \ldots,$$

where

$$I_k(2x) = x^k \sum_{r=0}^{\infty} \frac{x^{2r}}{r!\,\Gamma(k+r+1)}$$

is the modified Bessel function of the first kind and of order $k$.

*Hint.* One possible proof of Part (b) is based on the series expansion of the generating function of the random variable $\xi_1 - \xi_2$—see Sect. A.3.

**Problem 2.8.4.** Setting $m_1 = m_2 = 0$ in formula [$\underline{\mathrm{P}}$ §2.8, (4)], show that

$$f_{\xi/\eta}(z) = \frac{\sigma_1\sigma_2\sqrt{1-\rho^2}}{\pi(\sigma_2^2 z^2 - 2\rho\sigma_1\sigma_2 z + \sigma_1^2)}.$$

**Problem 2.8.5.** The *maximal correlation coefficient* between the random variables $\xi$ and $\eta$ is defined as the quantity $\rho^*(\xi,\eta) = \sup_{u,v} \rho(u(\xi), v(\eta))$, where the supremum is taken over all Borel functions $u = u(x)$ and $v = v(x)$, for which the correlation coefficient $\rho(u(\xi), v(\eta))$ is meaningful. Prove that the random variables $\xi$ and $\eta$ are independent if and only if $\rho^*(\xi,\eta) = 0$—see Problem 2.8.6 below.

*Hint.* The *necessity* part of the statement is obvious. To prove the *sufficiency* part, given two arbitrarily chosen sets $A$ and $B$, set $u(\xi) = I_A(\xi)$ and $v(\eta) = I_B(\eta)$ and show that $\sup_{u,v} \rho(u(\xi), v(\eta)) = 0$ implies

$$\mathsf{P}\{\xi \in A, \eta \in B\} - \mathsf{P}\{\xi \in A\}\mathsf{P}\{\eta \in B\} = \rho(I_A(\xi), I_B(\eta)) = 0.$$

As the sets $A$ and $B$ are arbitrarily chosen, the last relation guarantees that $\xi$ and $\eta$ are independent.

**Problem 2.8.6.** (*Continuation of Problem 2.8.5.*) Let $(\Omega, \mathscr{F}, P)$ be a probability space, let $\mathscr{F}_1 \subseteq \mathscr{F}$, and $\mathscr{F}_2 \subseteq \mathscr{F}$ be any two sub-$\sigma$-algebras of $\mathscr{F}$, and let $L^2(\Omega, \mathscr{F}_i, P)$, $i = 1, 2$, be the usual spaces of random variables with finite second moment.

Set

$$\rho^*(\mathscr{F}_1, \mathscr{F}_2) = \sup|\rho(\xi_1, \xi_2)|,$$

where $\rho(\xi_1, \xi_2)$ is the correlation coefficient between $\xi_1$ and $\xi_2$ and the supremum is taken with respect to all pairs of random variables $(\xi_1, \xi_2)$ with $\xi_i \in L^2(\Omega, \mathscr{F}_i, P)$, $i = 1, 2$.

(a) Prove that if $\mathscr{F}_1 = \sigma(X_1)$ and $\mathscr{F}_2 = \sigma(X_2)$, for some random variables $X_1$ and $X_2$ on $(\Omega, \mathscr{F}, P)$, then

$$\rho^*(\sigma(X_1), \sigma(X_2)) = |\rho(X_1, X_2)|.$$

(b) Let $\mathscr{F}_1 = \bigvee_{i \in I} \mathscr{A}_i \left(= \bigcup_{i \in I} \mathscr{A}_i\right)$ and $\mathscr{F}_2 = \bigvee_{i \in I} \mathscr{B}_i \left(= \bigcup_{i \in I} \mathscr{B}_i\right)$, where $I$ is some index set. Assuming that all $\sigma$-algebras $\sigma(\mathscr{A}_i, \mathscr{B}_i)$, $i \in I$, are *jointly independent* ($\sigma(\mathscr{A}_i, \mathscr{B}_i)$ stands for the $\sigma$-algebra generated by the sets $A_i \in \mathscr{A}_i$ and $B_i \in \mathscr{B}_i$), prove that

$$\rho^*\left(\bigvee_{i \in I} \mathscr{A}_i, \bigvee_{i \in I} \mathscr{B}_i\right) = \sup_{i \in I} \rho^*(\mathscr{A}_i, \mathscr{B}_i).$$

**Problem 2.8.7.** Let $(\Omega, \mathscr{F}, P)$ be any probability space and let $\mathscr{F}_1 \subseteq \mathscr{F}$ and $\mathscr{F}_2 \subseteq \mathscr{F}$ be any two sub-$\sigma$-algebras of $\mathscr{F}$. Define the following quantities, every one of which measures the degree of *mixing* between $\mathscr{F}_1$ and $\mathscr{F}_2$:

$$\alpha(\mathscr{F}_1, \mathscr{F}_2) = \sup\{|P(A \cap B) - P(A)P(B)| : A \in \mathscr{F}_1, B \in \mathscr{F}_2\};$$

$$\varphi(\mathscr{F}_1, \mathscr{F}_2) = \sup\{|P(B \mid A) - P(B)| : A \in \mathscr{F}_1, B \in \mathscr{F}_2, P(A) > 0\};$$

$$\psi(\mathscr{F}_1, \mathscr{F}_2) = \sup\left\{\left|\frac{P(A \cap B)}{P(A)P(B)} - 1\right| : A \in \mathscr{F}_1, B \in \mathscr{F}_2, P(A)P(B) > 0\right\}.$$

In addition, let

$$\beta(\mathscr{F}_1, \mathscr{F}_2) = \sup \frac{1}{2} \sum_{i=1}^{N} \sum_{j=1}^{M} |P(A_i \cap B_j) - P(A_i)P(B_j)|,$$

where the supremum is taken over all finite partitions $\{A_1, \ldots, A_N\}$ and $\{B_1, \ldots, B_M\}$, with $A_i \in \mathscr{F}_1$ and $B_j \in \mathscr{F}_2$, $1 \le i \le N$, $1 \le j \le M$ and $N \ge 1$, $M \ge 1$.

Verify the following inequalities, in which the quantity $\rho^*(\mathcal{F}_1, \mathcal{F}_2)$ is as defined in Problem 2.8.6:

$$\alpha(\mathcal{F}_1, \mathcal{F}_2) \leq \beta(\mathcal{F}_1, \mathcal{F}_2) \leq \varphi(\mathcal{F}_1, \mathcal{F}_2) \leq \psi(\mathcal{F}_1, \mathcal{F}_2),$$

and

$$\alpha(\mathcal{F}_1, \mathcal{F}_2) \leq \rho^*(\mathcal{F}_1, \mathcal{F}_2) \leq 2\varphi^{1/2}(\mathcal{F}_1, \mathcal{F}_2),$$

$$\rho^*(\mathcal{F}_1, \mathcal{F}_2) \leq 2\varphi^{1/2}(\mathcal{F}_1, \mathcal{F}_2)\varphi^{1/2}(\mathcal{F}_2, \mathcal{F}_1).$$

**Problem 2.8.8.** Assuming that $\tau_1, \ldots, \tau_k$ are independent and identically distributed random variables, all having exponential distribution with density

$$f(t) = \lambda e^{-\lambda t}, \quad t \geq 0,$$

prove that $\tau_1 + \cdots + \tau_k$ is distributed with density

$$\frac{\lambda^k t^{k-1} e^{-\lambda t}}{(k-1)!}, \quad t \geq 0,$$

and

$$P(\tau_1 + \cdots + \tau_k > t) = \sum_{i=0}^{k-1} e^{-\lambda t} \frac{(\lambda t)^i}{i!}.$$

**Problem 2.8.9.** Assuming that $\xi \sim \mathcal{N}(0, \sigma^2)$, prove that for every $p \geq 1$ one has

$$E|\xi|^p = C_p \sigma^p,$$

where

$$C_p = \frac{2^{p/2}}{\pi^{1/2}} \Gamma\left(\frac{p+1}{2}\right)$$

and $\Gamma(s) = \int_0^\infty e^{-x} x^{s-1}\, dx$ is Euler's Gamma function. In particular, for any integer $n \geq 1$ one can write (see Problem 2.6.36)

$$E\xi^{2n} = (2n-1)!!\, \sigma^{2n}.$$

**Problem 2.8.10.** Prove that if $\xi$ and $\eta$ are two independent random variables, such that the distribution of $\xi + \eta$ coincides with the distribution of $\xi$, then $\eta = 0$ (P-a. e.).

*Hint.* Use the fact that if $\eta_1, \ldots, \eta_n$, $n \geq 1$, are independent random variables, all having the same distribution as $\eta$, then, for any $n \geq 1$, the distribution of $\xi + \eta_1 + \cdots + \eta_n$ coincides with the distribution of $\xi$.

If $\xi$ and $\eta$ admit moments of all orders, then one can use the relationship between the semi-invariants $s_{\xi+\eta}^{(k)}$ and $s_{\xi}^{(k)}$, $k \geq 1$ (see [P §2.12]).

**Problem 2.8.11.** Suppose that the random point $(X, Y)$ is distributed uniformly in the unit disk $\{(x, y): x^2 + y^2 \leq 1\}$, let $W = X^2 + Y^2$, and set

$$U = X \sqrt{-\frac{2 \ln W}{W}}, \quad V = Y \sqrt{-\frac{2 \ln W}{W}}.$$

Prove that $U$ and $V$ are independent $\mathcal{N}(0, 1)$-distributed random variables.

**Problem 2.8.12.** Suppose that $U$ and $V$ are two independent random variables that are uniformly distributed in the interval $(0, 1)$, and set

$$X = \sqrt{-\ln V} \cos(2\pi U), \quad Y = \sqrt{-\ln V} \sin(2\pi U).$$

Prove that $X$ and $Y$ are independent $\mathcal{N}(0, 1)$-distributed random variables.

**Problem 2.8.13.** Consider some positive random variable $\mathbb{R}$, which is distributed according to *Rayley law*, i.e., has density

$$f_R(r) = \frac{r}{\sigma^2} \exp\left\{ -\frac{r^2}{2\sigma^2} \right\}, \quad r > 0,$$

with some $\sigma^2 > 0$, and suppose that the random variable $\theta$ is uniformly distributed in the interval $(\alpha, \alpha + 2\pi k)$, where $k \in \mathbb{N} = \{1, 2, \ldots\}$ and $\alpha \in [0, 2\pi)$.
   Prove that the random variables $X = R \cos \theta$ and $Y = R \sin \theta$ are independent and distributed with law $\mathcal{N}(0, \sigma^2)$.

**Problem 2.8.14.** Give an example of two Gaussian random variables, $\xi$ and $\eta$, for which $\xi + \eta$ is not Gaussian.

**Problem 2.8.15.** Let $X_1, \ldots, X_n$ be independent and identically distributed random variables with density $f = f(x)$ and let

$$\mathcal{R}_n = \max(X_1, \ldots, X_n) - \min(X_1, \ldots, X_n)$$

denote the "range" of the sample $(X_1, \ldots, X_n)$. Prove that the density of the random variable $\mathcal{R}_n$ is given by

$$f_{\mathcal{R}_n}(x) = n(n-1) \int_{-\infty}^{\infty} [F(y) - F(y - x)]^{n-2} f(y) f(y - x) \, dx, \quad x > 0,$$

where $F(y) = \int_{-\infty}^{y} f(z) \, dz$. In particular, if $X_1, \ldots, X_n$ are uniformly distributed in the interval $[0, 1]$, then one has

$$f_{\mathcal{R}_n}(x) = \begin{cases} n(n-1)x^{n-2}(1-x), & 0 \leq x \leq 1, \\ 0, & x < 0 \text{ or } x > 1. \end{cases}$$

**Problem 2.8.16.** Let $F(x)$ be any (cummulative) distribution function. Prove that for any $a > 0$ the functions

$$G_1(x) = \frac{1}{a} \int_x^{x+a} F(u)\, du \quad \text{and} \quad G_2(x) = \frac{1}{2a} \int_{x-a}^{x+a} F(u)\, du$$

are also (cummulative) distribution functions.

**Problem 2.8.17.** Suppose that $X$ is some exponentially distributed random variable with parameter $\lambda > 0$, i.e., $X$ has density $f_X(x) = \lambda e^{-\lambda x} I(x \geq 0)$.

(a) Find the density of the distribution law of the random variable $Y = X^{1/\alpha}$, $\alpha > 0$, which is known as the *Weibull distribution*.

(b) Find the density of the distribution law of the random variable $Y = \ln X$, which is known as the *double exponential law*.

(c) Prove that the integer part and the fractional part of the random variable $X$, i.e., $\lfloor X \rfloor$ and $\{X\} = X - \lfloor X \rfloor$, are independent random variables. Find the distribution of $\lfloor X \rfloor$ and $\{X\}$.

**Problem 2.8.18.** Let $X$ and $Y$ be any two random variables with joint density of the form $f(x, y) = g(\sqrt{x^2 + y^2})$.

Find the density of the joint distribution of the random variables $\rho = \sqrt{X^2 + Y^2}$ and $\theta = \tan^{-1}(Y/X)$, and prove that $\rho$ and $\theta$ are independent.

Setting $U = (\cos \alpha)X + (\sin \alpha)Y$ and $V = (-\sin \alpha)X + (\cos \alpha)Y$, $\alpha \in [0, 2\pi]$, prove that the joint density $U$ and $V$ coincides with $f(x, y)$. (This property reflects the fact that the distribution of the vector $(X, Y)$ is invariant under rotation in $\mathbb{R}^2$.)

**Problem 2.8.19.** Let $X_1, \ldots, X_n$ be independent and identically distributed random variables with *continuous* (cummulative) distribution function $F = F(x)$. As this assumption implies $P\{X_i = X_j\} = 0, i \neq j$ (see Problem 2.8.76 below), it follows that

$$P\{X_i = X_j \text{ for some } i \neq j\} = P\left[ \bigcup_{i<j} \{X_i = X_j\} \right] \leq \sum_{i<j} P\{X_i = X_j\} = 0.$$

Consequently, one can claim that *with probability 1* the numbers $X_1(\omega), \ldots, X_n(\omega)$ can be arranged (and in a unique way) in a strictly increasing sequence. The elements of this sequence, which we denote by $X_1^{(n)}(\omega), \ldots, X_n^{(n)}(\omega)$, are well-defined random variables that are commonly referred to as *order statistics*—see also [P §3.13] and Problem 1.12.8. Thus, with probability 1 we have

$$X_1^{(n)}(\omega) < \cdots < X_n^{(n)}(\omega)$$

and

$$X_1^{(n)}(\omega) = \min(X_1(\omega), \ldots, X_n(\omega)), \quad \ldots, \quad X_n^{(n)}(\omega) = \max(X_1(\omega), \ldots, X_n(\omega)).$$

In addition, we will suppose that the distribution $F = F(x)$ admits density $f = f(x)$.

Prove that:

(a) The density of $X_k^{(n)}$ is given by

$$n f(x) C_{n-1}^{k-1} [F(x)]^{k-1} [1 - F(x)]^{n-k}.$$

(b) The joint density, $f^{(n)}(x_1, \ldots, x_n)$, of $X_1^{(n)}, \ldots, X_n^{(n)}$ is given by

$$f^{(n)}(x_1, \ldots, x_n) = \begin{cases} n! \, f(x_1) \ldots f(x_n), & \text{if } x_1 < \cdots < x_n, \\ 0 & \text{in all other cases.} \end{cases}$$

(c) If $f(x) = I_{[0,1]}(x)$ (i.e., if the random variables $X_i$ are distributed uniformly in $[0, 1]$), then

$$EX_r^{(n)} = \frac{r}{n+1} \quad \text{and} \quad \text{cov}(X_r^{(n)}, X_p^{(n)}) = \frac{r(n-p+1)}{(n+1)^2(n+2)}, \quad r \le p.$$

**Problem 2.8.20.** Let $\xi_1, \ldots, \xi_n$ be independent and identically distributed random variables with normal distribution $\mathcal{N}(m, \sigma^2)$. The quantities $\bar{\xi}$ and $s^2$, given by

$$\bar{\xi} = \frac{1}{n} \sum_{i=1}^{n} \xi_i \quad \text{and} \quad s^2 = \frac{1}{n-1} \sum_{i=1}^{n} (\xi_i - \bar{\xi})^2, \quad n > 1,$$

are known, respectively, as *sample mean* and *sample variance* (for the sample $\xi_1, \ldots, \xi_n$).

Prove that:

(a) $Es^2 = \sigma^2$.

(b) The sample mean $\bar{\xi}$ and the sample variance $s^2$ are independent.

(c) $\bar{\xi}$ has normal $\mathcal{N}(m, \sigma^2/n)$-distribution, while $(n-1)s^2/\sigma^2$ has $\chi^2$-distribution with $(n-1)$ degrees of freedom.

**Problem 2.8.21.** Suppose that $X_1, \ldots, X_n$ are independent and identically distributed random variables, let $\nu$ be any random variable with values in the set $\{1, \ldots, n\}$, which is independent from $X_1, \ldots, X_n$, and set $S_\nu = X_1 + \cdots + X_\nu$. Prove that

$$DS_\nu = DX_1 E\nu + (EX_1)^2 D\nu, \qquad \frac{DS_\nu}{ES_\nu} = \frac{DX_1}{EX_1} + EX_1 \frac{D\nu}{E\nu}.$$

**Problem 2.8.22.** Let $M(s) = Ee^{sX}$ be the moment generating function for the random variables $X$ (see Problem 2.6.32). Prove that $P\{X \ge 0\} \le M(s)$ for any $s > 0$.

**Problem 2.8.23.** Let $X, X_1, \ldots, X_n$ be independent and identically distributed random variables and set $S_n = \sum_{i=1}^{n} X_i$, $S_0 = 0$, $\overline{M}_n = \max_{0 \leq j \leq n} S_j$ and $\overline{M} = \sup_{n \geq 0} S_n$. Prove that ("$\xi \overset{d}{=} \eta$" means that the distribution laws of $\xi$ and $\eta$ coincide):

(a) $\overline{M}_n \overset{d}{=} (\overline{M}_{n-1} + X)^+$, $n \geq 1$.

(b) If $S_n \to \infty$ (P-a. e.), then $\overline{M} \overset{d}{=} (\overline{M} + X)^+$.

(c) If $-\infty < EX < 0$ nd $EX^2 < \infty$, then

$$E\overline{M} = \frac{DX - D(S + X)^-}{-2EX}.$$

**Problem 2.8.24.** Let everything be as in the previous problem and let $\overline{M}(\varepsilon) = \sup_{n \geq 0}(S_n - n\varepsilon)$, for $\varepsilon > 0$. Prove that $\lim_{\varepsilon \downarrow 0} \varepsilon \overline{M}(\varepsilon) = (DX)/2$.

**Problem 2.8.25.** Suppose that $\xi$ and $\eta$ are two independent random variables with densities $f_\xi(x)$, $x \in \mathbb{R}$, and $f_\eta(y) = I_{[0,1]}(y)$, $y \in \mathbb{R}$ (i.e., $\eta$ is distributed uniformly in $[0, 1]$). Prove that in this special case formulas [P §2.8, (36) and (37)] can be written as

$$f_{\xi\eta}(z) = \begin{cases} \displaystyle\int_z^\infty \frac{f_\xi(x)\,dx}{x}, & z \geq 0, \\ 0, & z < 0, \end{cases}$$

and

$$f_{\xi/\eta}(z) = \begin{cases} \displaystyle\int_0^1 x f_\xi(zx)\,dx, & 0 \leq z \leq 1, \\ \displaystyle\frac{1}{z^2} \int_0^1 x f_\xi(x)\,dx, & z > 1, \\ 0, & z < 0. \end{cases}$$

In particular, prove that if $\xi$ is also uniformly distributed in $[0, 1]$, then

$$f_{\xi/\eta}(z) = \begin{cases} \displaystyle\frac{1}{2}, & 0 \leq z \leq 1, \\ \displaystyle\frac{1}{2z^2}, & z > 1, \\ 0, & z < 0. \end{cases}$$

**Problem 2.8.26.** Let $\xi$ and $\eta$ be two independent random variables that are exponentially distributed with the same parameter $\lambda > 0$.

(a) Prove that the random variable $\frac{\xi}{\xi+\eta}$ is distributed uniformly in $[0, 1]$.

(b) Prove that if $\xi$ and $\eta$ are two independent and exponentially distributed random variables with parameters, respectively, $\lambda_1$ nd $\lambda_2$, $\lambda_1 \neq \lambda_2$, then the density of $\xi + \eta$ is given by

$$f_{\xi+\eta}(z) = \frac{e^{-z/\lambda_1} - e^{-z/\lambda_2}}{\lambda_1 - \lambda_2} I_{(0,\infty)}(z).$$

**Problem 2.8.27.** Suppose that $\xi$ and $\eta$ are two independent standard normal (i.e., $\mathcal{N}(0,1)$) random variables and prove that:
   (a) Both $\xi/\eta$ and $\xi/|\eta|$ have Cauchy distribution with density $\frac{1}{\pi(1+x^2)}$, $x \in \mathbb{R}$.
   (b) $|\xi|/|\eta|$ has density $\frac{2}{\pi(1+x^2)}$, $x \geq 0$.

**Problem 2.8.28.** Let $X_1, \ldots, X_n$ be independent and exponentially distributed random variables with parameters, respectively, $\lambda_1, \ldots, \lambda_n$, and suppose that $\lambda_i \neq \lambda_j$, $i \neq j$. Setting $T_n = X_1 + \ldots + X_n$, prove that the probability $P\{T_n > t\}$ can be expressed in the form

$$P\{T_n > t\} = \sum_{i=1}^{n} a_{in} e^{-\lambda_i t}$$

and find the coefficients $a_{in}$, $i = 1, \ldots, n$. (Comp. with Problem 2.8.8.)

**Problem 2.8.29.** Let $\xi_1, \xi_2, \ldots$ be any sequence of random variables with $E\xi_n = 0$, $E\xi_n^2 = 1$ and let $S_n = \xi_1 + \ldots + \xi_n$. Prove that for any positive $a$ and $b$ one has

$$P\{S_n \geq a\,n + b \text{ for some } n \geq 1\} \leq \frac{1}{1 + a\,b}.$$

**Problem 2.8.30.** Suppose that the random variable $\xi$ takes values in the finite set $\{x_1, \ldots, x_k\}$. Prove that

$$\lim_{n \to \infty} (E\xi^n)^{1/n} = \max(x_1, \ldots, x_k).$$

**Problem 2.8.31.** Suppose that $\xi$ and $\eta$ are two independent random variables that take values in the set $\{1, 2, \ldots\}$ and are such that either $E\xi < \infty$, or $E\eta < \infty$. Prove that

$$E \min(\xi, \eta) = \sum_{k=1}^{\infty} P\{\xi \geq k\} P\{\eta \geq k\}.$$

**Problem 2.8.32.** Let $\xi_1$ and $\xi_2$ be any two independent and exponentially distributed random variables with parameters, respectively, $\lambda_1$ and $\lambda_2$. Find the distribution functions of the random variables $\frac{\xi_1}{\xi_1+\xi_2}$ and $\frac{\xi_1+\xi_2}{\xi_1}$.

**Problem 2.8.33.** Let $X$ and $Y$ be two random matrices and suppose that $\mathsf{E} Y Y^*$ is invertible. Prove the following   *matrix version of the Cauchy-Bunyakovsky inequality*:

$$(\mathsf{E} X Y^*)(\mathsf{E} Y Y^*)^{-1}(\mathsf{E} Y X^*) \le \mathsf{E} X X^*,$$

where the relation $\le$ is understood as the difference between the right and the left side being non-negative definite.

**Problem 2.8.34.** (*L. Shepp.*) Suppose that $X$ is a Bernoulli random variable with $\mathsf{P}\{X = 1\} = p, \mathsf{P}\{X = 0\} = 1 - p$.

(a) Prove that one can find a random variable $Y$ which is independent from $X$ and is such that $X + Y$ has symmetric distribution $(X + Y \overset{d}{=} -(X + Y))$.

(b) Among all random variables $Y$ that have the above property, find the one that has the smallest variance $\mathsf{D} Y$.

**Problem 2.8.35.** Suppose that the random variable $U$ is uniformly distributed in the interval $(0, 1)$. Prove that:

(a) Given any $\lambda > 0, -\frac{1}{\lambda} \ln U$ is exponentially distributed with parameter $\lambda$.

(b) $\tan \pi (U - \frac{1}{2})$ is distributed according to the Cauchy law with density $\frac{1}{\pi(1+x^2)}$, $x \in \mathbb{R}$.

(c) $\lfloor nU \rfloor + 1$ is distributed uniformly in the (discrete) set $\{1, 2, \ldots, n\}$.

(d) Given any $0 < q < 1$, the random variable $X = 1 + \left\lfloor \frac{\ln U}{\ln q} \right\rfloor$ has geometric distribution with $\mathsf{P}\{X = k\} = q^{k-1}(1 - q), k \ge 1$.

**Problem 2.8.36.** Give an example of a sequence of independent and identically distributed random variables $\{X_1, X_2, \ldots\}$ with $\sup_n \mathsf{E} X_n^p < \infty$, for all $p > 0$, but such that

$$\mathsf{P}\left\{ \sup_j X_{n_j} < \infty \right\} = 0$$

for any sub-sequence $\{n_1, n_2, \ldots\}$.

**Problem 2.8.37.** Let $\xi$ and $\eta$ be any two independent random variables with (cumulative) distribution functions $F = F(x)$ and $G = G(x)$. Since $\mathsf{P}\{\max(\xi, \eta) \le x\} = \mathsf{P}\{\xi \le x\}\mathsf{P}\{\eta \le x\}$, it is easy to see that the distribution function of $\max(\xi, \eta)$ is nothing but $F(x)G(x)$. Give an alternative proof of the last claim by identifying the event $\{\max(\xi, \eta) \le x\}$ with the union of the events $\{\xi \le x, \xi \ge \eta\}$ and $\{\eta \le x, \xi < \eta\}$, and by expressing the probabilities of these events in terms of the conditional probabilities (in the final stage use formula (68) from § 6).

**Problem 2.8.38.** Suppose that $\xi$ and $\eta$ are two independent random variables whose product $\xi \eta$ is distributed with Poisson law of parameter $\lambda > 0$. Prove that one of the variables $\xi$ and $\eta$ takes values in the set $\{0, 1\}$.

**Problem 2.8.39.** Prove that, given a standard normal, i.e., $\mathcal{N}(0, 1)$, random variable $\xi$, one has the following asymptotic result (see Problem 1.6.9):

$$\mathsf{P}\{\xi \geq x\} \sim \frac{\varphi(x)}{x}, \quad \text{as } x \to \infty, \quad \text{where} \quad \varphi(x) = \frac{1}{\sqrt{2\pi}} e^{-x^2/2}.$$

What is the analog of this result for a random variable $\gamma$ which has gamma-distribution (see [P §2.3, Table 3])?

**Problem 2.8.40.** Let $\xi$ be any random variable and let $M_a$ be the collection of all medians of $\xi$, as defined in part (a) of Problem 1.4.23. This object is meaningful for arbitrary random variables (Problem 1.4.23 refers to discrete random variables). Prove that for any $b \in \mathbb{R}$ and any $p \geq 1$ with $\mathsf{E}|\xi|^p < \infty$, one must have

$$|\mu - b|^p \leq 2\mathsf{E}|\xi - b|^p,$$

where $\mu = \mu(\xi) \in M_a$ is a median for $\xi$. In particular, if $\mathsf{E}|\xi|^2 < \infty$, then $|\mu - \mathsf{E}\xi| \leq \sqrt{2\mathsf{D}\xi}$.)

**Problem 2.8.41.** Suppose that $\xi$ and $\eta$ are two independent random variables with finite second moments. Prove that the random variables $\xi + \eta$ and $\xi - \eta$ are uncorrelated if and only if $\mathsf{D}\xi = \mathsf{D}\eta$.

**Problem 2.8.42.** Given any two $L^1$-functions, $f$ and $g$, their convolution, $f * g$, is defined as $f * g = \int_R f(y)g(x - y)\, dy$. Prove *Young's inequality*:

$$\int_{\mathbb{R}} |(f * g)(x)|\, dx \leq \int_{\mathbb{R}} |f(x)|\, dx \cdot \int_{\mathbb{R}} |g(x)|\, dx.$$

**Problem 2.8.43.** According to formula [P §2.8, (22)] the density, $f_\eta(y)$, of the random variable $\eta = \varphi(\xi)$ can be connected with the density, $f_\xi(x)$, of the random variable $\xi$ by the relation

$$f_\eta(y) = f_\xi(h(y)) \left| h'(y) \right|,$$

where $h(y) = \varphi^{-1}(y)$.

Suppose that $I$ is some open subset of $\mathbb{R}^n$ and that $y = \varphi(x)$ is some $\mathbb{R}^n$-valued function defined on $I$ (for $x = (x_1, \ldots, x_n) \in I$ and $y = (y_1, \ldots, y_n) \in \mathbb{R}^n$, the relation $y = \varphi(x)$ is understood as $y_i = \varphi_i(x_1, \ldots, x_n)$, $i = 1, \ldots, n$). Suppose that all derivatives $\frac{\partial \varphi_i}{\partial x_j}$ exist and are continuous and that $|J_\varphi(x)| > 0$, $x \in I$, where $J_\varphi(x)$ stands for the determinant of the Jacobian of $\varphi$, i.e.,

$$J_\varphi(x) = \det \left\| \frac{\partial \varphi_i}{\partial x_j}, \quad 1 \leq i, j \leq n \right\|.$$

Prove that if $\xi = (\xi_1, \ldots, \xi_n)$ is some $I$-valued random vector with density $f_\xi(x)$ and if $\eta = \varphi(\xi)$, then the density $f_\eta(y)$ is well defined on the set $\varphi(I) = \{y : y = \varphi(x), x \in I\}$ and can be written as

$$f_\eta(y) = f_\xi(h(y))|J_h(y)|,$$

where $h = \varphi^{-1}$ is the inverse of the function $\varphi$ (we have $|J_h(y)| > 0$, as long as $|J_\varphi(x)| > 0$).

*Hint.* Use the multivariate analog of the integration-by-substitution rule (Problem 2.6.74), with $g(x) = G(\varphi(x)) f_\xi(x)$, for some appropriate function $G$.

**Problem 2.8.44.** Let $\eta = A\xi + B$, where: $\xi = (\xi_1, \ldots, \xi_n)$, $\eta = (\eta_1, \ldots, \eta_n)$, $A$ is an $n \times n$-matrix with $|\det A| > 0$, and $B$ is an $n$-dimensional vector. Prove that

$$f_\eta(y) = \frac{1}{|\det A|} f_\xi(A^{-1}(y - B)).$$

*Hint.* Use the result established in Problem 2.8.43 with $\varphi(x) = Ax + B$ and prove that $|J_{\varphi^{-1}}(y)| = 1/|\det A|$.

**Problem 2.8.45.** (a) Let $\rho(\xi, \eta)$ be the correlation coefficient between two given random variables, $\xi$ and $\eta$. Prove that

$$\rho(c_1\xi + c_2, c_3\eta + c_4) = \rho(\xi, \eta) \cdot \text{sign}(c_1 c_3),$$

where $\text{sign}\, x = 1$ for $x > 0$, $\text{sign}\, x = 0$ for $x = 0$ and $\text{sign}\, x = -1$ for $x < 0$.

(b) Consider the random variables $\xi_1, \xi_2, \xi_3, \xi_4$ with correlation coefficients $\rho(\xi_i, \xi_j)$, $i \neq j$, and prove that

$$\rho(\xi_1 + \xi_2, \xi_3 + \xi_4) = [\rho(\xi_1, \xi_3) + \rho(\xi_1, \xi_4)] + [\rho(\xi_2, \xi_3) + \rho(\xi_2, \xi_4)].$$

**Problem 2.8.46.** Let $X = (X_1, \ldots, X_n)$ be any Gaussian random vector whose components are independent and identically distributed with $X_i \sim \mathcal{N}(0, \sigma^2)$, $i = 1, \ldots, n$. Consider the spherical coordinates, $\{R, \Phi_1, \ldots, \Phi_{n-1}\}$, of the vector $X = (X_1, \ldots, X_n)$; in other words, suppose that

$$X_1 = R \sin \Phi_1,$$

$$X_m = R \sin \Phi_m \cos \Phi_{m-1} \ldots \cos \Phi_1, \quad 2 \leq m \leq n - 1,$$

$$X_n = R \cos \Phi_{n-1} \cos \Phi_{n-2} \ldots \cos \Phi_1,$$

where $R \geq 0$, $\Phi_i \in [0, 2\pi)$, $1 \leq i \leq n - 1$. Prove that, for $r \geq 0$, $\varphi_i \in [0, 2\pi)$, $i = 1, \ldots, n - 1$, $n \geq 2$, the joint density, $f(r, \varphi_1, \ldots, \varphi_{n-1})$, of the random variables $(R, \Phi_1, \ldots, \Phi_{n-1})$ is given by

$$f(r, \varphi_1, \ldots, \varphi_{n-1}) = \frac{r^{n-1} \exp\left(-\frac{r^2}{2\sigma^2}\right)}{(2\pi)^{n/2}\sigma^n} \cos^{n-2}\varphi_1 \cos^{n-3}\varphi_2 \ldots \cos\varphi_{n-2},$$

where we set by convention $\varphi_0 = 0$.

**Problem 2.8.47.** Suppose that $X$ is a random variable with values in the interval $[0, 1]$ and such that the distribution of $X$ is given by the Cantor function (see § 3). Compute all moments $EX^n$, $n \geq 1$.

**Problem 2.8.48.** (a) Verify that each of the following functions is a (cummulative) distribution function:

$$F_G(x) = \exp(-e^{-x}), \qquad x \in \mathbb{R};$$

$$F_F(x) = \begin{cases} 0, & x < 0, \\ \exp(-x^{-\alpha}), & x \geq 0, \end{cases} \qquad \text{where } \alpha > 0;$$

$$F_W(x) = \begin{cases} \exp(-|x|^{\alpha}), & x < 0, \\ 1, & x \geq 0, \end{cases} \qquad \text{where } \alpha > 0.$$

These functions are known, respectively, as the *Gumbel's distribution*, or double exponential distribution; comp. with Problem 2.8.17 ($F_G(x)$), *Fréchet distribution* ($F_F(x)$), and *Weibull distribution* ($F_W(x)$).

These distributions are special cases of the following three types (everywhere below we suppose that $\mu \in \mathbb{R}$, $\sigma > 0$, and $\alpha > 0$):

Type 1 (Gumbel-type distributions):

$$F_G(x) = \exp\left\{-e^{-\frac{(x-\mu)^2}{\sigma}}\right\}.$$

Type 2 (Fréchet-type distribution):

$$F_F(x) = \begin{cases} 0, & x < \mu, \\ \exp\left\{-\left(\frac{(x-\mu)^2}{\sigma}\right)^{\alpha}\right\}, & x > \mu. \end{cases}$$

Type 3 (Weibull-type distribution):

$$F_W(x) = \begin{cases} \exp\left\{-\left(\frac{(x-\mu)^2}{\sigma}\right)^{\alpha}\right\}, & x \leq \mu, \\ 1, & x > \mu. \end{cases}$$

(b) Prove that if $X$ has Type 2 distribution, then the random variable

$$Y = \ln(X - \mu)$$

has Type 1 distribution. Similarly, if $X$ has Type 3 distribution, then

$$Y = -\ln(\mu - X)$$

also must have Type 1 distribution.

*Remark.* This explains why Type 1 distributions, which are often referred to as *extreme value distributions*, are fundamental in the "extreme value theory".

**Problem 2.8.49.** (*Factorial moments.*) Given some random variable $X$, its factorial moments are defined as

$$m_{(r)} = \mathsf{E}X(X-1)\ldots(X-r+1), \qquad r = 1, 2, \ldots,$$

i.e., $m_{(r)} = \mathsf{E}(X)_r$.

If $X$ has Poisson distribution law of parameter $\lambda$, then for $r = 3$ one has $m_{(3)} = \lambda$. Calculate $m_{(r)}$ for an arbitrary $r$.

**Problem 2.8.50.** Suppose that $\theta_1$ and $\theta_2$ are two independent random variables that are distributed uniformly in $[0, 2\pi)$, and let $X_1 = \cos \theta_1$ and $X_2 = \cos \theta_2$.
    Prove that

$$\frac{1}{2}(X_1 + X_2) \overset{\text{law}}{=} X_1 X_2$$

(recall that "$\overset{\text{law}}{=}$," or "$\overset{d}{=}$," means "*identical in law*").

**Problem 2.8.51.** The random variable $\theta$ is distributed uniformly in the interval $[0, 2\pi)$ and the random variable $C$ is distributed in the real line $\mathbb{R}$ according to the Cauchy law with density $\frac{1}{\pi(1+x^2)}$, $x \in \mathbb{R}$.
    (a) Prove that the random variables $\cos^2 \theta$ and $1/(1 + C^2)$ share the same distribution law (i.e., $\cos^2 \theta \overset{\text{law}}{=} 1/(1 + C^2)$).
    (b) Prove that $\cot \frac{\theta}{2} \overset{\text{law}}{=} C$.
    (c) Find the densities of the distribution laws of the random variables $\sin(\theta + \varphi)$, $\varphi \in \mathbb{R}$, and $\alpha \tan \theta$, $\alpha > 0$.

**Problem 2.8.52.** Let $\xi$ be any exponentially distributed random variable with $\mathsf{P}\{\xi \geq t\} = e^{-t}$, $t \geq 0$, and let $N$ be any standard normal random variable (i.e., $N \sim \mathcal{N}(0, 1)$), which is independent from $\xi$. Prove that

$$\xi \overset{\text{law}}{=} \sqrt{2\xi}\,|N|,$$

i.e., the distribution law of $\xi$ coincides with the distribution law of $\sqrt{2\xi}\,|N|$.

**Problem 2.8.53.** Suppose that $X$ is a random variable that takes values in the set $\{0, 1, \ldots, N\}$ and has *binomial moments* $b_0, b_1, \ldots, b_N$, given by $b_k = C_X^k = \frac{1}{k!}\mathsf{E}(X)_k \equiv \frac{1}{k!}\mathsf{E}X(X-1)\ldots(X-k+1)$—see Sect. A.3.

Prove that the moment generating function of $X$ is given by

$$G_X(s) = \mathbf{E}s^X = \sum_{k=0}^{N} b_k (s-1)^k = \sum_{n=0}^{N} s^n \left( \sum_{k=n}^{N} (-1)^{k-n} C_k^n b_k \right),$$

and that, consequently, for any $n = 0, 1, \ldots, N$, one has

$$\mathbf{P}\{X = n\} = \sum_{k=n}^{N} (-1)^{k-n} C_k^n b_k.$$

**Problem 2.8.54.** Suppose that $X$ and $Y$ are two independent random variables that are distributed uniformly in the interval $[0, 1]$. Prove that the random variable

$$Z = \begin{cases} X + Y, & 0 \le X + Y \le 1, \\ (X + Y) - 1, & 1 < X + Y \le 2, \end{cases}$$

is also uniformly distributed in $[0, 1]$.

**Problem 2.8.55.** Suppose that $X_1, \ldots, X_n$ are independent and identically distributed random variables that take the values 0, 1 and 2 with probability $1/3$ each. Find a general formula for the probabilities

$$P_n(k) = \mathbf{P}\{X_1 + \cdots + X_n = k\}, \qquad 0 \le k \le 2n$$

(for example, $P_n(0) = 3^{-n}$, $P_n(1) = n3^{-n}$, $P_n(2) = C_{n+1}^2 3^{-n}$, $P_n(5) = (C_{n+4}^5 - nC_{n+1}^2) \cdot 3^{-n}$, and so on).

**Problem 2.8.56.** The random variables $\xi$ and $\eta$ are such that $\mathbf{E}\xi^2 < \infty$ and $\mathbf{E}\eta^2 < \infty$. Prove that:
 (a) $\mathbf{D}(\xi \pm \eta) = \mathbf{D}\xi + \mathbf{D}\eta \pm 2\,\mathsf{cov}(\xi, \eta)$;
 (b) If, in addition, $\xi$ and $\eta$ are independent, then

$$\mathbf{D}(\xi\eta) = \mathbf{D}\xi \cdot \mathbf{D}\eta + \mathbf{D}\xi \cdot (\mathbf{E}\eta)^2 + \mathbf{D}\eta \cdot (\mathbf{E}\xi)^2.$$

(See Problem 2.8.69.)

**Problem 2.8.57.** The joint density, $f(x, y)$, for the pair of random variables $(X, Y)$, is said to be "spherically symmetric" if it can be expressed as

$$f(x, y) = g(x^2 + y^2),$$

for some choice of the probability density function $g = g(z)$, $z \ge 0$. Assuming that $R$ and $\theta$ represent the polar coordinates of $(X, Y)$, i.e., $X = R\cos\theta$, $Y = R\sin\theta$, prove that $\theta$ is uniformly distributed in $[0, 2\pi)$, while $R$ is distributed with density $h(r) = 2\pi r g(r^2)$.

**Problem 2.8.58.** Given a pair of random variables, $(X, Y)$, with density $f(x, y)$, consider the complex random variables

$$Z_t = Ze^{it}, \quad t \in \mathbb{R}, \quad \text{where } Z = X + iY.$$

Prove that in order to claim that the distribution law of $Z_t$ does not depend on $t \in \mathbb{R}$ it is necessary to assume that $f(x, y)$ has the form $f(x, y) = g(x^2 + y^2)$, where, just as in the previous problem, $g$ is some probability density function.

**Problem 2.8.59.** Let $\xi$ and $\eta$ be any two independent and exponentially distributed random variables with density $f(x) = \lambda e^{-\lambda x}$, $x > 0$. Prove that the random variables $\xi + \eta$ and $\frac{\xi}{\eta}$ are independent.

**Problem 2.8.60.** Suppose that $\xi$ and $\eta$ are two independent random variables with densities

$$f_\xi(x) = \frac{1}{\pi} \frac{1}{\sqrt{1 - x^2}}, \quad |x| < 1, \quad \text{and} \quad f_\eta(y) = \frac{y}{\sigma^2} e^{-\frac{y^2}{2\sigma^2}}, \quad y > 0, \quad \sigma > 0.$$

Prove that the random variable $\xi \eta$ is normally, $\mathcal{N}(0, \sigma^2)$-distributed.

**Problem 2.8.61.** Consider the random matrix $\|\xi_{ij}\|$ of size $n \times n$, whose (random) entries are such that $\mathsf{P}\{\xi_{ij} = \pm 1\} = 1/2$. Prove that the expected value and the variance of the determinant of this random matrix are equal, respectively, to $0$ and $n!$.

**Problem 2.8.62.** Suppose that $X_1, X_2, \ldots$ are independent random variables that are distributed uniformly in the interval $[0, 1]$. Prove that

$$\mathsf{E} \sum_{k=1}^{n} X_k^2 \left[ \sum_{k=1}^{n} X_k \right]^{-1} \to \frac{2}{3} \quad \text{as } n \to \infty.$$

**Problem 2.8.63.** Suppose that the random vector $X = (X_1, X_2, X_3)$ is distributed uniformly in the tetrahedron

$$\Sigma_3 = \{(x_1, x_2, x_3) : x_1 \geq 0, x_2 \geq 0, x_3 \geq 0 \ x_1 + x_2 + x_3 \leq c\},$$

where $c > 0$ is some fixed constant. Find the *marginal* distributions of the random vector $X = (X_1, X_2, X_3)$, associated with the components $X_1$ and $(X_1, X_2)$.

*Hint.* The density, $f(x_1, x_2, x_3)$, of the vector $X = (X_1, X_2, X_3)$ is equal to the constant $V^{-1}$, where $V = c^3/6$ is the volume of $\Sigma_3$. With this observation in mind, prove that the density of $(X_1, X_2)$ is given by $f(x_1, x_2) = 6(c - x_1 - x_2)/c$ and then calculate the density of $X_1$.

**Problem 2.8.64.** Let $X_1, \ldots, X_n$ be positive, independent and identically distributed random variables with $\mathsf{E}X_1 = \mu$, $\mathsf{E}X_1^{-1} = r$ and $S_m = X_1 + \cdots + X_n$, $1 \leq m \leq n$. Prove that:
(a) $\mathsf{E}S_n^{-1} \leq r$;
(b) $\mathsf{E}X_i S_n^{-1} = 1/n, i = 1, \ldots, n$;

(c) $\mathrm{E} S_m S_n^{-1} = m/n$, if $m \leq n$;
(d) $\mathrm{E} S_n S_m^{-1} = 1 + (n - m)\mathrm{E} S_m^{-1}$, if $m < n$.

**Problem 2.8.65.** (*Dirichlet distribution.*) In [P §2.3, Table 3] the beta-distribution, with parameters $\alpha > 0$ and $\beta > 0$, is defined as a probability distribution on $[0, 1]$ with density

$$f(x; \alpha, \beta) = \frac{x^{\alpha-1}(1 - x)^{\beta-1}}{\mathrm{B}(\alpha, \beta)},$$

where

$$\mathrm{B}(\alpha, \beta) = \int_0^1 x^{\alpha-1}(1 - x)^{\beta-1}\, dx$$

$$\left( = \frac{\Gamma(\alpha + \beta)}{\Gamma(\alpha)\Gamma(\beta)}, \quad \text{with } \Gamma(\alpha) = \int_0^\infty x^{\alpha-1} e^{-x}\, dx \right).$$

The *Dirichlet distribution*, is a multivariate analog of the beta-distribution and is defined as the probability distribution on the set

$$\Delta_{k-1} = \{(x_1, \ldots, x_{k-1}) : x_i \geq 0,\ 0 \leq x_1 + \ldots x_{k-1} \leq 1\}, \quad \text{for } k \geq 2,$$

given by the density

$$f(x_1, \ldots, x_{k-1}; \alpha_1, \ldots, \alpha_{k-1}, \alpha_k) =$$

$$= \frac{\Gamma(\alpha_1 + \ldots + \alpha_k)}{\Gamma(\alpha_1) \ldots \Gamma(\alpha_k)} x_1^{\alpha_1-1} \ldots x_{k-1}^{\alpha_k-1}(1 - (x_1 + \ldots + x_{k-1}))^{\alpha_k-1},$$

where $\alpha_i > 0$, $i = 1, \ldots, k$, are given parameters. Alternatively, the Dirichlet distribution can be defined on the simplex $\{(x_1, \ldots, x_k) : x_i \geq 0, \sum_{i=1}^k x_i = 1\}$ by specifying the "density"

$$\widetilde{f}(x_1, \ldots, x_k; \alpha_1, \ldots, \alpha_{k-1}, \alpha_k) = \frac{\Gamma(\alpha_1 + \ldots + \alpha_k)}{\Gamma(\alpha_1) \ldots \Gamma(\alpha_k)} x_1^{\alpha_1} \ldots x_k^{\alpha_k}$$

(the quotation marks around the word "density" are simply a reference to the fact that the function $\widetilde{f}(x_1, \ldots, x_k; \alpha_1, \ldots, \alpha_{k-1}, \alpha_k)$ does not represent a density relative to the Lebesgue measure in $\mathbb{R}^k$).

Suppose that all component of the random vector $X = (X_1, \ldots, X_k)$ are non-negative, i.e., $X_i \geq 0$, and are such that the sum $X_1 + \cdots + X_k = 1$ has Dirichlet distribution with density $f(x_1, \ldots, x_{k-1}; \alpha_1, \ldots, \alpha_{k-1}, \alpha_k)$ on $\Delta_{k-1}$ (in the sense that this function represents the joint density of the first $k - 1$ components, $X_1, \ldots, X_{k-1}$, of the vector $(X_1, \ldots, X_k)$, after eliminating the last component, $X_k$, from the relation $X_k = 1 - (X_1 + \ldots + X_{k-1})$).

(a) Prove that

$$EX_j = \frac{\alpha_j}{\sum_{i=1}^{k} \alpha_i}, \quad DX_j = \frac{\alpha_j \left( \sum_{i=1}^{k} \alpha_i - \alpha_j \right)}{\left( \sum_{i=1}^{k} \alpha_i \right)^2 \left( \sum_{i=1}^{k} \alpha_i + 1 \right)},$$

$$\mathrm{cov}(X_{j_1}, X_{j_2}) = -\frac{\alpha_{j_1} \alpha_{j_2}}{\left( \sum_{i=1}^{k} \alpha_i \right)^2 \left( \sum_{i=1}^{k} \alpha_i + 1 \right)}, \quad j_1 \neq j_2.$$

(b) Prove that for every choice of non-negative integer numbers, $r_1, \ldots, r_k$, one can write

$$EX_1^{r_1} \ldots X_k^{r_k} = \frac{\Gamma \left( \sum_{i=1}^{k} \alpha_i \right) \prod_{i=1}^{k} \Gamma(\alpha_i + r_i)}{\prod_{i=1}^{k} \Gamma(\alpha_i) \Gamma \left( \sum_{i=1}^{k} (\alpha_i + r_i) \right)}.$$

(c) Find the conditional density, $f_{X_k | X_1, \ldots, X_{k-1}}(x_k \mid x_1, \ldots, x_{k-1})$, of the random variables $X_k$, given $X_1, \ldots, X_{k-1}$.

**Problem 2.8.66.** The *concentration function* of a random variable $X$ is defined as

$$Q(X; l) = \sup_{x \in \mathbb{R}} P\{x < X \leq x + l\}, \quad l \geq 0.$$

Prove that:

(a) If $X$ and $Y$ are two independent random variables, then

$$Q(X + Y; l) \leq \min(Q(X; l), Q(Y; l)), \quad \text{for all } l \geq 0.$$

(b) There is a number $x_l^*$, for which one can write $Q(X; l) = P\{x_l^* < X \leq x_l^* + l\}$, and the distribution function of $X$ can be claimed to be continuous if and only if $Q(X; 0) = 0$.

Hint. (a) If $F_X$ and $F_Y$ stand for the distribution functions of $X$ and $Y$, then

$$P\{z < X + Y \leq z + l\} = \int_{-\infty}^{\infty} [F_X(z + l - y) - F_X(z - y)] \, dF_Y(y).$$

**Problem 2.8.67.** Suppose that $\xi \sim \mathcal{N}(m, \sigma^2)$, i.e., $\xi$ has normal distribution with parameters $m$ and $\sigma^2$, and consider the random variable $\eta = e^\xi$, which has *log-normal* distribution with density (see formula [P §2.8, (23)])

$$f_\eta(y) = \begin{cases} \frac{1}{\sqrt{2\pi}\,\sigma y} \exp\left[ -\frac{(m - \ln y)^2}{2\sigma^2} \right], & y > 0, \\ 0, & y \leq 0. \end{cases}$$

Given any $\alpha \in [-1, 1]$ define the function

$$f^{(\alpha)}(y) = \begin{cases} f_\eta(y)\{1 - \alpha \sin[\pi\sigma^{-2}(m - \ln y)]\}, & y > 0, \\ 0, & y \leq 0. \end{cases}$$

(a) Prove that $f^{(\alpha)}(y)$ is a probability density function, i.e., $f^{(\alpha)}(y) \geq 0$ and $\int_0^\infty f^{(\alpha)}(y)\,dy = 1$.

(b) Suppose that $\zeta$ is some random variable with density $f_\zeta(y) = f^{(\alpha)}(y)$, $\alpha \neq 0$, and prove that $\eta$ and $\zeta$ have identical moments of all orders: $E\eta^n = E\zeta^n$, $n \geq 1$. (This shows that the log-normal distribution admits moments of all orders, and yet this distribution is *not* uniquely determined by its moments.)

**Problem 2.8.68.** Let $(\xi_n)_{n\geq0}$ be any sequence of independent, identically, and symmetrically distributed random variables, and, given any $n \geq 1$, let $S_0 = 0$ and $S_n = \xi_1 + \ldots + \xi_n$. Define the respective sequences of partial maximums and partial minimums, $M = (M_n)_{n\geq0}$ and $m = (m_n)_{n\geq0}$, given by

$$M_n = \max(S_0, S_1, \ldots, S_n) \quad \text{and} \quad m_n = \min(S_0, S_1, \ldots, S_n).$$

As a generalization of Problem 1.10.7, prove that for any fixed $n$ one has

$$(M_n - S_n, S_n - m_n, S_n) \overset{\text{law}}{=} (-m_n, M_n, S_n) \overset{\text{law}}{=} (M_n, -m_n, S_n),$$

i.e., the joint distribution laws of the above triplets of random variables coincide.
    *Hint.* Use the following relation, which is easy to verify:

$$(S_n - S_{n-k}; k \leq n) \overset{\text{law}}{=} (S_k; k \leq n) \quad \text{for any } n \geq 1.$$

**Problem 2.8.69.** The random variables $\xi$ and $\eta$ are such that $D\xi < \infty$ and $D\eta < \infty$. Prove that

$$\text{cov}^2(\xi, \eta) \leq D\xi\, D\eta$$

and explain when does the identity in this relation hold.

**Problem 2.8.70.** Let $\xi_1, \ldots, \xi_n$ be independent and identically distributed random variables. Prove that

$$P\{\min(\xi_1, \ldots, \xi_n) = \xi_1\} = n^{-1}.$$

Show also that the random variables $\min(\xi_1, \ldots, \xi_n)$ and $I_{\{\xi_1 = \min(\xi_1, \ldots, \xi_n)\}}$ are independent.

**Problem 2.8.71.** Let $X$ be any random variable with distribution function $F = F(x)$ and let $C$ be any constant. Find the distribution functions for the following random variables:

$$X \vee C \equiv \max(X, C), \quad X \wedge C \equiv \min(X, C), \quad X^C = \begin{cases} X, & \text{if } |X| \leq C, \\ 0, & \text{if } |X| > C. \end{cases}$$

**Problem 2.8.72.** Let $X$ be any random variable, let $\lambda > 0$ and let $\varphi(x) = \dfrac{x}{1 + x}$. Prove the following inequalities:

$$E\left[\varphi(|X|^\lambda) - \varphi(x^\lambda)\right] \leq P\{|X| \geq x\} \leq \frac{E\varphi(|X|^\lambda)}{\varphi(x^\lambda)}.$$

**Problem 2.8.73.** Let $\xi$ and $\eta$ be two independent random variables that have gamma-distribution with parameters, respectively, $(\alpha_1, \beta)$ and $(\alpha_2, \beta)$ (see [$\underline{P}$ §2.3, Table 3]). Prove that:

(a) The random variables $\xi + \eta$ and $\dfrac{\xi}{\xi + \eta}$ are independent.

(b) The random variable $\dfrac{\xi}{\xi + \eta}$ has beta distribution with parameters $(\alpha_1, \alpha_2)$ (see also [$\underline{P}$ §2.3, Table 3]).

**Problem 2.8.74.** (*Bernoulli Scheme with random probability for success.*) Suppose that the random variables $\xi_1, \ldots, \xi_n$ and $\pi$ are chosen so that $\pi$ is uniformly distributed in $(0, 1)$, $\xi_i$, $i = 1, \ldots, n$, take values 1 and 0 with conditional probabilities

$$P(\xi_i = 1 \mid \pi = p) = p, \qquad P(\xi_i = 0 \mid \pi = p) = 1 - p,$$

and, furthermore, are conditionally independent, in the sense that (in what follows $x_i$, stands for a number that is either 0 or 1, for $i = 1, \ldots, n$)

$$P(\xi_1 = x_1, \ldots, \xi_n = x_n \mid \pi) = P(\xi_1 = x_1 \mid \pi) \ldots P(\xi_n = x_n \mid \pi).$$

Prove that:
(a) One has the identity

$$P\{\xi_1 = x_1, \ldots, \xi_n = x_n\} = \frac{1}{(n + 1)C_n^x},$$

where $x = x_1 + \ldots + x_n$.

(b) The random variable $S_n = \xi_1 + \ldots + \xi_n$ is uniformly distributed in the (discrete) set $\{0, 1, \ldots, n\}$.

(c) The conditional distributions $P(\pi \le p \mid \xi_1 = x_1, \ldots, \xi_n = x_n)$ and $P(\pi \le p \mid S_n = x_1 + \ldots + x_n)$ coincide, for any $p \in (0, 1)$.

(d) The conditional distribution $P(\pi \le p \mid S_n = x)$, where $x = x_1 + \ldots + x_n$, has density

$$f_{\pi \mid S_n}(p \mid x) = (n + 1)C_n^x p^x (1 - p)^{n-x},$$

and one has $E(\pi \mid S_n = x) = \dfrac{x + 1}{n + 2}$.

**Problem 2.8.75.** Let $\xi$ and $\eta$ be two non-negative, independent and identically distributed random variables with $P\{\xi = 0\} < 1$, and suppose that $\min(\xi, \eta)$ and $\xi/2$ have the same distribution. Prove that $\xi$ and $\eta$ must be exponentially distributed.

*Hint.* Consider the relation

$$(P\{\xi > x\})^2 = P\{\min(\xi, \eta) > x\} = P\{\xi > 2x\},$$

and conclude that $(P\{\xi > x\})^{2n} = P\{\xi > 2nx\}$. Then conclude that for every $a > 0$ and for every non-negative rational $x$ one has $P\{\xi > x\} = e^{-\lambda x/a}$, where $\lambda = -\ln P\{\xi > a\}$. Finally, conclude that $P\{\xi > x\} = e^{-\lambda x/a}$, for all $x \ge 0$.

**Problem 2.8.76.** Let $\xi$ and $\eta$ be two independent and identically distributed random variables with distribution function $F = F(x)$. Prove that

$$P\{\xi = \eta\} = \sum_{x \in \mathbb{R}} |F(x) - F(x-)|^2.$$

(Comp. with Problem 2.12.20.)

**Problem 2.8.77.** Consider the random variables $X_1, \ldots, X_n$ and prove the following "inclusion–exclusion" formula (for the *maximum* of several random variables—comp. with Problems 1.1.12 and 1.4.9):

$$\max(X_1, \ldots, X_n) = \sum_{i=1}^n X_i - \sum_{1 \le i_1 < i_2 \le n} \min(X_{i_1}, X_{i_2})$$

$$+ \sum_{1 \le i_1 < i_2 < i_3 \le n} \min(X_{i_1}, X_{i_2}, X_{i_3}) + \ldots + (-1)^{n+1} \min(X_1, \ldots, X_n).$$

By choosing the random variables $X_1, \ldots, X_n$ accordingly, prove the "inclusion–exclusion" formula for the probability $P(A_1 \cup \ldots \cup A_n)$ (see again Problems 1.1.12 and 1.4.9).

**Problem 2.8.78.** Let $\xi_1, \xi_2, \ldots$ be any sequence of independent and identically distributed Bernoulli random variables with $P\{\xi_1 = 1\} = P\{\xi_1 = -1\} = 1/2$ and let $z_n(\omega) = \dfrac{1}{2} \sum_{k=1}^n \dfrac{\xi_k(\omega)}{3^k}$ and $z_\infty(\omega) = \lim_{n \to \infty} z_n(\omega)$. Prove that the distribution function $F(x) = P\{z_\infty(\omega) \le x\}$ is the Cantor function (see [P §2.3]).

In particular, this means that $\mathrm{Law}(z_\infty)$ refers to a probability distribution concentrated on the Cantor set. The random variables $z_\infty = z_\infty(\omega)$ is an example of what is known as *fractal* random variable (its distribution is neither discrete nor absolutely continuous—see Problem 2.3.18).

**Problem 2.8.79.** Prove that in the binomial case (see [P §1.2, 1] and [P §2.3, Table 2]) the distribution function

$$B_n(m; p) \equiv \sum_{k=0}^{m} C_n^k \, p^k q^{n-k}, \qquad 0 \le m \le n,$$

can be expressed in terms of the (incomplete) *beta-function*:

$$B_n(m; p) = \frac{1}{B(m+1, n-m)} \int_p^1 x^m (1-x)^{n-m-1} \, dx,$$

where

$$B(p, q) = \int_0^1 x^{p-1} (1-x)^{q-1} \, dx$$

$$\left( = \frac{\Gamma(p)\Gamma(q)}{\Gamma(p+q)}, \text{ with } \Gamma(p) = \int_0^\infty x^{p-1} e^{-x} \, dx \right).$$

**Problem 2.8.80.** Prove that the Poisson distribution function $F(m; \lambda) = \sum_{k=0}^{m} \frac{e^{-\lambda}\lambda^k}{k!}$, $m = 0, 1, 2, \ldots$, can be expressed in terms of the (incomplete) *gamma-function* as

$$F(m; \lambda) = \frac{1}{m!} \int_\lambda^\infty x^m e^{-x} \, dx.$$

**Problem 2.8.81.** In addition to the mean and the variance, another important characteristics of the shape of the density $f = f(x)$ are the *"skewness"* parameter, given by

$$\alpha_3 = \frac{\mu_3}{\sigma^3},$$

and the *"kurtosis"* or *"peakedness"* parameter, given by

$$\alpha_4 = \frac{\mu_4}{\sigma^4},$$

where $\mu_k = \int (x - \mu)^k f(x) \, dx$, $\mu = \int x f(x) \, dx$, and $\sigma^2 = \mu_2$.

What are the values of the parameters $\alpha_3$ and $\alpha_4$ for the distributions listed in [P §2.3, Table 3]?

**Problem 2.8.82.** Suppose that $X$ is some binomial random variable with parameters $n$ and $p$ (see Table 2 in [P §2.3, 1]). Analogously to Problem 2.8.81), define the "skewness" parameter

$$\mathrm{skw}(X) = \alpha_3 \quad \left( = \frac{\mathsf{E}(X - \mathsf{E}X)^3}{(\mathsf{D}X)^{3/2}} \right),$$

and prove that (with $q = 1 - p$)

$$\mathrm{skw}(X) = \frac{q - p}{\sqrt{npq}}.$$

(If $0 < p < 1/2$, then $\mathrm{skw}(X) > 0$, in which case one says that the distribution has "long right tail".) Find also the value of the "kurtosis" parameter $\mathrm{kur}(X) = \alpha_4$ ($= \frac{\mathsf{E}(X - \mathsf{E}X)^4}{(\mathsf{D}X)^2}$).

**Problem 2.8.83.** Suppose that $\xi_1, \ldots, \xi_n$ are independent and identically distributed random variables with "skewness" parameter $\alpha_3$ ($= \mathrm{skw}(\xi_1)$) and with "kurtosis" parameter $\alpha_4$ ($= \mathrm{kur}(\xi_1)$). Prove that

$$\mathrm{skw}(\xi_1 + \ldots + \xi_n) = n^{-1/2} \mathrm{skw}(\xi_1)$$

and

$$\mathrm{kur}(\xi_1 + \ldots + \xi_n) = 3 + n^{-1} \{\mathrm{kur}(\xi_1) - 3\}.$$

**Problem 2.8.84.** The well known *binomial distribution* arises as the distribution law of the total number of "successes," $\nu$, in $n$ independent trials, with probability for success in each individual trial $0 \le p \le 1$. More precisely, this distribution can be identified with the collection of probabilities $\mathsf{P}_n\{\nu = r\} = C_n^r p^r q^{n-r}$, $r = 0, \ldots, n$, for some fixed integer $n$ and fixed $0 \le p \le 1$. The *negative binomial* distribution $\mathsf{P}^r\{\tau = k\}$ (a.k.a. the *Pascal distribution*) arises as the probability distribution of the trial, $\tau$, during which $r$-"successes" are observed for *the first time*. Prove that, for any $r = 1, 2, \ldots$ and any $k \ge r$, one has

$$\mathsf{P}^r(\tau = k) = C_{k-1}^{r-1} p^r q^{k-r}, \quad k = r, r+1, \ldots,$$

where $p$ is the probability for success in a single trial. The negative binomial distribution can be identified with the collection of all probabilities $\mathsf{P}^r\{\tau = k\}$, $k = r, r+1, \ldots$, for fixed $r$. Given any fixed $r$, prove that $\mathsf{E}^r \tau = rq/p$, where $q = 1 - p$.

**Problem 2.8.85.** The (discrete) random variable $\xi$, with values in the set $\{1, 2, \ldots\}$, is said to have a *discrete Pareto law* with parameter $\rho > 0$, if

$$\mathsf{P}\{\xi = k\} = \frac{c}{k^{\rho+1}}.$$

Prove that

$$c = \frac{1}{\zeta(\rho + 1)} \quad \text{and} \quad \mathsf{E}\xi = \frac{\zeta(\rho)}{\zeta(\rho + 1)},$$

where $\zeta(s) = \sum_{n=1}^{\infty} \frac{1}{n^s}$ stands for Riemann's zeta function (for a description of the continuous Pareto law see Problem 3.6.23).

**Problem 2.8.86.** Let $\xi_1, \ldots, \xi_n$ be independent and identically distributed random variables with distribution function $F(x \mid \theta)$, which depends on some (random) parameter $\theta$, with prior distribution $\Pi(\theta)$, known to be in some class $\mathcal{K}$. Let $\Pi(\theta \mid x_1, \ldots, x_n)$ be the posterior distribution, calculated from the Bayes formula, where $x_1, \ldots, x_n$ are the observed values of $\xi_1, \ldots, \xi_n$. If the posterior distribution also belongs to the class $\mathcal{K}$, we say that the distribution $\Pi(\theta)$ is the $\mathcal{K}$-conjugate of the distribution $F(x \mid \theta)$.

Prove that:

(a) If $F(\cdot \mid \theta) \sim \mathcal{N}(\theta, a_0^{-1})$ and $\Pi(\theta) \sim \mathcal{N}(m_0, b_0^{-1})$, then

$$\Pi(\theta \mid x_1, \ldots, x_n) \sim \mathcal{N}\left(\frac{b_0 m_0 + a_0(x_1 + \ldots + x_n)}{b_0 + n a_0}, \frac{1}{b_0 + n a_0}\right).$$

(b) If $F(\cdot \mid \theta) \sim \mathcal{N}(0, \theta^{-1})$ and $\Pi(\theta) \sim \Gamma(k; \lambda) \equiv$ gamma-distribution with density

$$\gamma_{(k;\lambda)}(x) = \frac{\lambda^k x^{k-1} e^{-\lambda x}}{\Gamma(k)} I_{[0,\infty)}(x),$$

where $k > 0$ and $\lambda > 0$, then

$$\Pi(\theta \mid x_1, \ldots, x_n) \sim \Gamma\left(k + \frac{1}{2}n; \lambda + \frac{1}{2}(x_1^2 + \ldots + x_n^2)\right).$$

(c) If $F(\cdot \mid \theta) \sim \exp(\theta) \equiv$ exponential distribution with parameter $\theta$, and $\Pi(\theta) \sim \Gamma(k; \lambda)$, then

$$\Pi(\theta \mid x_1, \ldots, x_n) \sim \Gamma(k + n; \lambda + (x_1 + \ldots + x_n)).$$

(d) If $F(\cdot \mid \theta) \sim \text{Poisson}(\theta)$ and $\Pi(\theta) \sim \Gamma(k; \lambda)$, then

$$\Pi(\theta \mid x_1, \ldots, x_n) \sim \Gamma(k + (x_1 + \ldots + x_n); \lambda + n).$$

(e) If $F(\cdot \mid \theta) \sim \text{Bernoulli}(\theta)$ and $\Pi(\theta) \sim B(k; L) \equiv$ beta-distribution with density

$$\beta_{(k;L)}(x) = \frac{x^{k-1}(1 - x)^{L-1}}{B(k; L)} I_{(0,1)}(x),$$

then

$$\Pi(\theta \mid x_1, \ldots, x_n) \sim B(k + (x_1 + \ldots + x_n); L + n - (x_1 + \ldots + x_n)).$$

**Problem 2.8.87.** Suppose that $X$ is a random variable with one of the following distributions: binomial, Poisson, geometric, negative-binomial, or Pareto. Find the probability of the event $\{X$ is even$\}$. If, for example, $X$ has geometric distribution with parameter $p$ (see [P §2.3, Table 2]), then $P\{X$ is even$\} = (1 - p)/(2 - p)$.

**Problem 2.8.88.** (*Exponentially distributed random variables.*) Let $\xi_1, \ldots, \xi_n$ be independent and exponentially distributed random variables with parameters, respectively, $\lambda_1, \ldots, \lambda_n$.

(a) Prove that $P\{\xi_1 < \xi_2\} = \lambda_1/(\lambda_1 + \lambda_2)$.

(b) Prove that $\min_{1 \le k \le n} \xi_k$ has exponential distribution with parameter $\lambda = \sum_{k=1}^{n} \lambda_k$ and conclude from part (a) that

$$P\left\{\xi_j = \min_{1 \le k \le n} \xi_k\right\} = \lambda_j \Big/ \sum_{k=1}^{n} \lambda_k.$$

(c) Assuming that $\lambda_i \ne \lambda_j$, $i \ne j$, find the density of the random variable $\xi_1 + \cdots + \xi_n$ (for the case $n = 2$, see Problem 2.8.26).

(d) Prove that $E \min(\xi_1, \xi_2) = 1/(\lambda_1 + \lambda_2)$ and find $E \max(\xi_1, \xi_2)$.

(e) Find the distribution density of the random variable $\xi_1 - \xi_2$.

(f) Prove that the random variables $\min(\xi_1, \xi_2)$ and $\xi_1 - \xi_2$ are independent.

## 2.9 Construction of Stochastic Processes with a Given System of Finite-Dimensional Distributions

**Problem 2.9.1.** Let $\Omega = [0, 1]$, let $\mathscr{F}$ be the class of Borel sets in $[0, 1]$, and let P stand for the Lebesgue measure on $[0, 1]$. Prove that $(\Omega, \mathscr{F}, P)$ is a *universal probability space*, in the sense that, given *any* distribution functions $F(x)$, one can construct on $(\Omega, \mathscr{F}, P)$ a random variable $\xi = \xi(\omega)$, $\omega \in \Omega$, whose distribution function, $F_\xi(x) = P\{\xi \le x\}$, coincides with $F(x)$.

*Hint.* Set $\xi(\omega) = F^{-1}(\omega)$, where $F^{-1}(\omega) = \sup\{x: F(x) < \omega\}$, $0 < \omega < 1$, ($\xi(0)$ and $\xi(1)$ may be chosen arbitrarily).

**Problem 2.9.2.** Verify the consistency of the families of probability distributions described in the corollaries to [P §2.9, Theorems 1 and 2].

**Problem 2.9.3.** Prove that Corollary 2 to [P §2.9, Theorem 2] can be derived from [P §2.9, Theorem 1].

*Hint.* Show that the measures defined in [P §2.9, (16)] form a consistent family of (finite-dimensional) distributions.

**Problem 2.9.4.** Consider the random variables $T_n$, $n \ge 1$, from [P §2.9, 4] and let $F_n$, $n \ge 1$, denote their respective distribution functions. Prove that $F_{n+1}(t) = \int_0^t F_n(t - s) \, dF(s)$, $n \ge 1$, where $F_1 = F$.

**Problem 2.9.5.** Prove that $P\{N_t = n\} = F_n(t) - F_{n+1}(t)$ (see [$\underline{P}$ §2.9, (17)]).

**Problem 2.9.6.** Prove that the renewal function $m(t)$ from [$\underline{P}$ §2.9, **4**] satisfies what is known as the *recovery equation*:

$$m(t) = F(t) + \int_0^t m(t - x)\, dF(x). \qquad (*)$$

**Problem 2.9.7.** Prove that the function defined by formula [$\underline{P}$ §2.9, (20)] is the only solution to equation $(*)$, within the class of functions that are bounded on every finite interval.

**Problem 2.9.8.** Let $T$ be an arbitrary set.

(a) Suppose that for every $t \in T$ there is a probability space $(\Omega_t, \mathscr{F}_t, P_t)$, and let $\Omega = \prod_{t \in T} \Omega_t$ and $\mathscr{F} = \otimes_{t \in T} \mathscr{F}_t$. Prove, that there is a unique probability measure $P$, defined on the $(\Omega, \mathscr{F})$, for which the following independence property holds:

$$P\left(\prod_{t \in T} B_t\right) = \prod_{t \in T} P(B_t),$$

where $B_t \in \mathscr{F}_t, t \in T$, and $B_t = \Omega_t$ for all but finitely many indices $t \in T$.

*Hint.* Define $P$ on the some appropriate algebra and use the argument of the proof of the Ionescu-Tulcea Theorem.

(b) Suppose that for every $t \in T$ there is a measurable space $(E_t, \mathscr{E}_t)$ and a probability measure $P_t$ defined on that space. Prove the following result, which is due to Łomnicki and Ulam: there is a probability space $(\Omega, \mathscr{F}, P)$ and *independent* random elements $(X_t)_{t \in T}$, such that each $X_t$ is $\mathscr{F}/\mathscr{E}_t$-measurable and $P\{X_t \in B\} = P_t(B), B \in \mathscr{E}_t$.

## 2.10   Various Types of Convergence of Sequences of Random Variables

**Problem 2.10.1.** By using [$\underline{P}$ §2.10, Theorem 5] prove that in [$\underline{P}$ §2.6, Theorems 3 and 4] one can replace "convergence almost surely" with "convergence in probability".

*Hint.* If $\xi_n \overset{P}{\to} \xi$, $|\xi_n| \leq \eta$, $E\eta < \infty$, and $E|\xi_n - \xi| \not\to 0$, then one can find some $\varepsilon > 0$ and a sub-sequence $(n_k)_{k \geq 1}$, such that $E|\xi_{n_k} - \xi| > \varepsilon$ and $\xi_{n_k} \overset{P}{\to} \xi$. Furthermore, according to [$\underline{P}$ §2.10, Theorem 5], one can find such a sub-sequence $(k_l)_{l \geq 1}$, that $\xi_{n_{k_l}} \not\to \xi$ (P-a. e.). The next step is to use [$\underline{P}$ §2.6, Theorem 3] to find a contradiction to the assumption $E|\xi_n - \xi| \not\to 0$.

**Problem 2.10.2.** Prove that the space $L^\infty$ is complete.

*Hint.* Take a sequence $(\xi_k)_{k\geq 1}$, which is fundamental in $L^\infty$, in the sense that $\|\xi_m - \xi_n\|_{L^\infty} \leq a_n$, for $n \leq m$, with $a_n \to 0$ as $n \to \infty$, and set

$$\xi(\omega) = \begin{cases} \overline{\lim}_n \, \xi_n(\omega), & \text{if } \overline{\lim}_n \, \xi_n(\omega) < \infty, \\ 0, & \text{if } \overline{\lim}_n \, \xi_n(\omega) = \infty. \end{cases}$$

Prove that, as defined above, $\xi(\omega)$ is a well defined random variable and, furthermore, $\|\xi - \xi_n\|_{L^\infty} \leq a_n \to 0$ as $n \to \infty$.

**Problem 2.10.3.** Prove that if $\xi_n \xrightarrow{P} \xi$ and, at the same time, $\xi_n \xrightarrow{P} \eta$, then $\xi$ and $\eta$ are *equivalent*, in the sense that $P\{\xi \neq \eta\} = 0$.

**Problem 2.10.4.** Let $\xi_n \xrightarrow{P} \xi$ and $\eta_n \xrightarrow{P} \eta$ and suppose that the random variables $\xi$ and $\eta$ are equivalent. Prove that for any $\varepsilon > 0$ one has

$$P\{|\xi_n - \eta_n| \geq \varepsilon\} \to 0 \text{ as } n \to \infty.$$

**Problem 2.10.5.** Let $\xi_n \xrightarrow{P} \xi$ and $\eta_n \xrightarrow{P} \eta$. Prove that if $\varphi = \varphi(x, y)$ is some continuous function, then $\varphi(\xi_n, \eta_n) \xrightarrow{P} \varphi(\xi, \eta)$. (*Slutsky's lemma.*)

*Hint.* Given some $\varepsilon > 0$, choose $c > 0$ so that

$$P\{|\xi_n| > c\} < \varepsilon, \quad P\{|\eta_n| > c\} < \varepsilon, \quad n \geq 1,$$
$$P\{|\xi| > c\} < \varepsilon, \quad P\{|\eta| > c\} < \varepsilon.$$

As the function $\varphi = \varphi(x, y)$ is continuous, it must be uniformly continuous on the compact $[-c, c] \times [-c, c]$. Therefore one can find some $\delta > 0$ so that for any $x, y \in [-c, c]$ with $\rho(x, y) < \delta$, one has $|\varphi(x) - \varphi(y)| \leq \varepsilon$ $(\rho(x, y) = \max(|x^1 - y^1|, |x^2 - y^2|), x = (x^1, x^2), y = (y^1, y^2))$. Finally, consider the estimate

$$P\{|\varphi(\xi_n, \eta_n) - \varphi(\xi, \eta)| > \varepsilon\} \leq P\{|\xi_n| > c\} + P\{|\eta_n| > c\}$$
$$+ P\{|\xi| > c\} + P\{|\eta| > c\} + P\{|\xi_n - \xi| > \delta\} + P\{|\eta_n - \eta| > \delta\},$$

and prove that for large $n$ the right side does not exceed $6\varepsilon$.

**Problem 2.10.6.** Let $(\xi_n - \xi)^2 \xrightarrow{P} 0$. Prove that $\xi_n^2 \xrightarrow{P} \xi^2$.

**Problem 2.10.7.** Prove that if $\xi_n \xrightarrow{d} C$, where $C$ is some constant, then the convergence must hold also in probability; in other words

$$\xi_n \xrightarrow{d} C \implies \xi_n \xrightarrow{P} C.$$

*Hint.* For a given $\varepsilon > 0$, consider the function $f_\varepsilon(x) = (1 - \frac{|x-c|}{\varepsilon})^+$ and notice that

$$P\{|\xi_n - c| \le \varepsilon\} \ge E f_\varepsilon(\xi_n) \to E f_\varepsilon(c) = 1.$$

**Problem 2.10.8.** Let the sequence $(\xi_n)_{n\ge 1}$ be such that $\sum_{n=1}^{\infty} E|\xi_n|^p < \infty$, for some $p > 0$. Prove that $\xi_n \to 0$ (P-a. e.).
   *Hint.* Use Chebyshev's inequality and the Borel-Cantelli lemma.

**Problem 2.10.9.** Let $(\xi_n)_{n\ge 1}$ be a sequence of identically distributed random variables. Prove the following implications:

$$E|\xi_1| < \infty \iff \sum_{n=1}^{\infty} P\{|\xi_1| > \varepsilon n\} < \infty, \ \varepsilon > 0 \iff$$

$$\iff \sum_{n=1}^{\infty} P\left\{\left|\frac{\xi_n}{n}\right| > \varepsilon\right\} < \infty, \ \varepsilon > 0 \implies \frac{\xi_n}{n} \to 0 \quad \text{(P-a. e.)}.$$

*Hint.* Use the following easy to verify inequalities:

$$\varepsilon \sum_{n=1}^{\infty} P\{|\xi_1| > \varepsilon n\} \le E|\xi_1| \le \varepsilon + \varepsilon \sum_{n=1}^{\infty} P\{|\xi_1| > \varepsilon n\}.$$

**Problem 2.10.10.** Let $(\xi_n)_{n\ge 1}$ be some sequence of random variables and let $\xi$ be a random variable.
   (a) Prove that if $P\{|\xi_n - \xi| \ge \varepsilon \text{ i. o.}\} = 0$ for every $\varepsilon > 0$, then $\xi_n \to \xi$ (P-a. e.).
   (b) Prove that if one can find a sub-sequence $(n_k)$, such that $\xi_{n_k} \to \xi$ (P-a. e.) and $\max_{n_{k-1} < l \le n_k} |\xi_l - \xi_{n_{k-1}}| \to 0$ (P-a. e.) for $k \to \infty$, then $\xi_n \to \xi$ (P-a. e.).
   (c) Prove that if $\xi_n \to \xi$ (P-a. e.), then $P\{|\xi_n - \xi| \ge \varepsilon \text{ i. o.}\} = 0$, for every $\varepsilon > 0$. (This is the converse of property (a).)

**Problem 2.10.11.** Define the distance, $d(\xi, \eta)$, between two random variables, $\xi$ and $\eta$, as

$$d(\xi, \eta) = E\frac{|\xi - \eta|}{1 + |\xi - \eta|},$$

and prove that the function $d = d(\cdot, \cdot)$ defines a metric in the space of all equivalence classes of random variables (on a given probability space) for the relation "identity almost everywhere." Prove that convergence in probability is equivalent to convergence in the metric $d(\cdot, \cdot)$.
   *Hint.* Check the triangle inequality and convince yourself that

$$\frac{\varepsilon}{1 + \varepsilon} P\{|\xi_n - \xi| > \varepsilon\} \le E\frac{|\xi_n - \xi|}{1 + |\xi_n - \xi|} \le \varepsilon P\{|\xi_n - \xi| \le \varepsilon\} + P\{|\xi_n - \xi| > \varepsilon\},$$

for every $\varepsilon > 0$.

**Problem 2.10.12.** Prove that the topology of convergence almost surely is not metrizable.

*Hint.* Suppose that there is metric, $\rho$, which defines convergence almost surely, and consider some sequence $(\xi_n)_{n \geq 1}$, chosen so that $\xi_n \xrightarrow{\text{P}} 0$, but $\xi_n \nrightarrow 0$ (P-a. e.). Then, for some $\varepsilon > 0$, one can find a sub-sequence $(n_k)_{k \geq 1}$ so that $\rho(\xi_{n_k}, 0) > \varepsilon$ and, at the same time, $\xi_{n_k} \xrightarrow{\text{P}} 0$. Finally, by using [P §2.10, Theorem 5] one can find a contradiction to the claim that convergence in the metric $\rho$ is the same as convergence almost surely.

**Problem 2.10.13.** Prove that if $X_1 \leq X_2 \leq \ldots$ and $X_n \xrightarrow{\text{P}} X$, then one also has $X_n \to X$ (P-a. e.).

**Problem 2.10.14.** Let $(X_n)_{n \geq 1}$ be a sequence random variables. Prove that:

(a) $X_n \to 0$ (P–a. e.) $\Longrightarrow S_n/n \to 0$ (P–a. e.), where, as usual, $S_n = X_1 + \cdots + X_n$.

(b) $X_n \xrightarrow{L^p} 0 \Longrightarrow S_n/n \xrightarrow{L^p} 0$, if $p \geq 1$ and, in general,

$$X_n \xrightarrow{L^p} 0 \nRightarrow \frac{S_n}{n} \xrightarrow{L^p} 0.$$

(c) In general, $X_n \xrightarrow{\text{P}} 0$ does not imply the convergence $S_n/n \xrightarrow{\text{P}} 0$ (comp. with the last statement in Problem 2.10.34).

(d) $S_n/n \to 0$ (P-a. e.) if and only if $S_n/n \xrightarrow{\text{P}} 0$ and $S_{2^n}/2^n \to 0$ (P-a. e.).

**Problem 2.10.15.** Let $(\Omega, \mathscr{F}, \text{P})$ be a probability space, on which one has the convergence $X_n \xrightarrow{\text{P}} X$. Prove that if P is an atomic measure, then $X_n \to X$ also with probability 1 (for the definition of atomic measure, see Problem 2.3.35).

**Problem 2.10.16.** According to part (a) in the Borel–Cantelli lemma (the "first Borel–Cantelli lemma"), if $\sum_{n=1}^{\infty} \text{P}(|\xi_n| > \varepsilon) < \infty$ for some $\varepsilon > 0$, then $\xi_n \to 0$ (P-a. e.). Give an example of a sequence $\{\xi_n\}$ for which $\xi_n \to 0$ (P-a. e.), and yet $\sum_{n=1}^{\infty} \text{P}(|\xi_n| > \varepsilon) = \infty$, for some $\varepsilon > 0$.

**Problem 2.10.17.** (*On part (b) in the Borel–Cantelli lemma; i.e., on the "second Borel–Cantelli lemma."*) Let $\Omega = (0, 1)$, $\mathscr{B} = \mathscr{B}((0, 1))$, and let P stand for the Lebesgue measure. Consider the events $A_n = (0, 1/n)$ and prove that $\sum \text{P}(A_n) = \infty$, even though every $\omega \in (0, 1)$ can belong only to *finitely many* sets $A_1, \ldots, A_{[1/\omega]}$, i.e., $\text{P}\{A_n \text{ i. o.}\} = 0$.

**Problem 2.10.18.** Prove that in the second Borel-Cantelli lemma, instead of requiring that the events $A_1, A_2, \ldots$ are independent, it is enough to require only that these events are *pair-wise independent*, in that $\text{P}(A_i \cap A_j) - \text{P}(A_i)\text{P}(A_j) = 0, i \neq j$; in fact, it is enough to require only that $A_1, A_2, \ldots$ are *pair-wise negatively correlated*, in that $\text{P}(A_i \cap A_j) - \text{P}(A_i)\text{P}(A_j) \leq 0, i \neq j$.

**Problem 2.10.19.** (*On the second Borel–Cantelli lemma.*) Prove the following variants of the second Borel–Cantelli lemma: given an arbitrary sequence of (not necessarily independent) events $A_1, A_2, \ldots$, one can claim that:

(a) If

$$\sum_{n=1}^{\infty} P(A_n) = \infty \quad \text{and} \quad \liminf_n \frac{\sum_{i,k=1}^{n} P(A_i A_k)}{\left[\sum_{k=1}^{n} P(A_k)\right]^2} = 1,$$

then (Erdös and Rényi [37]) $P(A_n \text{ i.o.}) = 1$.

(b) If

$$\sum_{n=1}^{\infty} P(A_n) = \infty \quad \text{and} \quad \liminf_n \frac{\sum_{i,k=1}^{n} P(A_i A_k)}{\left[\sum_{k=1}^{n} P(A_k)\right]^2} = L,$$

then (Kochen and Stone [64], Spitser [125]) $L \geq 1$ and $P(A_n \text{ i.o.}) = 1/L$.

(c) If

$$\sum_{n=1}^{\infty} P(A_n) = \infty \quad \text{and} \quad \liminf_n \frac{\sum_{1 \leq i < k \leq n} [P(A_i A_k) - P(A_i)P(A_k)]}{\left[\sum_{k=1}^{n} P(A_k)\right]^2} \leq 0,$$

then (Ortega and Wschebor [92]) $P(A_n \text{ i.o.}) = 1$.

(d) If $\sum_{n=1}^{\infty} P(A_n) = \infty$ and

$$\alpha_H = \liminf_n \frac{\sum_{1 \leq i < k \leq n} [P(A_i A_k) - H P(A_i)P(A_k)]}{\left[\sum_{k=1}^{n} P(A_k)\right]^2},$$

where $H$ is an arbitrary constant, then (Petrov [95]) $P(A_n \text{ i.o.}) \geq \frac{1}{H + 2\alpha_H}$ and $H + 2\alpha_H \geq 1$.

**Problem 2.10.20.** Let $A_1, A_2, \ldots$ be some sequence of independent events and suppose that $\sum_{n=1}^{\infty} P(A_n) < \infty$. Prove that for $S_n = \sum_{k=1}^{n} I(A_k)$ the following stronger version of the "second Borel-Cantelli lemma" is in force:

$$\lim_n \frac{S_n}{E S_n} = 1 \quad (\text{P-a. e.}).$$

**Problem 2.10.21.** Let $(X_n)_{n \geq 1}$ and $(Y_n)_{n \geq 1}$ be any two sequences of random variables with identical finite-dimensional distributions, i.e., $F_{X_1, \ldots, X_n} = F_{Y_1, \ldots, Y_n}$, $n \geq 1$, and suppose that $X_n \xrightarrow{P} X$. Prove that there is a random variable $Y$, whose distribution is identical to the distribution of $X$ (notation: $\text{Law}(X) = \text{Law}(Y)$, or $X \stackrel{\text{law}}{=} Y$, or $X \stackrel{d}{=} Y$), for which one can claim that $Y_n \xrightarrow{P} Y$.

**Problem 2.10.22.** Let $(X_n)_{n \geq 1}$ be a sequence independent random variables with $X_n \xrightarrow{P} X$, for some random variable $X$. Prove that $X$ must be a *degenerate* random variable.

**Problem 2.10.23.** Prove that for every sequence of random variables, $\xi_1, \xi_2, \ldots$, it is possible to find a sequence of *constants*, $a_1, a_2, \ldots$, so that $\xi_n / a_n \to 0$ (P-a. e.).

**Problem 2.10.24.** Let $\xi_1, \xi_2, \ldots$ be a sequence random variables and let $S_n = \xi_1 + \cdots + \xi_n$, $n \geq 1$. Prove that the set $\{S_n \to \}$, i.e., the set of all $\omega \in \Omega$, for which the series $\sum_{k \geq 1} \xi_k(\omega)$ converges, can be represented in the form:

$$\{S_n \to \} = \bigcap_{N \geq 1} \bigcup_{m \geq 1} \bigcap_{k \geq m} \left\{ \sup_{l \geq k} |S_l - S_k| \leq N^{-1} \right\}.$$

Similarly, the set $\{S_n \nrightarrow \}$, on which the series $\sum_{k \geq 1} \xi_k(\omega)$ diverges, can be represented in the form

$$\{S_n \nrightarrow \} = \bigcup_{N \geq 1} \bigcap_{m \geq 1} \bigcup_{k \geq m} \left\{ \sup_{l \geq k} |S_l - S_k| > N^{-1} \right\}.$$

**Problem 2.10.25.** Consider the probability space $(\Omega, \mathscr{F}, \mathsf{P})$, in which the sample space $\Omega$ is *at most countable*, and prove that if $\xi_n \overset{\mathsf{P}}{\to} \xi$, then $\xi_n \to \xi$ (P-a. e.).

**Problem 2.10.26.** Give an example of a sequence of random variables, such that with probability 1 one has $\limsup \xi_n = \infty$, $\liminf \xi_n = -\infty$, but, nevertheless, one can find a random variable $\eta$ with $\xi_n \overset{\mathsf{P}}{\to} \eta$.

**Problem 2.10.27.** Prove the following version of the *the 0–1 law* (comp. with the 0–1 law of [ $\underline{\text{P}}$ §4.1]): if the events $A_1, A_2, \ldots$ are pairwise independent, then

$$\mathsf{P}\{A_n \text{ i.o.}\} = \begin{cases} 0, & \text{if } \sum \mathsf{P}(A_n) < \infty, \\ 1, & \text{if } \sum \mathsf{P}(A_n) = \infty. \end{cases}$$

**Problem 2.10.28.** Let $A_1, A_2, \ldots$ be an arbitrary sequence of events, such that $\lim_n \mathsf{P}(A_n) = 0$ and $\sum_n \mathsf{P}(A_n \cap \overline{A}_{n+1}) < \infty$. Prove that $\mathsf{P}\{A_n \text{ i.o.}\} = 0$.

**Problem 2.10.29.** Prove, that if $\sum_n \mathsf{P}\{|\xi_n| > n\} < \infty$, then $\limsup_n (|\xi_n|/n) \leq 1$ (P-a. e.).

**Problem 2.10.30.** Suppose that $\xi_n \downarrow \xi$ (P-a. e.), $\mathsf{E}|\xi_n| < \infty$, $n \geq 1$, and $\inf_n \mathsf{E}\xi_n > -\infty$. Prove that $\xi_n \overset{L^1}{\longrightarrow} \xi$, i.e., $\mathsf{E}|\xi_n - \xi| \to 0$.

**Problem 2.10.31.** In conjunction with the second Borel-Cantelli lemma, prove that $\mathsf{P}\{A_n \text{ i.o.}\} = 1$ if and only if $\sum_n \mathsf{P}(A \cap A_n) = \infty$, for every set $A$ with $\mathsf{P}(A) > 0$.

**Problem 2.10.32.** Suppose that the events $A_1, A_2, \ldots$ are independent and chosen so that $\mathsf{P}(A_n) < 1$, for all $n \geq 1$. Prove that $\mathsf{P}\{A_n \text{ i.o.}\} = 1$ if and only if $\mathsf{P}(\bigcup A_n) = 1$.

**Problem 2.10.33.** Let $X_1, X_2, \ldots$ be any sequence of independent random variables with $P\{X_n = 0\} = 1/n$ and $P\{X_n = 1\} = 1 - 1/n$. Set $E_n = \{X_n = 0\}$. By using the properties $\sum_{n=1}^{\infty} P(E_n) = \infty$, $\sum_{n=1}^{\infty} P(\overline{E}_n) = \infty$, conclude that $\lim_n X_n$ does not exist (P-a. e.).

**Problem 2.10.34.** Let $X_1, X_2, \ldots$ be any sequence of random variables. Prove that $X_n \xrightarrow{P} 0$ if and only if

$$E\frac{|X_n|^r}{1 + |X_n|^r} \to 0, \quad \text{for some } r > 0.$$

In particular, if $S_n = X_1 + \cdots + X_n$, then

$$\frac{S_n - ES_n}{n} \xrightarrow{P} 0 \iff E\frac{(S_n - ES_n)^2}{n^2 + (S_n - ES_n)^2} \to 0.$$

Show also that, given any sequence of random variables $X_1, X_2, \ldots$, one can claim that

$$\max_{1 \leq k \leq n} |X_k| \xrightarrow{P} 0 \implies \frac{S_n}{n} \xrightarrow{P} 0.$$

**Problem 2.10.35.** Let $X_1, X_2, \ldots$ be any sequence of independent and identically distributed Bernoulli random variables with $P\{X_k = \pm 1\} = 1/2$. Setting $U_n = \sum_{k=1}^{n} \frac{X_k}{2^k}$, $n \geq 1$, prove that $U_n \to U$ (P-a. e.), where $U$ is some random variable, which is distributed uniformly on $[-1, +1]$.

**Problem 2.10.36.** (*Egoroff's Theorem.*) Let $(\Omega, \mathscr{F}, \mu)$ be any measurable space, endowed with a finite measure $\mu$, and let $f_1, f_2, \ldots$ be some sequence of Borel functions, which converges in measure $\mu$ to the Borel function $f$, i.e., $f_n \xrightarrow{\mu} f$. *Egoroff's Theorem* states that for every *given* $\varepsilon > 0$ it is possible to find a set $A_\varepsilon \in \mathscr{F}$, with $\mu(A_\varepsilon) < \varepsilon$, such that $f_n(\omega) \to f(\omega)$ uniformly for all $\omega \in \overline{A}_\varepsilon$, where $\overline{A}_\varepsilon = \Omega \setminus A_\varepsilon$ is the complement of $A_\varepsilon$. Prove this statement.

**Problem 2.10.37.** (*Luzin's theorem.*) Let $(\Omega, \mathscr{F}, P) = ([a, b], \mathscr{F}, \lambda)$, where $\lambda$ stands for the Lebesgue measure on $[a, b]$ and $\mathscr{F}$ is the collection of all Lebesgue sets. Let $f = f(x)$ be any finite $\mathscr{F}$-measurable function. Prove *Luzin's Theorem*: for every given $\varepsilon > 0$ one can find a *continuous* function $f_\varepsilon = f_\varepsilon(x)$, such that

$$P\{x \in [a, b] : f(x) \neq f_\varepsilon(x)\} < \varepsilon.$$

**Problem 2.10.38.** The statement of Egoroff's Theorem leads naturally to the notion of *almost uniform convergence*. We say that the sequence of functions $f_1, f_2, \ldots$ converges *almost uniformly* to the function $f$, if, for *every* $\varepsilon > 0$, it is possible to find a set $A_\varepsilon \in \mathscr{F}$ with $\mu(A_\varepsilon) < \varepsilon$, so that $f_n(\omega) \to f(\omega)$ uniformly for all $\omega \in \overline{A}_\varepsilon$ (notation: $f_n \rightrightarrows f$).

Prove that the almost uniform convergence $f_n \Rightarrow f$ implies both convergence in measure $(f_n \overset{\mu}{\to} f)$ and convergence almost surely $(f_n \overset{\mu\text{-a.e.}}{\to} f)$.

**Problem 2.10.39.** Let $X_1, X_2, \ldots$ be a sequence random variables and let $\{X_n \not\to\}$ denote the set of those $\omega \in \Omega$ for which $X_n(\omega)$ does not converge as $n \to \infty$. Prove that

$$\{X_n \not\to\} = \bigcup_{p<q} \{\lim\inf X_n \le p < q \le \lim\sup X_n\},$$

where the union is taken over all pairs of rational numbers, $(p, q)$, with $p < q$.

**Problem 2.10.40.** Let $X_1, X_2, \ldots$ be any sequence of random variables defined on some complete probability space, which converges with probability 1 to the random variable $X$. Show that the $\sigma$-algebras $\sigma(X_1, X_2, \ldots)$ and $\sigma(X_1, X_2, \ldots, X)$, generated, respectively, by the random elements $(X_1, X_2, \ldots)$ and $(X_1, X_2, \ldots, X)$ (see [P § 2.5]) coincide.

**Problem 2.10.41.** Let $X_1, X_2, \ldots$ be any sequence of independent and identically distributed random variables, such that their distribution function $F = F(x)$ satisfies the condition

$$\lim_{x \to \infty} x^2 [1 - F(x)] = 0.$$

Prove that

$$\frac{1}{\sqrt{n}} \max_{1 \le i \le n} X_i \overset{P}{\to} 0 \qquad \text{as} \quad n \to \infty.$$

**Problem 2.10.42.** Let $\xi_1, \xi_2, \ldots$ be any sequence of independent and identically distributed random variables with $E\xi_1 = \mu$, $D\xi_1 = \sigma^2 < \infty$ and $P\{\xi_1 = 0\} = 0$. Prove that

$$\frac{\sum_{k=1}^{n} \xi_k}{\sum_{k=1}^{n} \xi_k^2} \overset{P}{\to} \frac{\mu}{\mu^2 + \sigma^2} \qquad \text{as} \quad n \to \infty.$$

**Problem 2.10.43.** Suppose that $\xi_n \overset{P}{\to} \xi$, $\eta_n \overset{P}{\to} \eta$ and $P\{\xi_n \le \eta_n\} = 1$, $n \ge 1$. Prove that $P\{\xi \le \eta\} = 1$.

**Problem 2.10.44.** Let $\xi_1, \xi_2, \ldots$ be any sequence of non-negative random variables and suppose that the $\sigma$-algebras $\mathscr{F}_1, \mathscr{F}_2, \ldots$ are such that $E(\xi_n \mid \mathscr{F}_n) \overset{P}{\to} 0$. Prove that $\xi_n \overset{P}{\to} 0$.

**Problem 2.10.45.** Let $\xi_n \overset{d}{\to} \xi$ and $c_n \xi_n \overset{d}{\to} \xi$ ("$\overset{d}{\to}$" means convergence in distribution), where $\xi$ is some non-degenerate random variable and $c_n > 0$. Prove that $c_n \to 1$.

**Problem 2.10.46.** Let $A_1, A_2, \ldots$ be any sequence of random events. Setting $A = \overline{\lim}_n A_n$, prove that, if $\sum_{n=1}^{\infty} P(A_n) = \infty$, then the following relation, known as the *Kochen–Stone inequality* (see [64]), must hold:

$$P(A) \ge \overline{\lim_n} \frac{\left(\sum_{m=1}^{n} P(A_m)\right)^2}{\sum_{k=1}^{n} \sum_{l=1}^{n} P(A_k \cap A_l)}.$$

**Problem 2.10.47.** Let $\xi_1, \xi_2, \ldots$ be any sequence of independent and identically distributed random variables with $E|\xi_1| < \infty$. Given some positive constant, $\alpha$, set $A_n = \{|\xi_n| > \alpha n\}$, $n \geq 1$, and prove that $P(\overline{\lim} A_n) = 0$.

**Problem 2.10.48.** Prove that in the space of continuous functions, $C$, there is no metric $\rho$ for which the convergence $\rho(f_n, f) \to 0$ is equivalent to the point-wise the convergence $f_n \to f$ (comp. with Problem 2.10.12).

**Problem 2.10.49.** Assuming that $c > 0$ is an arbitrary constant, give an example of a sequence of random variables $\xi, \xi_1, \xi_2, \ldots$, such that $E\xi_n = -c$ for all $n \geq 1$, $\xi_n(\omega) \to \xi(\omega)$ in point-wise sense, and yet $E\xi = c$.

**Problem 2.10.50.** For each of the three definitions in Problem 1.4.23 find the median $\mu_n = \mu(\xi_n)$ of the random variables $\xi_n = I_{\{\eta \leq (-1)^n - 1/n\}}$, where $\eta$ is a standard Gaussian random variable.

**Problem 2.10.51.** Let $\mu_n = \mu(\xi_n)$ be the uniquely defined medians (see Problem 1.4.23) of the random variables $\xi_n$, $n \geq 1$, which converge almost surely to a random variable $\xi$. Give an example showing that, in general, $\lim_n \mu(\xi_n)$ may not exist.

**Problem 2.10.52.** Let $\xi_1, \xi_2, \ldots$ be any sequence of independent, non-negative and identically distributed non-degenerate random variables with $E\xi_1 = 1$. Setting $T_n = \prod_{k=1}^{n} \xi_k$, $n \geq 1$, prove that $T_n \to 0$ (P-a. e.).

**Problem 2.10.53.** Let $\xi_1, \xi_2, \ldots$ be any sequence of independent and identically distributed random variables and let $S_n = \xi_1 + \cdots + \xi_n$, $n \geq 1$. Prove that:

(a)   $E\xi_1^+ = \infty$ and $E\xi_1^- = \infty \implies \dfrac{S_n}{n} \to +\infty$ (P-a. e.);

(b)   $E|\xi_1| < \infty$ and $E\xi_1 \neq 0 \implies \dfrac{\max(|\xi_1|, \ldots, |\xi_n|)}{|S_n|} \to 0$ (P-a. e.);

(c)   $E\xi_1^2 < \infty \implies \dfrac{\max(|\xi_1|, \ldots, |\xi_n|)}{\sqrt{n}} \to 0$ (P-a. e.).

**Problem 2.10.54.** Let $\xi_1, \xi_2, \ldots$ be any sequence of random variables and let $\xi$ be a random variable. Prove that for every $p \geq 1$ the following conditions are equivalent:
(a) $\xi_n \xrightarrow{L^p} \xi$ (i.e., $E|\xi_n - \xi|^p \to 0$);
(b) $\xi_n \xrightarrow{P} \xi$ and the family $\{|\xi_n|^p, n \geq 1\}$ is uniformly integrable.

**Problem 2.10.55.** Let $\xi_1, \xi_2, \ldots$ be some sequence of independent and identically distributed random variables, chosen so that $P\{\xi_1 > x\} = e^{-x}$, $x \geq 0$ (i.e., each random variable is exponentially distributed). Prove that

$$P\{\xi_n > \alpha \ln n, \text{ i. o. }\} = \begin{cases} 1, & \text{if } \alpha \leq 1, \\ 0, & \text{if } \alpha > 1. \end{cases}$$

Convince yourself that the above statement can be further refined as follows:

$$P\{\xi_n > \ln n + \alpha \ln \ln n, \text{ i. o. }\} = \begin{cases} 1, & \text{if } \alpha \leq 1, \\ 0, & \text{if } \alpha > 1, \end{cases}$$

and, in general, for every $k \geq 1$ one has

$$P\{\xi_n > \ln n + \ln \ln n + \ldots + \underbrace{\ln \ldots \ln n}_{k \text{ times}} + \alpha \underbrace{\ln \ldots \ln n}_{k+1 \text{ times}} \text{ i.o.}\} = \begin{cases} 1, & \text{if } \alpha \leq 1, \\ 0, & \text{if } \alpha > 1. \end{cases}$$

**Problem 2.10.56.** Prove the following generalization of [P §2.10, Theorem 3], which is concerned with situations where convergence in $L^1$ comes as a consequence of convergence a. e.: if $\xi$ is a random variable and $(\xi_n)_{n \geq 1}$ is some sequence of random variables, chosen so that that $E|\xi| < \infty$, $E|\xi_n| < \infty$, and $\xi_n \to \xi$ (P-a. e.), then $E|\xi_n - \xi| \to 0$ if and only if $E|\xi_n| \to E|\xi|$ as $n \to \infty$. This statement is known as *Scheffe's lemma*. (Comp. with the statement in Problem 2.6.19).

**Problem 2.10.57.** Let $\xi_1, \xi_2, \ldots$ be any sequence of positive, independent and identically distributed random variables that share one and the same density $f = f(x)$, with $\lim_{x \downarrow 0} f(x) = \lambda > 0$. Prove that

$$n \min(\xi_1, \ldots, \xi_n) \xrightarrow{d} \eta,$$

where $\eta$ is an exponentially distributed random variable with parameter $\lambda$.

**Problem 2.10.58.** Prove that if one of the conditions (i), (ii), or (iii) in the assumptions of Problem 2.6.33 holds, then

$$E|X_n|^p \to E|X|^p, \quad \text{for all } 0 < p \leq r.$$

**Problem 2.10.59.** Let $(\xi_n)_{n \geq 1}$ be any sequence of independent and normally distributed random variables, i.e., $\xi_n \sim \mathcal{N}(\mu_n, \sigma_n^2)$. Prove that the series $\sum_{n \geq 1} \xi_n^2$ converges in $L^1$ if and only if

$$\sum_{n \geq 1} (\mu_n^2 + \sigma_n^2) < \infty.$$

Show also that when the above condition holds the series $\sum_{n \geq 1} \xi_n^2$ converges in $L^p$ for *all* $p \geq 1$.

*Hint.* To prove the second statement, one has to establish that

$$\left\| \sum_{n \geq 1} \xi_n^2 \right\|_p < \infty, \quad \text{for all } p \geq 1.$$

**Problem 2.10.60.** Let $X_1, X_2, \ldots$ be independent random variables that are uniformly distributed on the interval $[0, 1]$. Setting $Y_n = X_1 \ldots X_n$, $n \geq 1$, consider the series $\sum_{n=1}^{\infty} z^n Y_n$ and prove that its radius of convergence, $R = R(\omega)$, equals the constant $e$ with probability 1.
   *Hint.* Use the relation $1/R = \overline{\lim}_n |Y_n|^{1/n}$.

**Problem 2.10.61.** Let $(\Omega, \mathscr{F}, \mathsf{P}) = ([0, 1), \mathscr{B}([0, 1)), \lambda)$, where $\lambda$ denotes the Lebesgue measure, and let $\omega = (a_1, a_2, \ldots)$ be the continued fraction expansion of $\omega \in [0, 1)$ (in particular, $a_n = a_n(\omega)$ are integer numbers)—see [2]. Prove that as $n \to \infty$ one has

$$\lambda\{\omega : a_n(\omega) = k\} \to \frac{1}{\ln 2} \ln\left[\frac{1 + 1/k}{1 + 1/(k + 1)}\right].$$

*Remark.* Discussion of the origins of this problem and various approaches to its solution can be found in the "Essay on the history of probability theory" in the book "Probability-2" (see [121]) and on p. 101 in Arnold's book [3].

**Problem 2.10.62.** Let $X_1, X_2, \ldots$ be independent and identically distributed random variables with $\mathsf{P}\{X_1 = 0\} = \mathsf{P}\{X_1 = 2\} = 1/2$. Prove that
   (a) The series $\sum_{n=1}^{\infty} \frac{X_n}{3^n}$ converges almost surely to some random variable $X$.
   (b) The distribution function of the random variable $X$ is the *Cantor* function (see [$\underline{P}$ §2.3]).

## 2.11   Hilbert Spaces of Random Variables with Finite Second Moments

**Problem 2.11.1.** Prove that if $\xi = \text{l.i.m. } \xi_n$, then $\|\xi_n\| \to \|\xi\|$.

**Problem 2.11.2.** Prove that if $\xi = \text{l.i.m. } \xi_n$ and $\eta = \text{l.i.m. } \eta_n$, then $(\xi_n, \eta_n) \to (\xi, \eta)$.

**Problem 2.11.3.** Prove that the norm $\| \cdot \|$ satisfies the "parallelogram law:"

$$\|\xi + \eta\|^2 + \|\xi - \eta\|^2 = 2(\|\xi\|^2 + \|\eta\|^2).$$

**Problem 2.11.4.** Let $\{\xi_1, \dots, \xi_n\}$ be any family of *orthogonal* random variables. Prove that

$$\left\| \sum_{i=1}^{n} \xi_i \right\|^2 = \sum_{i=1}^{n} \|\xi_i\|^2.$$

This property is known as the "Pythagorean theorem."

**Problem 2.11.5.** Let $\xi_1, \xi_2, \dots$ be any sequence of orthogonal random variables and let $S_n = \xi_1 + \cdots + \xi_n$. Prove that if $\sum_{n=1}^{\infty} \mathsf{E}\xi_n^2 < \infty$, then one can find a random variable $S$ with $\mathsf{E}S^2 < \infty$, so that l.i.m. $S_n = S$, i.e., $\|S_n - S\|^2 \equiv \mathsf{E}|S_n - S|^2 \to 0$ as $n \to \infty$.

*Hint.* According to Problem 2.11.4 one must have

$$\|S_{n+k} - S_n\|^2 = \sum_{m=n+1}^{n+k} \|\xi_m\|^2.$$

**Problem 2.11.6.** Prove that Rademacher's functions $R_n$ can be defined by the relation

$$R_n(x) = \text{sign}\,(\sin 2^n \pi x), \quad 0 \le x \le 1, \, n = 1, 2, \dots$$

**Problem 2.11.7.** Prove, that for any $\xi \in L^2(\Omega, \mathscr{F}, \mathsf{P})$ and for any sub-$\sigma$-algebra $\mathscr{G} \subseteq \mathscr{F}$ one has

$$\|\xi\| \ge \|\mathsf{E}(\xi \mid \mathscr{G})\|,$$

with equality taking place if and only if $\xi = \mathsf{E}(\xi \mid \mathscr{G})$ (P-a. e.).

**Problem 2.11.8.** Prove that if $X, Y \in L^2(\Omega, \mathscr{F}, \mathsf{P})$, $\mathsf{E}(X \mid Y) = Y$ and $\mathsf{E}(Y \mid X) = X$, then $X = Y$ (P-a. e.). In fact, the assumption $X, Y \in L^2(\Omega, \mathscr{F}, \mathsf{P})$ can be relaxed to $X, Y \in L^1(\Omega, \mathscr{F}, \mathsf{P})$, but under this weaker assumption the property $X = Y$ (P-a. e.) is much harder to establish—see Problem 2.7.24.

**Problem 2.11.9.** Suppose that $\mathscr{F}$ is a $\sigma$-algebra and that $(\mathscr{G}_n^{(1)})$, $(\mathscr{G}_n^{(2)})$ and $(\mathscr{G}_n^{(3)})$ are three sequences of sub-$\sigma$-algebras that are contained in $\mathscr{F}$ and are chosen so that

$$\mathscr{G}_n^{(1)} \subseteq \mathscr{G}_n^{(2)} \subseteq \mathscr{G}_n^{(3)}, \quad \text{for every } n.$$

Then suppose that $\xi$ is some $\mathscr{F}$-measurable and bounded random variable, for which one can find a random variable $\eta$ with

$$\mathsf{E}(\xi \mid \mathscr{G}_n^{(1)}) \overset{\mathsf{P}}{\to} \eta \quad \text{and} \quad \mathsf{E}(\xi \mid \mathscr{G}_n^{(3)}) \overset{\mathsf{P}}{\to} \eta.$$

Prove, that when the above conditions hold one must also have $\mathsf{E}(\xi \mid \mathscr{G}_n^{(2)}) \overset{\mathsf{P}}{\to} \eta$.

**Problem 2.11.10.** Let $x \rightsquigarrow f(x)$ be any Borel-measurable function, which is defined on $[0, \infty)$ and is such that $\int_0^\infty e^{-\lambda x} f(x)\, dx = 0$, for any $\lambda > 0$. Prove that $f = 0$ almost surely relative to the Lebesgue measure on $[0, \infty)$.

**Problem 2.11.11.** Suppose that the random variable $\eta$ is uniformly distributed in $[-1, 1]$ and let $\xi = \eta^2$. Prove that:

(a) The optimal (in terms of the mean-square distance) estimate for $\xi$ given $\eta$, and for $\eta$ given $\xi$, can be expressed, respectively, as

$$\mathsf{E}(\xi \mid \eta) = \eta^2 \quad \text{and} \quad \mathsf{E}(\eta \mid \xi) = 0.$$

(b) The respective optimal linear estimates can be expressed as

$$\widehat{\mathsf{E}}(\xi \mid \eta) = 1/3 \quad \text{and} \quad \widehat{\mathsf{E}}(\eta \mid \xi) = 0.$$

## 2.12 Characteristic Functions

**Problem 2.12.1.** Let $\xi$ and $\eta$ be two independent random variables and suppose that $f(x) = f_1(x) + if_2(x)$, $g(x) = g_1(x) + ig_2(x)$, where $f_k(x), g_k(x), k = 1, 2$, are Borel functions. Prove that if $\mathsf{E}|f(\xi)| < \infty$ and $\mathsf{E}|g(\xi)| < \infty$, then

$$\mathsf{E}|f(\xi)g(\eta)| < \infty$$

and

$$\mathsf{E}f(\xi)g(\eta) = \mathsf{E}f(\xi) \cdot \mathsf{E}g(\eta).$$

(Recall that by definition $\mathsf{E}f(\xi) = \mathsf{E}f_1(\xi) + i\mathsf{E}f_2(\xi)$, $\mathsf{E}|f(\xi)| = \mathsf{E}(f_1^2(\xi) + f_2^2(\xi))^{1/2}$.)

**Problem 2.12.2.** Let $\xi = (\xi_1, \ldots, \xi_n)$ and $\mathsf{E}\|\xi\|^n < \infty$, where $\|\xi\| = \sqrt{\sum \xi_i^2}$. Prove that

$$\varphi_\xi(t) = \sum_{k=0}^{n} \frac{i^k}{k!} \mathsf{E}(t, \xi)^k + \varepsilon_n(t)\|t\|^n,$$

where $t = (t_1, \ldots, t_n)$, $(t, \xi) = t_1\xi_1 + \ldots + t_n\xi_n$, and $\varepsilon_n(t) \to 0$ as $\|t\| \to 0$.

*Hint.* The proof should be analogous to the one in the one-dimensional case, after replacing $t\xi$ with $(t, \xi)$.

**Problem 2.12.3.** Prove [P §2.12, Theorem 2] for $n$-dimensional distribution functions of the form $F = F_n(x_1, \ldots, x_n)$ and $G = G_n(x_1, \ldots, x_n)$.

**Problem 2.12.4.** Let $F = F(x_1, \ldots, x_n)$ be any multivariate distribution function and let $\varphi = \varphi(t_1, \ldots, t_n)$ be the associated characteristic function. By using the notation from equation [P §2.3, (12)], prove the *multivariate conversion formula*:

$$P(a, b] = \lim_{c \to \infty} \frac{1}{(2\pi)^n} \int_{-c}^{c} \cdots \int_{-c}^{c} \prod_{k=1}^{n} \frac{e^{-it_k a_k} - e^{-it_k b_k}}{it_k} \varphi(t_1, \ldots, t_n) \, dt_1 \ldots dt_n.$$

(In the above formula it is assumed that the set $(a, b]$, where $a = (a_1, \ldots, a_n)$ and $b = (b_1, \ldots, b_n)$, is a *continuity interval* for the function $P(a, b]$, in the sense that for all $k = 1, \ldots, n$ the marginal distribution functions $F_k(x_k)$, obtained from $F(x_1, \ldots, x_n)$ by setting all arguments except $x_k$ to $+\infty$, are continuous at the points $a_k, b_k$.)

**Problem 2.12.5.** Let $\varphi_k(t), k \geq 1$, be any sequence of characteristic functions and let $\lambda_k, k \geq 1$, be any sequence of non-negative numbers with $\sum \lambda_k = 1$. Prove that $t \rightsquigarrow \sum \lambda_k \varphi_k(t)$ must be a characteristic function.

**Problem 2.12.6.** Assuming that $\varphi(t)$ is a characteristic function, is it true that $\operatorname{Re} \varphi(t)$ and $\operatorname{Im} \varphi(t)$ are also characteristic functions?

*Hint.* Let $\varphi = \varphi(t)$ be the characteristic function for some distribution $P$. To answer the question regarding $\operatorname{Re} \varphi(t)$, consider the distribution $Q$ with $Q(A) = \frac{1}{2}[P(A) + P(-A)]$, where $-A = \{-x : x \in A\}$. To answer the question regarding $\operatorname{Im} \varphi(t)$, consider the characteristic function $\varphi(t) \equiv 1$.

**Problem 2.12.7.** Let $\varphi_1, \varphi_2, \varphi_3$ be any three characteristic functions with $\varphi_1 \varphi_2 = \varphi_1 \varphi_3$. Can one conclude that $\varphi_2 = \varphi_3$?

**Problem 2.12.8.** Prove the formulas for the characteristic functions listed in [P §2.12, Tables 4 and 5].

*Hint.* The characteristic functions for the first five discrete distributions can be obtained with elmentary calculations.

In the case of the negative binomial distribution $(C_{k-1}^{r-1} p^r q^{k-r}, k = r, r+1, \ldots$ and $r = 1, 2, \ldots)$, notice that for $|z| < 1$ one has

$$\sum_{k=r}^{\infty} C_{k-1}^{r-1} z^{k-r} = (1-z)^{-r}.$$

In the case of the characteristic function $\varphi(t)$, associated with the normal distribution $\mathcal{N}(m, \sigma^2)$, notice that with $m = 0$ and $\sigma^2 = 1$, according to the general theory of functions of complex variables, one must have

$$\varphi(t) = \frac{1}{\sqrt{2\pi}} \int_{\mathbb{R}} e^{itx} e^{-\frac{x^2}{2}} \, dx = e^{\frac{t^2}{2}} \frac{1}{\sqrt{2\pi}} \int_{\mathbb{R}} e^{-\frac{(x-it)^2}{2}} \, dx = e^{\frac{t^2}{2}} \int_L f(z) \, dz,$$

where $f(z) = \frac{1}{\sqrt{2\pi}} e^{-\frac{(z-it)^2}{2}}$, $L = \{z : \operatorname{Im} z = 0\}$, and

$$\int_L f(z) \, dz = \int_{L'} f(z) \, dz = \frac{1}{\sqrt{2\pi}} \int_{\mathbb{R}} e^{-\frac{y^2}{2}} \, dy = 1,$$

where $L' = \{z : \operatorname{Im} z = t\}$.

The characteristic function of the gamma-distribution can be computed in a similar fashion.

As for the characteristic function $\varphi(t)$, associated with the Cauchy distribution, notice that for $t > 0$ one has

$$\varphi(t) = \int_{\mathbb{R}} \frac{e^{itx}\,\theta}{\pi(x^2 + \theta^2)}\,dx = \int_L f(z)\,dz\,,$$

where $L = \{z : \operatorname{Im} z = 0\}$ and $f(z) = \frac{e^{itz}}{\pi(z^2 + \theta^2)}$. By the Cauchy's residue theorem and the Jordan lemma (see [47, vol. 1]) one has

$$\int_L f(z)\,dz = 2\pi i \operatorname*{res}_{i\theta} f = e^{-t\theta}$$

Similarly, for $t < 0$ one can prove that $\varphi(t) = e^{t\theta}$, so that $\varphi(t) = e^{-\theta|t|}$ for any real $t$.

**Problem 2.12.9.** Let $\xi$ be any *integer-valued* random variable and let $\varphi_\xi(t)$ be its characteristic function. Prove that

$$\mathsf{P}\{\xi = k\} = \frac{1}{2\pi} \int_{-\pi}^{\pi} e^{-ikt} \varphi_\xi(t)\,dt, \quad k = 0, \pm 1, \pm 2, \dots\,.$$

**Problem 2.12.10.** Consider the space $L^2 = L^2([-\pi, \pi])$, endowed with the Borel $\sigma$-algebra $\mathscr{B}[-\pi, \pi]$ and the Lebesgue measure, and prove that the collection of functions $\left\{\frac{1}{\sqrt{2\pi}} e^{i\lambda n}, n = 0, \pm 1, \pm 2, \dots\right\}$ forms an orthonormal basis in that space.

*Hint.* Use the following steps:
(a) For a given $\varepsilon > 0$ find a constant $c > 0$ such that

$$\|\varphi - f\|_{L^2} < \varepsilon,$$

where $f(x) = \varphi(x)I(|\varphi(x)| \le c)$.

(b) By using Lusin's theorem (see Problem 2.10.37), find a *continuous* function $f_\varepsilon(x)$ such that $|f_\varepsilon(x)| \le c$ and

$$\mu\{x \in [-\pi, \pi] : f_\varepsilon(x) \ne f(x)\} < \varepsilon\,,$$

so that $\|f - f_\varepsilon\|_{L^2} \le 2c\sqrt{\varepsilon}$.

(c) Find a continuous function $\widetilde{f}_\varepsilon(x)$ with the property $\widetilde{f}_\varepsilon(-\pi) = \widetilde{f}_\varepsilon(\pi)$ and $\|\widetilde{f} - f_\varepsilon\|_{L^2} \le \varepsilon$.

(d) By using the Weierstrass theorem find a function $\overline{f}_\varepsilon(x) = \sum_{k=-n}^{n} a_k e^{ikx}$ with the property

$$\sup_{x\in[-\pi,\pi]} |\widetilde{f}_\varepsilon(x) - \overline{f}_\varepsilon(x)| \leq \varepsilon,$$

which implies $\|\widetilde{f}_\varepsilon - \overline{f}_\varepsilon\|_{L^2} \leq \varepsilon$.

Conditions (a)–(d) above imply that the collection of all finite sums of the form $\sum_{k=-n}^{n} a_k e^{ikx}$ is everywhere dense in $L^2$, i.e., the system $\left\{ \frac{1}{\sqrt{2\pi}} e^{i\lambda n}, n = 0, \pm 1, \pm 2, \ldots \right\}$ forms and orthonormal basis.

**Problem 2.12.11.** In the statement of the Bochner–Khinchin theorem it is assumed that the function under consideration, $\varphi(t)$, is *continuous*. Prove the following result (due to F. Riesz), which shows to what extent it may be possible to remove the continuity assumption from the Bochner–Khinchin theorem.

Let $\varphi = \varphi(t)$ be any complex-valued and Borel-measurable function with the property $\varphi(0) = 1$. Then one can claim that the function $\varphi = \varphi(t)$ is positive definite if and only if it coincides with some characteristic function Lebesgue-almost everywhere on the real line.

**Problem 2.12.12.** Which of the functions

$$\varphi(t) = e^{-|t|^\alpha}, \quad 0 \leq \alpha \leq 2, \qquad \varphi(t) = e^{-|t|^\alpha}, \quad \alpha > 2,$$

$$\varphi(t) = (1 + |t|)^{-1}, \qquad \varphi(t) = (1 + t^4)^{-1},$$

$$\varphi(t) = \begin{cases} 1 - |t|^3, & |t| \leq 1, \\ 0, & |t| > 1, \end{cases} \qquad \varphi(t) = \begin{cases} 1 - |t|, & |t| \leq 1/2, \\ 1/(4|t|), & |t| > 1/2, \end{cases}$$

can be claimed to be a characteristic function?

*Hint.* In order to demonstrate that some of the above functions are not characteristic, use [P §2.12, Theorem 1] and also the inequalities established in Problem 2.12.21 below.

**Problem 2.12.13.** Prove that the function $t \rightsquigarrow \varphi(t)$, given by

$$\varphi(t) = \begin{cases} \sqrt{1 - t^2}, & |t| \leq 1, \\ 0, & |t| > 1, \end{cases}$$

cannot be identified with the characteristic function of any random variable.

Can one make the same claim about the function $t \rightsquigarrow \varphi(t) = \frac{\sin t}{t}$?

**Problem 2.12.14.** Prove that if the function $t \rightsquigarrow \varphi(t)$ is a characteristic, then so is also the function $t \rightsquigarrow |\varphi(t)|^2$.

**Problem 2.12.15.** Prove that if the function $t \rightsquigarrow \varphi(t)$ is characteristic, then so is also the function $t \rightsquigarrow e^{\lambda(\varphi(t)-1)}$ for every $\lambda \geq 0$. Can one claim that the function $t \rightsquigarrow \varphi(t) \equiv e^{\lambda(e^{-|t|}-1)}$ is characteristic?

**Problem 2.12.16.** Prove that if $t \rightsquigarrow \varphi(t)$ is a characteristic function, then the following functions must be characteristic, too:

$$t \rightsquigarrow \int_0^1 \varphi(ut)\, du, \quad t \rightsquigarrow \int_0^\infty e^{-u}\varphi(ut)\, du.$$

**Problem 2.12.17.** Prove that for every $n \geq 1$ the function

$$\varphi_n(t) \equiv \frac{e^{it} - \sum_{k=0}^{n-1}(it)^k/k!}{(it)^n/n!}$$

can be identified with the characteristic function of some random variable.

**Problem 2.12.18.** Let $\varphi_{X_n}(t)$ be the characteristic function of the random variable $X_n$, which is uniformly distributed in the interval $(-n, n)$. Prove that

$$\lim_{n\to\infty} \varphi_{X_n}(t) = \begin{cases} 1, & t = 0, \\ 0, & t \neq 0. \end{cases}$$

**Problem 2.12.19.** Let $(m^{(n)})_{n\geq 1}$ be the sequence of all moments of the random variable $X$, which has distribution function $F = F(x)$, i.e., $m^{(n)} = \int_{-\infty}^\infty x^n\, dF(x)$.

Prove that if the series $\sum_{n=1}^\infty \frac{m^{(n)}}{n!} s^n$ converges absolutely for some $s > 0$, then the sequence $(m^{(n)})_{n\geq 1}$ uniquely defines the distribution function $F = F(x)$.

**Problem 2.12.20.** Let $F = F(x)$ be any distribution function and let $\varphi(t) = \int_{-\infty}^\infty e^{itx}\, dF(x)$ be its characteristic function. Prove that

$$\lim_{c\to\infty} \frac{1}{2c} \int_{-c}^c e^{-itx}\varphi(t)\, dt = F(x) - F(x-)$$

and

$$\lim_{c\to\infty} \frac{1}{2c} \int_{-c}^c |\varphi(t)|^2\, dt = \sum_{x\in\mathbb{R}} [F(x) - F(x-)]^2.$$

In particular, the distribution function $F = F(x)$ can be claimed to be continuous if and only if its characteristic function $\varphi(t)$ satisfies the condition

$$\lim_{c\to\infty} \frac{1}{c} \int_{-c}^c |\varphi(t)|^2\, dt = 0.$$

**Problem 2.12.21.** Prove that any characteristic function $\varphi = \varphi(t)$ must satisfy the following inequalities:

$$1 - \operatorname{Re} \varphi(nt) \leq n[1 - (\operatorname{Re} \varphi(t))^n] \leq n^2[1 - \operatorname{Re} \varphi(t)], \quad n = 0, 1, 2, \ldots; \qquad (*)$$

$$|\operatorname{Im} \varphi(t)|^2 \leq \frac{1}{2}[1 - \operatorname{Re} \varphi(2t)]; \quad 1 - \operatorname{Re} \varphi(2t) \geq 2(\operatorname{Re} \varphi(t))^2;$$

$$|\varphi(t) - \varphi(s)|^2 \leq 4\varphi(0)|1 - \varphi(t-s)|; \quad 1 - |\varphi(2t)|^2 \leq 4[1 - |\varphi(t)|^2];$$

$$|\varphi(t) - \varphi(s)|^2 \leq 2[1 - \operatorname{Re} \varphi(t-s)];$$

$$\frac{1}{2h} \int_{t-h}^{t+h} \varphi(u) \, du \leq (1 + \operatorname{Re} \varphi(h))^{1/2}, \quad t > 0.$$

(The last two relations are known as *the Raikov inequalities*.)

   Hint. The proof is based on the relation $\varphi(t) = \int_{-\infty}^{\infty} e^{itx} dF(x)$ (and the associated relations for $\operatorname{Re} \varphi(t)$ and $\operatorname{Im} \varphi(t)$). Thus, for example, in order to prove the inequality

$$1 - \operatorname{Re} \varphi(2t) \leq 4[1 - \operatorname{Re} \varphi(t)] \qquad (**)$$

(a special case of $(*)$ with $n = 2$) it is enough to notice that

$$1 - \operatorname{Re} \varphi(2t) = \int_{-\infty}^{\infty} (1 - \cos 2tx) \, dF(x) \quad \text{and} \quad 1 - \cos 2tx \leq 4(1 - \cos tx).$$

**Problem 2.12.22.** Suppose that the characteristic function $\varphi = \varphi(t)$ is such that $\varphi(t) = 1 + f(t) + o(t^2)$ as $t \to 0$; where $f(t) = -f(-t)$. Prove that $\varphi(t) \equiv 1$.
   Hint. Use the relation $(**)$ in the previous problem.

**Problem 2.12.23.** Let $\varphi = \varphi(t)$ be the characteristic function of some random variable $X$, which has distribution function $F = F(x)$.
   (a) Prove that $\int_{-\infty}^{\infty} |x| \, dF(x) < \infty$ if and only if $\int_{-\infty}^{\infty} \dfrac{1 - \operatorname{Re} \varphi(t)}{t^2} \, dt < \infty$ and that these conditions imply

$$\mathsf{E}|X| \equiv \int_{-\infty}^{\infty} |x| \, dF(x) = \frac{1}{\pi} \int_{-\infty}^{\infty} \frac{1 - \operatorname{Re} \varphi(t)}{t^2} \, dt = \frac{2}{\pi} \int_{0}^{\infty} \frac{1 - \operatorname{Re} \varphi(t)}{t^2} \, dt.$$

   (b) Prove that if $\int_{-\infty}^{\infty} |x| \, dF(x) < \infty$ then one has

$$\mathsf{E}|X| \equiv \int_{-\infty}^{\infty} |x| \, dF(x) = -\frac{1}{\pi} \int_{-\infty}^{\infty} \frac{\operatorname{Re} \varphi'(t)}{t} \, dt.$$

(See Problem 2.12.37.)

*Hint.* (a) Use the following easy to check formula

$$|x| = \frac{1}{\pi} \int_{-\infty}^{\infty} \frac{1 - \cos xt}{t^2} \, dt.$$

(b) Use the fact that $|x| = x \operatorname{sign} x$, where

$$\operatorname{sign} x = \begin{cases} 1, & x > 0, \\ 0, & x = 0, \\ -1, & x < 0, \end{cases}$$

in conjunction with the relation

$$\operatorname{sign} x = \frac{1}{\pi} \int_{-\infty}^{\infty} \frac{\sin xt}{t} \, dx.$$

**Problem 2.12.24.** Consider a characteristic function of the form $\varphi(t) = 1 + O(|t|^\alpha)$ for $t \to 0$, where $\alpha \in (0, 2]$. Prove that if $\xi$ is a random variable with characteristic function $\varphi(t)$, then the following property must hold:

$$P\{|\xi| > x\} = O(x^{-\alpha}) \text{ as } x \to 0.$$

**Problem 2.12.25.** Let $X$ and $Y$ be any two independent and identically distributed random variables with vanishing means and standard deviations equal to 1. By using characteristic functions prove that if the distribution of the random variable $(X + Y)/\sqrt{2}$ coincides with the distribution of $X$ and $Y$, then $X$ and $Y$ must be Gaussian.

**Problem 2.12.26.** The *Laplace Transform* of a non-negative random variable $X$, with distribution function $F = F(x)$, is defined (see Problem 2.6.32) as the function $\widehat{F} = \widehat{F}(\lambda)$, $\lambda \geq 0$, given by

$$\widehat{F}(\lambda) = \mathsf{E} e^{-\lambda X} = \int_{[0,\infty)} e^{-\lambda x} \, dF(x), \quad \text{for } \lambda \geq 0.$$

Prove the following criterion, which is due to S.N. Bernstein: the function $\widehat{F} = \widehat{F}(\lambda)$, defined on $(0, \infty)$, is the Laplace transform of some distribution function $F = F(x)$ on $[0, \infty)$, if and only if $\widehat{F}$ is *completely monotone*, in the sense that all derivatives $\widehat{F}^{(n)}(\lambda)$, $n \geq 0$, exist and satisfy $(-1)^n \widehat{F}^{(n)}(\lambda) \geq 0$.

**Problem 2.12.27.** Suppose that the distribution function $F = F(x)$ admits density $f = f(x)$, has characteristic function $\varphi = \varphi(t)$, and suppose that at least one of the following conditions holds:

(a) $\int_{-\infty}^{\infty} |\varphi(t)| \, dt < \infty$     or     (b) $\int_{-\infty}^{\infty} f^2(x) \, dx < \infty.$

Prove *Parseval's formula*:

$$\int_{-\infty}^{\infty} f^2(x)\,dx = \frac{1}{2\pi}\int_{-\infty}^{\infty}|\varphi(t)|^2\,dt \quad (<\infty).$$

(Comp. with Parseval's idenity—see [P §2.11, (14)].)

**Problem 2.12.28.** Prove that if the distribution function $F = F(x)$ has density $f = f(x)$, then its characteristic function $\varphi = \varphi(t)$ must be such that $\varphi(t) \to 0$ as $t \to \infty$.

**Problem 2.12.29.** Let $F = F(x)$ and $\widetilde{F} = \widetilde{F}(x)$ be any two distribution functions on $(\mathbb{R}, \mathscr{B}(\mathbb{R}))$ and let $\varphi(t)$ and $\widetilde{\varphi}(t)$ be their respective characteristic functions. Prove *Parseval's relation*: for every $t \in \mathbb{R}$ one has

$$\int \widetilde{\varphi}(x-t)\,dF(x) = \int e^{-ity}\varphi(y)\,d\widetilde{F}(y). \qquad (*)$$

In particular, if $\widetilde{F}$ is the distribution function associated with the normal distribution law $\mathscr{N}(0,\sigma^2)$, then

$$\int_{-\infty}^{\infty} e^{-\frac{\sigma^2(x-t)^2}{2}}\,dF(x) = \frac{1}{\sqrt{2\pi\sigma^2}}\int_{-\infty}^{\infty} e^{-ity}e^{-\frac{y^2}{2\sigma^2}}\varphi(y)\,dy. \qquad (**)$$

(Comp. with the result in Problem 2.12.40.)

**Problem 2.12.30.** By using $(**)$ in the previous problem, conclude that if the distribution functions $F_1$ and $F_2$ share the same characteristic function, then one must have $F_1 = F_2$. (Comp. with the result in Problem 2.12.41.)

**Problem 2.12.31.** By using Parseval's relation $(*)$ in Problem 2.12.29, prove the following result: if $\varphi_\xi(t)$ is the characteristic function of the random variable $\xi$, then the Laplace transform of the random variable $|\xi|$ is given by the formula

$$\mathsf{E}e^{-\lambda|\xi|} = \int_{-\infty}^{\infty} \frac{\lambda}{\pi(\lambda^2+t^2)}\varphi(t)\,dt, \quad \lambda > 0.$$

(Comp. with the statement in Problem 2.12.23.)

**Problem 2.12.32.** Let $F = F(x)$ be any distribution function and let $\varphi(t) = \int_{-\infty}^{\infty} e^{itx}\,dF(x)$ be its characteristic function. According to [P §2.12, Theorem 3-b)], the property $\int_{-\infty}^{\infty}|\varphi(t)|\,dt < \infty$ guarantees the existence of a *continuous* density $f(x)$. Give an example of a distribution functions $F = F(x)$ which admits a continuous density, and yet $\int_{-\infty}^{\infty}|\varphi(t)|\,dt = \infty$.

**Problem 2.12.33.** Let $\varphi(t) = \int_{\mathbb{R}} e^{itx} \, dF(x)$ be some characteristic function. According to [P §2.12, Theorem 1], if $\int_{\mathbb{R}} |x| \, dF(x) < \infty$, then $\varphi(t)$ must be differentiable. By using appropriate examples, prove that, in general, the converse statement does not hold. Prove that, in fact, it is possible to find a characteristic function $\varphi(t)$, which is infinitely differentiable, and yet $\int_{\mathbb{R}} |x| \, dF(x) = \infty$.

**Problem 2.12.34.** (*The "inversion formula."*) By using the argument in the proof of [P §2.12, Theorem 3], prove that, for *any* distribution function $F = F(x)$ and *any* $a < b$, the following general "inversion formula" is in force:

$$\lim_{c \to \infty} \frac{1}{2\pi} \int_{-c}^{c} \frac{e^{-ita} - e^{-itb}}{it} \varphi(t) \, dt = \frac{1}{2}[F(b) + F(b-)] - \frac{1}{2}[F(a) + F(a-)].$$

**Problem 2.12.35.** (a) Prove that the probability distribution with density

$$f(x) = \frac{1 - \cos x}{\pi x^2}, \quad x \in \mathbb{R},$$

has characteristic function given by

$$\varphi(t) = \begin{cases} 1 - |t|, & |t| \le 1, \\ 0, & |t| > 1. \end{cases}$$

(b) What is the characteristic function of the distribution with density

$$f(x) = \frac{1 - \cos \pi x}{\pi^2 x^2}, \quad x \in \mathbb{R}?$$

(c) Prove that the characteristic functions of the probability densities

$$f_1(x) = \frac{1}{\pi \cosh x} \quad \text{and} \quad f_2(x) = \frac{1}{2 \cosh^2 x}, \quad x \in \mathbb{R},$$

are given, respectively, by

$$\varphi_1(t) = \frac{1}{\cosh \frac{1}{2}\pi t} \quad \text{and} \quad \varphi_2(t) = \frac{\pi t}{2 \sinh \frac{1}{2}\pi t},$$

where $\cosh y = (e^y + e^{-y})/2$ and $\sinh y = (e^y + e^{-y})/2$.

(d) Find the probability distributions associated with the following characteristic functions:

$$\frac{1 + it}{1 + t^2}, \quad \frac{1 - it}{1 + t^2}, \quad \cos \frac{t}{2}, \quad \frac{2}{3e^{it} - 1}, \quad \frac{1}{2}e^{-it} + \frac{1}{3} + \frac{1}{6}e^{2it}.$$

**Problem 2.12.36.** Let $m^{(k)} = \int_{\mathbb{R}} x^k \, dF(x)$, $k \geq 1$, be the moments of the probability distribution $F = F(x)$. Prove that

$$\int_{\mathbb{R}} \cosh(ax) \, dF(x) = \sum_k \frac{a^{2k}}{(2k)!} \, m^{(2k)}.$$

**Problem 2.12.37.** Suppose that, just as in Problem 2.12.23, $\varphi = \varphi(t)$ is the characteristic function of some random variable $X$, which has distribution function $F = F(x)$. Prove that

$$\int_{-\infty}^{\infty} |x|^\beta \, dF(x) < \infty \quad \text{for } \beta \in (0, 2)$$

if and only if

$$\int_{-\infty}^{\infty} \frac{1 - \operatorname{Re} \varphi(t)}{|t|^{1+\beta}} \, dt < \infty,$$

in which case

$$\mathsf{E}|X|^\beta \equiv \int_{-\infty}^{\infty} |x|^\beta \, dF(x) = C_\beta \int_{-\infty}^{\infty} \frac{1 - \operatorname{Re} \varphi(t)}{|t|^{1+\beta}} \, dt,$$

where

$$C_\beta = \left[ \int_{-\infty}^{\infty} \frac{1 - \cos t}{|t|^{1+\beta}} \, dt \right]^{-1} = \frac{\Gamma(1 + \beta)}{\pi} \sin \frac{\beta \pi}{2}.$$

*Hint.* Use the relation

$$|x|^\beta = C_\beta \int_{-\infty}^{\infty} \frac{1 - \cos xt}{|t|^{1+\beta}} \, dt.$$

**Problem 2.12.38.** Prove the statement in Problem 2.8.27 by calculating the characteristic functions of the random variables $\frac{\xi}{\eta}$, $\frac{\xi}{|\eta|}$, $\frac{|\xi|}{|\eta|}$, and the characteristic function of the random variable $C$ that has Cauchy distribution with density $\frac{1}{\pi(1+x^2)}$, $x \in \mathbb{R}$.

**Problem 2.12.39.** (*Non-uniqueness in the problem of moments.*) It was shown in [P §2.12, **9**] that it is possible to find two *different* distribution functions that, nevertheless, have identical moments of all orders $n \geq 1$. Here is one such construction in terms of densities.

Let $\eta$ be any standard normally distributed random variable ($\eta \sim \mathcal{N}(0, 1)$) and let $\xi = e^{\eta}$. Prove that:

(a) The density $f_\xi(x)$ is given by the formula

$$f_\xi(x) = \frac{1}{\sqrt{2\pi}} x^{-1} e^{-\frac{(\ln x)^2}{2}}, \qquad x > 0$$

(comp. with [P §2.8, (23)]).

(b) The function

$$g(x) = f_\xi(x)[1 + \sin(2\pi \ln x)], \qquad x > 0,$$

is such that $g(x) \geq 0$ and $\int_0^\infty g(x)\, dx = 1$.

(c) For all $n \geq 1$ one has

$$\int_0^\infty x^n f_\xi(x)\, dx = \int_0^\infty x^n g(x)\, dx.$$

**Problem 2.12.40.** Let $\xi$ and $\eta$ be two independent random variables, such that $\eta$ has standard normal distribution (i.e., $\eta \sim \mathcal{N}(0, 1)$) and let $f = f(x)$ be any bounded Borel function with compact support. Prove that for every $\sigma > 0$ one has

$$\mathsf{E}\, f\!\left(\xi + \frac{1}{\sigma}\eta\right) = \frac{1}{2\pi} \int_{-\infty}^\infty e^{-\frac{t^2}{2\sigma^2}} \varphi_\xi(t)\widehat{f}(-t)\, dt, \qquad (*)$$

where $\varphi_\xi(t) = \mathsf{E} e^{it\xi}$ and $\widehat{f}(t) = \int_{-\infty}^\infty e^{it\xi} f(x)\, dx$. (Comp. with the result in Problem 2.12.29.) Formulate an analogous result for multivariate random variables $\xi$ and $\eta$.

**Problem 2.12.41.** By using the relation $(*)$ in the previous problem, prove that the characteristic function $\varphi_\xi(t)$ of any random variable $\xi$ completely determines the probability distribution of $\xi$. (Comp. with [P §2.12, Theorem 2].)

*Hint.* Convince yourself that, under the assumptions of the previous problem, the relation $(*)$ implies that

$$\mathsf{E}\, f(\xi) = \lim_{\sigma \to \infty} \frac{1}{2\pi} \int_{-\infty}^\infty e^{-\frac{t^2}{2\sigma^2}} \varphi_\xi(t)\widehat{f}(-t)\, dt, \qquad (**)$$

and conclude (using the fact the $f = f(x)$ is an arbitrary bounded function with compact support), that the characteristic function $\varphi_\xi(t)$ indeed uniquely determines the distribution of the random variable $\xi$. Verify that the relation $(**)$ holds also for multivariate random variables, $\xi$ and obtain a multivariate analog of the relation $(*)$.

**Problem 2.12.42.** Let $\varphi = \varphi(t)$ be a characteristic function and suppose that for some $b > 0$ and some $0 < a < 1$ one has

$$\varphi(t) \le a, \quad \text{for any} \quad |t| \ge b.$$

Show the *Cramér's inequality*: for any $|t| < b$ one has

$$|\varphi(t)| \le 1 - (1 - a^2) \frac{t^2}{8b^2}.$$

*Hint.* Use the inequality $1 - |\varphi(2t)|^2 \le 4(1 - |\varphi(t)|^2)$ from Problem 2.12.21.

**Problem 2.12.43.** (*Addendum to the inequalities in Problem 2.12.21.*) Let $F = F(x)$ be any distribution function and let $\varphi = \varphi(t)$ be its characteristic function. Show the *von Bahr–Esseen inequality*:

$$|1 - \varphi(t)| \le C_r \beta^{(r)} |t|^r, \quad \text{for every} \quad 1 \le r \le 2,$$

where $\beta^{(r)} = \int_{-\infty}^{\infty} |x|^r \, dF(x)$ and $C_r$ is some constant.

**Problem 2.12.44.** For integer numbers $n \ge 1$ the moments $m^{(n)} = \mathsf{E}X^n$ and the absolute moments $\beta_n = \mathsf{E}|X|^n$ of the random variable $X$ can be expressed in terms of derivatives of order at most $n$ of the characteristic function $\varphi(t) = \mathsf{E}e^{itX}, t \in \mathbb{R}$ (see formula [$\underline{P}$ §2.12, (13)] or (c) in Problem 2.12.23). In order to obtain similar representation for the moments $m^{(\alpha)} = \mathsf{E}X^\alpha$ and $\beta_\alpha = \mathsf{E}|X|^\alpha$ for arbitrary $\alpha > 0$, one must resort to fractional derivatives, as explained below.

Let $\alpha = n + a$, for some integer number $n$ and some $0 < a < 1$. The *fractional derivative* $D^{(\alpha)} f(t) \ (= \dfrac{d^\alpha}{dt^\alpha} f(t))$ of the function $f = f(t), t \in \mathbb{R}$, is defined as the function

$$\frac{a}{\Gamma(1 - a)} \int_{-\infty}^{t} \frac{f^{(n)}(t) - f^{(n)}(s)}{(t - s)^{1+\alpha}} \, ds,$$

assuming that the integral in the above expression is well defined for any $t \in \mathbb{R}$. In particular, if $f(t) = \varphi(t) \ (= \int_{-\infty}^{\infty} e^{itx} dF(x))$ is some characteristic function, then

$$
\begin{aligned}
D^{n+a}\varphi(t)\Big|_{t=0} &= -\frac{1}{\Gamma(-a)} \int_0^\infty \frac{f^{(n)}(0) - f^{(n)}(-u)}{u^{1+a}} \, du \\
&= \frac{(-1)^{n+1}}{\Gamma(-a)} \left\{ \int_0^\infty \left[ \int_{-\infty}^\infty x^n (1 - \cos ux) \, dF(x) \right. \right. \quad\quad (*) \\
&\qquad\qquad \left. \left. + i \int_{-\infty}^\infty x^n \sin ux \, dF(x) \right] \frac{du}{u^{1+a}} \right\}.
\end{aligned}
$$

Prove that for even numbers $n$ the absolute moments $\beta_{n+a}$ are finite, i.e., $\beta_{n+a} < \infty$, if and only if:

(i) $\beta_n < \infty$;

(ii) $\mathrm{Re}[D^{n+a}\varphi(t)|_{t=0}]$ exists.

Prove that when these conditions hold one must have

$$\beta_{n+a} = \frac{1}{\cos\frac{a\pi}{2}} \, \mathrm{Re}\left[(-1)^{n/2} D^\alpha \varphi(t)\big|_{t=0}\right].$$

*Hint.* Use $(*)$ and the fact that for every $0 < b < 2$ one has the following formula:

$$\int_0^\infty \frac{1 - \cos u}{u^{1+b}} \, du = -\Gamma(-b) \cos\frac{\pi b}{2}.$$

*Remark.* A detailed discussion of the calculation of $\mathsf{E}|X|^{n+a}$, for arbitrary $n \geq 0$ and $0 < a < 1$, can be found in the book [84].

**Problem 2.12.45.** Prove that the following inequality is in force for every characteristic function $\varphi = \varphi(t)$ and every $u$ and $s$:

$$|\varphi(u + s)| \geq |\varphi(u)| \cdot |\varphi(s)| - [1 - |\varphi(u)|^2]^{1/2}[1 - |\varphi(s)|^2]^{1/2}.$$

*Hint.* Use Bochner–Khinchin's theorem (see [P §2.12, 6]).

**Problem 2.12.46.** Suppose that $(\xi, \eta)$ is a pair of random variables with joint density

$$f(x, y) = \frac{1}{4}\{1 + xy(x^2 - y^2)\}I(|x| < 1, |y| < 1).$$

Prove that $\xi$ and $\eta$ are two *dependent* random variables with densities

$$f_\xi(x) = \frac{1}{2}I(|x| < 1), \qquad f_\eta(y) = \frac{1}{2}I(|y| < 1).$$

Show also that the characteristic function, $\varphi_{\xi+\eta}(t)$, of the sum $\xi + \eta$, equals the *product* of the characteristic functions $\varphi_\xi(t)$ and $\varphi_\eta(t)$, i.e., $\varphi_{\xi+\eta}(t) = \varphi_\xi(t)\,\varphi_\eta(t)$.

**Problem 2.12.47.** Let $\xi_1, \xi_2, \ldots$ be a sequence of independent and identically distributed random variables that take the values $0, 1, \ldots, 9$ with probability $1/10$ and let

$$X_n = \sum_{k=1}^n \frac{\xi_k}{10^k}.$$

Prove that the sequence $(X_n)_{n\geq 1}$ converges not only in distribution, but also almost surely to a random variable that is uniformly distributed in the interval $[0, 1]$.

*Hint.* Use the method of characteristic functions.

## 2.13   Gaussian Systems of Random Variables

**Problem 2.13.1.** Prove that, given any Gaussian system of random variables, $(\xi, \eta_1, \ldots, \eta_n)$, the conditional expectation $\mathsf{E}(\xi \mid \eta_1, \ldots, \eta_n)$ coincides with the conditional expectations in wide sense $\widehat{\mathsf{E}}(\xi \mid \eta_1, \ldots, \eta_n)$.

**Problem 2.13.2.** Let $(\xi, \eta_1, \ldots, \eta_k)$ be a Gaussian system. Describe the structure of the conditional expectations $\mathsf{E}(\xi^n \mid \eta_1, \ldots, \eta_k), n \geq 1$, as functions of the random variables $\eta_1, \ldots, \eta_k$.

**Problem 2.13.3.** Let $X = (X_k)_{1 \leq k \leq n}$ and $Y = (Y_k)_{1 \leq k \leq}$ be two Gaussian random sequences with $\mathsf{E}X_k = \mathsf{E}Y_k, \mathsf{D}X_k = \mathsf{D}Y_k, 1 \leq k \leq n$, and

$$\mathrm{cov}(X_k, X_l) \leq \mathrm{cov}(Y_k, Y_l), \quad 1 \leq k, l \leq n.$$

Prove the *Slepyan's inequality*: for every $x \in \mathbb{R}$ one has

$$\mathsf{P}\left\{ \sup_{1 \leq k \leq n} X_k < x \right\} \leq \mathsf{P}\left\{ \sup_{1 \leq k \leq n} Y_k < x \right\}.$$

**Problem 2.13.4.** Let $\xi_1, \xi_2, \xi_3$ be three independent standard Gaussian random variables, i.e., $\xi_i \sim \mathcal{N}(0, 1), i = 1, 2, 3$. Prove that

$$\frac{\xi_1 + \xi_2 \xi_3}{\sqrt{1 + \xi_3^2}} \sim \mathcal{N}(0, 1).$$

(This gives rise to the interesting problem of describing the family of all *nonlinear* transformations of a given family of independent Gaussian random variables, $\xi_1, \ldots, \xi_n, n \geq 2$, that yield a Gaussian distribution.)

**Problem 2.13.5.** In the context of [P §2.13], prove that the "matrix" $R = (r(s, t))_{s,t \in \mathfrak{A}}$, associated with the function $r(s, t)$ from [P §2.13, (25), (29) and (30)], is non-negative definite.

**Problem 2.13.6.** Let $A$ be any matrix of order $m \times n$. We say that the matrix $A^{\oplus}$, of order $n \times m$, is the *pseudo-inverse* of the matrix $A$, if one can find two matrices, $U$ and $V$, such that

$$AA^{\otimes}A = A, \quad A^{\oplus} = UA^* = A^*V.$$

Prove that the matrix $A^{\oplus}$, defined by the above conditions, exists and is unique.

**Problem 2.13.7.** Prove that formulas (19) and (20) in the Theorem of the Normal Correlation (Theorem 2 in [P §2.13, **4**]) remains valid in the case of a *degenerate* matrix $\mathsf{D}_{\xi\xi}$, provided that the inverse $\mathsf{D}_{\xi\xi}^{-1}$ is replaced by the pseudo-inverse $\mathsf{D}_{\xi\xi}^{\oplus}$.

**Problem 2.13.8.** Let $(\theta, \xi) = (\theta_1, \ldots, \theta_k; \xi_1, \ldots, \xi_l)$ be a Gaussian vector and suppose that the matrix $\Delta \equiv D_{\theta\theta} - D_{\xi\xi}^{\oplus} D_{\theta\xi}^*$ is non-degenerate. Prove that the conditional distribution function $P(\theta \le a \mid \xi) = P(\theta_1 \le a_1, \ldots, \theta_k \le a_k \mid \xi)$ admits density given by (P-a. e.)

$$p(a_1, \ldots, a_k \mid \xi) = \frac{|\Delta|^{-1/2}}{(2\pi)^{k/2}} \exp\left\{-\frac{1}{2}(a - E(\theta \mid \xi))^* \Delta^{-1}(a - E(\theta \mid \xi))\right\}.$$

**Problem 2.13.9.** Let $\xi$ and $\eta$ be two independent standard Gaussian random variables (i.e., Gaussian random variables with vanishing mean and standard deviation equal to 1).

(a) Prove that the random variables $\xi + \eta$ and $\xi - \eta$ are also independent and Gaussian.

(b) By using (a) and the result in Problem 2.8.27, prove that

$$C \stackrel{\text{law}}{=} \frac{\xi + \eta}{\xi - \eta} \stackrel{\text{law}}{=} \frac{1 + \frac{\eta}{\xi}}{1 - \frac{\eta}{\xi}} \stackrel{\text{law}}{=} \frac{1 + C}{1 - C} \stackrel{\text{law}}{=} \frac{1}{C},$$

where $C$ is a random variable with Cauchy density $\frac{1}{\pi(1+x^2)}$ (recall that "$\stackrel{\text{law}}{=}$" stands for "equality in distribution").

**Problem 2.13.10.** (*S. N. Bernstein.*) Let $\xi$ and $\eta$ be any two independent and identically distributed random variables with finite variance. Prove that if $\xi + \eta$ and $\xi - \eta$ are independent, then $\xi$ and $\eta$ must be *Gaussian*. (For a generalization of this result, see the *Darmois–Skitovich Theorem* stated in Problem 2.13.44.)

*Hint.* Use the following line of reasoning (by $\varphi_\zeta(t)$ we denote the characteristic function of the random variable $\zeta$):

(a) From $\varphi_\xi(t) = \varphi_{\frac{\xi+\eta}{2}}(t)\, \varphi_{\frac{\xi-\eta}{2}}(t) = \varphi_{\xi+\eta}(t/2) + \varphi_{\xi-\eta}(t/2)$ conclude that

$$\varphi_\xi(t) = \left(\varphi_\xi\left(\frac{t}{2}\right)\right)^2 \left|\varphi_\eta\left(\frac{t}{2}\right)\right|^2, \quad t \in \mathbb{R},$$

and, analogously,

$$\varphi_\eta(t) = \left(\varphi_\eta\left(\frac{t}{2}\right)\right)^2 \left|\varphi_\xi\left(\frac{t}{2}\right)\right|^2, \quad t \in \mathbb{R}.$$

(b) By using (a) conclude that $|\varphi_\xi(t)| = |\varphi_\eta(t)|$ and that $|\varphi_\eta(t)| = \left|\varphi_\xi\left(\frac{t}{2}\right)\right|^4$.

(c) By using (b) conclude that $\varphi_\eta(t) \neq 0$ for every $t \in R$, so that one can define the function $f(t) = \ln |\varphi_\eta(t)|$, with $f(t) = 4f(t/2), t \in \mathbb{R}$.

(d) From $E\eta^2 < \infty$ conclude that $\varphi_\eta(t) \in C^2(\mathbb{R})$ and by using (c) conclude that

$$f''(t) = f''\left(\frac{t}{2}\right) = \cdots = f''\left(\frac{t}{2^k}\right) \to f''(0), \quad t \in \mathbb{R},$$

so that $f''(t) = \text{const}.$

(e) By using (d) conclude that $f(t) = at^2 + bt + c$, which, in conjunction with (c), gives $f(t) = at^2$.

(f) By using (e) conclude that $\varphi_\eta(t) = e^{i\alpha(t) + at^2}$, where the function $\alpha(t)$ should be continuous as long as $\varphi_\eta(t)$ is continuous.

(g) Convince yourself that $\alpha(t)$ has the property

$$\alpha(t) = 2\alpha\left(\frac{t}{2}\right), \quad t \in \mathbb{R}.$$

(h) By using the relation $\mathsf{E}\eta^2 < \infty$ conclude that $\varphi_\eta(t)$ is differentiable at 0 and by using (g) conclude that as $k \to \infty$ one must have

$$\frac{\alpha(t)}{t} = \frac{\alpha(t/2^k)}{t/2^k} \to \alpha'(0), \quad t \neq 0,$$

which shows that $\alpha(t) = \alpha'(0)t$.

As a result, $\varphi_\eta(t) = e^{i\alpha'(0)t + at^2}$, i.e., $\eta$ has Gaussian distribution. With a similar line of reasoning one can show that the random variable $\xi$ is also Gaussian.

**Problem 2.13.11.** (*Mercer Theorem.*) Let $r = r(s, t)$ be any continuous covariance function defined on the rectangle $[a, b] \times [a, b]$, where $-\infty < a < b < \infty$. Prove that the equation

$$\lambda \int_a^b r(s, t)u(t)\, dt = u(s), \quad a \leq s \leq b$$

admits a continuous solution, $u(t)$, for infinitely many values $\lambda = \lambda_k > 0$, $k \geq 1$, and the respective system of solutions $\{u_k = u_k(s), k \geq 1\}$ forms a complete orthonormal system in $L^2(a, b)$, such that

$$r(s, t) = \sum_{k=1}^{\infty} \frac{u_k(s)u_k(t)}{\lambda_k},$$

where the series converges absolutely and uniformly on $[a, b] \times [a, b]$.

**Problem 2.13.12.** Let $X = \{X_t, t \geq 0\}$ be any Gaussian process with $\mathsf{E}X_t = 0$ and with covariance function $r(s, t) = e^{-|t-s|}$, $s, t \geq 0$. Given any $0 < t_1 < \cdots < t_n$, let $f_{t_1,\ldots,t_n}(x_1, \ldots, x_n)$ denote the (joint) density of the random variables $X_{t_1}, \ldots, X_{t_n}$. Prove, that this density admits the following representation:

$$f_{t_1,\ldots,t_n}(x_1, \ldots, x_n) = \left[(2\pi)^n \prod_{i=2}^n \left(1 - e^{2(t_{i-1} - t_i)}\right)\right]^{-1/2}$$

$$\times \exp\left\{-\frac{x_1^2}{2} - \frac{1}{2}\sum_{i=2}^n \frac{(x_i - e^{(t_{i-1} - t_i)}x_{i-1})^2}{1 - e^{2(t_{i-1} - t_i)}}\right\}.$$

**Problem 2.13.13.** Let $f = \{f_n = f_n(u), n \geq 1; u \in [0, 1]\}$ be a complete orthonormal (for the Lebesgue measure on $[0, 1]$) system of $L^2$-functions and let $(\xi_n)_{n\geq 1}$ be any sequence of independent and identically distributed $\mathcal{N}(0, 1)$-random variables. Prove that the process $B_t = \sum_{n\geq 1} \xi_n \int_0^t f_n(u) \, du$, $0 \leq t \leq 1$ is a Brownian motion.

**Problem 2.13.14.** Prove, that if $B^\circ = (B_t^\circ)_{0\leq t\leq 1}$ is a Brownian bridge process, then the process $B = (B_t)_{t\geq 0}$ given by $B_t = (1 + t)B_{t/(1+t)}^\circ$ is a Brownian motion.

**Problem 2.13.15.** Verify that if $B = (B_t)_{t\geq 0}$ is a Brownian motion, then each of the following processes is also a Brownian motion:

$$B_t^{(1)} = -B_t, \ t \geq 0;$$

$$B_t^{(2)} = tB_{1/t}, \ t > 0, \qquad B_0^{(2)} = 0;$$

$$B_t^{(3)} = B_{t+s} - B_s, \quad s > 0, \ t \geq 0;$$

$$B_t^{(4)} = B_T - B_{T-t} \quad \text{for } 0 \leq t \leq T, T > 0;$$

$$B_t^{(5)} = \frac{1}{a}B_{a^2t}, \quad a > 0, \ t \geq 0 \quad \text{(scaling property).}$$

**Problem 2.13.16.** Let $B^\mu = (B_t + \mu t)_{t\geq 0}$ be a Brownian motion with drift.
(a) Find the distribution of the random variables $B_{t_1}^\mu + B_{t_2}^\mu$, for $t_1 < t_2$.
(b) Calculate $\mathsf{E}B_{t_0}^\mu B_{t_1}^\mu$ and $\mathsf{E}B_{t_0}^\mu B_{t_1}^\mu B_{t_2}^\mu$, for $t_0 < t_1 < t_2$.

**Problem 2.13.17.** Consider the process $B^\mu$ from the previous problem and calculate the conditional distributions

$$\mathsf{P}(B_{t_2}^\mu \in \cdot \mid B_{t_1}^\mu), \quad \text{for } t_1 < t_2 \text{ and } t_1 > t_2,$$

and

$$\mathsf{P}(B_{t_2}^\mu \in \cdot \mid B_{t_0}^\mu, B_{t_1}^\mu), \quad \text{for } t_0 < t_1 < t_2.$$

**Problem 2.13.18.** Let $B = (B_t)_{t\geq 0}$ be a Brownian motion process. Prove that the process $Y = (Y_t)_{t\in\mathbb{R}}$, given by $Y_t = e^{-t}B_{e^{2t}}$, is an *Ornstein–Uhlenbeck* process, i.e., a Gauss–Markov process with $\mathsf{E}Y_t = 0$ and $\mathsf{E}Y_s Y_t = e^{-|t-s|}$.

**Problem 2.13.19.** Let $Y = (Y_t)_{t\in\mathbb{R}}$ be an Ornstein–Uhlenbeck process. Prove that the process

$$B_t^\circ = \begin{cases} \sqrt{t(1-t)} \, Y_{\frac{1}{2} \ln \frac{t}{1-t}}, & 0 < t < 1, \\ 0, & t = 0, 1, \end{cases}$$

is a Brownian bridge.

**Problem 2.13.20.** Let $\xi_0, \xi_1, \xi_2, \dots$ be independent and identically distributed standard Gaussian (i.e., $\mathcal{N}(0, 1)$) random variables. Prove that the series

$$B_t^\circ = \sum_{k=1}^{\infty} \xi_k \frac{\sqrt{2}\sin k\pi t}{k\pi}, \qquad 0 \le t \le 1,$$

defines a Brownian bridge, while, just as the series in [ P §2.13, (26)], the series

$$B_t = \xi_0 t + \sum_{k=1}^{\infty} \xi_k \frac{\sqrt{2}\sin k\pi t}{k\pi}, \qquad 0 \le t \le 1,$$

defines a Brownian motion.

**Problem 2.13.21.** Give a detailed proof of the fact that the processes $(B_t)_{0 \le t \le 1}$, defined in [ P §2.13, (26) and (28)], and the process

$$B_t = \sqrt{2} \sum_{n=1}^{\infty} \xi_n \frac{1 - \cos n\pi t}{n\pi},$$

where $\xi_n$, $n \ge 1$, are chosen as in [ P §2.13, (26) and (28)], are all Brownian motions.

**Problem 2.13.22.** Let $X = (X_k)_{1 \le k \le n}$ be any Gaussian sequence, let

$$m = \max_{1 \le k \le n} \mathsf{E}X_k, \qquad \sigma^2 = \max_{1 \le k \le n} \mathsf{D}X_k,$$

and suppose that

$$\mathsf{P}\Big\{ \max_{1 \le k \le n} (X_k - \mathsf{E}X_k) \ge a \Big\} \le 1/2, \qquad \text{for some } a.$$

Prove the following inequality, which is due to *E. Borel*:

$$\mathsf{P}\Big\{ \max_{1 \le k \le n} X_k > x \Big\} \le 2\Psi\Big( \frac{x - m - a}{\sigma} \Big),$$

where $\Psi(x) = (2\pi)^{-1/2} \int_x^{\infty} e^{-y^2/2} \, dy$.

**Problem 2.13.23.** Let $(X, Y)$ be any bi-variate Gaussian random variable with $\mathsf{E}X = \mathsf{E}Y = 0$, $\mathsf{E}X^2 > 0$, $\mathsf{E}Y^2 > 0$, and with correlation coefficient $\rho = \frac{\mathsf{E}XY}{\sqrt{\mathsf{E}X^2\mathsf{E}Y^2}}$.

(a) Prove that the variables $X$ and $Z = (Y - \rho X)/\sqrt{1 - \rho^2}$ are independent and normally distributed.

(b) Prove that

$$\mathsf{P}\{XY < 0\} = 1 - 2\mathsf{P}\{X > 0, Y > 0\} = \pi^{-1} \arccos \rho,$$

and conclude that

$$P\{X > 0, Y > 0\} = P\{X < 0, Y < 0\} = \frac{1}{4} + \frac{1}{2\pi} \arcsin \rho,$$

$$\frac{\partial}{\partial \rho} P\{X > 0, Y > 0\} = \frac{1}{2\pi \sqrt{1 - \rho^2}},$$

$$P\{X > 0, Y < 0\} = P\{X < 0, Y > 0\} = \frac{1}{4} - \frac{1}{2\pi} \arcsin \rho.$$

(c) Let $Z = \max(X, Y)$, where $EX^2 = EY^2 = 1$. Prove that

$$EZ = \sqrt{\frac{1 - \rho}{\pi}}, \quad EZ^2 = 1.$$

(d) Prove that for arbitrary $a$ and $b$ one has the following inequalities:

$$(1 - \Phi(a))(1 - \Phi(c)) \leq P\{X > a, Y > b\}$$

$$\leq (1 - \Phi(a))(1 - \Phi(c)) + \frac{\rho\varphi(b)(1 - \Phi(d))}{\varphi(a)},$$

where $c = (b - a\rho)/\sqrt{1 - \rho^2}$, $d = (a - b\rho)/\sqrt{1 - \rho^2}$ and $\varphi(x) = \Phi'(x)$ is the standard normal density.

   *Hint.* Property (b) can be derived from property (a).

**Problem 2.13.24.** Let $Z = XY$, where $X \sim \mathcal{N}(0, 1)$ and $P\{Y = 1\} = P\{Y = -1\} = \frac{1}{2}$. Prove that $Z \sim \mathcal{N}(0, 1)$, find the distribution of the pairs $(X, Z)$ and $(Y, Z)$, and find the distribution of the random variable $X + Z$. Convince yourself that $X$ and $Z$ are uncorrelated and yet dependent.

**Problem 2.13.25.** Let $\xi$ be any standard normal random variable, i.e., $\xi \sim \mathcal{N}(0, 1)$, and let

$$\eta_\alpha = \begin{cases} \xi, & \text{if } |\xi| \leq \alpha, \\ -\xi, & \text{if } |\xi| > \alpha. \end{cases}$$

Prove that $\eta_\alpha \sim \mathcal{N}(0, 1)$ and that with $\alpha$ chosen so that

$$\int_0^\alpha x^2 f_\xi(x)\, dx = \frac{1}{4} \quad \left( f_\xi(x) = \frac{1}{\sqrt{2\pi}} e^{-\frac{x^2}{2}} \right),$$

the variables $\xi$ and $\eta_{1/4}$ are uncorrelated and yet dependent Gaussian random variables (comp. with [$\underline{P}$ §2.13, Theorem 1-a)]).

**Problem 2.13.26.** Let $\xi$ and $\eta$ be two normally distributed random variables with $E\xi = E\eta = 0$, $E\xi^2 = E\eta^2 = 1$ and $E\xi\eta = \rho$. Prove that:

(a)   $E\max(\xi, \eta) = \sqrt{(1 - \rho)/\pi}$;

(b)   $E(\xi \mid \eta) = \rho\eta$,   $D(\xi \mid \eta) = 1 - \rho^2$;

(c)   $E(\xi \mid \xi + \eta = z) = z/2$,   $D(\xi \mid \xi + \eta = z) = (1 - \rho)/2$;

(d)   $E(\xi + \eta \mid \xi > 0, \eta > 0) = 2\sqrt{2/\pi}$.

Give the analogs of the above formulas for the case where $D\xi = \sigma_1^2$ and $D\eta = \sigma_2^2$, for arbitrary $\sigma_1 > 0$ and $\sigma_2 > 0$.

**Problem 2.13.27.** Let $\left(\begin{smallmatrix} X \\ Y \end{smallmatrix}\right)$ be any bi-variate Gaussian random variable with co-variance matrix

$$\operatorname{cov}(X, Y) = \begin{pmatrix} \sigma^2 & \sigma^2 \\ \sigma^2 & \sigma^2 \end{pmatrix}.$$

Write $\left(\begin{smallmatrix} X \\ Y \end{smallmatrix}\right)$ in the form

$$\begin{pmatrix} X \\ Y \end{pmatrix} = Q \begin{pmatrix} \xi \\ \eta \end{pmatrix},$$

where $Q$ is an orthogonal matrix and $\xi$ and $\eta$ are two independent Gaussian random variables.

**Problem 2.13.28.** Let $\xi = (\xi_1, \ldots, \xi_n)$ be any non-degenerate Gaussian vector with vanishing mean and with covariance matrix $R = \|E\xi_i\xi_j\|$, and suppose that $\lambda_1, \ldots, \lambda_n$ are the eigenvalues of the matrix $R$. Prove that the characteristic function, $\varphi(t)$, of the random variable $\xi_1^2 + \ldots + \xi_n^2$ coincides with the characteristic function of a random variable of the form $\lambda_1\eta_1^2 + \ldots + \lambda_n\eta_n^2$, where $\eta_1, \ldots, \eta_n$ are independent standard Gaussian random variables ($\eta_k \sim \mathcal{N}(0, 1)$), and, furthermore, one has

$$\varphi(t) = \prod_{j=1}^{n} |1 - 2it\lambda_j|^{-1/2}.$$

**Problem 2.13.29.** Let $\xi_1, \ldots, \xi_n$, $n \geq 2$, be any set of independent and identically distributed random variables. Prove that the distribution of the vector $(\xi_1, \ldots, \xi_n)$ is rotation invariant if and only if each of the variables $\xi_1, \ldots, \xi_n$ is normally distributed with vanishing mean.

*Hint.* Use characteristic functions.

**Problem 2.13.30.** (*Statistics of the normal distribution $\mathcal{N}(m,\sigma^2)$: part I.*) Suppose that $\xi_1,\ldots\xi_n$, $n \geq 2$, are independent and identically distributed normal, $\mathcal{N}(m,\sigma^2)$, random variables. Prove that the variables

$$\bar{\xi} = \frac{1}{n}\sum_{k=1}^{n}\xi_k \quad \text{and} \quad s_1^2 = \frac{1}{n-1}\sum_{k=1}^{n}(\xi_k - \bar{\xi})^2$$

are independent and

$$(n-1)s_1^2 \overset{d}{=} \sum_{k=1}^{n-1}(\xi_k - m)^2.$$

*Hint.* Use the statement in the previous problem.

**Problem 2.13.31.** (*Statistics of the normal distribution $\mathcal{N}(m,\sigma^2)$: part II.*) Let $\xi_1\ldots,\xi_n$ be any set of independent and identically distributed random variables with normal distribution $\mathcal{N}(m,\sigma^2)$, and let $x = (x_1,\ldots,x_n)$ be some sample of observations over $\xi = (\xi_1,\ldots,\xi_n)$, $n \geq 1$.

(a) Prove that the pairs of statistics

$$T_1(x) = \sum_{i=1}^{n}x_i, \qquad T_2(x) = \sum_{i=1}^{n}x_i^2$$

and

$$\bar{x} = \frac{1}{n}\sum_{i=1}^{n}x_i, \qquad s^2(x) = \frac{1}{n}\sum_{i=1}^{n}(x_i - \bar{x})^2$$

are *sufficient*.

(b) Convince yourself that

$$s^2(x) = \frac{1}{n}\sum_{i=1}^{n}x_i^2 - \bar{x}^2.$$

**Problem 2.13.32.** (*Statistics of the normal distribution $\mathcal{N}(m,\sigma^2)$: part III—m is unknown and $\sigma^2 = \sigma_0^2$.*) In this and the following problem it is assumed that $\xi_1,\ldots,\xi_n$ is a set of independent and identically distributed $\mathcal{N}(n,\sigma^2)$-random variables and the notation from Problem 2.13.30 (with $n \geq 2$) and from Problem 2.13.31 (with $n \geq 1$) is assumed.

Suppose that $m$ is unknown, $\sigma^2$ is known to be $\sigma^2 = \sigma_0^2$.

(a) Prove that, for $\bar{\xi} = \frac{1}{n}\sum_{i=1}^{n}\xi_i\ (=\frac{1}{n}T_i(\xi))$, one has

$$\mathsf{E}\bar{\xi} = m \quad \text{(unbiased estimate)} \quad \text{and} \quad \mathsf{D}\bar{\xi} = \frac{\sigma_0^2}{n}.$$

(b) Prove that (for $\sigma^2 = \sigma_0^2$) the *sample mean* $\bar{x}$ is an *effective estimate*, i.e., unbiased estimate with minimal dispersion. For that purpose, prove that in this case the unbiased estimate, $T(x)$, for the parameter $m$ satisfies the *Rao–Cramér's inequality*:

$$\mathsf{D}T \geq \frac{1}{n\mathsf{E}\left(\frac{\partial \ln p_{(m,\sigma_0^2)}(\xi)}{\partial m}\right)} = \frac{1}{\left(\frac{n}{\sigma_0^2}\right)},$$

where

$$p_{(m,\sigma_0^2)}(x) = \frac{1}{\sqrt{2\pi\sigma_0^2}} e^{-\frac{(x-m)^2}{2\sigma_0^2}}.$$

(c) Prove that the variable

$$\frac{\bar{\xi} - m}{\left(\frac{\sigma_0}{\sqrt{n}}\right)}$$

has a standard normal, i.e., $\mathcal{N}(0, 1)$, distribution, and, furthermore, if $\lambda(\varepsilon)$ is chosen so that

$$1 - \varepsilon = \frac{1}{\sqrt{2\pi}} \int_{-\lambda(\varepsilon)}^{\lambda(\varepsilon)} e^{-t^2/2}\, dt \quad (= 2\Phi(\lambda(\varepsilon)) - 1),$$

where $0 < \varepsilon < 1$, then the interval

$$\left(\bar{x} - \frac{\sigma_0}{\sqrt{n}}\lambda(\varepsilon), \bar{x} + \frac{\sigma_0}{\sqrt{n}}\lambda(\varepsilon)\right)$$

is a *confidence interval* for $m$ with *confidence level* $1 - \varepsilon$, i.e., the "probability for cover," satisfies

$$P_{(m,\sigma_0^2)}\left\{\bar{\xi} - \frac{\sigma_0}{\sqrt{n}}\lambda(\varepsilon) \leq m \leq \bar{\xi} + \frac{\sigma_0}{\sqrt{n}}\lambda(\varepsilon)\right\} = 1 - \varepsilon,$$

where $P_{(m,\sigma_0^2)}$ stands for the probability law with density $p_{(m,\sigma_0^2)}$. (Comp. with the Definition in [P §1.7, 2].)

**Problem 2.13.33.** (*Statistics of the normal distribution* $\mathcal{N}(m,\sigma^2)$: *part IV*—$m = m_0$, *but* $\sigma^2$ *is unknown.*)

If $m$ is known ($m = m_0$), then it is natural to estimate $\sigma^2$ not by the variable $s^2(x) = \frac{1}{n}\sum_{i=1}^n (x_i - \bar{x})^2$, but, rather, by the variable

$$s_0^2(x) = \frac{1}{n}\sum_{i=1}^n (x_i - m_0)^2.$$

(a) Prove that

$$\mathsf{E}s_0^2(\xi) = \sigma^2 \quad \text{(unbiased estimate)} \quad \text{and} \quad \mathsf{D}s_0^2(\xi) = \frac{2\sigma^4}{n}.$$

(b) Prove that the *sample dispersion* $s_0^2(x)$ (with $m = m_0$) is an *effective* estimate of the variable $\sigma^2$, i.e., unbiased estimate with a minimal dispersion. To this end, prove that the Rao–Cramér inequality for the unbiased estimate $T(x)$ of the variable $\sigma^2$ has the form:

$$\mathsf{D}T \geq \frac{1}{n\mathsf{E}\left(\frac{\partial \ln p_{(m_0,\sigma^2)}(\xi)}{\partial \sigma^2}\right)} = \frac{1}{\left(\frac{n}{2\sigma^4}\right)}.$$

*Remark.* As for the accuracy of the estimate $s_0^2(x)$, one can construct a confidence interval for $\sigma^2$ by using the following considerations.

Given $x = (x_1, \ldots, x_n)$, let

$$\chi_n^2(x) = \sum_{i=1}^{n} \left(\frac{x_i - m_0}{\sigma}\right)^2.$$

Since

$$\chi_n^2(\xi) \overset{d}{=} \sum_{i=1}^{n} \eta_i^2 \quad (= \chi_n^2),$$

according to [$\underline{\mathsf{P}}$ §2.8, (34)], the variable $\chi_n^2(\xi)$ has $\chi^2$-distribution with $n$ degrees of freedom; more specifically, it has density $(x \geq 0)$

$$f_{\chi_n^2}(x) = \frac{x^{\frac{n}{2}-1}e^{-x/2}}{2^{n/2}\Gamma(n/2)}$$

(see also [$\underline{\mathsf{P}}$ §2.3, Table 3]). Since, at the same time, one can write

$$s_0^2(x) = \frac{\chi_n^2(x)\sigma^2}{n},$$

one must have

$$\mathsf{P}_{(m_0,\sigma_0^2)}\left\{\frac{s_0^2(x)n}{\sigma^2} \leq x\right\} = \int_0^x f_{\chi_n^2}(t)\,dt.$$

For this reason, given any $0 < \varepsilon < 1$, it is possible to find a $\lambda'(\varepsilon)$ and $\lambda''(\varepsilon)$ so that

$$\int_0^{\lambda'(\varepsilon)} f_{\chi_n^2}(t)\,dt = \frac{\varepsilon}{2} \quad \text{and} \quad \int_{\lambda''(\varepsilon)}^{\infty} f_{\chi_n^2}(t)\,dt = \frac{\varepsilon}{2}.$$

Consequently,

$$\int_{\lambda'(\varepsilon)}^{\lambda''(\varepsilon)} f_{\chi_n^2}(t)\, dt = 1 - \varepsilon.$$

Furthermore, the interval

$$\left( \frac{s_0^2(x)n}{\lambda''(\varepsilon)}, \frac{s_0^2(x)n}{\lambda'(\varepsilon)} \right)$$

is a confidence interval for $\sigma^2$ with confidence level $(1 - \varepsilon)$, since

$$\left\{ \frac{s_0^2(x)n}{\lambda''(\varepsilon)} \le \sigma^2 \le \frac{s_0^2(x)n}{\lambda'(\varepsilon)} \right\} = \{\lambda'(\varepsilon) \le \chi_n^2(x) \le \lambda''(\varepsilon)\}.$$

Finally, we note that the choice of $\varepsilon > 0$ does not determine uniquely $a'(\varepsilon)$ and $a''(\varepsilon)$ from the relation

$$\int_{a'(\varepsilon)}^{a''(\varepsilon)} f_{\chi_n^2}(t)\, dt = 1 - \varepsilon.$$

(c) How should one choose $a'(\varepsilon)$ and $a''(\varepsilon)$ in order to define the narrowest possible confidence interval for $\sigma^2$ with confidence level $(1 - \varepsilon)$? Are these values for $a'(\varepsilon)$ and $a''(\varepsilon)$ going to be the same as $\lambda'(\varepsilon)$ and $\lambda''(\varepsilon)$?

**Problem 2.13.34.** (*Statistics of the normal distribution* $\mathcal{N}(m, \sigma^2)$: *part V—m is unknown and* $\sigma^2$ *is unknown*).

(a) Prove that in this case, for any $n > 1$, the *unbiased* estimates for $m$ and $\sigma^2$ are given by

$$\bar{x} = \frac{1}{n} \sum_{i=1}^{n} x_i \quad \text{and} \quad s_1^2(x) \equiv \frac{n}{n-1} s^2(x) = \frac{1}{n-1} \sum_{i=1}^{n} (x_i - \bar{x})^2.$$

(b) Prove that the statistics

$$t_{n-1}(x) = \frac{\bar{x} - m}{\left( \frac{s_1(x)}{\sqrt{n}} \right)}$$

has Student distribution with $n - 1$ degrees of freedom—see [P §2.3, Table 3].

*Hint.* Write the variables $t_{n-1}(x)$ in the form

$$t_{n-1}(x) = \frac{\frac{\bar{x}-m}{\sigma} \sqrt{n}}{\left( \frac{s_1(x)}{\sigma} \right)}.$$

and notice that:

(i) The numerator in the last expression has standard normal, $\mathcal{N}(0, 1)$, distribution.

(ii) The denominator $\frac{s_1(\xi)}{\sigma}$ has the same distribution as the random variable $\sqrt{\frac{1}{n-1}\chi_{n-1}^2}$, where $\chi_{n-1}^2 \overset{d}{=} \sum_{i=1}^{n-1} \eta_i^2$ and $\eta_1, \ldots, \eta_{n-1}$ are independent standard normal, $\mathcal{N}(0, 1)$, random variables.

(iii) The variables $\frac{\bar{x}-m}{\sigma}\sqrt{n}$ and $\frac{s_1(\xi)}{\sigma}$ are independent.

The desired statement with regard to the variables $t_{n-1}(\xi)$ follows from (i), (ii), (iii), and the formula [ P §2.8, (38)].

(c) By taking into account that the variable $t_{n-1}(x) = \frac{\bar{x}-m}{\left(\frac{s_1}{\sqrt{n}}\right)}$ has Student distribution, construct confidence intervals for the parameter $m$ with confidence level $1 - \varepsilon$.

(d) Prove that the variable $(n-1)\left(\frac{s_1}{\sigma}\right)^2$ has $\chi^2$-distribution with $(n-1)$ degrees of freedom and, by using this property, construct a confidence interval for the parameter $\sigma$ with confidence level $(1 - \varepsilon)$.

**Problem 2.13.35.** Suppose that $\varphi(t)$ is the characteristic function from Problem 2.13.28 and prove that for every choice of $0 < a_1 < \ldots < a_n$ and $p_k \geq 0$, $1 \leq k \leq n$, with $\sum_{k=1}^{n} p_k = 1$, the function

$$\psi(t) = \sum_{k=1}^{n} p_k \varphi\left(\frac{t}{a_k}\right)$$

is characteristic.

**Problem 2.13.36.** Consider the Gaussian sequences $X = (X_n)_{n \geq 0}$, with covariance function of the form

$$e^{-|i-j|} \quad \text{or} \quad \min(i, j) \; (= 2^{-1}(|i| + |j| - |i - j|)), \quad i, j = 0, 1, 2, \ldots .$$

What structural properties (such as independent increments, stationarity, Markovian, etc.) does this sequence have?

**Problem 2.13.37.** Let $N$ be a standard Gaussian random variable ($N \sim \mathcal{N}(0, 1)$). Prove that for any $\alpha < 1$ one has

$$E\frac{1}{|N|^\alpha} = \frac{1}{\sqrt{2^\alpha \pi}}\Gamma\left(\frac{1}{2} - \frac{\alpha}{2}\right).$$

**Problem 2.13.38.** Let $X$ and $Y$ be two independent standard normal ($\mathcal{N}(0, 1)$) random variables. Prove that

$$E\left(\frac{1}{(X^2 + Y^2)^{p/2}}\right) < \infty$$

if and only if $p < 2$.

**Problem 2.13.39.** Let everything be as in Problem 2.13.38 and suppose that

$$T = \frac{1}{2}(X^2 + Y^2) \quad \text{and} \quad g = \frac{X^2}{X^2 + Y^2}.$$

Prove that:
  (a) $T$ and $g$ are independent;
  (b) $T$ has exponential distribution ($P\{T > t\} = e^{-t}, t > 0$);
  (c) $g$ has arcsin-distribution (with density $\frac{1}{\pi\sqrt{x(1-x)}}$, $x \in (0, 1)$).

**Problem 2.13.40.** Let $B = (B_t)_{t \geq 0}$ be a Brownian motion and let

$$T_a = \inf\{t \geq 0 : B_t = a\}$$

be the first passage time to level $a > 0$, with the understanding that $T_a = \infty$, if the set in the right side of the last relation is empty.

By using *the reflection principle*, i.e., the property $P\{\sup_{s \leq t} B_s > a\} = 2P\{B_t \geq a\}$ (see [17], [103]), prove that the density $p_a(t) = \frac{\partial P\{T_a \leq t\}}{\partial t}$, $t > 0$, is given by the formula

$$p_a(t) = \frac{a}{\sqrt{2\pi t^3}} e^{-a^2/(2t)}.$$

*Hint.* Use the fact that $P\{T_a \leq t\} = 2P\{B_t \geq a\}$.

**Problem 2.13.41.** Let $T = T_1$, where $T_a$ is the first passage time defined in the previous problem. Prove that

$$T \overset{\text{law}}{=} N^{-2},$$

where $N$ is a standard normal ($N \sim \mathcal{N}(0, 1)$) random variable. In addition, prove that the Laplace transform of $T$ is given by

$$\mathsf{E}e^{-\frac{\lambda^2}{2}T} = \mathsf{E}e^{-\frac{\lambda^2}{2} \cdot \frac{1}{N^2}} = e^{-\lambda}, \quad \lambda \geq 0,$$

while the Fourier transform of $T$ is given by

$$\mathsf{E}e^{itT} = \mathsf{E}e^{it\frac{1}{N^2}} = \exp\left\{-|t|^{1/2}\left(1 - i\frac{t}{|t|}\right)\right\}, \quad t \in \mathbb{R}.$$

(The above relations may be viewed as a *constructive* definition of the random variable $1/N^2$, which has a stable distribution with parameters $\alpha = \frac{1}{2}$, $\beta = 0$, $\theta = -1$, and $d = 1$ (see [$\underline{P}$ §3.6, (9) and (10)]).

**Problem 2.13.42.** Let $X$ and $Y$ be two independent normally distributed $(\mathcal{N}(0, \sigma^2))$ random variables.

(a) Prove that the variables

$$\frac{2XY}{\sqrt{X^2 + Y^2}} \quad \text{and} \quad \frac{X^2 - Y^2}{\sqrt{X^2 + Y^2}}$$

are independent and normally distributed with mean 0 and dispersion $1/2$.

(b) Conclude that

$$\frac{X}{Y} - \frac{Y}{X} \stackrel{\text{law}}{=} 2C,$$

where $C$ is a Cauchy random variable with density $1/(\pi(1 + x^2))$, $x \in \mathbb{R}$, and that

$$C - \frac{1}{C} \stackrel{\text{law}}{=} 2C.$$

(c) Generalize this result by showing that for every $a > 0$ one has

$$C - \frac{a}{C} \stackrel{\text{law}}{=} (1 + a)C.$$

(d) Prove that the variables $X^2 + Y^2$ and $\dfrac{X}{\sqrt{X^2 + Y^2}}$ are independent.

*Hint.* (a) Use the representation for the variables $X$ and $Y$ obtained in Problem 2.8.13.

(b) Use the result in Problem 2.8.27 (a).

(c) For the proof it suffices to show that if $f(x) = \dfrac{x - ax^{-1}}{1 + a}$, then for any bounded function $g(x)$ the integrals $\int_{-\infty}^{\infty} g(f(x)) \dfrac{dx}{1 + x^2}$ and $\int_{-\infty}^{\infty} g(x) \dfrac{dx}{1 + x^2}$ coincide.

**Problem 2.13.43.** Prove that for any $0 < H \le 1$ the function

$$R(s, t) = \frac{1}{2}\left(t^{2H} + s^{2H} - |t - s|^{2H}\right), \quad s, t \ge 0$$

is non-negative definite (see formula [P §2.13, (24)]) and that, therefore, one can construct a Gaussian process $B^H = (B_t^H)_{t \ge 0}$ with mean 0 and covariance function $R(s, t)$. (By using Kolmogorov's criterion—see, for example, [17]—it is possible to show that, in fact, $B^H = (B_t^H)_{t \ge 0}$ may be chosen to have continuous sample paths. Such a process is commonly referred to as a *fractal Brownian motion* with *Hurst parameter $H$*.)

Convince yourself that for $H > 1$ the function $R(s, t)$ is *not* non-negative definite.

**Problem 2.13.44.** (*The Darmois–Skitovich theorem.*) Let $\xi_1, \ldots, \xi_n$ be independent and identically distributed random variables and let $a_1, \ldots, a_n$ and $b_1, \ldots, b_n$ be some non-zero constants. Prove that the following characterization holds: if the random variables $\sum_{i=1}^{n} a_i \xi_i$ and $\sum_{i=1}^{n} b_i \xi_i$ are independent, then the variables $\xi_1, \ldots, \xi_n$ must have normal distribution. (With $n = 2$ and with $a_1 = a_2 = 1$, $b_1 = 1$ and $b_2 = -1$ this is nothing but the Bernstein theorem from Problem 2.13.10.)

**Problem 2.13.45.** Let $\xi_1, \xi_2, \ldots$ be any sequence of independent standard normal ($\mathcal{N}(0, 1)$) random variables. Prove that as $n \to \infty$ the random variables

$$X_n = \sqrt{n} \, \frac{\sum_{i=1}^{n} X_i}{\sum_{i=1}^{n} X_i^2} \qquad Y_n = \frac{\sum_{i=1}^{n} X_i}{\left( \sum_{i=1}^{n} X_i^2 \right)^{1/2}}$$

converge in distribution to a standard normal ($\mathcal{N}(0, 1)$) random variable.

**Problem 2.13.46.** Let $(X, Y)$ be any pair of Gaussian random variables with $\mathsf{E}X = \mathsf{E}Y = 0$, $\mathsf{D}X = \mathsf{D}Y = 1$, and with correlation coefficient $\rho_{X,Y}$. Prove that the correlation coefficient $\rho_{\Phi(X), \Phi(Y)}$ of the variables $\Phi(X)$ and $\Phi(Y)$, where $\Phi(x) = (2\pi)^{-1/2} \int_{-\infty}^{x} e^{-y^2/2} \, dy$, is given by the formula

$$\rho_{\Phi(X), \Phi(Y)} = \frac{6}{\pi} \arcsin \frac{\rho_{X,Y}}{2}.$$

**Problem 2.13.47.** Let $(X, Y, Z)$ be any 3-dimensional Gaussian random vector with $\mathsf{E}X = \mathsf{E}Y = \mathsf{E}Z = 0$, $\mathsf{D}X = \mathsf{D}Y = \mathsf{D}Z = 1$ and with correlation coefficients $\rho(X, Y) = \rho_1$, $\rho(X, Z) = \rho_2$, $\rho(Y, Z) = \rho_3$. Prove that (comp. with statement (b) in Problem 2.13.23)

$$\mathsf{P}\{X > 0, Y > 0, Z > 0\} = \frac{1}{8} + \frac{1}{4\pi}\{\arcsin \rho_1 + \arcsin \rho_2 + \arcsin \rho_3\}.$$

*Hint.* Let $A = \{X > 0\}$, $B = \{Y > 0\}$, $C = \{Z > 0\}$. Then, for $p = \mathsf{P}(A \cap B \cap C)$, by the "inclusion–exclusion formula" (Problem 1.1.12), one has

$$1 - p = \mathsf{P}(A \cup B \cup C) = \big[\mathsf{P}(A) + \mathsf{P}(B) + \mathsf{P}(C)\big]$$
$$- \big[\mathsf{P}(A \cap B) + \mathsf{P}(A \cap C) + \mathsf{P}(B \cap C)\big] + p.$$

Finally, use the result in Problem 2.13.23(b).

**Problem 2.13.48.** Prove that the Laplace transform, $\mathsf{E}e^{-\lambda R^2}$, $\lambda > 0$, of the square of the "span", of the Brownian bridge $B^\circ = (B_t^\circ)_{0 \le t \le 1}$, namely, the quantity

$$R = \sqrt{\frac{2}{\pi}} \left( \max_{0 \le t \le 1} B_t^\circ - \min_{0 \le t \le 1} B_t^\circ \right),$$

is given by the formula

$$\mathsf{E}e^{-\lambda R^2} = \left( \frac{\sqrt{\lambda \pi}}{\sinh \sqrt{\lambda \pi}} \right)^2 .$$

**Problem 2.13.49.** (*O. V. Viskov.*) Let $\eta$ and $\zeta$ be any two independent standard normal ($\mathscr{N}(0, 1)$) random variables. Prove that:

(a) For any given function $f = f(z)$, $z \in \mathbb{C}$, with $\mathsf{E}|f(x + (\eta + i\zeta))| < \infty$, the following "averaging" property is in force

$$f(x) = \mathsf{E}f(x + (\eta + i\zeta)).$$

(b) For any Hermite polynomial $\mathrm{He}_n(x)$, $n \geq 0$ (see p. 380 in the Appendix) the following representation is in force

$$\mathrm{He}_n(x) = \mathsf{E}(x + i\zeta)^n.$$

# Chapter 3
# Topology and Convergence in Spaces of Probability Measures: The Central Limit Theorem

## 3.1 Weak Convergence of Probability Measures and Distributions

**Problem 3.1.1.** We say that the function $F = F(x)$, defined on $\mathbb{R}^m$, is *continuous at the point* $x \in \mathbb{R}^m$ if, for every $\varepsilon > 0$, one can find a $\delta > 0$, such that $|F(x) - F(y)| < \varepsilon$ for all $y \in \mathbb{R}^m$ that satisfy

$$x - \delta e < y < x + \delta e,$$

where $e = (1, \ldots, 1) \in \mathbb{R}^m$. The sequence of distribution functions $(F_n)_{n \geq 1}$ is said to *converge in general* to the distribution function $F$ (notation: $F_n \Rightarrow F$) if $F_n(x) \to F(x)$ as $n \to \infty$, for any $x \in \mathbb{R}^m$ at which the function $F = F(x)$ is continuous.

Prove that the statement in [P §3.1, Theorem 2] also holds for the spaces $\mathbb{R}^m$, $m > 1$ (see Remark 1 after [P §3.1, Theorem 1]).

*Hint.* In the context of [P §3.1], it is enough to show only the equivalence (1) $\Leftrightarrow$ (4). To prove the implication (1) $\Rightarrow$ (4), suppose that $x \in \mathbb{R}^m$ is a continuity point for $F$, and convince yourself that if $\partial(-\infty, x]$ is the boundary of the set $(-\infty, x] = (-\infty, x_1] \times \cdots \times (-\infty, x_m]$, then $P(\partial(-\infty, x]) = 0$, so that $P_n((-\infty, x]) \to P((-\infty, x])$, i.e., $F_n(x) \to F(x)$. The proof of the implication (4) $\Rightarrow$ (1) in the $m$-dimensional case is analogous to the one-dimensional argument in the proof of [P §3.1, Theorem 2].

**Problem 3.1.2.** Prove that in the spaces $\mathbb{R}^m$ the class of "elementary" sets, $\mathscr{K}$, is a *convergence defining* class.

**Problem 3.1.3.** Let $E$ be one of the space $\mathbb{R}^\infty$, $C$ or $D$ (see [P §2.2]). The sequence of probability measures $(P_n)_{n \geq 1}$ (defined on the Borel $\sigma$-algebra, $\mathscr{E}$, generated by the open sets in the respective space) *converges in general*, in the sense of finite-dimensional distributions, to the probability measure $P$ (notation: $P_n \overset{f}{\Rightarrow} P$), if $P_n(A) \to P(A)$ as $n \to \infty$, for all *cylindrical sets* $A$ with $P(\partial A) = 0$.

A.N. Shiryaev, *Problems in Probability*, Problem Books in Mathematics, DOI 10.1007/978-1-4614-3688-1_3,
© Springer Science+Business Media New York 2012

Prove that in the case of the space $\mathbb{R}^\infty$ one has

$$(P_n \overset{f}{\Rightarrow} P) \iff (P_n \Rightarrow P). \tag{$*$}$$

Can one make the same statement for the spaces $C$ and $D$?

*Hint.* The implication $\Leftarrow$ in $(*)$ is straight-forward. Therefore it is enough to prove only that $(P_n \overset{f}{\Rightarrow} P) \Rightarrow (P_n \to P)$. Let $f$ be any bounded ($|f| \leq c$) function from the space $C(\mathbb{R}^\infty)$. Given any $m \in \mathbb{N} = \{1, 2, \dots\}$, define the functions $f_m \colon \mathbb{R}^\infty \to \mathbb{R}$ by

$$f_m(x_1, \dots, x_m, x_{m+1}, \dots) = f_m(x_1, \dots, x_m, 0, 0, \dots).$$

Clearly, one has $f_m \in C(\mathbb{R}^\infty)$, $|f_m| \leq c$ and $f_m(x) \to f(x)$, for every $x \in \mathbb{R}^\infty$. Next, consider the sets

$$A_m = \Big\{ x \in \mathbb{R}^\infty : |f_m(x) - f(x)| \leq \varepsilon \Big\},$$

and convince yourself that the following estimate holds for all sufficiently large $n$ and $m$:

$$\left| \int_{\mathbb{R}^\infty} f_m \, dP_n - \int_{\mathbb{R}^\infty} f \, dP_n \right| \leq \varepsilon P_n(A_m) + 2c P(\overline{A}_m) \leq \varepsilon + 4 c \varepsilon.$$

Then notice that $\int_{\mathbb{R}^\infty} f_m \, dP_n \to \int_{\mathbb{R}^\infty} f_m \, dP$ for every $m$ and by using the above estimate prove that

$$\left| \overline{\lim_n} \int_{\mathbb{R}^\infty} f \, dP_n - \int_{\mathbb{R}^\infty} f_m \, dP \right| \leq \varepsilon + 4 c \varepsilon,$$

$$\left| \underline{\lim_n} \int_{\mathbb{R}^\infty} f \, dP_n - \int_{\mathbb{R}^\infty} f_m \, dP \right| \leq \varepsilon + 4 c \varepsilon.$$

for all sufficiently large $m$. The Lebesgue dominated convergence theorem yields $\int_{\mathbb{R}^\infty} f_m \, dP \to \int_{\mathbb{R}^\infty} f \, dP$, and the previous two inequalities yield:

$$\left| \overline{\lim_n} \int_{\mathbb{R}^\infty} f \, dP_n - \int_{\mathbb{R}^\infty} f \, dP \right| \leq \varepsilon + 4 c \varepsilon;$$

$$\left| \underline{\lim_n} \int_{\mathbb{R}^\infty} f \, dP_n - \int_{\mathbb{R}^\infty} f \, dP \right| \leq \varepsilon + 4 c \varepsilon.$$

Since $\varepsilon > 0$ is arbitrarily chosen, it follows that

$$\int_{\mathbb{R}^\infty} f \, dP_n \to \int_{\mathbb{R}^\infty} f \, dP, \qquad n \to \infty.$$

**Problem 3.1.4.** Let $F$ and $G$ be any two distribution functions on the real line and let

$$L(F, G) = \inf \left\{ h > 0: F(x - h) - h \le G(x) \le F(x + h) + h \right\}$$

be the *Lévy distance* between them. Prove that the convergence in the *Lévy metric*, $L(\cdot, \cdot)$, is equivalent to the *convergence in general*, i.e.

$$(F_n \Rightarrow F) \iff (L(F_n, F) \to 0).$$

*Hint.* The implication $(L(F_n, F) \to 0) \Rightarrow (F_n \Rightarrow F)$ follows directly from the definition. The inverse implication can be established by contradiction, i.e., by showing that $F_n \Rightarrow F$, while, at the same time, $L(F_n, F) \not\to 0$, leads to a contradiction.

**Problem 3.1.5.** Suppose that $F_n \Rightarrow F$ and that the distribution function $F$ is *continuous*. Prove that the functions $F_n(x)$ converge *uniformly* to $F(x)$ as $n \to \infty$ (comp. with Problem 1.6.8):

$$\sup_x |F_n(x) - F(x)| \to 0, \quad n \to \infty.$$

*Hint.* Choose an arbitrary $\varepsilon > 0$ and let $m > 1/\varepsilon$. Taking into account that $F$ is continuous, choose the points $x_1, \ldots, x_{m-1}$ so that $F(x_i) = \frac{i}{m}$ and $|F_n(x_i) - F(x_i)| < \varepsilon$, $i = 1, \ldots, m - 1$, for any sufficiently large $n$. Conclude that for any $x \in [x_k, x_{k+1}]$ (with the understanding that $x_0 = -\infty$ and $x_m = \infty$) one must have

$$F_n(x) - F(x) \le F_n(x_{k+1}) - F(x_k) \le F(x_{k+1}) + \varepsilon - F(x_k) = \varepsilon + \frac{1}{m} < 2\varepsilon .$$

Analogously, $F(x) - F_n(x) < 2\varepsilon$ and, therefore, $|F_n(x) - F(x)| < 2\varepsilon$ for all $x \in \mathbb{R}$.

**Problem 3.1.6.** Prove the statement formulated in Remark 1 after Theorem 1 in [P §3.1].

**Problem 3.1.7.** Prove the equivalence of conditions (I*)—(IV*), formulated in Remark 2 after Theorem 1 in [P §3.1].

**Problem 3.1.8.** Prove that $P_n \xrightarrow{w} P$ ($\xrightarrow{w}$ stands for "weakly converges to") if and only if every sub-sequence, $(P_{n'})$, of the sequence $(P_n)$ contains a sub-sub-sequence $(P_{n''})$ with the property $P_{n''} \xrightarrow{w} P$.

*Hint.* The *necessity* part is obvious. For the *sufficiency* part, it is enough to notice that if $P_n \xrightarrow{w} P$, then one can find: some continuous and bounded function $f$, some $\varepsilon > 0$, and some sub-sequence $(n')$, so that

$$\left| \int_E f \, dP_{n'} - \int_E f \, dP \right| > \varepsilon.$$

By using this property, one can show that the existence of a sub-sub-sequence $(n'') \subseteq (n')$ with $P_{n''} \xrightarrow{w} P$ leads to a contradiction.

**Problem 3.1.9.** Give an example of probability measures $P$, $P_n$, $n \geq 1$, on $(\mathbb{R}, \mathscr{B}(\mathbb{R}))$, such that $P_n \overset{w}{\to} P$, and, at the same time, it is not true that $P_n(B) \to P(B)$ *for all* Borel sets $B \in \mathscr{B}(\mathbb{R})$.

**Problem 3.1.10.** Give an example of distribution functions $F = F(x)$, $F_n = F_n(x)$, $n \geq 1$, such, that $F_n \overset{w}{\to} F$, but $\sup_x |F_n(x) - F(x)| \not\to 0$, $n \to \infty$.

**Problem 3.1.11.** In many probability theory texts, the implication (4) $\Rightarrow$ (3) in [P §3.1, Theorem 2], concerning the convergence of the distribution functions $F_n$, $n \geq 1$, to the distribution function $F$, is attributed to E. Helly and H. E. Bray. Prove one more time the following statements:

(a) *Helly–Bray Lemma.* If $F_n \Rightarrow F$ (see Definition 1), then

$$\lim_n \int_a^b g(x)\, dF_n(x) = \int_a^b g(x)\, dF(x),$$

where $a$ and $b$ are any two continuity points for the distribution function $F = F(x)$, and $g = g(x)$ is any continuous function on the interval $[a, b]$.

(b) *Helly–Bray Theorem.* If $F_n \Rightarrow F$, then

$$\lim_n \int_{-\infty}^{\infty} g(x)\, dF_n(x) = \int_{-\infty}^{\infty} g(x)\, dF(x),$$

for any bounded and continuous function $g = g(x)$ defined on the real line $\mathbb{R}$.

**Problem 3.1.12.** Suppose that $F_n \Rightarrow F$ and that for some $b > 0$ the sequence $\left( \int |x|^b\, dF_n(x) \right)_{n \geq 1}$ happens to be bounded. Prove that:

$$\lim_n \int |x|^a\, dF_n(x) = \int |x|^a\, dF(x), \quad 0 \leq a \leq b;$$

$$\lim_n \int x^k\, dF_n(x) = \int x^k\, dF(x) \quad \text{for every } k = 1, 2, \ldots, [b], k \neq b.$$

**Problem 3.1.13.** Let $F_n \Rightarrow F$ and let $\mu = \mathrm{med}(F)$ and $\mu_n = \mathrm{med}(F_n)$ denote, respectively, the medians of the distributions $F$ and $F_n$, $n \geq 1$ (see Problem 1.4.5). Assuming that the medians $\mu$ and $\mu_n$ are uniquely defined for all $n \geq 1$, prove that $\mu_n \to \mu$.

**Problem 3.1.14.** Suppose that the distribution function $F$ is uniquely determined by its moments $m^{(k)} = \int_{-\infty}^{\infty} x^k\, dF(x)$, $k = 1, 2, \ldots$, and let $(F_n)_{n \geq 1}$ be any sequence of distribution functions, such that

$$m_n^{(k)} = \int_{-\infty}^{\infty} x^k\, dF_n(x) \to m^{(k)} = \int_{-\infty}^{\infty} x^k\, dF(x), \quad k = 1, 2, \ldots .$$

Prove that $F_n \Rightarrow F$.

**Problem 3.1.15.** Let $\mu$ be any $\sigma$-finite measure on the Borel $\sigma$-algebra, $\mathscr{E}$, for some metric space $(E, \rho)$. Prove that for every $B \in \mathscr{E}$ one has

$$\mu(B) = \sup\{\mu(F); \ F \subseteq B, \ F \text{ is closed}\} = \inf\{\mu(G); \ G \supseteq B, \ G \text{ is open})\}.$$

**Problem 3.1.16.** Prove that a sequence of distribution functions, $F_n, n \geq 1$, defined on the real line $\mathbb{R}$, converges weakly to the distribution functions $F$ ($F_n \overset{w}{\to} F$) if and only if there is a set $D$ which is everywhere dense in $\mathbb{R}$ and is such that $F_n(x) \to F(x)$ for every $x \in D$.

**Problem 3.1.17.** Suppose that the functions $g(x)$ and $(g_n(x))_{n \geq 1}$, $x \in \mathbb{R}$, are continuous and have the properties:

$$\sup_{x,n} |g_n(x)| \leq c < \infty;$$

$$\limsup_{n} \ _{x \in B} |g_n(x) - g(x)| = 0,$$

for every bounded interval $B = [a, b]$.
  Prove that the convergence of distribution functions $F_n \Rightarrow F$ implies

$$\lim_{n} \int_{\mathbb{R}} g_n(x) \, dF_n(x) = \int_{\mathbb{R}} g(x) \, dF(x).$$

By constructing appropriate examples, prove that, in general, the *point-wise* convergence $g_n(x) \to g(x)$, $x \in \mathbb{R}$, is not enough to guarantee the above convergence.

**Problem 3.1.18.** Suppose that the following convergence of distribution functions takes place: $F_n \Rightarrow F$ as $n \to \infty$.
  (a) By constructing appropriate examples, prove that, in general,

$$\int_{\mathbb{R}} x \, dF_n(x) \nrightarrow \int_{\mathbb{R}} x \, dF(x).$$

  (b) Prove that if $\sup_n \int_{\mathbb{R}} |x|^k \, dF_n \leq c < \infty$, for some $k \geq 1$, then for all $1 \leq l \leq k - 1$ one must have

$$\int_{\mathbb{R}} x^l \, dF_n(x) \to \int_{\mathbb{R}} x^l \, dF(x).$$

**Problem 3.1.19.** As a generalization of the previous problem, prove that if $f = f(x)$ is some continuous function, not necessarily bounded, but such that

$$\lim_{|x| \to \infty} \frac{|f(x)|}{g(x)} = 0,$$

for some positive function $g = g(x)$ with $\sup_n \int_{\mathbb{R}} g(x)\,dF_n(x) \le c < \infty$, then

$$\int_{\mathbb{R}} f(x)\,dF_n(x) \to \int_{\mathbb{R}} f(x)\,dF(x).$$

## 3.2  Relative Compactness and Tightness of Families of Probability Distributions

**Problem 3.2.1.** Prove Theorems 1 and 2 from [P §3.2] for the spaces $\mathbb{R}^n$, $n \ge 2$.

**Problem 3.2.2.** Let $P_\alpha$ be a Gaussian measure on the real line, with parameters $m_\alpha$ and $\sigma_\alpha^2$, for every $\alpha \in \mathfrak{A}$. Prove that the family $\mathscr{P} = \{P_\alpha; \alpha \in \mathfrak{A}\}$ is tight if and only if there are constants, $a$ and $b$, for which one can write

$$|m_\alpha| \le a, \ \sigma_\alpha^2 \le b, \ \alpha \in \mathfrak{A}.$$

*Hint.* The *sufficiency* statement follows from the fact that for every $\alpha \in \mathfrak{A}$ one can find a random variable $\eta_\alpha \sim \mathcal{N}(0, 1)$, such that $\xi_\alpha = m_\alpha + \sigma_\alpha \eta_\alpha$. With this observation in mind, one can conclude that $\mathsf{P}\{|\xi_\alpha| \ge n\} \le \mathsf{P}\{|\eta_\alpha| \le \frac{n-a}{b}\}$. As a result, the family $\{P_\alpha\}$ must be tight. The *necessity* statement can be established by contradiction.

**Problem 3.2.3.** Give examples of tight and non-tight families of probability measures $\mathscr{P} = \{P_\alpha; \alpha \in \mathfrak{A}\}$, defined on the measure space $(\mathbb{R}^\infty, \mathscr{B}(\mathbb{R}^\infty))$.
  *Hint.* Consider the following families of measures:
  (a) $\{P_\alpha\}$, where $P_\alpha \equiv P$ is such that

$$P(A) = \begin{cases} 1, & \text{if } (0, 0, \dots) \in A, \\ 0, & \text{if } (0, 0, \dots) \notin A; \end{cases}$$

  (b) $\{P_n, n \in \mathbb{N}\}$, where $P_n$ is a probability measure concentrated at the point $x_n = (n, 0, 0, \dots)$.

**Problem 3.2.4.** Let $P$ be a probability measure, defined on the Borel $\sigma$-algebra, $\mathscr{E}$, in some metric space $(E, \rho)$. We say that the measure $P$ is *tight* (comp. with [P §3.2, Definition 2]), if for any $\varepsilon > 0$ one can find a compact set $K \subseteq E$, such that $P(K) \ge 1 - \varepsilon$. Prove the following result, known as "Ulam theorem": *every probability measure $P$, defined on the Borel $\sigma$-algebra in some Polish space (i.e., some complete and separable metric space) is automatically tight.*

**Problem 3.2.5.** Suppose that $X = \{X_\alpha \in \mathbb{R}^d; \alpha \in \mathfrak{A}\}$ is some family of random vectors in $\mathbb{R}^d$, chosen so that $\sup_\alpha \mathsf{E}\|X_\alpha\|^r < \infty$ for some $r > 0$. Setting $P_\alpha = \mathrm{Law}(X_\alpha)$, $\alpha \in \mathfrak{A}$, show that family $\mathscr{P} = \{P_\alpha; \alpha \in \mathfrak{A}\}$ is tight.

**Problem 3.2.6.** The family of random vectors $\{\xi_t \in \mathbb{R}^n; t \in T\}$ is said to be *tight*, if

$$\lim_{a \to \infty} \sup_{t \in T} P\{\|\xi_t\| > a\} = 0.$$

(a) Prove that $\{\xi_k \in \mathbb{R}^n; k \geq 0\}$ is tight if and only if

$$\lim_{a \to \infty} \overline{\lim_{k \to \infty}} \, P\{\|\xi_k\| > a\} = 0.$$

(b) Prove that the family of non-negative random variables $\{\xi_k; k \geq 0\}$ is tight if and only if

$$\lim_{\lambda \downarrow 0} \overline{\lim_k}[1 - Ee^{-\lambda \xi_k}] = 0.$$

**Problem 3.2.7.** Let $(\xi_k)_{k \geq 0}$ be any sequence of random vectors in $\mathbb{R}^n$, and suppose that $\xi_k \xrightarrow{d} \xi$, i.e., the distributions $F_k$ of the vectors $\xi_k$ converge *weakly* (equivalently, converge *essentially*) to the distribution $F$ of some random vector $\xi$. Prove that family $\{\xi_k; k \geq 0\}$ is tight.

**Problem 3.2.8.** Let $(\xi_k)_{k \geq 0}$ be any tight sequence of random variables and suppose that the sequence $(\eta_k)_{k \geq 0}$ is such that $\eta_k \xrightarrow{P} 0$ as $k \to \infty$. Conclude from these conditions that $\xi_k \eta_k \xrightarrow{P} 0$ as $k \to \infty$.

**Problem 3.2.9.** Let $X_1, X_2, \ldots$ be any *infinite* sequence of exchangeable random variables (for a definition, see Problem 2.5.4) and suppose that the variables $X_i$ take only the values 0 or 1.

Prove the following result: there is a probability distribution function $G = G(\lambda)$ on the interval $[0, 1]$, such that, for every $0 \leq k \leq n$, and every $n \geq 1$, one has

$$P\{X_1 = 1, \ldots, X_k = 1, X_{k+1} = 0, \ldots, X_n = 0\} = \int_0^1 \lambda^k (1 - \lambda)^{n-k} \, dG(\lambda).$$

(This is a special case of B. de Finetti's Theorem, according to which the distribution law of every *infinite* sequence of exchangeable random variables can be identified with the distribution law of a (convex) mixture of infinite sequences of independent and identically distributed random variables—see [1] and [29].)

*Hint.* Consider the event

$$A_k = \{X_1 = 1, \ldots, X_k = 1, X_{k+1} = 0, \ldots, X_n = 0\}$$

and write the probability $P(A_k)$ in the form

$$P(A_k) = \sum_{j=0}^m P(A_k \mid S_m = j)P\{S_m = j\}, \tag{*}$$

where $m \geq n$ and $S_m = \sum_{i=1}^{m} X_i$. Next, by using the exchangeability property, prove that the right side of $(*)$ may be re-written as

$$
E \prod_{i=0}^{k-1} (mY_m - i) \times \prod_{j=0}^{n-k-1} (m(1 - Y_m) - j) \times \frac{1}{\prod_{l=0}^{n-1}(m - l)} ,
$$

where $Y_m = S_m / m$. (Notice that for large $m$ this expression is close to $E[Y_m^k(1 - Y_m)^{n-k}]$.) Finally, pass to the limit as $m \to \infty$ and conclude that the limit can be expressed as $\int_0^1 \lambda^k (1 - \lambda)^{n-k} \, dG(\lambda)$, where $G(\lambda)$ is some distribution function on the interval $[0, 1]$.

**Problem 3.2.10.** Let $\xi_1, \ldots, \xi_n$ be any sequence of exchangeable random variables, which take the values 0 and 1. Prove that:

(a) $P(\xi_i = 1 \mid S_n) = \dfrac{S_n}{n}$, where $S_n = \xi_1 + \ldots + \xi_n$;

(b) $P(\xi_i = 1, \xi_j = 1 \mid S_n) = \dfrac{S_n(S_n - 1)}{n(n - 1)}$, where $i \neq j$.

**Problem 3.2.11.** As a generalization of [P §1.11, Theorem 2], prove that if $\eta_1, \ldots, \eta_n$ is some set of exchangeable random variables with values in $\{0, 1, 2, \ldots\}$, and if $S_k = \eta_1 + \ldots + \eta_k$, $1 \leq k \leq n$, then

$$
P(S_k < k \text{ for all } 1 \leq k \leq n \mid S_n) = \left(1 - \frac{S_n}{n}\right)^+ .
$$

## 3.3   The Method of Characteristic Functions for Establishing Limit Theorems

**Problem 3.3.1.** Prove the statement in [P §3.3, Theorem 1] in the case of the spaces $\mathbb{R}^n$, $n \geq 2$.

*Hint.* The proof is analogous to the one-dimensional case, except for [P §3.3, Lemma 3]. The multidimensional analog of this lemma can be stated in the form:

$$
\int_A dF(x) \leq \frac{k}{a^n} \int_B (1 - \operatorname{Re} \varphi(t)) \, dt,
$$

where

$$
A = \left\{x \in \mathbb{R}^n : |x_1| \leq \tfrac{1}{a}, \ldots, |x_n| \leq \tfrac{1}{a}\right\},
$$

$$
B = \left\{t \in \mathbb{R}^n : 0 \leq t_1 \leq a, \ldots, 0 \leq t_n \leq a\right\}.
$$

**Problem 3.3.2.** (*The law of large numbers.*)

(a) Let $\xi_1, \xi_2, \ldots$ be any sequence of independent random variables with finite expected values $\mathsf{E}|\xi_n|$ and dispersions $\mathsf{D}\xi_n \leq K$, $n \geq 1$. Prove that the law of large numbers holds: for every $\varepsilon > 0$

$$\mathsf{P}\left\{ \left| \frac{\xi_1 + \ldots + \xi_n}{n} - \frac{\mathsf{E}(\xi_1 + \ldots + \xi_n)}{n} \right| \geq \varepsilon \right\} \to 0 \quad \text{as } n \to \infty. \qquad (*)$$

(b) Let $\xi_1, \xi_2, \ldots$ be any sequence of random variables with finite expected values $\mathsf{E}|\xi_n|$, dispersions $\mathsf{D}\xi_n \leq K$, $n \geq 1$, and covariances $\mathrm{cov}(\xi_i, \xi_j) \leq 0$, $i \neq j$. Prove that the law of large numbers $(*)$ holds.

*Hint.* To prove (a) and (b), use Chebyshev's inequality.

(c) (S. N. Bernstein.) Let $\xi_1, \xi_2, \ldots$ be any sequence of random variables with finite expected values $\mathsf{E}|\xi_n|$ and dispersions $\mathsf{D}\xi_n \leq K$, $n \geq 1$, and suppose that the covariances are such that $\mathrm{cov}(\xi_i, \xi_j) \to 0$ as $|i - j| \to \infty$. Prove that when these conditions are satisfied the law of large numbers $(*)$ holds.

*Hint.* Convince yourself that under the specified conditions one has

$$\mathsf{D}(\xi_1 + \ldots + \xi_n)/n \to 0 \quad \text{as } n \to \infty.$$

(d) Let $\xi_1, \xi_2, \ldots$ be independent and identically distributed random variables, let $\mu_n = \mathsf{E}[\xi_1 I(|\xi_1| \leq n)]$, and suppose that

$$\lim_{x \to \infty} x \mathsf{P}\{|\xi_1| > x\} = 0.$$

Prove the following version of the law of large numbers:

$$\frac{S_n}{n} - \mu_n \overset{\mathsf{P}}{\to} 0,$$

where, as usual, $S_n = \xi_1 + \ldots + \xi_n$. (See also Problem 3.3.20.)

*Hint.* Given some $s > 0$, set $\xi_i^{(s)} = \xi_i I(|\xi_i| \leq s)$ and $m_n^{(s)} = \mathsf{E}[\xi_1^{(s)} + \ldots + \xi_n^{(s)}]$, and prove that

$$\mathsf{P}\left\{ |\xi_1 + \ldots + \xi_n - m_n^{(s)}| > t \right\} \leq t^{-2} \mathsf{D}\left( \xi_1^{(s)} + \ldots + \xi_n^{(s)} \right)$$

$$+ \mathsf{P}\{\xi_1 + \ldots + \xi_n \neq \xi_1^{(s)} + \ldots + \xi_n^{(s)}\}.$$

By using this estimate, convince yourself (setting $s = n$ and $t = \varepsilon n$, $\varepsilon > 0$) that

$$\mathsf{P}\left\{ \left| \frac{\xi_1 + \ldots + \xi_n}{n} - \mathsf{E}\xi_1 I(|\xi_1| \leq n) \right| > \varepsilon \right\}$$

$$\leq \frac{2}{\varepsilon^2 n} \int_0^n x \mathsf{P}\{|\xi_1| > x\} dx + n \mathsf{P}\{|\xi_1| > n\},$$

which leads to the desired property.

**Problem 3.3.3.** In the setting of Theorem 1, prove that the family $\{\varphi_n, n \geq 1\}$ is *uniformly equicontinuous* and the convergence $\varphi_n \to \varphi$ is uniform on every finite interval.

*Hint.* The uniform equicontinuity of the family $\{\varphi_n, n \geq 1\}$ means that for every $\varepsilon > 0$ one can find a $\delta > 0$, such that, for every $n \geq 1$ and every $s, t$ with $|t - s| < \delta$, one has $|\varphi_n(t) - \varphi_n(s)| < \varepsilon$.

Assuming that $F_n \overset{w}{\to} F$, the Prokhorov Theorem (see [P §3.2, Theorem 1]) implies that, given any $\varepsilon > 0$, one can find some $a > 0$ so that $\int_{|x| \geq a} dF_n < \varepsilon$, $n \geq 1$. Consequently,

$$|\varphi_n(t+h) - \varphi_n(t)| \leq \int_{|x| \leq a} |e^{itx} - 1| \, dF_n + 2\varepsilon,$$

from where the desired uniform equicontinuity property easily follows. By using this property one can prove that

$$\sup_{t \in [a,b]} |\varphi_n(t) - \varphi(t)| \to 0 \quad \text{as } n \to \infty,$$

for every finite interval $[a, b]$.

**Problem 3.3.4.** Let $\xi_n$, $n \geq 1$, be any sequence of random variables with characteristic functions $\varphi_{\xi_n}(t)$, $n \geq 1$. Prove that $\xi_n \overset{d}{\to} 0$ if and only if $\varphi_{\xi_n}(t) \to 1$ as $n \to \infty$, in some neighborhood of the point $t = 0$.

*Hint.* For the proof of the sufficiency part, consider using Lemma 3, according to which the family of measures $\{\text{Law}(\xi_n), n \geq 1\}$ is tight.

**Problem 3.3.5.** Let $X_1, X_2, \ldots$ be independent and identically distributed random vectors in $\mathbb{R}^k$ with vanishing mean and with (finite) covariance matrix $\Gamma$. Prove that

$$\frac{X_1 + \cdots + X_n}{\sqrt{n}} \overset{d}{\to} \mathcal{N}(0, \Gamma).$$

(Comp. with Theorem 3.)

*Hint.* According to Problem 3.3.1, it is enough to prove that, for every $t \in \mathbb{R}^k$, one has

$$\mathsf{E}e^{i(t,\xi_n)} \to \mathsf{E}e^{i(t,\xi)} \quad \text{as } n \to \infty,$$

where $\xi_n = n^{-1/2}(X_1 + \cdots + X_n)$ and $\xi \sim \mathcal{N}(0, \Gamma)$.

**Problem 3.3.6.** Let $\xi_1, \xi_2, \ldots$ and $\eta_1, \eta_2, \ldots$ be two sequences of random variables, chosen so that $\xi_n$ and $\eta_n$ are independent for every $n$, and suppose that $\xi_n \overset{d}{\to} \xi$ and $\eta_n \overset{d}{\to} \eta$ as $n \to \infty$, where $\xi$ and $\eta$ are also independent.

(a) Prove that the sequence of bi-variate random variables $(\xi_n, \eta_n)$ converges in distribution to $(\xi, \eta)$.

(b) Let $f = f(x, y)$ be any continuous function. Verify that the sequence $f(\xi_n, \eta_n)$ converges in distribution to $f(\xi, \eta)$.

*Hint.* The convergence $(\xi_n, \eta_n) \xrightarrow{d} (\xi, \eta)$ obtains from the statement in Problem 3.3.1. In order to establish the convergence $f(\xi_n, \eta_n) \xrightarrow{d} f(\xi, \eta)$, consider the composition $\varphi \circ f \colon \mathbb{R}^2 \to \mathbb{R}$, where $\varphi \colon \mathbb{R} \to \mathbb{R}$ is some continuous and bounded function.

**Problem 3.3.7.** By constructing an appropriate example, prove that in part (2) of [P §3.3, Theorem 1] the continuity condition at 0 for the limiting characteristic function $\varphi(t) = \lim_n \varphi_n(t)$ cannot be weakened in general. (In other words, if $\varphi(t)$ is not continuous at 0, then it is possible that $\varphi_n(t) \to \varphi(t)$, but there is no function $F$ for which $F_n \xrightarrow{w} F$.) Convince yourself by way of example that if the continuity at 0 for the limiting function $\varphi(t)$ fails, then the family of probability distributions $\{P_n, n \geq 1\}$, with characteristic functions $\varphi_n(t)$, $n \geq 1$, may no longer be tight.

*Hint.* Take $F_n$ to be the distribution function of a Gaussian random variable with mean 0 and dispersion $n$.

**Problem 3.3.8.** As an extension to inequality [P §3.3, (4)] from [P §3.3, Lemma 3], prove that if $\xi$ is a random variable with characteristic function $\varphi(t)$, then:

(a) For any $a > 0$ one has

$$P\{|\xi| \leq a^{-1}\} \leq \frac{2}{a} \int_{|t| \leq a} |\varphi(t)| \, dt.$$

(b) For any positive $b$ and $\delta$ one has

$$P\{|\xi| \geq b\} \leq \frac{\left(1 + \frac{2\pi}{b\delta}\right)^2}{\delta} \int_0^\delta [1 - \operatorname{Re} \varphi(t)] \, dt.$$

(c) If $\xi$ is a non-negative random variable and $\psi(a) = Ee^{-a\xi}$, $a \geq 0$, is its Laplace transform, then

$$P\{\xi \geq a^{-1}\} \leq 2(1 - \psi(a)).$$

**Problem 3.3.9.** Suppose that $\xi, \xi_1, \xi_2, \ldots$ is some sequence of random vectors in $\mathbb{R}^n$. Prove that $\xi_k \xrightarrow{d} \xi$ as $k \to \infty$ if and only if for any vector $t \in \mathbb{R}^n$ one has the following convergence of the respective scalar products

$$(\xi_k, t) \xrightarrow{d} (\xi, t).$$

(This result is the basis for the *Cramér–Wold method*, which comes down to replacing the test for convergence in distribution of random vectors from $\mathbb{R}^n$ to the test for convergence in distribution of certain scalar random variables.)

**Problem 3.3.10.** As a continuation to Theorem 2, which is known as *Khinchin law of large numbers* (or *Khinchin criterion*), prove the following statement.

Let $\xi_1, \xi_2, \ldots$ be some sequence of independent and identically distributed random variables and let $S_n = \xi_1 + \ldots + \xi_n$, $0 < p < 2$. Then there is a constant, $c \in \mathbb{R}$, for which

$$n^{-1/p} S_n \xrightarrow{\mathsf{P}} c,$$

if and only if one can claim that as $r \to 0$:

(a) $r^p \mathsf{P}\{|\xi_1| > r\} \to 0$ and $c = 0$, if $p < 1$;
(b) $r \mathsf{P}\{|\xi_1| > r\} \to 0$ and $\mathsf{E}[\xi_1 I(|\xi_1| \le r)] \to c$, if $p = 1$;
(c) $r^p \mathsf{P}\{|\xi_1| > r\} \to 0$ and $\mathsf{E}\xi_1 = c = 0$, if $p > 1$.

**Problem 3.3.11.** Let $\xi_1, \xi_2, \ldots$ be a sequence of independent and identically distributed random variables and let $S_n = \xi_1 + \ldots + \xi_n$. Prove that the variables $n^{-1/2} S_n$ converge in probability as $n \to \infty$ if and only if $\mathsf{P}\{\xi_1 = 0\} = 1$.

**Problem 3.3.12.** Let $F(x)$ and $(F_n(x))_{n \ge 1}$ be some distribution functions and let $\varphi(t)$ and $(\varphi_n(t))_{n \ge 1}$ be their respective characteristic functions. Prove that if

$$\sup_t |\varphi_n(t) - \varphi(t)| \to 0,$$

then

$$\sup_x |F_n(x) - F(x)| \to 0.$$

**Problem 3.3.13.** Let $\xi_1, \xi_2, \ldots$ be independent and identically distributed random variables with distribution function $F = F(x)$ and let $S_n = \xi_1 + \cdots + \xi_n$, $n \ge 1$.

Prove the following version of *the law of large numbers* (due to A. N. Kolmogorov): for the existence of a sequence of numbers $(a_n)_{n \ge 1}$, such that

$$\frac{S_n}{n} - a_n \xrightarrow{\mathsf{P}} 0 \quad \text{as } n \to \infty, \tag{$*$}$$

it is *necessary and sufficient* that

$$n \mathsf{P}\{|\xi_1| > n\} \to 0 \quad \text{as } n \to \infty, \tag{$**$}$$

or, equivalently, that

$$x[1 - F(x) - F(-x)] \to 0 \quad \text{as } x \to \infty.$$

Furthermore, when these conditions hold one has $a_n - \mathsf{E}(\xi_1 I(|\xi_1| \le n)) \to 0$ as $n \to \infty$. (The existence of a sequence $(a_n)_{n \ge 1}$ for which the property $(*)$ holds is known as "*stability of the sequence* $\left(\frac{S_n}{n}\right)_{n \ge 1}$ *in the sense of Kolmogorov*".)

**Problem 3.3.14.** In the context of the previous problem, prove that if $\mathsf{E}|\xi_1| < \infty$ then the condition $(**)$ holds and it is possible to take $a_n \equiv m$, where $m = \mathsf{E}\xi_1$.

(Comp. with [P §3.3, Theorem 2], *Khinchin criterion* for the law of large numbers, and Problems 3.3.10 and 3.3.13.)

**Problem 3.3.15.** Let $\xi_1, \xi_2, \ldots$ be any sequence of independent and identically distributed random variables that take values $\pm 3, \pm 4, \ldots$ with probabilities

$$P\{\xi_1 = \pm x\} = \frac{c}{2x^2 \ln x}, \qquad x = 3, 4, \ldots ,$$

where the normalizing constant $c$ is given by

$$c = \left( \sum_{x=3}^{\infty} \frac{1}{x^2 \ln x} \right)^{-1}.$$

Prove that in this case $E|\xi_1| = \infty$, but condition $(**)$ from Problem 3.3.13 holds and it is possible to take $a_n \equiv 0$, i.e., with this choice one has $\frac{S_n}{n} \xrightarrow{P} 0$.

*Remark.* As the random variables $\xi_1, \xi_2, \ldots$ do not possess finite first moments $(E|\xi_i| = \infty)$, it is not possible to formulate the law of large numbers in the sense of Khinchin $(n^{-1} S_n \to m$, where $m = E\xi_1$—see [P §3.3, Theorem 2]). Nevertheless the random variables $\xi_1, \xi_2, \ldots$ exhibit stability in the sense of Kolmogorov (see Problem 3.3.13), in that

$$\frac{S_n}{n} \xrightarrow{P} \widetilde{m} \ (= 0),$$

where $\widetilde{m} = \widetilde{E}\xi_1$ is *the generalized expected value*, which was defined by A. N. Kolmogorov (see [66, Chap. VI., §4]) by the formula

$$\widetilde{E}\xi_1 = \lim_{n \to \infty} E\big(I(|\xi_1| \leq n)\xi_1\big).$$

Later A. N. Kolmogorov called this generalized expected value the "*A-integral*". (It is common in analysis to say that the function $f = f(x)$, $x \in \mathbb{R}$, is *A-integrable*, if:

   (i) $f$ belongs to the space $L^1$ in weak sense (i.e., $\lim_n n\lambda\{x : |f(x)| > n\} \to 0$); and

   (ii) The limit $\lim_n \int_{\{x:|f(x)|\leq n\}} f(x)\lambda(dx)$ exists, where $\lambda$ is the Lebesgue measure on $(\mathbb{R}, \mathscr{B}(\mathbb{R}))$.

Usually this integral is denoted by $(A) \int f(x)\lambda(dx)$. One must be aware that many of the usual properties of the Lebesgue integral—the additivity property, for example—may not hold for the $A$-integral.)

**Problem 3.3.16.** Let $\xi_1, \xi_2, \ldots$ be a sequence of independent random variables (with finite expected values), such that

$$\frac{1}{n^{1+\delta}} \sum_{i=1}^{n} E|\xi_i|^{1+\delta} \to 0 \qquad \text{as } n \to \infty ,$$

for some $\delta \in (0, 1)$. Prove that this "$(1 + \delta)$-condition" guarantees that the *law of large numbers* is in force, i.e.,

$$\frac{1}{n} \sum_{i=1}^{n} (\xi_i - \mathsf{E}\xi_i) \xrightarrow{\mathsf{P}} 0 \quad \text{as } n \to \infty.$$

**Problem 3.3.17.** (*Restatement of [P §3.3, Theorem 3] for the case of non-identically distributed random variables.*) By using the continuity theorem ([P §3.3, Theorem 1]) and the method of characteristic functions, the central limit theorem was established in Theorem 3 in the case of independent and *identically* distributed random variables. By using the same method, prove the central limit theorem for the case of independent but *not necessarily identically distributed* random variables by using the following scheme.

Let $A_1, A_2, \ldots$ be a sequence independent events, chosen so that $\mathsf{P}(A_n) = 1/n$ (for examples of such events, see Problem 2.4.21). Setting $\xi = I_{A_n}$ and $S_n = \xi_1 + \ldots + \xi_n$, prove that

$$\mathsf{E}S_n = \sum_{k \leq n} \frac{1}{k} \quad (\sim \ln n \quad \text{as } n \to \infty),$$

$$\mathsf{D}S_n = \sum_{k \leq n} \frac{1}{k}\left(1 - \frac{1}{k}\right) \quad (\sim \ln n \quad \text{as } n \to \infty).$$

Next, consider the characteristic functions $\varphi_n(t)$ of the random variables $\frac{S_n - \mathsf{E}S_n}{\sqrt{\mathsf{D}S_n}}$, $n \geq 1$, and prove that $\varphi_n(t) \to e^{-t^2/2}$. Finally, conclude that *the central limit theorem* ([P §3.3, Theorem 1]) holds: as $n \to \infty$ one has

$$\mathsf{P}\left\{\frac{S_n - \mathsf{E}S_n}{\sqrt{\mathsf{D}S_n}} \leq x\right\} \to \Phi(x), \quad x \in \mathbb{R}.$$

**Problem 3.3.18.** As a supplement to inequality [P §3.3, (4)] from [P §3.3, Lemma 3], show that the following *double-sided inequality* holds for any $a > 0$:

$$(1 - \sin 1) \int_{|x| \geq 1/a} dF(x) \leq \frac{1}{a} \int_0^a [1 - \operatorname{Re} \varphi(t)] \, dt \leq 2 \int_{|x| \geq \sqrt{1/a}} dF(x) + \frac{a}{2}.$$

*Hint.* To prove the right inequality, write $1 - \operatorname{Re} \varphi(t)$ in the form

$$1 - \operatorname{Re} \varphi(t) = \int_R (1 - \cos tx) \, dF(x) =$$

$$= \int_{|x| \geq \sqrt{1/a}} (1 - \cos tx) \, dF(x) + \int_{|x| < \sqrt{1/a}} (1 - \cos tx) \, dF(x),$$

estimate the above integrals in the obvious way, and, just as in Lemma 3, use Fubini's theorem.

**Problem 3.3.19.** Let $(\xi_n)_{n \geq 1}$ be a sequence of independent random variables, distributed according to the Cauchy law with density

$$\frac{\theta}{\pi(\theta^2 + x^2)}, \quad \theta > 0, \quad x \in \mathbb{R}.$$

Prove that the distributions $F_n$ of the random variables $\frac{1}{n} \max_{i \leq n} \xi_i$ converge weakly to the Fréchet distribution with parameter $\alpha = 1$ (see Problem 2.8.48), i.e., the distribution of a random variable of the form $1/T_c$, where $T_c$ has exponential distribution with parameter $c = \theta/\pi$:

$$\mathsf{P}\left\{\frac{1}{T_c} \leq x\right\} = e^{-c/x}, \quad x > 0.$$

**Problem 3.3.20.** (*Continuity theorem for discrete random variables.*) Let $\xi, \xi_1, \xi_2, \ldots$ be a sequence of random variables taking integer values $k = 0, 1, 2, \ldots$ and let

$$G(s) = \sum_{k=0}^{\infty} \mathsf{P}\{\xi = k\} s^k \quad \text{and} \quad G_n(s) = \sum_{k=0}^{\infty} \mathsf{P}\{\xi_n = k\} s^k$$

be the generating functions, respectively, of the variables $\xi$ and $\xi_n$, $n \geq 1$.
   Prove that

$$\lim_n \mathsf{P}\{\xi_n = k\} = \mathsf{P}\{\xi = k\}, \quad k = 0, 1, 2, \ldots,$$

if and only if
$$\lim_n G_n(s) = G(s), \quad s \in [0, 1).$$

**Problem 3.3.21.** Prove the statement in Problem 2.10.35 by using the method of characteristic functions.
   *Hint.* The characteristic function of the random variable $U$, which is uniformly distributed in the interval $[-1, 1]$, is the function $\frac{\sin t}{t}$.

## 3.4   The Central Limit Theorem for Sums of Independent Random Variables I. Lindeberg's Condition

**Problem 3.4.1.** Let $\xi_1, \xi_2, \ldots$ be a sequence of independent and identically distributed random variables with $\mathsf{E}\xi_1^2 < \infty$. Prove that (comp. with Problem 2.10.53)

$$\frac{\max(|\xi_1|, \ldots, |\xi_n|)}{\sqrt{n}} \xrightarrow{d} 0 \quad \text{as } n \to \infty.$$

*Hint.* Use the relation

$$P\left\{\frac{\max(|\xi_1|,\ldots,|\xi_n|)}{\sqrt{n}} \le \varepsilon\right\} = \left[P\{\xi_1^2 \le n\varepsilon^2\}\right]^n$$

and the fact that $n\varepsilon^2 P\{\xi_1 > n\varepsilon\} \to 0$ as $n \to \infty$.

**Problem 3.4.2.** Give a direct proof of the fact that in the Bernoulli scheme the variable $\sup_x |F_{T_n}(x) - \Phi(x)|$ has order $\frac{1}{\sqrt{n}}$ as $n \to \infty$.

**Problem 3.4.3.** Let $X_1, X_2, \ldots$ be any infinite sequence of exchangeable random variables (see Problem 2.5.4) with $\mathsf{E}X_n = 0$, $\mathsf{E}X_n^2 = 1$, $n \ge 1$, and let

$$\mathrm{cov}(X_1, X_2) = \mathrm{cov}(X_1^2, X_2^2). \tag{*}$$

Prove that the central limit theorem holds for any such sequence, i.e.,

$$\frac{1}{\sqrt{n}} \sum_{i=1}^n X_i \xrightarrow{d} \mathcal{N}(0, 1). \tag{**}$$

Conversely, if $\mathsf{E}X_n^2 < \infty$, $n \ge 1$, then $(**)$ implies $(*)$.

**Problem 3.4.4.** (a) (*The local limit theorem for random variables on a lattice.*) Let $\xi_1, \xi_2, \ldots$ be independent and identically distributed random variables with mean value $\mu = \mathsf{E}\xi_1$ and with dispersion $\sigma^2 = \mathsf{D}\xi_1$. Set $S_n = \xi_1 + \ldots + \xi_n$, $n \ge 1$, and suppose that the variables $\xi_1, \xi_2, \ldots$ take values on a lattice of step-size $h > 0$, i.e., take the values $a + hk$, $k = 0, \pm 1, \pm 2, \ldots$, for some $h > 0$.

Prove that as $n \to \infty$ one has

$$\sup_k \left| \frac{\sqrt{n}}{h} P\{S_n = an + hk\} - \frac{1}{\sqrt{2\pi}\sigma} \exp\left\{-\frac{(hk + an - n\mu)^2}{2\sigma^2 n}\right\} \right| \to 0.$$

(Comp. with the local limit theorem in [P §2.6].)

*Hint.* The proof can be carried out with the following line of reasoning, which involves characteristic functions. By Problem 2.12.9 one can write

$$P\{S_n = an + hk\} = P\{(S_n - an)h^{-1} = k\} = \frac{1}{2\pi} \int_{-\pi}^{\pi} e^{-iuk} e^{\frac{iuna}{h}} \left[\varphi\left(\frac{u}{h}\right)\right]^n du,$$

where $\varphi(u)$ stands for the characteristic function of the variable $\xi_1$. It is clear that

$$e^{-z^2/2} = \frac{1}{\sqrt{2\pi}} \int_{-\infty}^{\infty} e^{iuz} e^{-u^2/2} du,$$

and, therefore,

$$2\pi \left| \frac{\sqrt{n}\sigma}{n} P\{S_n = an + hk\} - \frac{1}{2\pi} e^{-z^2/2} \right| \le$$

$$\le \int_{|t|\le \frac{\pi\sqrt{n}\sigma}{h}} \left| \left[ \varphi\left(\frac{t}{\sigma\sqrt{n}}\right) \right]^n - e^{-t^2/2} \right| dt + \int_{|t|> \frac{\pi\sqrt{n}\sigma}{h}} e^{-t^2/2}\, dt.$$

The expression in the right side does not depend on $k$ and it only remains to show that as $n \to \infty$ this expression converges to 0.

(b) (*The local limit theorem for random variables with density.*) Let $\xi_1, \xi_2, \ldots$ be independent and identically distributed random variables with mean value $\mu = \mathsf{E}\xi_1$ and dispersion $\sigma^2 = \mathsf{D}\xi_1$. Suppose that the characteristic function $\varphi = \varphi(t)$ of the variable $\xi_1$ is integrable and, consequently, $\xi_1$ admits a probability density given by

$$f(x) = \frac{1}{2\pi} \int_{-\infty}^{\infty} e^{-itx} \varphi(t)\, dt$$

(see [$\underline{P}$ §2.12, Theorem 3]).

Let $f_n = f_n(x)$ denote the probability density function of the variable $S_n = \xi_1 + \ldots + \xi_n$, $n \ge 1$. Prove that as $n \to \infty$ one has

$$\sup_x \left| \sqrt{n}\, f_n(x) - \frac{1}{\sqrt{2\pi}\sigma} \exp\left\{ -\frac{(x - n\mu)^2}{2\sigma^2 n} \right\} \right| \to 0.$$

*Hint.* Follow the argument used in the case of lattice-valued random variables.

**Problem 3.4.5.** Let $X_1, X_2, \ldots$ be independent and identically distributed random variables with $\mathsf{E}X_1 = 0$ and $\mathsf{E}X_1^2 = 1$, and let $d_1, d_2, \ldots$ be any sequence of non-negative constants, such that $d_n = o(D_n)$, where $D_n^2 = \sum_{k=1}^{n} d_k^2$. Prove that the "weighted sequence" $d_1 X_1, d_2 X_2, \ldots$ satisfies the central limit theorem:

$$\frac{1}{D_n} \sum_{k=1}^{n} d_k X_k \xrightarrow{d} \mathcal{N}(0, 1).$$

**Problem 3.4.6.** Let $\xi_1, \xi_2, \ldots$ be independent and identically distributed random variables with $\mathsf{E}\xi_1 = 0$ and $\mathsf{E}\xi_1^2 = 1$ and suppose that $(\tau_n)_{n\ge 1}$ is some sequence of random variables with values in the set $\{1, 2, \ldots\}$, chosen so that $\tau_n/n \xrightarrow{\mathsf{P}} c$, where $c > 0$ is some fixed constant. Setting $S_n = \xi_1 + \ldots + \xi_n$, prove that

$$\mathrm{Law}\left(\tau_n^{-1/2} S_{\tau_n}\right) \to \Phi,$$

i.e., $\tau_n^{-1/2} S_{\tau_n} \xrightarrow{d} \xi$, where $\xi \sim \mathcal{N}(0, 1)$. (Note that the sequences $(\tau_n)_{n\ge 1}$ and $(\xi_n)_{n\ge 1}$ are not assumed to be independent.)

**Problem 3.4.7.** Let $\xi_1, \xi_2, \ldots$ be independent and identically distributed random variables with $E\xi_1 = 0$ and $E\xi_1^2 = 1$. Prove that

$$\mathrm{Law}\left(n^{-1/2} \max_{1 \le m \le n} S_m\right) \to \mathrm{Law}(|\xi|), \quad \text{where } \xi \sim \mathcal{N}(0, 1);$$

in other words, for any $x > 0$ one has

$$P\left\{n^{-1/2} \max_{1 \le m \le n} S_m \le x\right\} \to \sqrt{\frac{2}{\pi}} \int_0^x e^{-y^2/2}\, dy \quad \left(= \frac{1}{\sqrt{2}} \mathrm{erf}(x)\right).$$

*Hint.* First prove the statement for symmetric *Bernoulli* random variables $\xi_1, \xi_2, \ldots$ with $P\{\xi_n = \pm 1\} = 1/2$, and then use – or, better yet, prove, which is non-trivial – the fact that the limiting distribution would be the same for any sequence $\xi_1, \xi_2, \ldots$ with the specified properties. (The *independence of the limiting distribution* from the particular choice of the independent and identically distributed random variables $\xi_1, \xi_2, \ldots$, with $E\xi_n = 0$, $E\xi_n^2 = 1$, is known as "invariance principle"; see, for example, [10] and [17].)

**Problem 3.4.8.** In the context of the previous problem (and hint) prove that

$$P\left\{n^{-1/2} \max_{1 \le m \le n} |S_m| \le x\right\} \to H(x), \quad x > 0,$$

where

$$H(x) = \frac{4}{\pi} \sum_{k=0}^{\infty} \frac{(-1)^k}{2k + 1} \exp\left\{-\frac{(2k + 1)^2 \pi^2}{8x^2}\right\}.$$

**Problem 3.4.9.** Let $X_1, X_2, \ldots$ be independent random variables with

$$P\{X_n = \pm n^{\alpha}\} = \frac{1}{2n^{\beta}}, \quad P\{X_n = 0\} = 1 - \frac{1}{n^{\beta}}, \quad \text{where } 2\alpha > \beta - 1.$$

Prove that in this case the Lindeberg condition holds if and only if $0 \le \beta < 1$.

**Problem 3.4.10.** Let $X_1, X_2, \ldots$ be independent random variables chosen so that $|X_n| \le C_n$ (P-a. e.) and let $C_n = o(D_n)$, where

$$D_n^2 = \sum_{k=1}^{n} E(X_k - EX_k)^2 \to \infty.$$

Prove that

$$\frac{S_n - ES_n}{D_n} \xrightarrow{d} \mathcal{N}(0, 1), \quad \text{where } S_n = X_1 + \cdots + X_n.$$

**Problem 3.4.11.** Let $X_1, X_2, \ldots$ be any sequence of independent random variables with $\mathsf{E}X_n = 0$ and $\mathsf{E}X_n^2 = \sigma_n^2$. In addition, suppose that this sequence satisfies the central limit theorem and has the property

$$\mathsf{E}\left(D_n^{-1/2} \sum_{i=1}^n X_i\right)^k \to \frac{(2k)!}{2^k k!} \quad \text{for some } k \geq 1 .$$

Prove that *Lindeberg's condition of order $k$* holds, i.e.,

$$\sum_{j=1}^n \int_{\{|x|>\varepsilon\}} |x|^k \, dF_j(x) = o(D_n^k), \quad \varepsilon > 0 .$$

Note that the usual "Lindeberg condition" is of order $k = 2$—see [$\underline{\mathsf{P}}$ §3.4, (1)].

**Problem 3.4.12.** Let $X = X(\lambda)$ and $Y = Y(\mu)$ be two independent random variables having Poisson distribution with parameters, respectively, $\lambda > 0$ and $\mu > 0$. Prove that

$$\frac{(X(\lambda) - \lambda) - (Y(\mu) - \mu)}{\sqrt{X(\lambda) + Y(\mu)}} \xrightarrow{d} \mathscr{N}(0, 1) \quad \text{as } \lambda \to \infty, \mu \to \infty.$$

**Problem 3.4.13.** Given any $n \geq 1$, suppose that the random vector

$$(X_1^{(n)}, \ldots, X_{n+1}^{(n)})$$

is uniformly distributed on the unit sphere in $\mathbb{R}^{n+1}$. Prove the following statement, due to H. Poincaré: for every $x \in \mathbb{R}$,

$$\lim_{n\to\infty} \mathsf{P}\left\{\sqrt{n}\, X_{n+1}^{(n)} \leq x\right\} = \frac{1}{\sqrt{2\pi}} \int_{-\infty}^x e^{-u^2/2} \, du.$$

**Problem 3.4.14.** Let $\xi_1, \xi_2, \ldots$ be a sequence of independent and $\mathscr{N}(0, 1)$-distributed random variables. Setting $S_n = \xi_1 + \ldots + \xi_n, n \geq 1$, find the limiting probability distribution (as $n \to \infty$) of the random variables

$$\frac{1}{n} \sum_{k=1}^n |S_{k-1}| (\xi_k^2 - 1), \quad n \geq 1.$$

**Problem 3.4.15.** Let $\xi_1, \xi_2, \ldots$ be a symmetric Bernoulli scheme (i.e., a sequence of independent and identically distributed random variables with $\mathsf{P}\{\xi_1 = 1\} = \mathsf{P}\{\xi_1 = -1\} = 1/2$) and let $S_0 = 0$ and $S_k = \xi_1 + \ldots + \xi_k, k \geq 1$.

Define the continuous processes $X^{(2n)} = (X_t^{(2n)})_{0 \le t \le 1}$ so that

$$X_t^{(2n)} = \frac{S_{2nt}}{\sqrt{2n}},$$

where, given any $u \ge 0$, $S_u$ is defined by way of linear interpolation from the nearest integer values.

*Hint.* Prove the following—difficult but important—statements:

(a) The distributions $P^{2n} = \mathrm{Law}(X_t^{(2n)}, 0 \le t \le 1)$ converge (in terms of finite dimensional distributions and in terms of the weak convergence of distributions on the metric space $C$, endowed with the uniform distance) to the distribution law $P = \mathrm{Law}(B_t, 0 \le t \le 1)$ of the Brownian motion $B = (B_t)_{0 \le t \le 1}$. (The statement about the weak convergence in $C$ is a special case of the *Donsker–Prokhorov invariance principle*—see [P §7.8, 1].)

(b) The conditional distributions $Q^{2n} = \mathrm{Law}(X_t^{(2n)}, 0 \le t \le 1 \mid X_1^{2n})$ converge (in the same sense as in (a)) to the distribution $Q = \mathrm{Law}(B_t^\circ, 0 \le t \le 1)$ of the Brownian bridge $B^\circ = (B_t^\circ)_{0 \le t \le 1}$.

*Hint.* Use the same line of reasoning as in the derivation of Kolmogorov's limiting distribution in [P §3.13]. For more details see the books [10] and [17].

**Problem 3.4.16.** Conclude from the results in the previous problem (and compare these results with the statements in Problems 3.4.7 and 3.4.8) the following limiting relations: for any $x > 0$ one has:

(a$_1$)  $\quad P\left\{ \dfrac{1}{\sqrt{2n}} \max_{0 \le k \le 2n} S_k \le x \right\} \to P\left\{ \max_{0 \le t \le 1} B_t \le x \right\}$  $\quad (= P\{|B_1| \le x\});$

(a$_2$)  $\quad P\left\{ \dfrac{1}{\sqrt{2n}} \max_{0 \le k \le 2n} |S_k| \le x \right\} \to P\left\{ \max_{0 \le t \le 1} |B_t| \le x \right\};$

and

(b$_1$)  $\quad P\left( \dfrac{1}{\sqrt{2n}} \max_{0 \le k \le 2n} S_k \le x \mid S_{2n} = 0 \right) \to P\left\{ \max_{0 \le t \le 1} B_t^\circ \le x \right\};$

(b$_2$)  $\quad P\left( \dfrac{1}{\sqrt{2n}} \max_{0 \le k \le 2n} |S_k| \le x \mid S_{2n} = 0 \right) \to P\left\{ \max_{0 \le t \le 1} |B_t^\circ| \le x \right\}.$

**Problem 3.4.17.** As a continuation of Problems 3.4.15 and 3.4.16, verify the following relations:

(a)  $\quad P\left\{ \dfrac{1}{\sqrt{2n}} \left[ \max_{0 \le k \le 2n} S_k - \min_{0 \le k \le 2n} S_k \right] \le x \right\} \to P\left\{ \max_{0 \le t \le 1} B_t - \min_{0 \le t \le 1} B_t \le x \right\};$

and

(b)   $P\left(\dfrac{1}{\sqrt{2n}}\left[\max_{0\le k\le 2n} S_k - \min_{0\le k\le 2n} S_k\right] \le x \,\middle|\, S_{2n} = 0\right)$

$$\to P\left\{\max_{0\le t\le 1} B_t^\circ - \min_{0\le t\le 1} B_t^\circ \le x\right\}.$$

**Problem 3.4.18.** Assuming that $N \in [0, \infty)$ and $\lambda \in (0, \infty)$, prove that

$$\lim_{n\to\infty} e^{-\lambda n} \sum_{k\le nN} \frac{(\lambda n)^k}{k!} = \begin{cases} 1, & \text{if } N > \lambda, \\ 1/2, & \text{if } N = \lambda, \\ 0, & \text{if } N < \lambda. \end{cases}$$

Show also that

$$\lim_{n\to\infty} \left(\sum_{k\ge nN} \frac{(\lambda n)^k}{k!}\right)^{1/n} = \begin{cases} e^\lambda, & \text{if } N \le \lambda, \\ e^{-N \ln \frac{N}{\lambda}+N}, & \text{if } N > \lambda. \end{cases}$$

*Hint.* Let $(\xi_n)_{n\ge 1}$ be any sequence of independent Poisson random variables with expected value $\lambda$, i.e., $P\{\xi_n = k\} = e^{-\lambda}\lambda^k/k!, k \ge 0$. Convince yourself that

$$P\left\{\frac{\xi_1 + \ldots + \xi_n}{n} \le N\right\} = e^{-n\lambda} \sum_{k\le nN} \frac{(n\lambda)^k}{k!},$$

and then use the central limit theorem.

**Problem 3.4.19.** Prove that

$$\frac{1}{n!} \int_0^{n+1} x^n e^{-x}\, dx \to \frac{1}{2} \quad \text{as } n \to \infty,$$

and, more generally, that

$$\lim_n \frac{1}{\Gamma(n+1)} \int_0^{y\sqrt{n+1}+(n+1)} x^n e^{-x}\, dx = \Phi(y), \quad y \ge 0.$$

Show also that

$$1 + \frac{n+1}{1!} + \frac{(n+1)^2}{2!} + \ldots + \frac{(n+1)^n}{n!} \sim \frac{1}{2}e^{n+1} \quad \text{as } n \to \infty.$$

*Hint.* Use the result from Problem 2.8.80 and the statement in the previous problem (in the case $N = \lambda$).

**Problem 3.4.20.** Let $\xi_1, \xi_2, \ldots$ be any sequence of independent and identically distributed random variables such that the expected value $\mu = \mathsf{E}\xi_1$ is well defined and $\sigma^2 = \mathsf{D}\xi_1 < \infty$. Setting $S_0 = 0$ and $S_n = \xi_1 + \ldots + \xi_n$, prove that the sequence of partial maxima $M_n = \max(S_0, S_1, \ldots, S_n)$, $n \geq 0$, satisfies the central limit theorem: if $0 < \mu < \infty$, then

$$\lim_n \mathsf{P}\left\{\frac{M_n - n\mu}{\sigma\sqrt{n}} \leq x\right\} = \varPhi(x), \quad x \in \mathbb{R},$$

where $\varPhi(x)$ is the distribution function of the standard normal distribution.

**Problem 3.4.21.** Let $\xi_1, \xi_n, \ldots$ be any sequence of independent and identically distributed random variables with $\mathsf{E}\xi_1 = 0$ and $\mathsf{E}\xi_1^2 = \sigma^2$. Set $S_n = \xi_1 + \ldots + \xi_n$, $n \geq 1$, and let $\{N_t, t \geq 0\}$ be any family of random variables with values in the set $\{1, 2, \ldots\}$, such that $N_t/t \overset{\mathsf{P}}{\to} \lambda$ as $t \to \infty$, $0 < \lambda < \infty$.

Prove the following version of the central limit theorem (due to F. J. Anscombe): as $t \to \infty$ one has

$$\mathsf{P}\left\{\frac{S_{N_t}}{\sigma\sqrt{N_t}} \leq x\right\} \to \varPhi(x) \quad \text{and} \quad \mathsf{P}\left\{\frac{S_{N_t}}{\sigma\sqrt{\lambda}\sqrt{t}} \leq x\right\} \to \varPhi(x).$$

*Hint.* For the sake of simplicity set $\sigma = 1$ and let $n_0 = \lfloor \lambda t \rfloor$. The expression $S_{N_t}/\sqrt{N_t}$ can now be written in the form

$$\frac{S_{N_t}}{\sqrt{N_t}} = \left(\frac{S_{n_0}}{\sqrt{n_0}} + \frac{S_{N_t} - S_{n_0}}{\sqrt{n_0}}\right)\sqrt{\frac{n_0}{N_t}}.$$

Since $\mathsf{P}\{S_{n_0}/\sqrt{n_0} \leq x\} \to \varPhi(x)$ and $n_0/N_t \overset{\mathsf{P}}{\to} 1$ as $t \to \infty$, it only remains to show that $(S_{N_t} - S_{n_0})/\sqrt{n_0} \overset{\mathsf{P}}{\to} 0$. For that purpose it is enough to write the probability $\mathsf{P}\{|S_{N_t} - S_{n_0}| > \varepsilon\sqrt{n_0}\}$ as the sum

$$\mathsf{P}\{|S_{N_t} - S_{n_0}| > \varepsilon\sqrt{n_0},\, N_t \in [n_1, n_2]\} + \mathsf{P}\{|S_{N_t} - S_{n_0}| > \varepsilon\sqrt{n_0},\, N_t \notin [n_1, n_2]\},$$

with $n_1 = \lfloor n_0(1 - \varepsilon^3) \rfloor + 1$, $n_2 = \lfloor n_0(1 + \varepsilon^3) \rfloor$. The convergence of the above probabilities to 0 can be established by using Kolmogorov's inequality (see [P §4.2]).

**Problem 3.4.22.** (*On the convergence of moments in the central limit theorem.*) Let $\xi_1, \xi_2, \ldots$ be any sequence of independent and identically distributed random variables with $\mathsf{E}\xi_1 = 0$ and $\mathsf{E}\xi_1^2 = \sigma^2 < \infty$. According to [P §3.3, Theorem 3] and part (b) of [P §3.4, Theorem 1], one has

$$\frac{S_n}{\sigma\sqrt{n}} \overset{d}{\to} N,$$

where $N$ is a standard normal ($\mathcal{N}(0, 1)$) random variable.

Prove that if $E|\xi_1|^r < \infty$, for some $r \geq 2$, then for any $0 < p \leq r$ the following convergence of moments takes place

$$E\left|\frac{S_n}{\sigma\sqrt{n}}\right|^p \rightarrow E|N|^p.$$

*Hint.* Prove that the family of random variables $\left\{\left|\frac{S_n}{\sigma\sqrt{n}}\right|^r : n \geq 1\right\}$ is uniformly integrable and then use the statement in part (b) of Problem 2.10.54.

**Problem 3.4.23.** Let $(\xi_n)_{n\geq1}$ be a sequence of independent and identically distributed standard normal random variables (i.e., $\xi_n \sim \mathcal{N}(0, 1)$) and suppose that the random variable $\xi \sim \mathcal{N}(0, 1)$ is independent from the sequence $(\xi_n)_{n\geq1}$.
Prove that the limit

$$\lim_n E\left|\frac{\xi_1 + \ldots + \xi_n}{\sqrt{n}} - \xi\right|$$

exists and equals $2/\sqrt{\pi}$.

*Hint.* Convince yourself that the family of random variables $\{S_n/\sqrt{n} - \xi : n \geq 1\}$, where $S_n = \xi_1 + \ldots + \xi_n$, is uniformly integrable.

**Problem 3.4.24.** Let P and Q be two probability measures on $(\Omega, \mathscr{F})$, chosen so that Q is absolutely continuous with respect to P ($Q \ll P$), and suppose that, relative to P, $X_1, X_2, \ldots$ is a sequence of independent and identically distributed random variables with $m = E_P X_i, \sigma^2 = E_P(X_i - m)^2$, where $E_P$ means expectation with respect to P.
According to [P §3.3, Theorem 3], the central limit theorem holds: as $n \rightarrow \infty$ one has

$$P\left\{\frac{1}{\sigma\sqrt{n}}\sum_{i=1}^n (X_i - m) \leq x\right\} \rightarrow \Phi(x), \qquad x \in R,$$

where $\Phi(x) = \frac{1}{\sqrt{2\pi}}\int_{-\infty}^x e^{-y^2/2}\,dy$.
Now consider the measure Q. Even if $Q \ll P$, the sequence $X_1, X_2, \ldots$ will not, in general, represent a sequence of independent random variables relative to Q. Prove that, nevertheless, the central limit theorem still holds in the following form, which is due to A. Rényi: as $n \rightarrow \infty$ one has

$$Q\left\{\frac{1}{\sigma\sqrt{n}}\sum_{i=1}^n (X_i - m) \leq x\right\} \rightarrow \Phi(x), \qquad x \in R.$$

*Hint.* One has to prove that if $f = f(x)$ is some bounded and continuous function, then

$$E_Q f(\widehat{S}_n) \rightarrow E_P f(N(0, 1)),$$

where $\widehat{S}_n = \frac{1}{\sigma\sqrt{n}}\sum_{i=1}^{n}(X_i - m)$ and $N(0,1)$ is a standard normal $(\mathcal{N}(0,1))$ random variable. For that purpose, consider the Radon–Nikodym density $D = \frac{dQ}{dP}$ and the random variables $D_k = \mathsf{E}_\mathsf{P}(D \mid \mathscr{F}_k)$, where $\mathscr{F}_k = \sigma(X_1, \ldots, X_k)$, and write $\mathsf{E}_\mathsf{Q} f(\widehat{S}_n)$ in the following form

$$\mathsf{E}_\mathsf{Q} f(\widehat{S}_n) = \mathsf{E}_\mathsf{P}[(D - D_k)f(\widehat{S}_n)] + \mathsf{E}_\mathsf{P}[D_k f(\widehat{S}_n)].$$

Then prove that $\lim_k \sup_n |\mathsf{E}_\mathsf{P}[(D - D_k)f(\widehat{S}_n)]| = 0$ and

$$\mathsf{E}_\mathsf{P}[D_k f(\widehat{S}_n)] \to \mathsf{E}_\mathsf{P} f(N) \quad \text{as } n \to \infty, \text{ for every } k \geq 1.$$

**Problem 3.4.25.** Let $\xi_1, \xi_2, \ldots$ be a sequence of independent and identically distributed random variables, such that

$$P\{\xi_1 > x\} = P\{\xi_1 < -x\}, \quad x \in R, \quad \text{and} \quad P\{|\xi_1| > x\} = x^{-2}, \quad x \geq 1.$$

Prove that as $n \to \infty$

$$P\left\{\frac{S_n}{\sqrt{n \ln n}} \leq x\right\} \to \Phi(x), \quad x \in R,$$

where $S_n = \xi_1 + \ldots + \xi_n$.

*Remark.* This problem shows that, after a suitable normalization, the distribution of the sums $S_n$ may converge to the standard normal distribution even if $\mathsf{E}\xi_1^2 = \infty$.

*Hint.* Consider the random variables $\xi_{nk} = \xi_k I(|\xi_k| \leq \sqrt{n} \ln\ln n)$ and convince yourself that:

(i) $\sum_{k=1}^{n} P\{\xi_{nk} \neq \xi_k\} \to 0$ as $n \to \infty$;

(ii) $\mathsf{E}\xi_{nk}^2 \sim \ln n$ as $n \to \infty$;

(iii) by Lindeberg's Theorem (Theorem 1) one has

$$\frac{1}{\sqrt{n \ln n}} \sum_{k=1}^{n} \xi_{nk} \overset{d}{\to} \mathcal{N}(0,1);$$

(iv) $P\left\{S_n \neq \sum_{k=1}^{n} \xi_{nk}\right\} \to 0$ as $n \to \infty$.

## 3.5   The Central Limit Theorem for Sums of Independent Random Variables II. Non-classical Conditions

**Problem 3.5.1.** Prove formula [P §3.5, (5)].
   *Hint.* By using the relations

$$\int_{\mathbb{R}} x^2 \, dF_{nk}(x) < \infty \quad \text{and} \quad \int_{\mathbb{R}} x^2 \, d\Phi_{nk}(x) < \infty,$$

conclude that the integrals in the left and the right sides of [P §3.5, (5)] are finite and then use the relation

$$\int_{-\infty}^{\infty} \left( e^{itx} - itx + \frac{t^2 x^2}{2} \right) d(F_{nk} - \Phi_{nk}) =$$

$$= \lim_{a \to \infty} \left( e^{itx} - itx + \frac{t^2 x^2}{2} \right) (F_{nk}(x) - \Phi_{nk}(x)) \Big|_{-a}^{a} -$$

$$- it \int_{-\infty}^{\infty} (e^{itx} - 1 - itx)[F_{nk}(x) - \Phi_{nk}(x)] \, dx.$$

**Problem 3.5.2.** Verify the relations [P §3.5, (10) and (12)].

**Problem 3.5.3.** Let $N = (N_t)_{t \geq 0}$ be the renewal process introduced in [P §2.9, 4] (i.e., $N_t = \sum_{n=1}^{\infty} I(T_n \leq t)$, $T_n = \sigma_1 + \cdots + \sigma_n$, where $\sigma_1, \sigma_2, \ldots$ is a sequence of independent and identically distributed positive random variables). Assuming that $\mu = \mathsf{E}\sigma_1 < \infty$ and $0 < \mathsf{D}\sigma_1 < \infty$, prove that the Central Limit Theorem holds:

$$\frac{N_t - t\mu^{-1}}{\sqrt{t\mu^{-3}\mathsf{D}\sigma_1}} \xrightarrow{d} N(0, 1),$$

where $N(0, 1)$ is a standard normal random variable.

## 3.6   Infinitely Divisible and Stable Distributions

**Problem 3.6.1.** Prove that if $\xi_n \xrightarrow{d} \xi$ and $\xi_n \xrightarrow{d} \eta$, then $\xi \overset{d}{=} \eta$.

**Problem 3.6.2.** Prove that if $\varphi_1$ and $\varphi_2$ are two infinitely divisible characteristic functions, then $\varphi_1 \cdot \varphi_2$ is also an infinitely divisible characteristic function.

**Problem 3.6.3.** Let $\varphi_n = \varphi_n(t)$, $n \geq 1$, be infinitely divisible characteristic functions and suppose that there is a characteristic function $\varphi = \varphi(t)$ for which one can claim that $\varphi_n(t) \to \varphi(t)$, for each $t \in \mathbb{R}$. Prove that $\varphi(t)$ must be infinitely divisible.

*Hint.* Use the fact that if $\varphi_n$ is infinitely divisible, then one can find some independent and identically distributed random variables $\xi_1, \ldots, \xi_n$, such that $S_n = \xi_1 + \cdots + \xi_n$ has characteristic function $\varphi_n$ and $S_n \overset{d}{\to} T$, where $T$ is infinitely divisible.

**Problem 3.6.4.** Prove that the characteristic function of an infinitely divisible distribution cannot be equal to 0 (see also Problem 3.6.12).

*Hint.* The required statement follows directly from the Kolmogorov–Lévy–Khinchin formula, but one can give an independent proof by using the following argument: if $\varphi(t)$ is the characteristic function of some infinitely divisible distribution, then for every $n \geq 1$ one can find a characteristic function $\varphi_n(t)$, such that $\varphi(t) = (\varphi_n(t))^n$, and, setting $\psi_n(t) = |\varphi_n(t)|^2$, prove that the function $\psi(t) = \lim_n \psi_n(t)$ must be identically 1.

**Problem 3.6.5.** Prove that the gamma-distribution is infinitely divisible but is not stable.

*Hint.* The proof can be constructed by analogy to the following line of reasoning. A random variables $\xi$, which is distributed according to the Poisson law $P\{\xi = k\} = e^{-k}/k!$, must be infinitely divisible (see Problem 2.8.3). However, such a random variable *does not* have a stable distribution. Indeed, assuming that $\xi_1 + \xi_2 \overset{d}{=} a\xi + b$, where $a > 0, b \in \mathbb{R}$, and $\xi_1$ and $\xi_2$ are two independent copies of $\xi$, argue that it must be that $a = 1$ and $b = 0$. This means that $\xi_1 + \xi_2 \overset{d}{=} \xi$, which is not possible.

**Problem 3.6.6.** Prove that for a stable random variable $\xi$ one must have $\mathsf{E}|\xi|^r < \infty$, for all $r \in (0, \alpha)$ and all $0 < \alpha < 2$.

*Hint.* By using the Lévy–Khinchin representation of the characteristic function $\varphi(t)$ of the stable random variable $\xi$, conclude that there exists some $\delta > 0$, such that for any $t \in (0, \delta)$ and any $\alpha < 2$ one has $\mathrm{Re}\,\varphi(t) \geq 1 - c|t|^\alpha$, where $c > 0$. By [$\underline{\mathrm{P}}$ §3.3, Lemma 3], for $\alpha \in (0, \delta)$ one has

$$ \mathsf{P}\left\{|\xi| \geq \frac{1}{a}\right\} \leq \frac{cK}{\alpha + 1} a^\alpha, $$

and therefore

$$ \mathsf{P}\{|\xi|^r \geq n\} \leq \frac{cK}{\alpha + 1} n^{-\alpha/r}, \qquad \mathsf{E}|\xi|^r \leq 1 + \sum_{n=1}^\infty \mathsf{P}\{|\xi|^r \geq n\} < \infty, $$

if $r \in (0, \alpha)$ and $0 < \alpha < 2$.

**Problem 3.6.7.** Prove that if $\xi$ is a stable random variable with parameter $0 < \alpha \leq 1$, then its characteristic function $\varphi(t)$ cannot be differentiable at $t = 0$.

**Problem 3.6.8.** Give a direct proof of the fact that the function $e^{-d|t|^\alpha}$ is a characteristic function $d \geq 0$ and $0 < \alpha \leq 2$, but not if $d > 0$ and $\alpha > 2$.

**Problem 3.6.9.** Let $(b_n)_{n \geq 1}$ be any sequence of real numbers, chosen so that for all $|t| < \delta$ and $\delta > 0$ the limit $\lim_n e^{itb_n}$ exists. Prove that $\overline{\lim}_n |b_n| < \infty$.

Hint. The statement can be proved by contradiction. Let $\overline{\lim}_n b_n = +\infty$. Switching, if necessary, to a subsequence, one can claim that $b_n \to \infty$ as $n \to \infty$. Then, setting $h(t) = \lim_n e^{itb_n}$ for $t \in [-\delta, \delta]$, one can write for any $[\alpha, \beta] \subseteq [-\delta, \delta]$

$$\int_{-\delta}^{\delta} I_{[\alpha,\beta]}(t) h(t) \, dt = \lim_n \int_{-\delta}^{\delta} I_{[\alpha,\beta]}(t) e^{itb_n} \, dt = 0.$$

By using the suitable sets principle (see [P §2.2]), it is possible to conclude that $\int_{-\delta}^{\delta} I_A(t) h(t) \, dt = 0$, for every Borel set $A \in \mathcal{B}([-\delta, \delta])$. Consequently, one must have $h(t) = 0$ for any $t \in [-\delta, \delta]$. At the same time, since $|e^{itb_n}| = 1$, one must have $|h(t)| = 1$, for any $t \in [-\delta, \delta]$. This contradiction shows that $\overline{\lim}_n |b_n| < \infty$.

**Problem 3.6.10.** Prove that the binomial, the uniform and the triangular distributions *are not* infinitely divisible. (Recall that the *triangular distribution* on $(-1, 1)$ has density $f(x) = (1 - |x|) I_{(-1,1)}(x)$.)

Prove the following more general statement: a *non-degenerate* distribution with finite support cannot be infinitely divisible.

**Problem 3.6.11.** Suppose that the distribution function $F$ and its characteristic function $\varphi$ admit the representations $F = F^{(n)} * \cdots * F^{(n)}$ ($n$ times) and $\varphi = [\varphi^{(n)}]^n$, for some distribution functions $F^{(n)}$ and their respective characteristic functions $\varphi^{(n)}$, $n \geq 1$. Prove that it is possible to find a (sufficiently "rich") probability space $(\Omega, \mathcal{F}, P)$ and random variables $T$ and $(\eta_k^n)_{k \leq n}$, $n \geq 1$, defined on that space ($T$ has distribution $F$, while $\eta_1^{(n)}, \ldots, \eta_n^{(n)}$ are independent and identically distributed with law $F^{(n)}$) and such that $T \overset{d}{=} \eta_1^{(n)} + \cdots + \eta_n^{(n)}$, $n \geq 1$.

**Problem 3.6.12.** Give examples of random variables that are not infinitely divisible, and yet their characteristic functions never vanish (see Problem 3.6.4).

**Problem 3.6.13.** Prove that:

(a) The function $\varphi = \varphi(t)$ is a characteristic function of an infinitely divisible distribution if and only if for every $n \geq 1$ one can claim that the $n^{\text{th}}$ root $\varphi^{1/n}(t) = e^{\frac{1}{n} \ln \varphi(t)}$ (here ln stands for the principle value of the logarithmic function) is a characteristic function.

(b) The product of finitely many characteristic functions associated with infinitely divisible distributions is an infinitely divisible characteristic function.

(c) If the characteristic functions $\varphi_n(t)$, $n \geq 1$, associated with some infinitely divisible distributions, converge in *point-wise sense* to the function $\varphi(t)$, which happens to be characteristic function, then $\varphi(t)$ must be the characteristic of some infinitely divisible distribution.

**Problem 3.6.14.** By using the results established in the previous problem and the fact that

$$\varphi(t) = \exp\{\lambda(e^{itu} - 1) + it\beta\}, \qquad \lambda > 0, \ u \in R, \ \beta \in R,$$

is known to be a characteristic function of an infinitely divisible distribution law (of Poisson type), prove that the following functions (studied by B. de Finetti) have the same property:

$$\varphi(t) = \exp\left\{\sum_{j=1}^{k} [\lambda_j (e^{itu_j} - 1) + it\beta_j]\right\}$$

and

$$\varphi(t) = \exp\left\{it\beta + \int_{-\infty}^{\infty} (e^{itu} - 1)\, dG(u)\right\},$$

where $G = G(u)$ is some bounded and increasing function.

**Problem 3.6.15.** Let $\varphi = \varphi(t)$ be the characteristic function of some distribution that has a finite second moment. Prove that $\varphi(t)$ can be a characteristic function of an infinitely divisible distribution law if and only if it admits the so-called *Kolmogorov representation*:

$$\varphi(t) = \exp \psi(t)$$

with

$$\psi(t) = itb + \int_{-\infty}^{\infty} (e^{itu} - 1 - itu)\frac{1}{u^2}\, dG(u),$$

where $b \in R$ and $G = G(u)$ is a non-decreasing and left-continuous function with $G(-\infty) = 0$ and $G(\infty) < \infty$ (comp. with de Finetti's function from the previous problem).

**Problem 3.6.16.** Prove that if $\varphi(t)$ is the characteristic function of some infinitely divisible distribution, then for every $\lambda \geq 0$ the function $\varphi^{\lambda}(t)$ is characteristic.

**Problem 3.6.17.** (*On the Kolmogorv–Lévy–Khinchin representation.*) Let $h = h(x)$ be a *cutoff function*, defined for $x \in R$ (i.e, a bounded and continuous function chosen so that $h(x) = x$ in some neighborhood of $x = 0$).
    Prove that:
    (a) The Kolmogorov-Lévy-Khinchin representation [P §3.6, (2)] can be re-written in the form

$$\varphi(t) = \exp \psi_h(t)$$

with

$$\psi_h(t) = itb - \frac{t^2 c}{2} + \int_{-\infty}^{\infty} (e^{itx} - 1 - ith(x)) \, F(dx),$$

where $b = b(h) \in \mathbb{R}, c \geq 0$ and $F(dx)$ is a measure on $(\mathbb{R}, \mathscr{B}(\mathbb{R}))$ with $F(\{0\}) = 0$ and $\int (x^2 \wedge 1) \, F(dx) < \infty$.

(b) For two different cutoff functions $h$ and $h'$ the coefficients $b(h)$ and $b(h')$ are linked through the relation

$$b(h') = b(h) + \int (h'(x) - h(x)) \, F(dx).$$

(c) If $\varphi(t)$ corresponds to a distribution that has a finite second moment (comp. with Problem 3.6.15), then $\int_{-\infty}^{\infty} x^2 \, F(dx) < \infty$.

**Problem 3.6.18.** Prove that the probability distribution with density

$$f(x) = \frac{1}{\sqrt{2\pi x^3}} e^{-1/(2x)}, \quad x > 0,$$

is stable for $\alpha = \frac{1}{2}, \beta = 0$ and $\theta = -1$ (see formula [$\underline{P}$ §3.6, (10)]).

**Problem 3.6.19.** One says that random variable $\xi_m$ has *generalized Poisson distribution* with parameter $\lambda(\{x_m\}) > 0$, if

$$P\{\xi_m = kx_m\} = \frac{e^{-\lambda(\{x_m\})} \lambda^k(\{x_m\})}{k!}, \quad k = 0, 1, 2, \ldots,$$

where $x_m \in \mathbb{R} \setminus \{0\}$.

Let $\xi_1, \ldots, \xi_n$ be $n$ mutually independent random variables that share the above distribution. Let $\lambda = \lambda(dx)$ denote the measure on $\mathbb{R} \setminus \{0\}$ which is supported on the set $\{x_m : m = 1, \ldots, n\}$, consisting of $n$ different points and let $\lambda(\{x_m\})$ denote the probability mass of the point $x_m$. The probability distribution of the random variable $T_n = \xi_1 + \ldots + \xi_n$ is known as the *compound Poisson distribution*.

Prove that the characteristic function $\varphi_{T_n}(t)$ of such a random variable is given by:

$$\varphi_{T_n}(t) = \exp\left\{ \int_{\mathbb{R} \setminus \{0\}} (e^{itx} - 1) \, \lambda(dx) \right\}.$$

(It is clear from the above formula that the compound Poisson distribution is infinitely divisible. In conjunction with the Kolmogorov–Lévy–Khinchin formula [$\underline{P}$ §3.6, (2)] this illustrates the "generating role" that this distribution plays in the class of all infinitely divisible distributions. Formally this property can be stated as follows: *every infinitely divisible distribution is a (weak) limit of some sequence of compound Poisson distributions.*)

**Problem 3.6.20.** On the probability space $(\Omega, \mathscr{F}, P)$ consider the observation scheme consisting of the events $A^{(n)} = (A_{nk}, 1 \leq k \leq n), n \geq 1$, chosen so that for every $n$ the events $A_{n1}, \ldots, A_{nn}$ are independent. Let

$$\lim_n \max_{1 \leq k \leq n} P(A_{nk}) = 0$$

and

$$\lim_n \sum_{k=1}^n P(A_{nk}) = \lambda, \quad \lambda > 0.$$

Prove the "*rare events law:*" the sequence of random variables $\xi^{(n)} = \sum_{k=1}^n I(A_{nk})$ converges in distribution to a random variable $\xi$ that has Poisson distribution with parameter $\lambda > 0$.

**Problem 3.6.21.** Let $X$ and $Y$ be any two independent random variables, distributed with Poisson law of parameter $\lambda > 0$. Find the characteristic function $\varphi(t)$ of the random variable $X - Y$ which is often referred to as a *double-sided Poisson* random variable. Prove that the probability distribution of the random variable $X - Y$ is the *compound* Poisson distribution (see Problems 3.6.19 and 2.8.3).

**Problem 3.6.22.** Let $\xi^{(n)} = (\xi_{nk}, 1 \leq k \leq n), n \geq 1$, be an *observation series* of random variables, such that for any $n$ the variables $\xi_{n1}, \ldots, \xi_{nn}$ are independent. Let $\varphi_{nk} = \varphi_{nk}(t)$ denote the characteristic functions of the random variables $\xi_{nk}$. Prove that the following conditions are equivalent:

(a) $\lim_n \max_{1 \leq k \leq n} P\{|\xi_{nk}| > \varepsilon\} = 0$ (limiting, or asymptotic, negligibility of the series $\xi^{(n)}, n \geq 1$);

(b) $\lim_n \max_{1 \leq k \leq n} |1 - \varphi_{nk}(t)| = 0$ for every $t \in \mathbb{R}$.

**Problem 3.6.23.** The random variable $\xi$ is said to be distributed according to the (continuous) *Pareto law* (with parameters $\rho > 0, b > 0$), if its density is given by

$$f_{\rho,b}(x) = \frac{\rho b^\rho}{x^{\rho+1}} I(x \geq b).$$

Prove that this distribution is infinitely divisible.

*Remark.* The discrete Pareto Law is defined in Problem 2.8.85.

**Problem 3.6.24.** The random variable $\xi$ with values in $(0, \infty)$ is said to have a *logistic distribution* with parameters $(\mu, \rho)$, where $\mu \in \mathbb{R}$ and $\rho > 0$, if

$$P\{\xi \leq x\} = \frac{1}{1 + e^{-(x-\mu)/\rho}}, \quad x > 0.$$

Prove that this distribution is infinitely divisible.

## 3.7   Metrizability of the Weak Convergence

**Problem 3.7.1.** Prove that, in the case of the space $E = \mathbb{R}$, the Lévy–Prokhorov distance $L(P, \widetilde{P})$ between the distribution laws $P$ and $\widetilde{P}$ is *not smaller* than the Lévy distance $L(F, \widetilde{F})$ between the distribution functions $F$ and $\widetilde{F}$, associated with the laws $P$ and $\widetilde{P}$ (see Problem 3.1.4). By constructing appropriate examples, prove that the inequality between these two metrics can be strict.

  *Hint.* To prove that $L(F, \widetilde{F}) \leq L(P, \widetilde{P})$, it is enough to show that

$$L(F, \widetilde{F}) = \inf\{\varepsilon > 0 : P(D) \leq \widetilde{P}(D^{\varepsilon}) + \varepsilon \text{ and } \widetilde{P}(D) \leq P(D^{\varepsilon}) + \varepsilon$$
$$\text{for all sets } D \text{ of the form } (-\infty, x], x \in \mathbb{R}\}$$

and

$$L(P, \widetilde{P}) = \inf\{\varepsilon > 0 : P(D) \leq \widetilde{P}(D^{\varepsilon}) + \varepsilon \text{ and } \widetilde{P}(D) \leq P(D^{\varepsilon}) + \varepsilon$$
$$\text{for all closed sets } D \subseteq \mathbb{R}\}.$$

In order to obtain the strict inequality, take $P = \delta_0$ and $\widetilde{P} = \frac{1}{2}(\delta_{-1} + \delta_1)$, where $\delta_a$ is the measure concentrated at the point $a$:

$$\delta_a(A) = \begin{cases} 1, & \text{if } a \in A, \\ 0, & \text{if } a \notin A. \end{cases}$$

In this case $L(F, \widetilde{F}) = \frac{1}{2}$ and $L(P, \widetilde{P}) = 1$—prove these two identities.

**Problem 3.7.2.** Prove that formula [P §3.7, (19)] defines a metric in the space $BL$.
  *Hint.* To prove that $\|P - \widetilde{P}\|^*_{BL} = 0 \Rightarrow P = \widetilde{P}$ (the remaining properties of the metric are easy to verify), given any closed set $A$ and any $\varepsilon > 0$, consider the function $f_A^{\varepsilon}(x)$, defined by formula [P §3.7, (14)]. Since as $\varepsilon \downarrow 0$ one has

$$\int f_A^{\varepsilon}(x) P(dx) \to P(A) \quad \text{and} \quad \int f_A^{\varepsilon}(x) \widetilde{P}(dx) \to \widetilde{P}(A),$$

then $P(A) = \widetilde{P}(A)$ for every closed set $A$. Finally, consider the class $\mathcal{M} = \{A \in \mathcal{B}(E) : P(A) = \widetilde{P}(A)\}$ and by using the suitable sets principle and the $\pi$-$\lambda$-systems from [P §2.2], conclude that $\mathcal{M} = \mathcal{B}(E)$.

**Problem 3.7.3.** Prove the inequalities [P §3.7, (20), (21) and (22)].

**Problem 3.7.4.** Let $F = F(x)$ and $G = G(x)$, $x \in \mathbb{R}$, be any two distribution functions and suppose that $P_c$ and $Q_c$ are the intersecting points of their graphs with the graph of the line $x + y = c$. Prove that the Lévy distance $L(F, G)$ (see Problem 3.1.4) can be expressed as

$$L(F, G) = \sup_c \frac{\overline{P_c Q_c}}{\sqrt{2}},$$

where $\overline{P_c Q_c}$ is the length of the segment connecting the points $P_c$ and $Q_c$.

**Problem 3.7.5.** Prove that the space of all distribution functions is complete for the Lévy metric.

**Problem 3.7.6.** Consider the *Kolmogorov distance* between the distribution functions $F$ and $\widetilde{F}$, which is given by

$$K(F, \widetilde{F}) = \sup_x |F(x) - \widetilde{F}(x)|,$$

and let $L(F, \widetilde{F})$ be the Lévy distance from Problem 3.7.4. Prove that

$$L(F, \widetilde{F}) \le K(F, \widetilde{F})$$

and that, if the distribution function $\widetilde{F}$ is absolutely continuous, then

$$K(F, \widetilde{F}) \le \left(1 + \sup_x |\widetilde{F}'(x)|\right) L(F, \widetilde{F}).$$

**Problem 3.7.7.** Let $X$ and $\widetilde{X}$ be two random variables defined on one and the same probability space and let $F$ and $\widetilde{F}$ be their respective distribution functions. Prove that the Lévy distance $L(F, \widetilde{F})$ is subject to the following inequalities:

$$L(F, \widetilde{F}) \le d + \mathsf{P}\{|X - \widetilde{X}| > d\}, \quad \forall\, d > 0,$$

and

$$L(F, \widetilde{F}) \le (c + 1) e^{\frac{c}{c+1}} (\mathsf{E}|X - \widetilde{X}|^c)^{\frac{1}{c+1}}, \quad \forall\, c \ge 1.$$

**Problem 3.7.8.** By using the results in Problems 3.7.6 and 3.7.7, prove that if $X$ and $\widetilde{X}$ are two random variables defined on one and the same probability space, if $F$ and $\widetilde{F}$ denote their respective distribution functions, and if $\Phi = \Phi(x)$ stands for the distribution function of the standard normal law $\mathcal{N}(0, 1)$, then the following inequality holds for any $\sigma > 0$:

$$\sup_x \left| F(x) - \Phi\left(\frac{x}{\sigma}\right) \right| \le \left(1 + \frac{1}{\sqrt{2\pi\sigma^2}}\right) \left[ \sup_x \left| \widetilde{F}(x) - \Phi\left(\frac{x}{\sigma}\right) \right| + 2(\mathsf{E}|X - \widetilde{X}|^2)^{1/2} \right].$$

## 3.8   The Connection Between Almost Sure Convergence and Weak Convergence of Probability Measures (the "Common Probability Space" Method)

**Problem 3.8.1.** Prove that if $E$ is a separable metric space with metric $\rho(\cdot,\cdot)$, and $X(\omega)$ and $Y(\omega)$ are any two random elements in $E$, defined on the probability space $(\Omega, \mathscr{F}, \mathsf{P})$, then one can claim that $\rho(X(\omega), Y(\omega))$ is a real *random variable* on $(\Omega, \mathscr{F}, \mathsf{P})$.

*Hint.* Let $\{z_1, z_2, \dots\}$ be any countable and everywhere dense subset of $E$. Prove that for every $a > 0$

$$\{\omega : \rho(X(\omega), Y(\omega)) < a\} =$$

$$= \bigcap_{n=1}^{\infty} \bigcup_{m=1}^{\infty} \left( \left\{ \omega : \rho(X(\omega), z_m) < \frac{1}{n} \right\} \cap \left\{ \omega : \rho(Y(\omega), z_m) < a - \frac{1}{n} \right\} \right),$$

and, by using [$\underline{\mathsf{P}}$ §2.4, Lemma 1], conclude that $\rho(X(\omega), Y(\omega))$ must be a $\mathscr{F}$-measurable function on $\Omega$.

**Problem 3.8.2.** Prove, that the function $d_{\mathsf{P}}(X, Y)$, as defined in [$\underline{\mathsf{P}}$ §3.8, (2)], is a *metric* in the space of random elements in $E$.

*Hint.* The statement in the previous problem shows that the set $\{\rho(X, Y) < \varepsilon\}$ is measurable, and therefore $d_{\mathsf{P}}(X, Y)$ is a well-defined random variable. The proof of the fact that $d_{\mathsf{P}}(X, Y)$ actually represents a metric is straight-forward.

**Problem 3.8.3.** Prove the implication [$\underline{\mathsf{P}}$ §3.8, (5)].

**Problem 3.8.4.** Setting $\Delta_h = \{x \in E : h(x) \text{ is not } \rho\text{-continuous at the point } x\}$, prove that $\Delta_h \in \mathscr{E}$.

*Hint.* Let $\{a_1, a_2, \dots\}$ be any countable and everywhere dense subset of $E$. In order to prove that $\Delta_n \in \mathscr{E}$, it is enough to establish the following representation

$$\Delta_n = \bigcup_{n=1}^{\infty} \bigcap_{m=1}^{\infty} \bigcup_{k=1}^{\infty} A_{n,m,k} \,,$$

where the sets

$$A_{n,m,k} = \begin{cases} B_{1/m}(a_k), & \text{if one can find } y, z \in B_{1/m}(a_k) \\ & \text{so that } |h(y) - h(z)| > \frac{1}{n}, \\ \varnothing & \text{otherwise} \end{cases}$$

all belong to $\mathscr{E}$.

**Problem 3.8.5.** Suppose that $(\xi, \eta)$ and $(\tilde{\xi}, \tilde{\eta})$ are two identically distributed pairs of random variables, i.e., $(\xi, \eta) \overset{d}{=} (\tilde{\xi}, \tilde{\eta})$, with $\mathsf{E}|\xi| < \infty$. Prove that $\mathsf{E}(\xi \mid \eta) \overset{d}{=} \mathsf{E}(\tilde{\xi} \mid \tilde{\eta})$.

**Problem 3.8.6.** Let $\xi$ and $\eta$ be *any* two random elements (defined on a sufficiently rich probability space) which take values in the Borel space $(E, \mathscr{E})$ (see [P §2.7, Definition 9]). Prove that one can find a measurable function $f = f(x, y)$, defined on $E \times [0, 1]$ and taking values in $E$, and also a random variable $\alpha$, which is uniformly distributed in the interval $[0, 1]$, so that the following representation holds with probability 1

$$\xi = f(\eta, \alpha).$$

**Problem 3.8.7.** Let $\xi_1, \xi_2, \ldots$ be any sequence of independent and identically distributed random variables with $\mathsf{E}\xi_1 = 0$, $\mathsf{E}\xi_1^2 < \infty$ and let $X_n = \sum_{k=1}^{n} \xi_k$, $n \geq 1$. Prove the following result (known as *Skorokhod's embedding*): there is a probability space $(\widetilde{\Omega}, \widetilde{\mathscr{F}}, \widetilde{\mathsf{P}})$, on which one can construct a Brownian motion $\widetilde{B} = (\widetilde{B}_t)_{t \geq 0}$ and a sequence of stopping times $\widetilde{\tau} = (\widetilde{\tau}_k)_{k \geq 0}$ with $0 = \widetilde{\tau}_0 \leq \widetilde{\tau}_1 \leq \ldots$, so that

$$(X_n)_{n \geq 1} \overset{d}{=} (\widetilde{B}_{\widetilde{\tau}_n})_{n \geq 1}$$

and $\widetilde{\mathsf{E}}(\tau_n - \tau_{n-1}) = \mathsf{E}\xi_1^2$, $n \geq 1$. (As usual, the symbol "$\overset{d}{=}$" is understood to mean "identity in distribution".)

**Problem 3.8.8.** Let $F = F(x)$ be a distribution function on $\mathbb{R}$ and define its *inverse* $F^{-1}(u), 0 \leq u \leq 1$, by

$$F^{-1}(u) = \begin{cases} \inf\{x : F(x) > u\}, & u < 1, \\ \infty, & u = 1. \end{cases}$$

Prove that:

    (a) $\{x; F(x) > u\} \subseteq \{x; F^{-1}(u) \leq x\} \subseteq \{x; F(x) \geq u\}$;

    (b) $F(F^{-1}(u)) \geq u$, $F^{-1}(F(x)) \geq x$;

    (c) if the function $F = F(x)$ is continuous, then $F^{-1}(u) = \inf\{x; F(x) \geq u\}$, $F^{-1}(u) = \max\{x; F(x) = u\}$, $F(F^{-1}(u)) \geq u$, and $\{x; F(x) > u\} = \{x; F^{-1}(u) < x\}$;

    (d) $\inf\{x; F(x) \geq u\} = \sup\{x; F(x) < u\}$.

*Remark.* In Statistics the function $Q(u) = F^{-1}(u)$ is known as the *quantile function* of the distribution $F$.

**Problem 3.8.9.** Let $F = F(x)$ be any distribution function and let $F^{-1} = F^{-1}(u)$ be its inverse.

Prove that if $U$ is any random variable which is uniformly distributed on $[0, 1]$, then the distribution of the random variable $F^{-1}(U)$ is precisely $F$, i.e.,

$$\mathsf{P}\{F^{-1}(U) \leq x\} = F(x).$$

In addition, prove that if the distribution function $F = F(x)$, associated with the random variable $X$, happens to be *continuous*, then the random variable $F(X)$ must be uniformly distributed in $[0, 1]$.

*Remark.* If $C(u) = P\{U \leq u\}$ is the distribution function of the random variables $U$, which is uniformly distributed in $[0, 1]$, then one must have $C(F(x)) = F(x)$—see Problem 3.8.12.

**Problem 3.8.10.** Let $F(x, y)$ be the distribution function of the pair of random variables $(\xi, \eta)$ and let $F_1(x) = P\{\xi \leq x\}$ and $F_2(y) = P\{\eta \leq y\}$ be the distribution functions of $\xi$ and $\eta$. Prove the *Fréchet–Hoeffding inequality*:

$$\max(F_1(x) + F_2(y) - 1, 0) \leq F(x, y) \leq \min(F_1(x), F_2(y)), \quad \text{for all } x, y \in R.$$

**Problem 3.8.11.** Let $(U, V)$ be some random vector in $[0, 1]^2$ with distribution function

$$C(u, v) = P\{U \leq u, V \leq v\},$$

and suppose that $U$ and $V$ are both uniformly distributed in $[0, 1]$. Let $F_1(x)$ and $F_2(y)$, $x, y \in \mathbb{R}$, be any two continuous distribution functions.

Prove that the function

$$F(x, y) = C(F_1(x), F_2(y)), \qquad x, y \in R, \qquad (*)$$

is a bi-variate *distribution function* with marginal distributions $F_1(x)$ and $F_2(y)$.

*Remark.* For a *given* bi-variate distribution function $F(x, y)$, with marginal distributions $F_1(x)$ and $F_2(y)$, $x, y \in \mathbb{R}$, it is interesting to know how to construct the function $C(u, v)$ so that property $(*)$ holds. Functions that share this property and can be written as bi-variate distributions of the form $P\{U \leq u, V \leq v\}$, for some random variables $U$ and $V$ that take values in $[0, 1]$, were introduced by A. Sklar in 1959 under the name *copula*. His work [122] contains *existence and uniqueness* results for such functions. The next problem provides an example.

**Problem 3.8.12.** Consider the bi-variate distribution function

$$F(x, y) = \max(x + y - 1, 0),$$

where $0 \leq x, y \leq 1$.

Prove that the associated marginal distribution functions, $F_1(x)$ and $F_2(y)$, give the uniform distribution on $[0, 1]$.

Show also that for the copula

$$C(u, v) = \max(u + v - 1, 0), \qquad 0 \leq u, v \leq 1,$$

one must have

$$F(x, y) = C(F_1(x), F_2(y)).$$

*Remark.* Compare the statement in this problem with the one in Problem 3.8.9.

**Problem 3.8.13.** Suppose that the random variables $\xi$ and $\xi_1, \xi_2, \ldots$ are chosen so that $\xi_n \geq 0$ and $\mathrm{Law}(\xi_n) \to \mathrm{Law}(\xi)$. Prove that

$$\mathsf{E}\xi \leq \varliminf_n \mathsf{E}\xi_n.$$

*Hint.* Use [P §3.8, Theorem 1] and [P §2.6, Theorem 2] (Fatou's Lemma).

## 3.9   The Variation Distance Between Probability Measures. The Kakutani-Hellinger Distance and the Hellinger Integral. Applications to Absolute Continuity and Singularity of Probability Measures

**Problem 3.9.1.** Adopting the notation introduced in [P §3.9, Lemma 2], set

$$P \wedge \widetilde{P} = E_Q(z \wedge \tilde{z}),$$

where $z \wedge \tilde{z} = \min(z, \tilde{z})$. Prove that

$$\| P - \widetilde{P} \| = 2(1 - P \wedge \widetilde{P})$$

and conclude that $\mathscr{E}\mathrm{r}\,(P, \widetilde{P}) = P \wedge \widetilde{P}$ (for the definition of $\mathscr{E}\mathrm{r}\,(P, \widetilde{P})$ see [P §3.9, **1**]).
   *Hint.* Use the relation $a \wedge b = \frac{1}{2}(a + b - |a - b|)$.

**Problem 3.9.2.** Let $P$, $P_n$, $n \geq 1$, be probability measures on $(\mathbb{R}, \mathscr{B}(\mathbb{R}))$ with densities (relative to the Lebesgue measure) $p(x)$, $p_n(x)$, $n \geq 1$, and suppose that $p_n(x) \to p(x)$ for Lebesgue-almost every $x$. Prove that

$$\| P - P_n \| = \int_{-\infty}^{\infty} |p(x) - p_n(x)|\, dx \to 0 \quad \text{as } n \to \infty.$$

*Hint.* Consider the inequality

$$\int_{-\infty}^{\infty} |p_n(x) - p(x)|\, dx \leq \int_{\{|x| \leq a\}} |p_n(x) - p(x)|\, dx + \int_{\{|x| > a\}} p(x)\, dx$$

$$+ \int_{\{|x| > a\}} p_n(x)\, dx,$$

where $a > 0$ is chosen so that $\int_{\{|x|\le a\}} p(x)\,dx > 1 - \varepsilon$ for every $\varepsilon > 0$. By Fatou's lemma

$$\lim_n \int_{\{|x|\le a\}} p_n(x)\,dx \ge 1 - \varepsilon.$$

**Problem 3.9.3.** Let $P$ and $\widetilde{P}$ be any two probability measures. The *Kullback information* $K(P, \widetilde{P})$, which measures the "divergence" of $\widetilde{P}$ from $P$, is defined as

$$K(P, \widetilde{P}) = \begin{cases} E\left[\ln \frac{dP}{d\widetilde{P}}\right], & \text{if } P \ll \widetilde{P}, \\ \infty & \text{otherwise.} \end{cases}$$

Prove that

$$K(P, \widetilde{P}) \ge -2\ln(1 - \rho^2(P, \widetilde{P})) \ge 2\rho^2(P, \widetilde{P}),$$

where $\rho(P, \widetilde{P})$ is the Kakutani–Hellinger distance between the measures $P$ and $\widetilde{P}$.

   *Hint.* The second inequality follows from the relation $-\ln(1 - x) \ge x, 0 \le x \le 1$. To prove the first inequality, show that

$$-2\ln(1 - \rho^2(P, \widetilde{P})) = -2\ln E_P \sqrt{\frac{\widetilde{z}}{z}},$$

and then conclude from Jensen's inequality that

$$-2\ln E_P \sqrt{\frac{\widetilde{z}}{z}} \le K(P, \widetilde{P}).$$

**Problem 3.9.4.** Prove formulas [P §3.9, (11) and (12)].

**Problem 3.9.5.** Prove the two inequalities in [P §3.9, (24)].

   *Hint.* With $Q = \frac{1}{2}(P + \widetilde{P})$, $z = \frac{dP}{dQ}$, and $\widetilde{z} = \frac{d\widetilde{P}}{dQ}$, setting $y = z - 1$, one finds that $\widetilde{z} = 2 - z = 1 - y$ and that [P §3.9, (24)] can be written in the form

$$2(1 + E_Q f(y)) \le 2E_Q|y| \le \sqrt{c_\alpha(1 - E_Q f(y))},$$

where $f(y) = (1 + y)^\alpha (1 - y)^{1-\alpha}$, $y \in [-1, 1]$. By analyzing the functions $f'(y)$ and $f''(y)$ on the interval $(-1, 1)$, one can prove that:
   (a) $f = f(y)$ is concave on $[-1, 1]$ and $f(y) \ge 1 - |y|$;
   (b) $f(y) \le 1 + f'(0)y - \widetilde{c}_\alpha y^2$, $y \in [-1, 1]$, with $\widetilde{c}_\alpha = \alpha(1 - \alpha)/4$.
   Finally, the first inequality can be deduced from (a), while the second one can be deduced from (b).

**Problem 3.9.6.** Let $P, \widetilde{P}, Q$ be probability measures on $(\mathbb{R}, \mathscr{B}(\mathbb{R}))$ and let $P * Q$ and $\widetilde{P} * Q$ stand for the respective convolutions (see [P §2.8, **4**]). Show, that

$$\| P * Q - \widetilde{P} * Q \| \leq \| P - \widetilde{P} \|.$$

*Hint.* Use [P §3.9, Lemma 1].

**Problem 3.9.7.** Prove the relations (30) from [P §3.9, Example 2].
*Hint.* By using straight-forward calculation, prove that

$$H\left(\frac{1}{2}; P, \widetilde{P}\right) = \exp\left\{ -\frac{1}{2}\sum_{k=1}^{\infty}\left(\sqrt{\lambda_k} - \sqrt{\widetilde{\lambda}_k}\right)^2\right\}$$

and then use [P §3.9, Theorems 2 and 3].

**Problem 3.9.8.** Let $\xi$ and $\eta$ be any two random elements on $(\Omega, \mathscr{F}, \mathrm{P})$ with values in the measurable space $(E, \mathscr{E})$. Prove that

$$|\mathrm{P}\{\xi \in A\} - \mathrm{P}\{\eta \in A\}| \leq \mathrm{P}(\xi \neq \eta), \quad A \in \mathscr{E}.$$

*Hint.* Use the relation

$$|I(\xi \in A) - I(\eta \in A)| = |I(\xi \in A) - I(\eta \in A)|I(\xi \neq \eta).$$

**Problem 3.9.9.** The *Hellinger integral* of order $\alpha$ for the measures $P$ and $\widetilde{P}$ is defined by (see formula [P §3.9, (20)])

$$H(\alpha; P, \widetilde{P}) = \int_{\Omega} (dP)^{\alpha}(d\widetilde{P})^{1-\alpha}.$$

A useful tool in the study of many statistical experiments is what is known as the Hellinger transformation $H(\alpha; \mathscr{E})$, which is defined as follows:

Consider the *statistical experiment* $\mathscr{E} = (\Omega, \mathscr{F}; P_0, P_1, \ldots, P_k)$, which consists of the measurable space $(\Omega, \mathscr{F})$ and the finite family of probability measures $P_0, P_1, \ldots, P_k$ defined on that space.

In symbolic form the *Hellinger transformation* $H(\alpha; \mathscr{E})$ of the experiment $\mathscr{E}$ is defined by the formula:

$$H(\alpha; \mathscr{E}) = \int_{\Omega} (dP_0)^{\alpha_0} \ldots (dP_k)^{\alpha_k}, \tag{$*$}$$

where $\alpha = (\alpha_0, \ldots, \alpha_k)$ belongs to the symplex

$$\Sigma_{k+1} = \left\{\alpha = (\alpha_0, \ldots, \alpha_k) : \alpha_i \geq 0, \sum_{i=0}^{k}\alpha_i = 1\right\}.$$

Similarly to the case $k = 1$, give meaning to the "integral" in $(*)$ (by using the concept of "dominating measures") and prove the corresponding analog of Lemma 3.

**Problem 3.9.10.** Let $(\Sigma_k, \mathscr{B}(\Sigma_k))$ denote the simplex

$$\Sigma_k = \left\{ x = (x_1, \ldots, x_k) : x_i \geq 0, \sum_{i=0}^{k} x_i = 1 \right\},$$

equipped with the associated Borel $\sigma$-algebra $\mathscr{B}(\Sigma_k)$.

Let $\mu = \mu(dx)$ be any measure on $(\Sigma_k, \mathscr{B}(\Sigma_k))$, such that $\mu(\Sigma_k) < \infty$ and

$$\int_{\Sigma_k} x_i \, \mu(dx) = 1, \quad i = 1, \ldots, k.$$

(In the theory of statistical experiments measures, $\mu$, with the above properties are known as *standard measures*.)

In Mathematical Analysis the *Hellinger transformation*, $\mathbb{H}(\alpha; \mu)$, *of the measure* $\mu$ is defined by the formula

$$\mathbb{H}(\alpha; \mu) = \int_{\Sigma_k} x_1^{\alpha_1} \ldots x_k^{\alpha_k} \, \mu(dx),$$

for all $\alpha \in \Sigma_k$.

Prove the following statements:

(a) If $\mu_1$ and $\mu_2$ are two standard measures, such that $\mathbb{H}(\alpha; \mu_1) = \mathbb{H}(\alpha; \mu_2)$, for all $\alpha \in \Sigma_k$, then one must have $\mu_1 = \mu_2$.

(b) The sequence of standard measure $\mu_n$ converges weakly to the standard measure $\mu$ if and only if $\mathbb{H}(\alpha; \mu_n) \to \mathbb{H}(\alpha; \mu)$ as $n \to \infty$, for all $\alpha \in \Sigma_k$.

Let $\mathscr{E} = (\Omega, \mathscr{F}; P_0, P_1, \ldots, P_k)$ be some statistical experiment, let $Q$ be some probability measure that dominates the measures $P_0, P_1, \ldots, P_k$, and let

$$f_i = \frac{dP_i}{dQ}, \quad i = 0, 1, \ldots, k.$$

Setting

$$\mu(A) = Q\{\omega : (f_0(\omega), \ldots, f_k(\omega)) \in A\}, \quad A \in \mathscr{B}(\Sigma_{k+1}),$$

prove that $\mu$ is a *standard* probability measure on the space $(\Sigma_k, \mathscr{B}(\Sigma_k))$ that shares the property

$$H(\alpha; \mathscr{E}) = \mathbb{H}(\alpha; \mu).$$

**Problem 3.9.11.** Let $\mathscr{E} = (\Omega, \mathscr{F} : P_0, P_1, \ldots, P_k)$ be some statistical experiment, suppose that the measure $P_0$ dominates the measures $P_1, \ldots, P_k$, and let

$$z_i = \frac{dP_i}{dP_0}, \quad i = 1, \ldots, k.$$

In probability theory, the *Mellin transformation* of the experiment $\mathscr{E}$ is defined as a function of the argument $\beta \in \Delta_k$ given by

$$M(\beta; \mathscr{E}) = \int_\Omega z_1^{\beta_1} \ldots z_k^{\beta_k} P_0(d\omega) \quad (= E_0 z_1^{\beta_1} \ldots z_k^{\beta_k}),$$

where

$$\Delta_k = \left\{ \beta = (\beta_1, \ldots, \beta_k) : 0 \le \beta_i < 1, \sum_{i=1}^k \beta_i < 1 \right\}.$$

In mathematical analysis, the *Mellin transformation* $\mathsf{M}(\beta; \nu)$ of the measure $\nu$ is defined somewhat differently. Specifically, if

$$\mathbb{R}_+^k = \{x = (x_1, \ldots, x_k) : x_i \ge 0, i = 1, \ldots, k\},$$

and $\nu$ is a probability measure on $(\mathbb{R}_+^k, \mathscr{B}(\mathbb{R}_+^k))$, chosen so that

$$\int_{\mathbb{R}_+^k} x_i \, \nu(dx) \le 1$$

(a measures $\nu$ on $\mathbb{R}_+^k$ with this property is commonly referred to as *standard measure*), then one sets

$$\mathsf{M}(\beta; \nu) = \int_{\mathbb{R}_+^k} x_1^{\beta_1} \ldots x_k^{\beta_k} \nu(dx),$$

where $\beta = (\beta_1, \ldots, \beta_k) \in \Delta_k$.

Prove that:

(a) if $\nu_1$ and $\nu_2$ are any two standard measures for which $\mathsf{M}(\beta; \nu_1) = \mathsf{M}(\beta; \nu_2)$ for all $\beta \in \Delta_k$, then $\nu_1 = \nu_2$;

(b) the sequence of standard measures $(\nu_n)$ converges weakly to the standard measure $\nu$ if and only if

$$\mathsf{M}(\beta; \nu_n) \to \mathsf{M}(\beta; \nu), \quad \text{for all} \quad \beta \in \Delta_k;$$

(c)

$$M(\beta; \mathscr{E}) = \mathsf{M}(\beta; \nu).$$

**Problem 3.9.12.** Prove that if $\alpha = (\alpha_0, \ldots, \alpha_k) \in \Sigma_{k+1}$ and $\alpha_0 > 0$, then

$$H(\alpha; \mathscr{E}) = M(\beta; \mathscr{E}),$$

where $(\beta_1, \ldots, \beta_k) = (\alpha_1, \ldots, \alpha_k)$.

Convince yourself that if $\alpha = (\alpha_0, \ldots, \alpha_k) \in \Sigma_{k+1}$, with $\alpha_0 > 0$, and $L_i = \ln z_i$, $i = 1, \ldots, k$, then

$$H(\alpha; \mathscr{E}) = E_0 \exp \left\{ \sum_{i=1}^{k} \alpha_i L_i \right\},$$

i.e., the Hellinger transformation $H(\alpha; \mathscr{E})$ coincides with the Laplace transformation of the vector $(L_1, \ldots, L_k)$ with respect to the measure $P_0$.

**Problem 3.9.13.** Suppose that $P = \| p_{ij} \|, 1 \leq i, j \leq N < \infty$, is a stochastic matrix (see $[\underline{P} \ \S1.12]$). The variable

$$D(P) = \frac{1}{2} \sup_{i,j} \sum_{k=1}^{N} |p_{ik} - p_{jk}|$$

is known as the *Dobrushin ergodicity coefficient* of the matrix $P$.

Prove that:
(a) $D(P) = \sup_{i,j} \| p_{i \cdot} - p_{j \cdot} \|$;
(b) $D(P) = 1 - \inf_{i,j} \sum_{k=1}^{\infty} (p_{ik} \wedge p_{jk})$;
(c) if $P$ and $Q$ are any two stochastic matrices of the same dimension, then

$$D(PQ) \leq D(P)D(Q);$$

(d) if $\mu = (\mu_1, \ldots, \mu_N)$ and $\nu = (\nu_1, \ldots, \nu_N)$ are any two distributions, then

$$\| \mu P^n - \nu P^n \| \leq \| \mu - \nu \| \, (D(P))^n.$$

**Problem 3.9.14.** Suppose that $\xi$ and $\eta$ are two random variables with probability distributions $P$ and $Q$. Show the *coupling inequality*:

$$P\{\xi = \eta\} \leq 1 - \frac{1}{2} \| P - Q \|$$

and compare this relation with the statement in Problem 3.9.8. In particular, if $\xi$ and $\eta$ are two random variables with densities $p(x)$ and $q(x)$, then

$$P\{\xi = \eta\} \leq 1 - \frac{1}{2} \int_R |p(x) - q(x)| \, dx.$$

Give examples in which the above inequality turns into equality.

**Problem 3.9.15.** Let $X = (X_n)_{n \geq 0}$ and $Y = (Y_n)_{n \geq 0}$ be any two random sequences, defined on some probability space $(\Omega, \mathscr{F}, \mathsf{P})$. Let $\tau$ be a random moment, such that $X_n(\omega) = Y_n(\omega)$ for all $n \geq \tau(\omega)$ (the moment $\tau$ is sometimes referred to as the *coupling time*). Letting $P_n$ and $Q_n$ denote the probability distributions of the variables $X_n$ and $Y_n$, prove the *coupling inequality*:

$$\frac{1}{2}\|P_n - Q_n\| \leq \mathsf{P}\{\tau \geq n\}.$$

**Problem 3.9.16.** Let $f = f(x)$ and $g = g(x)$ be any two probability densities on $(\mathbb{R}, \mathscr{B}(\mathbb{R}))$. Prove that:

(a) $\displaystyle\int |f(x) - g(x)|\, dx = 2 \int (f(x) - g(x))^+ \, dx = 2 \int (g(x) - f(x))^+ \, dx;$

(b) $\displaystyle\left( \int \sqrt{f(x)g(x)}\, dx \right)^2 \leq 2 \int \min(f(x), g(x))\, dx;$

(c) $\int |f(x) - g(x)|\, dx \leq \sqrt{2K(f, g)}$, where $K(f, g) = \int f(x) \ln \frac{f(x)}{g(x)}\, dx$ is Kullback's information (see Problem 3.9.3) and the probability distribution $P_f$, associated with the density $f$, is assumed to be absolutely continuous with respect to the distribution $P_g$, associated with the density $g$;

(d) $\displaystyle\int \min(f(x), g(x))\, dx \geq \frac{1}{2}e^{-K(f,g)}.$

**Problem 3.9.17.** Suppose that the random vector $X = (X_1, \ldots, X_k)$ is uniformly distributed inside the set

$$T_k = \left\{ x = (x_1, \ldots, x_k), x_i \geq 0, \sum_{i=1}^{k} x_i \leq 1 \right\}.$$

Prove that the probability density of the random vector $X$ is given by the formula

$$f(x) = k!, \quad x \in T_k.$$

**Problem 3.9.18.** Let $X$ and $Y$ be any two random variables with $\mathsf{E}X^2 < \infty$ and $\mathsf{E}Y^2 < \infty$, let $\mathrm{cov}(X, Y) = \mathsf{E}(X - \mathsf{E}X)(Y - \mathsf{E}Y)$, and let $F(x, y)$, $F_1(x)$, and $F_2(y)$ denote, respectively, the distribution functions of the random elements $(X, Y)$, $X$, and $Y$.

Prove the *Hoeffding formula*:

$$\mathrm{cov}(X, Y) = \iint \big( F(x, y) - F_1(x)F_2(y) \big)\, dx\, dy.$$

## 3.10 Contiguity (Proximity) and Full Asymptotic Separation of Probability Measures

**Problem 3.10.1.** Let $P^n = P^n_1 \times \cdots \times P^n_n$ and $\widetilde{P}^n = \widetilde{P}^n_1 \times \cdots \times \widetilde{P}^n_n$, $n \geq 1$, where $P^n_k$ and $\widetilde{P}^n_k$ are *Gaussian measures* with parameters $(a^n_k, 1)$ and $(\tilde{a}^n_k, 1)$. Find conditions for $(a^n_k)$ and $(\tilde{a}^n_k)$ that ensure the relations $(\widetilde{P}^n) \lhd (P^n)$ and $(\widetilde{P}^n) \lhd (P^n)$.

*Hint.* Use direct calculation to show that

$$H(\alpha; P^n, \widetilde{P}^n) = \exp\left\{ -\frac{\alpha(1 - \alpha)}{2} \sum_{k=1}^{n} (a^n_k - \tilde{a}^n_k)^2 \right\}$$

and take into account the relations [P §3.10, (11) and (12)].

**Problem 3.10.2.** Let $P^n = P^n_1 \times \cdots \times P^n_n$ and $\widetilde{P}^n = \widetilde{P}^n_1 \times \cdots \times \widetilde{P}^n_n$, where $P^n_k$ and $\widetilde{P}^n_k$ are probability measures on $(\mathbb{R}, \mathscr{B}(\mathbb{R}))$, such that $P^n_k(dx) = I_{[0,1]}(x)\,dx$ and $\widetilde{P}^n_k(dx) = I_{[a_n, 1 + a_n]}(x)\,dx$, for some choice of $0 \leq a_n \leq 1$. Prove that $H(\alpha; P^n_k, \widetilde{P}^n_k) = 1 - a_n$ and

$$(\widetilde{P}^n) \lhd (P^n) \iff (P^n) \lhd (\widetilde{P}^n) \iff \varlimsup_n n a_n = 0,$$

$$(\widetilde{P}^n) \triangle (P^n) \iff \varlimsup_n n a_n = \infty.$$

**Problem 3.10.3.** Consider the structure $(\Omega, \mathscr{F}, (\mathscr{F}_n)_{n \geq 0})$, which consists of a measurable space $(\Omega, \mathscr{F})$ and a flow of $\sigma$-algebras $(\mathscr{F}_n)_{n \geq 0}$, chosen so that $\mathscr{F}_0 \subseteq \mathscr{F}_1 \subseteq \cdots \subseteq \mathscr{F}$. Set $\mathscr{F} = \sigma(\bigcup_n \mathscr{F}_n)$, suppose that $P$ and $\widetilde{P}$ are two probability measures on $(\Omega, \mathscr{F})$, and denote by $P_n = P|\mathscr{F}_n$ and $\widetilde{P}_n = \widetilde{P}|\mathscr{F}_n$ their respective restrictions to $\mathscr{F}_n$. Prove that

$$(P_n) \lhd (P_n) \iff P \ll P,$$

$$(\widetilde{P}_n) \lhd\rhd (P_n) \iff \widetilde{P} \sim P,$$

$$(\widetilde{P}_n) \lhd (P_n) \iff \widetilde{P} \perp P.$$

**Problem 3.10.4.** Consider the probability space $(\Omega, \mathscr{F}, \mathsf{P})$, in which $\Omega = \{-1, 1\}^\infty$ is the space of binary sequences $\omega = (\omega_1, \omega_2, \ldots)$ and the probability measure $\mathsf{P}$ is chosen so that $\mathsf{P}\{\omega : (\omega_1, \ldots, \omega_n) = (a_1, \ldots, a_n)\} = 2^{-n}$, for every $a_i = \pm 1$, $i = 1, \ldots, n$. Given any $n \geq 1$, let $\varepsilon_n(\omega) = \omega_n$. (In particular, under the measure $\mathsf{P}$, the sequence $\varepsilon = (\varepsilon_1, \varepsilon_2, \ldots)$ is a sequence of independent Bernoulli random variables with $\mathsf{P}\{\varepsilon_n = 1\} = \mathsf{P}\{\varepsilon_n = -1\} = \frac{1}{2}$.)

Next, define the sequence $S = (S_n)_{n \geq 0}$ according to the recursive rule $S_0 = 1$ and $S_n = S_{n-1}(1 + \rho_n)$, where $\rho_n = \mu_n + \sigma_n \varepsilon_n$, $\sigma_n > 0$, $\mu_n > \sigma_n - 1$. (In the context of financial mathematics the random variable $S_n > 0$ is usually interpreted as "*the price*" of a given security in period $n$—see [P §7.11].)

Let $P^n = P|\mathscr{F}_n$, where $\mathscr{F}_n = \sigma(\varepsilon_1, \ldots, \varepsilon_n)$. On the probability space $(\Omega, \mathscr{F})$ one can define a new measure $\widetilde{P}$ in such a way that under $\widetilde{P}$ the random variables $\varepsilon_1, \varepsilon_2, \ldots$ are again independent, but also share the property

$$\widetilde{P}\{\varepsilon_n = 1\} = \frac{1}{2}(1 + b_n), \qquad \widetilde{P}\{\varepsilon_n = -1\} = \frac{1}{2}(1 - b_n), \qquad \text{where } b_n = -\frac{\mu_n}{\sigma_n}.$$

Prove that the sequence $S = (S_n)_{n \geq 0}$ forms a martingale relative to the measure $\widetilde{P}$ (see [P §1.11] and [P §7.1]).

Setting $P^n = P|\mathscr{F}_n$, prove that

$$H(\alpha; \widetilde{P}^n, P^n) = \prod_{k=1}^{n} \left[ \frac{(1 + b_k)^\alpha + (1 - b_k)^\alpha}{2} \right].$$

Finally, by using [P §3.10, Theorem 1] conclude that

$$(P^n) \lhd (\widetilde{P}^n) \quad \Leftrightarrow \quad \sum_{k=1}^{\infty} b_k^2 < \infty.$$

(In the context of "large" financial markets the previous statement implies that the condition $\sum_{k=1}^{\infty} \left( \frac{\mu_k}{\sigma_k} \right)^2 < \infty$ is necessary and sufficient for the absence of *asymptotic arbitrage*—for more details see § 3, Chap. VI, in the book [120].)

**Problem 3.10.5.** Suppose that, unlike the security-pricing model discussed in the previous problem, one sets $S_0 = 1$ and $S_n = e^{h_1 + \ldots + h_n}$, $n \geq 1$, where $h_k = \mu_k + \sigma_k \varepsilon_k$, for some $\sigma_k > 0$ and some sequence $(\varepsilon_1, \varepsilon_2, \ldots)$ of independent and identically distributed *Gaussian* $(\mathscr{N}(0, 1))$ random variables.

With $\mathscr{F}_n = \sigma(\varepsilon_1, \ldots, \varepsilon_n)$ and $P^n = P|\mathscr{F}_n$, $n \geq 1$, prove that the sequence $(S_n)_{n \geq 0}$ forms a martingale (see [P §7.11]) relative to the measure $\widetilde{P}$, which is defined by $\widetilde{P}|\mathscr{F}_n = \widetilde{P}^n$, where $d\widetilde{P}^n = z_n \, dP^n$ with

$$z_n = \exp \left\{ -\sum_{k=1}^{n} \left( \frac{\mu_k}{\sigma_k} + \frac{\sigma_k}{2} \right) \varepsilon_k + \frac{1}{2} \sum_{k=1}^{n} \left( \frac{\mu_k}{\sigma_k} + \frac{\sigma_k}{2} \right)^2 \right\}.$$

Show also that

$$H(\alpha; \widetilde{P}^n, P^n) = \exp \left\{ -\frac{\alpha(1 - \alpha)}{2} \sum_{k=1}^{n} \left( \frac{\mu_k}{\sigma_k} + \frac{\sigma_k}{2} \right)^2 \right\}$$

and

$$(P^n) \lhd (\widetilde{P}^n) \quad \Leftrightarrow \quad \sum_{k=1}^{n} \left( \frac{\mu_k}{\sigma_k} + \frac{\sigma_k}{2} \right)^2 < \infty.$$

(The condition $\sum_{k=1}^{n} \left( \frac{\mu_k}{\sigma_k} + \frac{\sigma_k}{2} \right)^2$ guarantees the absence of *asymptotic arbitrage* in this financial market—see § 3c, Chap. VI, in the book [120].)

## 3.11  Rate of Convergence in the Central Limit Theorem

**Problem 3.11.1.** Prove the inequalities in [P §3.11, (8)].

**Problem 3.11.2.** Let $\xi_1, \xi_2, \ldots$ be a sequence of independent and identically distributed random variables with $\mathsf{E}\xi_k = 0$, $\mathsf{D}\xi_k = \sigma^2$ and $\mathsf{E}|\xi_1|^3 < \infty$. The following *non-uniform* estimate is well known:

$$|F_n(x) - \Phi(x)| \leq \frac{C\mathsf{E}|\xi_1|^3}{\sigma^3 \sqrt{n}} \times \frac{1}{(1 + |x|)^3} \quad \text{for all } -\infty < x < \infty.$$

Prove this result at least in the case of Bernoulli random variables. (The statements in this problem and in Problems 3.11.5–3.11.7 bellow are discussed, for example, in the book [94].)

**Problem 3.11.3.** Let $(\xi_k)_{k\geq 1}$ be a sequence of independent and identically distributed random variables that take two values $\pm 1$ with equal probability $(1/2)$. Setting $\varphi_2(t) = \mathsf{E}e^{it\xi_1} = \frac{1}{2}(e^{it} + e^{-it})$ and $S_k = \xi_1 + \cdots + \xi_k$, show, following Laplace, that

$$\mathsf{P}\{S_{2n} = 0\} = \frac{1}{\pi} \int_0^{\pi} \varphi_2^n(t)\, dt \sim \frac{1}{\sqrt{\pi n}} \quad \text{as } n \to \infty.$$

**Problem 3.11.4.** Let $(\xi_k)_{k\geq 1}$ be a sequence of independent and identically distributed random variables, taking $2a + 1$ integer values $0, \pm 1, \ldots, \pm a$, and set $\varphi_{2a+1}(t) = \mathsf{E}e^{it\xi_1} = \frac{1}{1+2a}(1 + 2\sum_{k=1}^{a} \cos tk)$.
Just as in the previous problem, prove—again, following Laplace—that

$$\mathsf{P}\{S_n = 0\} = \frac{1}{\pi} \int_0^{\pi} \varphi_{2a+1}^n(t)\, dt \sim \frac{\sqrt{3}}{\sqrt{2\pi(a + 1)n}} \quad \text{as } n \to \infty.$$

In particular, for $a = 1$, i.e., in the special case where $\xi_k$ takes the values $-1, 0, 1$, one must have

$$\mathsf{P}\{S_n = 0\} \sim \frac{\sqrt{3}}{2\sqrt{\pi n}} \quad \text{as } n \to \infty.$$

**Problem 3.11.5.** Prove that if $F = F(x)$ and $G = G(x)$ are two distribution functions, associated with two *integer-valued* random variables, and if $f(t)$ and $g(t)$ denote their respective characteristic functions, then

$$\sup_x |F(x) - G(x)| \leq \frac{1}{4} \int_{-\pi}^{\pi} \left| \frac{f(t) - g(t)}{t} \right| dt.$$

**Problem 3.11.6.** Prove that if $F$ and $G$ are two distribution functions, $f(t)$ and $g(t)$ are their respective characteristic functions, and $L(F, G)$ is the Lévy distance between $F$ and $G$ (see Problem 3.1.4), then for every $T \geq 2$ one must have

$$L(F, G) \leq \frac{1}{\pi} \int_0^T \left| \frac{f(t) - g(t)}{t} \right| dt + 2e \frac{\ln T}{T}.$$

**Problem 3.11.7.** Let $F_n(x)$ be the distribution function of the normalized sum $\frac{1}{\sigma\sqrt{n}} \sum_{i=1}^n \xi_i$ of some finite collection of independent and identically distributed random variables, such that $\mathsf{E}\xi_i = 0$, $\mathsf{E}\xi_i^2 = \sigma^2 > 0$ and $\mathsf{E}|\xi_i|^3 = \beta_3 < \infty$. Setting $\rho = \frac{\beta_3}{\sigma^3}$, prove that

$$\varlimsup_n \inf_{(a,\sigma)} \sqrt{n} \left| F_n(x) - \Phi\left( \frac{x - \widetilde{a}}{\widetilde{\sigma}} \right) \right| \leq \frac{\rho}{\sqrt{2\pi}}.$$

## 3.12   Rate of Convergence in the Poisson Limit Theorem

**Problem 3.12.1.** Prove that with $\lambda_k = -\ln(1 - p_k)$ the variation distance $\|B(p_k) - \Pi(\lambda_k)\|$ satisfies the relation

$$\|B(p_k) - \Pi(\lambda_k)\| = 2(1 - e^{-\lambda_k} - \lambda_k e^{-\lambda_k}) \ (\leq \lambda_k^2)$$

and, therefore, $\|B - \Pi\| \leq \sum_{k=1}^n \lambda_k^2$.

*Hint.* The inequality $\|B(p_k) - \Pi(\lambda_k)\| \leq \lambda_k^2$ follows from the formula

$$\|B(p_k) - \Pi(\lambda_k)\| = |(1 - p_k) - e^{-\lambda_k}| + |p_k - \lambda_k e^{-\lambda_k}| +$$

$$+ e^{-\lambda_k} \sum_{i=2}^{\infty} \frac{\lambda_k^i}{i!} = 2(1 - e^{-\lambda_k} - \lambda_k e^{-\lambda_k})$$

and the fact that $2(1 - e^{-x} - xe^{-x}) \leq x^2$, for $x \geq 0$.

**Problem 3.12.2.** Prove the relations [$\underline{P}$ §3.12, (9) and (10)].

**Problem 3.12.3.** Let $\xi_1, \ldots, \xi_n$ be independent Bernoulli random variables that take the values 1 and 0 with probabilities $P\{\xi_k = 1\} = p_k$, $P\{\xi_k = 0\} = 1 - p_k$, $1 \le k \le n$. Given any $0 \le t \le 1$ and $\lambda > 0$, set $\xi_0 = 0$,

$$S_n(t) = \sum_{k=0}^{\lfloor nt \rfloor} \xi_k,$$

$$P_k^{(n)}(t) = P\{S_n(t) = k\}, \quad \pi_k(t) = \frac{(\lambda t)^k e^{-\lambda t}}{k!}, \qquad k = 0, 1, 2, \ldots,$$

and

$$A_n(t) = \sum_{k=0}^{\lfloor nt \rfloor} p_k \quad (= \mathsf{E} S_n(t)).$$

Prove that the probabilities $P_k^{(n)}(t)$ and $\pi_k(t)$ satisfy the following relations:

$$P_0^{(n)}(t) = 1 - \int_0^t P_0^{(n)}(s-) \, dA_n(s),$$

$$P_k^{(n)}(t) = -\int_0^t \left[ P_k^{(n)}(s-) - P_{k-1}^{(n)}(s-) \right] dA_n(s), \quad k \ge 1,$$

$$(*)$$

and

$$\pi_0(t) = \int_0^t \pi_0(s-) \, d(\lambda s),$$

$$\pi_k(t) = \int_0^t \left[ \pi_k(s-) - \pi_{k-1}(s-) \right] d(\lambda s), \quad k \ge 1.$$

$$(**)$$

**Problem 3.12.4.** By using the relations $(*)$ and $(**)$ in the previous problem, prove that

$$\sum_{k=0}^{\infty} \left| P_k^{(n)}(t) - \pi_k(t) \right| \le 2 \int_0^t \sum_{k=0}^{\infty} \left| P_k^{(n)}(s-) - \pi_k(s-) \right| d(\lambda s)$$

$$+ (2 + 4A_n(t)) \max_{0 \le s \le t} |A_n(s) - \lambda s|. \qquad (***)$$

**Problem 3.12.5.** By using the Gronwall–Bellman inequality (see Problem 2.6.51) and the notation adopted in Problem 3.12.3, conclude from $(***)$ that

$$\sum_{k=0}^{\infty} \left| P_k^{(n)}(t) - \pi_k(t) \right| \le e^{2\lambda t} + (2 + 4A_n(t)) \max_{0 \le s \le t} |A_n(s) - \lambda s|.$$

Then conclude from the last relation that

$$\sum_{k=0}^{\infty} |P\{S_n(1) = k\} - \pi_k(1)| \leq \left(2 + 4\sum_{k=1}^{n} p_k\right) e^{2\lambda} \min_i \sup_{0 \leq s \leq 1} \left|\sum_{k=0}^{\lfloor ns \rfloor} p_{i_k} - \lambda s\right|,$$

where the min is taken with respect to all permutations $i = (i_1, \ldots, i_n)$ of the numbers $(1, \ldots, n)$ and $p_{i_0} = 0$.

By using the above inequality, prove that if $\sum_{k=1}^{n} p_k = \lambda$ then

$$\sum_{k=0}^{\infty} \left|P\left\{\sum_{l=1}^{n} \xi_l = k\right\} - \frac{\lambda^k e^{-\lambda}}{k!}\right| \leq C(\lambda) \min_i \sup_{0 \leq s \leq 1} \left|\sum_{k=0}^{\lfloor ns \rfloor} p_{i_k} - \lambda s\right| \leq C(\lambda) \max_{1 \leq k \leq n} p_k,$$

where $C(\lambda) = (2 + 4\lambda)e^{2\lambda}$.

## 3.13  The Fundamental Theorems of Mathematical Statistics

**Problem 3.13.1.** Prove formula [ P §3.13, (18)].

**Problem 3.13.2.** By using the notation adopted in [ P §3.13, **4** ], prove that the convergence $P^{(N)} \xrightarrow{w} P$ (in $(D, \mathscr{D}, \rho)$) implies the convergence $f(X^{(N)}) \xrightarrow{d} f(X)$.

**Problem 3.13.3.** Verify the implication [ P §3.13, (22)].

**Problem 3.13.4.** Let $\xi_1, \xi_2 \ldots$ and $\eta_1, \eta_2, \ldots$ be two sequences of independent and identically distributed random variables with continuous distribution functions, respectively, $F = F(x)$ and $G = G(x)$. Consider the empirical distribution functions

$$F_N(x; \omega) = \frac{1}{N} \sum_{k=1}^{N} I(\xi_k(\omega) \leq x) \quad \text{and} \quad G_N(x; \omega) = \frac{1}{N} \sum_{k=1}^{N} I(\eta_k(\omega) \leq x)$$

and set

$$D_{N,M}(\omega) = \sup_x |F_N(x; \omega) - G_M(x; \omega)|$$

and

$$D_{N,M}^{+}(\omega) = \sup_x (F_N(x; \omega) - G_M(x; \omega)).$$

In the case of *two samples*, of the type described above, it is well known that

$$\lim_{N,M \to \infty} P\left\{\sqrt{\frac{NM}{N+M}} D_{N,M}(\omega) \leq y\right\} = K(y), \quad y > 0, \qquad (*)$$

where $K = K(y)$ denotes the *Kolmogorov distribution* (see [ P §3.13, **5** ]).

By following the ideas on which the proof of [$\underline{P}$ §3.13, (25)] is based, sketch the main steps in the proof of (∗) and the proof of [$\underline{P}$ §3.13, (27) and (28)].

**Problem 3.13.5.** Consider the "*omega-square statistics*"

$$\omega_N^2(\omega) = \int_{-\infty}^{\infty} |F_N(x;\omega) - F(x)|^2 \, dF(x), \qquad (**)$$

associated with the continuous distribution function $F = F(x)$. Prove that, similarly to the statistics $D_N(\omega)$ and $D_N^+(\omega)$, the distribution of the statistics $\omega_N^2(\omega)$ is one and the same for all *continuous* distribution functions $F = F(x)$. Show also that

$$\mathsf{E}\omega_N^2(\omega) = \frac{1}{6N}, \quad \mathsf{D}\omega_N^2(\omega) = \frac{4N-3}{180N^3}.$$

**Problem 3.13.6.** Let $\xi_1, \xi_2, \ldots$ be a sequence of independent and identically distributed random variables, chosen so that $\mathsf{E}\xi_1 = 0$ and $\mathsf{D}\xi_1 = 1$. Setting

$$\mathscr{R}_n = \max_{k \le n} \left( S_k - \frac{k}{n} S_n \right) - \min_{k \le n} \left( S_k - \frac{k}{n} S_n \right),$$

prove that

$$\frac{\mathscr{R}_n}{\sqrt{n}} \overset{d}{=} \max_{\{t=k/n:k=0,1,\ldots,n\}} |B_t - tB_1| - \min_{\{t=k/n:k=0,1,\ldots,n\}} |B_t - tB_1|$$

$$\overset{d}{=} \max_{\{t=k/n:k=0,1,\ldots,n\}} B_t^\circ - \min_{\{t=k/n:k=0,1,\ldots,n\}} B_t^\circ,$$

where $B = (B_t)_{t \le 1}$ is a Brownian motion, $B^\circ = (B_t^\circ)_{t \le 1}$ is a Brownian bridge and, as usual, "$\overset{d}{=}$" stands for "identity in distribution."
   Show also that

$$\mathsf{E}\mathscr{R}_n \sim \sqrt{\tfrac{\pi}{2}n} \quad \text{and} \quad \mathsf{D}\mathscr{R}_n \sim \left( \tfrac{\pi^2}{6} - \tfrac{\pi}{2} \right) n.$$

(Comp. with Problem 2.13.48.)

**Problem 3.13.7.** Let $F = F(x)$ and $G = G(x)$ be any two distribution functions, let $F^{-1}(t) = \inf\{x : F(x) > t\}$ and $G^{-1}(t) = \inf\{x : G(x) > t\}$, let $\mathfrak{F}_2$ stand for the space of all distribution functions $F$ with $\int_{-\infty}^{\infty} x^2 \, dF(x) < \infty$, and let

$$d_2(F, G) = \left( \int_0^1 |F^{-1}(t) - G^{-1}(t)|^2 \, dt \right)^{1/2}, \quad F, G \in \mathfrak{F}_2.$$

(a) Prove that the function $d_2 = d_2(F, G)$, which is known as *the Wasserstein metric*, is indeed a metric and the space $\mathfrak{F}_2$ is complete for the metric $d_2$.

(b) Let $\xi_1, \ldots, \xi_n$ be independent and identically distributed random variables, which share one and the same distribution function $F \in \mathfrak{F}_2$, and let $\widehat{F}_n$ be the empirical distribution function associated with the sample $\xi_1, \ldots, \xi_n$. Prove that one has (P-a. e.)

$$d_2(F, \widehat{F}_n) \to 0 \quad \text{as } n \to \infty.$$

(c) Prove that for any $F, G \in \mathfrak{F}_2$ the following (*coupling-type*) relation is in force

$$d_2(F, G) = \inf \mathsf{E}(\xi - \eta)^2,$$

where the inf is taken over all possible pairs of random variables $(\xi, \eta)$, chosen so that $\xi$ has distribution function $F \in \mathfrak{F}_2$ and $\eta$ has distribution function $G \in \mathfrak{F}_2$.

**Problem 3.13.8.** Let $\mathfrak{F}_1$ stand for the space of all distribution functions $F$ with $\int_{-\infty}^{\infty} |x| \, dF(x) < \infty$ and let

$$d_1(F, G) = \int_0^1 |F^{-1}(t) - G^{-1}(t)| \, dt, \quad F, G \in \mathfrak{F}_1.$$

(a) Prove that the function $d_1 = d_1(F, G)$, which is known as *the Dobrushin metric*, is indeed a metric and the space $\mathfrak{F}_1$ is complete for the metric $d_1$.

(b) Prove that if $F, F_1, F_2, \ldots \in \mathfrak{F}_1$, then $d_1(F, F_n) \to 0$ as $n \to \infty$ if and only if $F_n \Rightarrow F$ and $\int |x| \, dF_n(x) \to \int |x| \, dF(x)$ (the symbol "$\Rightarrow$" stands for "converges essentially" —see [P §3.1]).

(c) Prove that for any $F, G \in \mathfrak{F}_1$ the following (*coupling-type*) relation is in force

$$d_1(F, G) = \inf \mathsf{E}|\xi - \eta|,$$

where the infimum is taken over all possible pairs of random variables $(\xi, \eta)$, chosen so that $\xi$ has distribution function $F \in \mathfrak{F}_1$ and $\eta$ has distribution function $G \in \mathfrak{F}_1$.

(d) Let $\xi_1, \ldots, \xi_n$ be independent and identically distributed random variables, which share one and the same distribution function $F \in \mathfrak{F}_1$ and let $\widehat{F}_n$ is the empirical distribution function associated with the sample $\xi_1, \ldots, \xi_n$. Prove that one has (P-a. e.)

$$d_1(F, \widehat{F}_n) \to 0 \quad \text{as } n \to \infty.$$

**Problem 3.13.9.** Let $F = F(x)$, $x \in \mathbb{R}$, be the distribution function of some random variable $X$ and let $F^{-1} = F^{-1}(u)$, $u \in [0, 1]$, be the inverse of $F$, as defined in Problem 3.8.8. Given any $0 < p < 1$, the quantity $x_p = F^{-1}(p)$ is known as the *p-quantile* of the random variable $X$, or, equivalently, of the distribution function $F = F(x)$. (The quantity $F^{-1}(1/2)$ is often referred as the "median," while $F^{-1}(1/4)$ and $F^{-1}(3/4)$ are commonly referred to, respectively, as the "lower quantile" and the "upper quantile".)

Give the conditions under which the *p*-quantile $x_p$ can be characterized as the unique root of the equation $F(x) = p$.

**Problem 3.13.10.** Let $X_1, \ldots, X_n, \ldots$ be independent and identically distributed random variables, which share one and the same distribution function $F = F(x)$, and let

$$\widehat{F}_n(x) = F_n(x; \omega) = \frac{1}{n} \sum_{k=1}^{n} I(X_k(\omega) \leq x)$$

be the empirical distribution function constructed from the sample $X_1, \ldots, X_n$ (see formula [P §3.13, (1)]).

Prove that if $\widehat{X}_1^{(n)}, \ldots, \widehat{X}_n^{(n)}$ are the ordered statistics constructed from the observations $X_1, \ldots, X_n$ (in Problems 1.12.8 and 2.8.19 these statistics are denoted by $X_1^{(n)}, \ldots, X_n^{(n)}$), then the empirical distribution function $\widehat{F}_n = \widehat{F}_n(x)$ admits the following representation:

$$\widehat{F}_n(x) = \begin{cases} 0, & \text{if } x < \widehat{X}_1^{(n)}, \\ k/n, & \text{if } \widehat{X}_k^{(n)} \leq x < \widehat{X}_{k+1}^{(n)}, \quad k = 1, \ldots, n-1, \\ 1, & \text{if } x \geq \widehat{X}_n^{(n)}. \end{cases}$$

**Problem 3.13.11.** Let everything be as in the previous problem, let $\varkappa_p$ be the $p$-quantile of the distribution function $F = F(x)$ and let $\widehat{\varkappa}_p(n) = \widehat{F}_n^{-1}(p)$ be the $p$-quantile of the distribution $\widehat{F}_n = \widehat{F}_n(x)$. Prove that if $\varkappa_p$ is the unique value with the property $F(\varkappa_p-) \leq p \leq F(\varkappa_p)$, then as $n \to \infty$ one has

$$\widehat{\varkappa}_p(n) \to \varkappa_p \quad \text{(P-a. e.)}.$$

*Hint.* Notice that $\widehat{\varkappa}_p(n) = \widehat{X}_{\lceil np \rceil}^{(n)}$ and convince yourself that for every $\varepsilon > 0$ one has

$$\mathsf{P}\left\{ \varliminf \widehat{X}_{\lceil np \rceil}^{(n)} > \varkappa_p - \delta \right\} = \mathsf{P}\left\{ \varlimsup \widehat{X}_{\lceil np \rceil}^{(n)} < \varkappa_p + \delta \right\} = 1,$$

where, just as before, $\lceil x \rceil$ stands for the smallest integer that is greater than or equal to $x$.

**Problem 3.13.12.** Let $X_1, X_2, \ldots$ be independent and identically distributed random variables that share one and the same continuous distribution function $F = F(x)$. In addition, suppose that the following conditions hold: for any given $0 < p < 1$ the equation $F(x) = p$ has unique solution $\varkappa_p$; the derivative $F'(x)$ exists and is continuous at the point $x = \varkappa_p$ and, furthermore, $F'(\varkappa_p) > 0$. Let $\widehat{X}_{\lceil np \rceil}^{(n)}$ denote the $p$-quantile in the sample.

Prove that as $n \to \infty$ the random variables $\sqrt{n}(\widehat{X}_{\lceil np \rceil}^{(n)} - \varkappa_p)$ converge in distribution to a Gaussian random variable $N$ that has zero mean and dispersion $p(1-p)(F'(\varkappa_p))^2$, i.e.,

$$\sqrt{n}\left(\widehat{X}_{\lceil np \rceil}^{(n)} - \varkappa_p\right) \overset{\text{law}}{\to} N.$$

*Hint.* Suppose that the random variables $\xi_1, \ldots, \xi_n$ are independent and uniformly distributed in the interval $[0, 1]$, and let $\widehat{\xi}_1^{(n)}, \ldots, \widehat{\xi}_n^{(1)}$ denote the associated ordered statistics. In order to prove the required statement, notice first that the variables $\widehat{X}_{\lceil np \rceil}^{(n)} - \varkappa_p$ and $F^{-1}(\widehat{\xi}_{\lceil np \rceil}^{(n)}) - F^{-1}(p)$ coincide in distribution, and then use the statement in [P §3.13, Lemma 2] and the Central Limit Theorem in terms of Lindeberg's conditions (see [P §3.4, Theorem 1]).

# Chapter 4
# Sequences and Sums of Independent Random Variables

## 4.1 0–1 Laws

**Problem 4.1.1.** Prove the Corollary to Theorem 1 in [ P §4.1].

*Hint.* Use the fact that the distribution function of the variable $\eta$ takes only the values 0 and 1.

**Problem 4.1.2.** Prove that if $(\xi_n)_{n\geq 1}$ is some sequence of independent random variables, then the random variables $\overline{\lim}\, \xi_n$ and $\underline{\lim}\, \xi_n$ are degenerate (i.e., have vanishing dispersion).

*Hint.* Show first that $\overline{\lim}\, \xi_n$ and $\underline{\lim}\, \xi_n$ are $\mathcal{X}$-measurable, where $\mathcal{X}$ is the associated tail $\sigma$-algebra

**Problem 4.1.3.** Let $(\xi_n)_{n\geq 1}$ be any sequence of independent random variables, let $S_n = \xi_1 + \ldots + \xi_n$ and suppose that the constants $b_n$ are chosen so that $0 < b_n \uparrow \infty$. Prove that the random variables $\overline{\lim}\, \frac{S_n}{b_n}$ and $\underline{\lim}\, \frac{S_n}{b_n}$ are degenerate (i.e., have vanishing dispersion).

*Hint.* Fix some integer $N$ in the set $\{1, 2, \ldots\}$ and set

$$\widetilde{S}_n = \begin{cases} 0, & n \leq N, \\ S_n - S_N, & n > N. \end{cases}$$

By using the property $\overline{\lim}_n \frac{S_n}{b_n} = \overline{\lim}_n \frac{\widetilde{S}_n}{b_n}$ conclude that the variable $\overline{\lim}_n \frac{S_n}{b_n}$ must be measurable for $\bigcap_{n=N}^{\infty} \mathscr{F}_n$ and, therefore, since $N$ is arbitrarily chosen, must be measurable also for $\mathcal{X}$.

**Problem 4.1.4.** Let $S_n = \xi_1 + \ldots + \xi_n$, $n \geq 1$, and $\mathcal{X}(S) = \bigcap \mathscr{F}_n^{\infty}(S)$, $\mathscr{F}_n^{\infty}(S) = \sigma\{\omega: S_n, S_{n+1}, \ldots\}$. Prove that all events in the tail $\sigma$-algebra $\mathcal{X}(S)$ are trivial.

**Problem 4.1.5.** Let $(\xi_n)_{n\geq 1}$ be any sequence of random variables. Prove that $\{\overline{\lim}\, \xi_n \geq c\} \supseteq \overline{\lim}\{\xi_n \geq c\}$ for every constant $c$.

A.N. Shiryaev, *Problems in Probability*, Problem Books in Mathematics,
DOI 10.1007/978-1-4614-3688-1_4,
© Springer Science+Business Media New York 2012

*Hint.* It is enough to notice that

$$\overline{\lim_n}\{\xi_n \geq c\} = \{\omega : \xi_n(\omega) \geq c \text{ i.o.}\}.$$

**Problem 4.1.6.** Give examples of tail events $A$ (i.e., events in the $\sigma$-algebra $\mathscr{X} = \bigcap_{n=1}^{\infty} \mathscr{F}_n^{\infty}$, where $\mathscr{F}_n^{\infty} = \sigma(\xi_n, \xi_{n+1}, \ldots)$, for some sequence of random variables $(\xi_n)_{n \geq 1}$) that have the property $0 < P(A) < 1$.

**Problem 4.1.7.** Let $\xi_1, \xi_2, \ldots$ be any sequence of independent random variables with $E\xi_n = 0$ and $E\xi_n^2 = 1$, $n \geq 1$, for which the central limit theorem holds, i.e., $P\{S_n/\sqrt{n} \leq x\} \to \Phi(x)$, $x \in \mathbb{R}$, where $S_n = \xi_1 + \ldots + \xi_n$). Prove that

$$\overline{\lim_{n \to \infty}} \, n^{-1/2} S_n = +\infty \quad (\text{P-a. e.}).$$

In particular, this property must hold when $\xi_1, \xi_2, \ldots$ are independent and *identically* distributed with $E\xi_1 = 0$ and $E\xi_1^2 = 1$.

**Problem 4.1.8.** Let $\xi_1, \xi_2, \ldots$ be any sequence of independent and identically distributed random variables with $E|\xi_1| > 0$. Prove that

$$\overline{\lim_{n \to \infty}} \left| \sum_{k=1}^{n} \xi_k \right| = +\infty \quad (\text{P-a. e.}).$$

**Problem 4.1.9.** Let $\xi_1, \xi_2, \ldots$ be any sequence of independent and identically distributed random variables with $E\xi_1 = 0$ and $E|\xi_1| > 0$ and let $S_n = \xi_1 + \ldots + \xi_n$. Prove that (P-a. e.)

$$\overline{\lim_{n \to \infty}} \, n^{-1/2} S_n = +\infty \quad \text{and} \quad \underline{\lim_{n \to \infty}} \, n^{-1/2} S_n = -\infty.$$

(Comp. with the statements in Theorem 2 and Problem 4.1.7.)

**Problem 4.1.10.** Let $\mathscr{F}_1, \mathscr{F}_2, \ldots$ be any sequence of independent $\sigma$-algebras and let $\mathscr{G} = \bigcap_{n=1}^{\infty} \sigma\left(\bigcup_{j \geq n} \mathscr{F}_j\right)$. Prove that every set $G \in \mathscr{G}$ satisfies the "zero-one" law: $P(G)$ is either 0 or 1.

**Problem 4.1.11.** Let $A_1, A_2, \ldots$ be some sequence of independent random events, chosen so that $P(A_n) < 1$, $n \geq 1$, and $P\left(\bigcup_{n=1}^{\infty} A_n\right) = 1$. Show that $P(\overline{\lim} A_n) = 1$.

**Problem 4.1.12.** Let $A_1, A_2, \ldots$ be any sequence of independent random events and let $p_n = P(A_n)$, $n \geq 1$. The "zero-one" law implies that the probabilities $P(\overline{\lim} A_n)$ and $P(\underline{\lim} A_n)$ must equal either zero or one. Give conditions, expressed in terms of the probabilities $p_n$, $n \geq 1$, which guarantee that: (a) $P(\underline{\lim} A_n) = 0$; (b) $P(\underline{\lim} A_n) = 1$; (c) $P(\overline{\lim} A_n) = 0$; and (d) $P(\overline{\lim} A_n) = 1$.

**Problem 4.1.13.** Let $\xi_1, \xi_2, \ldots$ be any sequence of *non-degenerate* and identically distributed random variables and let $S_n = \xi_1 + \ldots + \xi_n$. Prove that:
   (a) $P\{S_n \in A \text{ i.o.}\} = 0$ or 1 for every Borel set $A \in \mathscr{B}(\mathbb{R})$.

___(b) Only the following two relations are possible: either $\overline{\lim}\, S_n = \infty$ (P-a. e.), or $\overline{\lim}\, S_n = -\infty$ (P-a. e.); furthermore,

$$P\{\overline{\lim}\, S_n = \infty\} = 1, \quad \text{if} \quad \sum_{n=1}^{\infty} \frac{1}{n} P\{S_n > 0\} = \infty,$$

$$P\{\overline{\lim}\, S_n = -\infty\} = 1, \quad \text{if} \quad \sum_{n=1}^{\infty} \frac{1}{n} P\{S_n > 0\} < \infty.$$

(c) If the distribution of the variables $\xi_n$ is symmetric, then $\overline{\lim}\, S_n = \infty$ and $\underline{\lim}\, S_n = -\infty$ (P-a. e.).

**Problem 4.1.14.** According to the corollary to [P §4.1, Theorem 1], every random variable $\eta$, which is measurable for the *tail* $\sigma$-algebra $\mathscr{X}$, associated with some sequence of independent (say, relative to some measure P) random variables $\xi_1, \xi_2, \ldots$, must be constant P-a. e., i.e., $P\{\eta = C_P\} = 1$, for some constant $C_P$. Let Q be another probability measure, relative to which the variables $\xi_1, \xi_2, \ldots$ are also independent. Then it must be the case that $Q\{\eta = C_Q\} = 1$, for some constant $C_Q$. Can one claim that the constant $C_Q$ must coincide with the constant $C_P$?

**Problem 4.1.15.** Let $S_m = \xi_1 + \ldots + \xi_m, m \geq 1$, where $\xi_1, \xi_2, \ldots$ is some sequence of independent Bernoulli random variables, such that $P\{\xi_i = 1\} = P\{\xi_i = -1\} = 1/2, i \geq 1$. Let $\sigma_0 = \inf\{n \geq 1 : S_n = 0\}$, with the understanding that $\sigma_0 = \infty$ if $S_n \neq 0$ for all $n \geq 1$. Prove that the random walk $(S_m)_{m \geq 0}$, which starts from 0 ($S_0 = 0$), is recurrent, in that $P\{\sigma_0 < \infty\} = 1$. By using this property argue that $P\{S_n = 0 \text{ i.o. }\} = 1$.

*Hint.* Use the result established in Problem 1.5.7, according to which

$$P\{S_1 \ldots S_{2m} \neq 0\} = \left(\frac{1}{2}\right)^{2m} C_{2m}^m, \quad \text{for every } m \geq 1.$$

**Problem 4.1.16.** Let $\xi_1, \xi_2, \ldots$ be any sequence of independent and identically distributed random variables with $E|\xi_i| < \infty$. Assuming that $E\xi_i = 0$ and setting $S_n = \xi_1 + \ldots + \xi_n, n \geq 1$, prove that one has (P-a. e.)

$$\lim_{n \to \infty} |S_n| < \infty.$$

**Problem 4.1.17.** Let $X = (X_1, X_2, \ldots)$ be any *infinite* sequence of exchangeable random variables (for the definition of "exchangeable," see Problem 2.5.4), let $\mathscr{X}_n = \sigma(X_n, X_{n+1}, \ldots)$ and let $\mathscr{X} = \bigcap_n \mathscr{X}_n$ be the "tail" $\sigma$-algebra, associated with the sequence $X$. Prove that for every bounded Borel function $g = g(x)$ one must have (P-a. e.)

$$E[g(X_1) \mid \mathscr{X}] = E[g(X_1) \mid \mathscr{X}_2].$$

Show also that the random variables $X_1, X_2, \ldots$ are conditionally independent relative to the "tail" $\sigma$-algebra $\mathscr{X}$.

**Problem 4.1.18.** Let $(X_1, \ldots, X_N)$ be any *Gaussian* vector of exchangeable random variables. Prove that there is a vector $(\varepsilon_1, \ldots, \varepsilon_N)$, of independent standard normal random variables $(\varepsilon_i \sim \mathscr{N}(0,1))$, for which one can write

$$X_n \stackrel{\text{law}}{=} a + b\varepsilon_n + c \sum_{i=1}^{N} \varepsilon_i, \quad 1 \le n \le N,$$

for some choice of the constants $a, b$ and $c$.

**Problem 4.1.19.** Let $(X_1, X_2, \ldots)$ be any *infinite* Gaussian sequence of exchangeable random variables. Prove that one can find a sequence $(\varepsilon_0, \varepsilon_1, \ldots)$, that consists of independent and identically distributed Gaussian random variables $\varepsilon_i \sim \mathscr{N}(0,1), i \ge 0$, so that

$$X_n \stackrel{\text{law}}{=} a + b\varepsilon_0 + c\varepsilon_n, \quad n \ge 1.$$

**Problem 4.1.20.** Let $\xi_1, \xi_2, \ldots$ be any sequence of independent random variables with exponential distribution $\mathsf{P}\{\xi_i > x\} = e^{-x}, x \ge 0$. Consider the event $A_n = \{\xi_n \ge h(n)\}, n \ge 1$, where $h(n)$ is any of the functions $(c \ln n), (\ln n + c \ln \ln n)$, or $(\ln n + \ln \ln n + c \ln \ln \ln n)$.
   Prove that

$$\mathsf{P}(A_n \text{ i.o.}) = \begin{cases} 0, & \text{if } c > 1, \\ 1, & \text{if } c \le 1. \end{cases}$$

*Hint.* Use the Borel–Cantelli lemma.

**Problem 4.1.21.** Let $\xi_1, \xi_2, \ldots$ be any sequence of independent and identically distributed Bernoulli random variables with $\mathsf{P}\{\xi_n = 1\} = \mathsf{P}\{\xi_n = 0\} = 1/2$, $n \ge 1$. Consider the the events

$$A_n = \{\xi_{n+1} = 1, \ldots, \xi_{n+[\log_2 \log_2 n]} = 1\}, \quad n \ge 4.$$

(a) Prove that $\mathsf{P}(A_n \text{ i.o.}) = 1$.
   *Hint.* Consider first the sequence the events $A_{2^m}, m \ge 2$.
(b) Calculate the probability $\mathsf{P}(B_n \text{ i.o.})$, where

$$B_n = \{\xi_{n+1} = 1, \ldots, \xi_{n+[\log_2 n]} = 1\}, \quad n \ge 2.$$

**Problem 4.1.22.** Let $A_1, A_2, \ldots$ be some sequence of independent events and let

$$B_{\le x} = \left\{ \omega; \lim_n \frac{1}{n} \sum_{k=1}^{n} I_{A_k} \le x \right\}, \quad x \in R.$$

Prove that for every $x \in \mathbb{R}$ one has

$$P(B_{\leq x}) = 0 \text{ or } 1.$$

## 4.2    Convergence of Series of Random Variables

**Problem 4.2.1.** Let $\xi_1, \xi_2, \dots$ be any sequence of independent random variables and let $S_n = \xi_1 + \dots + \xi_n$. By using the "three series theorem" prove that:

(a) If $\sum \xi_n^2 < \infty$ (P-a. e.), then the series $\sum \xi_n$ converges with Probability 1 if and only if the series $\mathsf{E}\xi_i\, I(|\xi_i| \leq 1)$ converges.

(b) If the series $\sum \xi_n$ converges (P-a. e.), then $\sum \xi_n^2 < \infty$ (P-a. e.), if and only if

$$\sum (\mathsf{E}|\xi_n|\, I(|\xi_n| \leq 1))^2 < \infty.$$

**Problem 4.2.2.** Let $\xi_1, \xi_2, \dots$ be any sequence of independent random variables. Prove that $\sum \xi_n^2 < \infty$ (P-a. e.), if and only if

$$\sum \mathsf{E}\left[\frac{\xi_n^2}{1 + \xi_n^2}\right] < \infty.$$

*Hint.* Use the "three series theorem" and notice that

$$\sum \mathsf{E}\left[\frac{\xi_n^2}{1 + \xi_n^2}\right] < \infty \iff \left[\sum \mathsf{E}\xi_n^2 I(|\xi_n| \leq 1) < \infty \quad \sum \mathsf{P}\{|\xi_n| > 1\} < \infty\right].$$

**Problem 4.2.3.** Let $\xi_1, \xi_2, \dots$ be any sequence of independent random variables. Prove that the following three conditions are equivalent:

1. The series $\sum \xi_n$ converges *with Probability* 1.
2. Series $\sum \xi_n$ converges *in probability*.
3. Series $\sum \xi_n$ converges *in distribution*.

*Hint.* Consider proving the implications $(1) \Rightarrow (3) \Rightarrow (2) \Rightarrow (1)$. The first implication follows from [ P §2.10, Theorem 2]. The implication $(3) \Rightarrow (2)$ can be proved by contradiction by using the Prokhorov Theorem. To prove the implication $(2) \Rightarrow (1)$, show first that the following inequality holds for arbitrary $m < n$ and $C > 0$:

$$\mathsf{P}\left\{\max_{m \leq k \leq n} |S_k - S_m| > 2C\right\} \leq 2 \max_{m \leq k \leq n} \mathsf{P}\{|S_n - S_k| > C\}.$$

If the series $\sum \xi_n$ converges in probability, then for every $\varepsilon > 0$ one can find an integer $m \in \mathbb{N} = \{1, 2, \dots\}$, so that the following inequality holds for every $n \geq m$:

$$\max_{m \leq k \leq n} \mathsf{P}\{|S_n - S_k| > \varepsilon\} < \varepsilon.$$

Finally, conclude from the last relation that the series $\sum \xi_n$ converges with Probability 1.

**Problem 4.2.4.** By providing appropriate examples, prove that, in general, in [P §4.2, Theorems 1 and 2] one cannot remove the uniform boundedness requirement, i.e., the condition: for every given $n \geq 1$ one has $\mathsf{P}\{|\xi_n| \leq c_n\} = 1$ for some appropriate constant $c_n > 0$.

*Hint.* Consider the sequence of independent random variables $\xi_1, \xi_2, \ldots$ chosen so that

$$\mathsf{P}\{\xi_n = 0\} = 1 - \frac{2}{n^2}, \quad \mathsf{P}\{\xi_n = n\} = \mathsf{P}\{\xi_n = -n\} = \frac{1}{n^2}, \quad n \geq 1.$$

**Problem 4.2.5.** Let $\xi_1, \ldots, \xi_n$ be independent and identically distributed random variables with $\mathsf{E}\xi_1 = 0$ and $\mathsf{E}\xi_1^2 < \infty$, and let $S_k = \xi_1 + \ldots + \xi_k, k \leq n$. Prove the following *one-sided analog* (due to A. V. Marshall) of the *Kolmogorov's inequality* [P §4.2, (2)]:

$$\mathsf{P}\left\{\max_{1 \leq k \leq n} S_k \geq \varepsilon\right\} \leq \frac{\mathsf{E}S_n^2}{\varepsilon^2 + \mathsf{E}S_n^2}, \quad \varepsilon \geq 0.$$

**Problem 4.2.6.** Let $\xi_1, \xi_2, \ldots$ be any sequence of random variables. Prove that if $\sum_{n \geq 1} \mathsf{E}|\xi_n| < \infty$, then the series $\sum_{n \geq 1} \xi_n$ converges absolutely with Probability 1.

**Problem 4.2.7.** Let $\xi_1, \xi_2, \ldots$ be any sequence of independent and symmetrically distributed random variables. Prove that

$$\mathsf{E}\left[\left(\sum_n \xi_n\right)^2 \wedge 1\right] \leq \sum_n \mathsf{E}(\xi_n^2 \wedge 1).$$

**Problem 4.2.8.** Let $\xi_1, \xi_2, \ldots$ be any sequence of independent random variables with finite second moments. Prove that the series $\sum \xi_n$ converges in $L^2$ if and only if the series $\sum \mathsf{E}\xi_n$ and $\sum \mathsf{D}\xi_n$ both converge.

**Problem 4.2.9.** Let $\xi_1, \xi_2, \ldots$ be any sequence of independent random variables and suppose that the series $\sum \xi_n$ converges (P-a. e). Prove that the value of the sum $\sum \xi_n$ does not depend on the order of summation (P-a. e.) if and only if $\sum |\mathsf{E}(\xi_n; |\xi_n| \leq 1)| < \infty$.

**Problem 4.2.10.** Let $\xi_1, \xi_2, \ldots$ be any sequence of independent random variables with $\mathsf{E}\xi_n = 0, n \geq 1$, and suppose that

$$\sum_{n=1}^{\infty} \mathsf{E}[\xi_n^2 I(|\xi_n| \le 1) + |\xi_n| I(|\xi_n| > 1)] < \infty.$$

Prove that the series $\sum_{n=1}^{\infty} \xi_n$ converges (P-a. e.).

**Problem 4.2.11.** Let $A_1, A_2, \dots$ be any sequence of independent events with $\mathsf{P}(A_n) > 0, n \ge 1$, and suppose that $\sum_{n=1}^{\infty} \mathsf{P}(A_n) = \infty$. Show that

$$\sum_{j=1}^{n} I(A_j) \Big/ \sum_{j=1}^{n} \mathsf{P}(A_j) \to 1 \quad \text{as } n \to \infty \text{ (P-a. e.).}$$

**Problem 4.2.12.** Let $\xi_1, \xi_2, \dots$ be any sequence of independent random variables with mean values $\mathsf{E}\xi_n$ and dispersions $\sigma_n^2$, chosen so that $\lim_{n} \mathsf{E}\xi_n = c$ and $\sum_{n=1}^{\infty} \sigma_n^{-2} = \infty$. Prove that

$$\sum_{j=1}^{n} \frac{\xi_j}{\sigma_j^2} \Big/ \sum_{j=1}^{n} \frac{1}{\sigma_j^2} \to c \quad \text{as } n \to \infty \text{ (P-a. e.).}$$

**Problem 4.2.13.** Let $\xi_1, \xi_2, \dots$ be any sequence of independent and identically exponentially distributed random variables, so that $\mathsf{P}\{\xi_1 > x\} = e^{-\lambda}, x \ge 0$.

Prove that if the positive numbers $a_n, n \ge 1$, are chosen so that the series $\sum_{n \ge 1} a_n$ converges, then the series $\sum_{n \ge 1} a_n \xi_n$ converges with Probability 1 and also in $L^p$-sense for every $p \ge 1$.

**Problem 4.2.14.** Let $(T_1, T_2, \dots)$ be the moments of jumps for some Poisson process (see [P §7.10]) and let $\alpha \in (0, 1)$. Prove that the series $\sum_{i=1}^{\infty} T_i^{-1/\alpha}$ converges with Probability 1.

**Problem 4.2.15.** Let $\xi_1, \xi_2, \dots$ be any sequence of independent random variables, chosen so that $\xi_n$ is uniformly distributed in $\left[-\frac{1}{n}, \frac{1}{n}\right], n \ge 1$. Prove that (P-a.e.):
  (a) the series $\sum_n \xi_n$ converges;
  (b) $\sum_n |\xi_n| = +\infty$.
  *Hint.* Use the two-series and three-series theorems of A. Khinchin and A. N. Kolmogorov ([P §4.2, Theorem 2] and [P §4.2, Theorem 3]).

**Problem 4.2.16.** The three-series theorem ([P §4.2, Theorem 3]) guarantees that, if $\xi_1, \xi_2, \dots$ is any sequence of independent random variables, then the series $\sum_{n \ge 1} \xi_n$ converges (P-a. e.), if one can find a constant $c > 0$, for which the following three series happen converge (with $\xi_n^c = \xi_n I(|\xi_n| \le c)$):

$$\sum_{n \ge 1} \mathsf{E}\xi_n^c, \quad \sum_{n \ge 1} \mathsf{D}\xi_n^c, \quad \sum_{n \ge 1} \mathsf{P}\{|\xi_n| > c\}.$$

By using appropriate examples, prove that if any one of the above series fails to converge for some $c > 0$, then the convergence (P-a. e.) of the series $\sum_{n \geq 1} \xi_n$ may not hold.

**Problem 4.2.17.** Let $\xi_1, \xi_2, \ldots$ be any sequence of random variables, chosen so that $\sum_{k=1}^{\infty} \mathsf{E}|\xi_k|^r < \infty$, for some $r > 0$. Prove that $\xi_n \to 0$ as $n \to \infty$ with Probability 1.

**Problem 4.2.18.** Let $\xi_1, \xi_2, \ldots$ be any sequence of independent Bernoulli random variables with $\mathsf{P}\{\xi_k = 1\} = \mathsf{P}\{\xi_k = -1\} = \frac{1}{2}$, $k \geq 1$. Prove that the random variable $\sum_{k=1}^{\infty} \frac{\xi_k}{2^k}$ is well defined (P-a.e.) and is uniformly distributed in $[-1, 1]$.

**Problem 4.2.19.** Let $\xi_1, \xi_2, \ldots$ be any sequence of independent and symmetrically distributed random variables. Prove that the following conditions are equivalent:

1. The series $\sum \xi_n$ converges with Probability 1.
2. $\sum \xi_n^2 < \infty$, P-a. e.
3. $\sum \mathsf{E}(\xi_n^2 \wedge 1) < \infty$.

**Problem 4.2.20.** Let $\xi$ be any random variable and let $\overline{\xi}$ denote its symmetrization, i.e., $\overline{\xi} = \xi - \widetilde{\xi}$, where $\widetilde{\xi}$ is independent of $\xi$ and has the same distribution as $\xi$. (We suppose that the probability space is sufficiently rich to support both $\widetilde{\xi}$ and $\xi$.) Let $\mu = \mu(\xi)$ denote the median of the random variable $\xi$, defined by $\max(\mathsf{P}\{\xi > \mu\}, \mathsf{P}\{\xi < \mu\}) \leq \frac{1}{2}$ (comp. with Problem 1.4.23). Prove that for every $a \geq 0$ one

$$\mathsf{P}\{|\xi - \mu| > a\} \leq 2\mathsf{P}\{|\overline{\xi}| > a\} \leq 4\mathsf{P}\left\{|\xi| > \frac{a}{2}\right\}.$$

**Problem 4.2.21.** Let $\xi_1, \xi_2, \ldots$ be any sequence of independent random variables with

$$\mathsf{P}\{\xi_n = 1\} = 2^{-n}, \quad \mathsf{P}\{\xi_n = 0\} = 1 - 2^{-n}.$$

Prove that the series $\sum_{n=1}^{\infty} \xi_n$ converges with Probability 1 and the following relations hold:

$$\mathsf{P}\left\{\sum_{n=1}^{\infty} \xi_n = 0\right\} = \prod_{n=1}^{\infty}(1 - 2^{-n}) > 0$$

and

$$\mathsf{P}\left\{\sum_{n=1}^{\infty} \xi_n = 1\right\} = \sum_{n=1}^{\infty} \frac{2^{-n}}{1 - 2^{-n}} \cdot \prod_{n=1}^{\infty}(1 - 2^{-n}).$$

**Problem 4.2.22.** Suppose that $\xi_1, \xi_2, \ldots$ is some sequence of independent random variables and let $S_m = \xi_1 + \ldots \xi_m$, $m \geq 1$. Prove *Etemadi's inequality*: for every $\varepsilon > 0$ and every integer $n \geq 1$ one has

$$\mathsf{P}\left\{\max_{1\leq m\leq n}|S_m| > 4\varepsilon\right\} \leq 4 \max_{1\leq m\leq n}\mathsf{P}\{|S_m| > \varepsilon\}.$$

(This inequality may be used in the proof of the implication (2) $\Rightarrow$ (1) in Problem 4.2.3.)

**Problem 4.2.23.** Let $\xi_1, \ldots, \xi_n$ be independent random variables with $\mathsf{E}\xi_k = 0$, chosen so that for any given $h > 0$ one has $\mathsf{E}e^{h\xi_k} < \infty$, $k = 1, \ldots, n$. Setting $S_k = \xi_1 + \ldots + \xi_k$, $1 \leq k \leq n$, prove the *exponential analog* of Kolomogorov inequality: for every $\varepsilon > 0$ one has

$$\mathsf{P}\left\{\max_{1\leq k\leq n} S_k \geq \varepsilon\right\} \leq e^{-h\varepsilon}\mathsf{E}e^{hS_n}.$$

*Hint.* Just as in the proof of the (usual) Kolomogorov inequality, one must introduce the sets $A = \{\max_{1\leq k\leq n} S_k \geq \varepsilon\}$ and $A_k = \{S_i < \varepsilon, 1 \leq i \leq k-1, S_k \geq \varepsilon\}$, $1 \leq k \leq n$, and, by using Jensen's inequality, establish the following relations:

$$\mathsf{E}e^{hS_n} \geq \mathsf{E}e^{hS_n}I_A = \sum_{k=1}^{n}\mathsf{E}e^{S_n}I_{A_k} \geq \ldots \geq e^{h\varepsilon}\mathsf{P}(A).$$

**Problem 4.2.24.** Let $Y$ be a random variable and let $(Y_n)_{n\geq 1}$ be a sequence random variables, such that $Y_n \overset{d}{\to} Y$ as $n \to \infty$ ("$\overset{d}{\to}$" means convergence in distribution). In addition, suppose that $\{N_t; t \geq 0\}$ is some family of positive integer-valued random variables, which are *independent* of $(Y_n)_{n\geq 1}$, and are such that $N_t \overset{\mathsf{P}}{\to} \infty$ as $t \to \infty$. Prove that $Y_{N_t} \overset{d}{\to} Y$ as $t \to \infty$.

*Hint.* Use the method of characteristic functions.

**Problem 4.2.25.** Let $Y$ be a random variable, let $(Y_n)_{n\geq 1}$ be some sequence of random variables, chosen so that

$$Y_n \to Y \quad \text{as} \quad n \to \infty \text{ (P-a.e.)},$$

and let $\{N_t, t \geq 0\}$ be some family of positive integer-valued random variables. (Unlike in Problem 4.2.24, the independence between $(Y_n)_{n\geq 1}$ and $\{N_t, t \geq 0\}$ is no longer assumed.)

Prove the following properties:
(a) If $N_t \to \infty$ (P-a.e.), then $Y_{N_t} \to Y$ as $t \to \infty$ P-a.e.
(b) If $N_t \to N$ (P-a.e.), then $Y_{N_t} \to Y_N$ as $t \to \infty$, P-a.e.
(c) If $N_t \overset{\mathsf{P}}{\to} \infty$, then $Y_{N_t} \overset{\mathsf{P}}{\to} Y$ as $t \to \infty$.

*Hint.* For the proof of (c), use the fact that a sequence that converges in probability must contain a sub-sequence that converges almost surely.

**Problem 4.2.26.** Let $\xi_1, \xi_2, \ldots$ be any sequence of independent Bernoulli random variables with $\mathsf{P}\{\xi_n = \pm 1\} = 1/2, n \geq 1$. Prove that the random variable $X = \sum_{n=1}^{\infty} \frac{\xi_n}{n}$ is well defined and its distribution function admits a probability density.

**Problem 4.2.27.** Let $\xi_1, \xi_2, \ldots$ be some sequence of independent Bernoulli random variables with $\mathsf{P}\{\xi_n = 0\} = \mathsf{P}\{\xi_n = 1\} = 1/2, n \geq 1$. Let $a_n > 0, b_n > 0$, $a_n + b_n = 1, n \geq 1$, and let

$$X_n = 2a_n^{\xi_n} b_n^{1-\xi_n}.$$

Prove that the following statements are equivalent:

1. $\prod_{n=1}^{\infty} X_n$ converges almost surely (i.e., $\lim_N \prod_{n=1}^{N} X_n$ exists and does not vanish almost surely);
2. $\prod_{n=1}^{\infty}(2 - X_n)$ converges almost surely;
3. $\prod_{n=1}^{\infty} a_n b_n$ converges.

*Hint.* To prove that (3) $\Rightarrow$ (1), consider the quantities $\mathsf{E} \ln X_n$ and $\mathsf{D} \ln X_n$ and use the Three-Series Theorem.

**Problem 4.2.28.** Let $\xi_1, \xi_2, \ldots$ be any sequence of independent and identically distributed random variables with (Cauchy) density $f(x) = \frac{1}{\pi(1+x^2)}, x \in \mathbb{R}$. Prove that there is no constant $m$ for which the property $\frac{1}{n} \sum_{i=1}^{n} \xi_i \xrightarrow{\mathsf{P}} m$ can hold.

**Problem 4.2.29.** Let $\xi_1, \xi_2, \ldots$ be any sequence of independent and identically distributed random variables with $\mathsf{E}\xi_i = \mu$ and $\mathsf{D}\xi_i < \infty$. Prove that

$$\frac{1}{C_n^2} \sum_{1 \leq i < j \leq n} \xi_i \xi_j \xrightarrow{\mathsf{P}} \mu \quad \text{as} \quad n \to \infty,$$

where, as usual, $C_n^2$ stands for the number of combinations $n$ choose 2 ($= n(n-1)/2$).

**Problem 4.2.30.** Let $\xi_1, \xi_2, \ldots$ be any sequence of independent Bernoulli random variables with $\mathsf{P}\{\xi_n = 0\} = \mathsf{P}\{\xi_n = 1\} = 1/2, n \geq 1$. Given any $n \geq 1$, let $Z_n$ denote the length of the maximal block inside the set of values $\xi_n, \ldots, \xi_n$, that contains only 1's. Prove that

$$\lim_n \frac{Z_n}{\ln n} = 1 \quad (\text{P-a. e.}).$$

*Hint.* Prove that with Probability 1 $\underline{\lim}_n Z_n / \ln n \geq 1$ and $\overline{\lim}_n Z_n / \ln n \leq 1$.

**Problem 4.2.31.** Let $\xi_1, \xi_2, \ldots$ be any sequence of independent Bernoulli random variables with $\mathsf{P}\{\xi_n = 1\} = p_n$ and $\mathsf{P}\{\xi_n = 0\} = 1 - p_n, n \geq 1$.

(a) Prove that if $\sum_{k=1}^{\infty} p_k p_{k+1} < \infty$, then the series $\sum_{k=1}^{\infty} \xi_k \xi_{k+1}$ converges with Probability 1.

(b) Prove *the Persi Diaconis Theorem*: if $p_n = 1/n$ for any $n \geq 1$, then the random variable $S = \sum_{n=1}^{\infty} \xi_n \xi_{n+1}$ has Poisson distribution of parameter $\lambda = 1$.

## 4.3 The Strong Law of Large Numbers

**Problem 4.3.1.** Prove that $\mathsf{E}\xi^2 < \infty$ if and only if $\sum_{n=1}^{\infty} n\mathsf{P}\{|\xi| > n\} < \infty$.
*Hint.* Prove that

$$\sum_{n=1}^{\infty} n\mathsf{P}\{|\xi| > n\} \leq \mathsf{E}\xi^2 \leq 1 + 4\sum_{n=1}^{\infty} n\mathsf{P}\{|\xi| > n\}.$$

**Problem 4.3.2.** Assuming that $\xi_1, \xi_2, \ldots$ is some sequence of independent and identically distributed random variables, prove the *Marcinkiewicz–Zygmund strong law of large numbers*: if $\mathsf{E}|\xi_1|^\alpha < \infty$, for some $0 < \alpha < 1$, then $\frac{S_n}{n^{1/\alpha}} \to 0$ (P-a. e.), and if $\mathsf{E}|\xi_1|^\beta < \infty$ for some $1 \leq \beta < 2$, then $\frac{S_n - n\mathsf{E}\xi_1}{n^{1/\beta}} \to 0$ (P-a. e.).

**Problem 4.3.3.** Let $\xi_1, \xi_2, \ldots$ be any sequence of independent and identically distributed random variables with $\mathsf{E}|\xi_1| = \infty$. Prove that the following relation holds for any sequence of real numbers $(a_n)_{n \geq 1}$:

$$\varlimsup_n \left| \frac{S_n}{n} - a_n \right| = \infty \quad \text{(P-a. e.)}.$$

**Problem 4.3.4.** Can one claim that all rational numbers in the interval $[0, 1)$ are normal, in the context of Example 2 in [ P §4.3, 4 ]?

**Problem 4.3.5.** Consider the *decimal* expansions $\omega = 0.\omega_1\omega_2\ldots$ of the numbers $\omega \in [0, 1)$.
(a) Formulate the decimal-expansions analog of the strong law of large numbers, formulated in [ P §4.3, 4 ] for binary expansions.
(b) In terms of decimal expansions, are the rational numbers normal, in the sense that $\frac{1}{n} \sum_{k=1}^{n} I(\xi_k(\omega) = i) \to \frac{1}{10}$ (P-a. e.) as $n \to \infty$, for any $i = 0, 1, \ldots, 9$?
(c) Prove the Champernowne's proposition: the number

$$\omega = 0.123456789101112\ldots,$$

where the (decimal) expansion consists of all positive integers (written as decimals) arranged in an increasing order, is *normal*, as a decimal expansion—see [ P §4.3, Example 2].

**Problem 4.3.6.** (N. Etemadi) Prove that the statement in [P §4.3, Theorem 3] remains in force even if the "independence" of the random variables $\xi_1, \xi_2, \ldots$ is replaced with "*pairwise* independence".

**Problem 4.3.7.** Prove that under the conditions in [P §4.3, Theorem 3] one can also claim convergence in the first-order mean: $E\left|\frac{S_n}{n} - m\right| \to 0$ as $n \to \infty$.

**Problem 4.3.8.** Let $\xi_1, \xi_2, \ldots$ be independent and identically distributed random variables with $E|\xi_1|^2 < \infty$. Prove that

$$n\, P\{|\xi_1| \geq \varepsilon \sqrt{n}\} \to 0 \quad \text{and} \quad \frac{1}{\sqrt{n}} \max_{k \leq n} |\xi_k| \xrightarrow{P} 0.$$

(Comp. with Problem 2.10.41.)

**Problem 4.3.9.** Construct a sequence of independent random variables

$$\xi_1, \xi_2, \ldots,$$

with the property that $\lim_{n \to \infty} \frac{1}{n}(\xi_1 + \cdots + \xi_n)$ exists as a "limit in probability" but not as a "limit almost surely".

*Hint.* Consider the independent random variables $\xi_1, \xi_2, \ldots$, chosen so that

$$P\{\xi_n = 0\} = 1 - \frac{1}{n \ln n}, \qquad P\{\xi_n = \pm n\} = \frac{1}{2n \ln n}.$$

By using the second Borel–Cantelli lemma, in conjunction with the fact that $ES_n^2 \leq n^2/\ln n$ and that $\sum_{n=1}^{\infty} P\{|\xi_n| \geq n\} = 1$, conclude that $P\{|\xi_n| \geq n \text{ i.o.}\} = 1$.

**Problem 4.3.10.** Let $\xi_1, \xi_2, \ldots$ be a sequence of independent random variables, chosen so that $P\{\xi_n = \pm n^a\} = 1/2$. Prove that this sequence satisfies the strong law of large numbers if and only if $a < 1/2$.

**Problem 4.3.11.** Prove that the *Kolmogorov strong law of large numbers* (Theorem 3) can be formulated in the following equivalent form: for any sequence of independent and identically distributed random variables $\xi_1, \xi_2, \ldots$ one has

$$E|\xi_1| < \infty \iff n^{-1} S_n \to E\xi_1 \quad (\text{P-a.e.}),$$

$$E|\xi_1| = \infty \iff \overline{\lim}\, n^{-1} S_n = +\infty \quad (\text{P-a.e.}).$$

In addition, prove that the first relation remains valid even if "independent" is replaced by "*pair-wise* independent".

**Problem 4.3.12.** Let $\xi_1, \xi_2, \ldots$ be independent and identically distributed random variables. Prove that

$$E \sup_n \left| \frac{\xi_n}{n} \right| < \infty \iff E|\xi_1| \ln^+ |\xi_1| < \infty.$$

**Problem 4.3.13.** Let $\xi_1, \xi_2, \ldots$ be independent and identically distributed random variables and let $S_n = \xi_1 + \cdots + \xi_n$, $n \geq 1$. Prove that for any given $\alpha \in (0, 1/2]$ one of the following properties holds:
  (a) $n^{-\alpha} S_n \to \infty$ (P-a. e.);
  (b) $n^{-\alpha} S_n \to -\infty$ (P-a. e.);
  (c) $\overline{\lim} \, n^{-\alpha} S_n = \infty$, $\underline{\lim} \, n^{-\alpha} S_n = -\infty$ (P-a. e.).

**Problem 4.3.14.** Let $\xi_1, \xi_2, \ldots$ be independent and identically distributed random variables and let $S_0 = 0$ and $S_n = \xi_1 + \ldots + \xi_n$, $n \geq 1$. Prove that:
  (a) If $\varepsilon > 0$ then

$$\sum_{n=1}^{\infty} P\{|S_n| \geq n\varepsilon\} < \infty \iff E\xi_1 = 0, \ E\xi_1^2 < \infty.$$

  (b) If $E\xi_1 < 0$, then for any $p > 1$ one has

$$E(\sup_{n \geq 0} S_n)^{p-1} < \infty \iff E(\xi_1^+)^p < \infty.$$

  (c) If $E\xi_1 = 0$ and $1 < p \leq 2$, then there is a constant $C_p$, for which the following relations are in force:

$$\sum_{n=1}^{\infty} P\left\{ \max_{k \leq n} S_k \geq n \right\} \leq C_p E|\xi_1|^p, \quad \sum_{n=1}^{\infty} P\left\{ \max_{k \leq n} |S_k| \geq n \right\} \leq 2C_p E|\xi_1|^p.$$

  (d) If $E\xi_1 = 0$, $E\xi_1^2 < \infty$ and $M(\varepsilon) = \sup_{n \geq 0}(S_n - n\varepsilon)$, $\varepsilon > 0$, then

$$\lim_{\varepsilon \to \infty} \varepsilon M(\varepsilon) = \sigma^2/2.$$

**Problem 4.3.15.** (*On [P §4.3, Theorem 2].*) Let $\xi_1, \xi_2, \ldots$ be independent random variables, chosen so that

$$P\{\xi_n = 1\} = P\{\xi_n = -1\} = \frac{1}{2}(1 - 2^{-n}),$$
$$P\{\xi_n = 2^n\} = P\{\xi_n = -2^n\} = 2^{-(n+1)}.$$

Prove that $\sum_{n=1}^{\infty} \frac{D\xi_n}{n^2} = \infty$ (comp. with [$\underline{P}$ §4.3, (3)]), but nevertheless one has (P-a. e.)

$$\frac{\xi_1 + \ldots + \xi_n}{n} \to 0,$$

i.e., the strong law of large numbers holds, in that [$\underline{P}$ §4.3, (4)] holds (notice that $E\xi_n = 0, n \geq 1$).

**Problem 4.3.16.** Let $\xi_1, \xi_2, \ldots$ be independent and identically distributed random variables with $E|\xi_1| = \infty$. Prove that at least one of the following two properties must be satisfied:

$$P\left\{\overline{\lim_n} \frac{1}{n} \sum_{k=1}^{n} \xi_k = +\infty\right\} = 1 \quad \text{or} \quad P\left\{\underline{\lim_n} \frac{1}{n} \sum_{k=1}^{n} \xi_k = -\infty\right\} = 1.$$

**Problem 4.3.17.** As a generalization of the Kolomogorov strong law of large numbers [$\underline{P}$ §4.3, Theorem 2] prove the following result, which is due to M. Loève: if $\xi_1, \xi_2, \ldots$ are independent random variables, chosen so that

$$\sum_{n=1}^{\infty} \frac{E|\xi_n|^{\alpha_n}}{n^{\alpha_n}} < \infty,$$

where $0 < \alpha_n \leq 2$, and, moreover, $E\xi_n = 0$ for $1 \leq \alpha_n \leq 2$, then $\frac{1}{n} \sum_{i=1}^{n} \xi_n \to 0$ almost everywhere.

**Problem 4.3.18.** Give an example of a sequence $\xi_1, \xi_2, \ldots$ of independent random variables such that $E\xi_n = 0, n \geq 1$, and

$$\frac{1}{n} \sum_{i=1}^{n} \xi_i \to -\infty \quad \text{(P-a. e.)}.$$

*Hint.* Choose, for example, the random variables $\xi_n$ so that $P\{\xi_n = -n\} = 1 - n^{-2}$ and $P\{\xi_n = n^3 - n\} = n^{-2}, n \geq 1$.

**Problem 4.3.19.** Let $\xi_1, \xi_2, \ldots$ be any sequence of independent random variables, such that $E\xi_k = 0, k \geq 1$. Setting

$$\xi_k^{(n)} = \begin{cases} \xi_k, & \text{if } |\xi_k| \leq n, \\ 0, & \text{if } |\xi_k| > n, \end{cases}$$

prove the following version of *the law of large numbers*, which is due to A. N. Kolmogorov: in order to claim that

$$\frac{1}{n} \sum_{k=1}^{n} \xi_k \xrightarrow{P} 0 \quad \text{as } n \to \infty,$$

it is necessary and sufficient that the following relations hold as $n \to \infty$,

$$\sum_{k=1}^{n} P\{|\xi_k| > n\} \to 0,$$

$$\frac{1}{n} \sum_{k=1}^{n} E\xi_k^{(n)} \to 0,$$

$$\frac{1}{n^2} \sum_{k=1}^{n} D\xi_k^{(n)} \to 0.$$

By using appropriate examples, prove that the last condition (as a necessary condition) cannot be replaced by the condition

$$\frac{1}{n^2} \sum_{k=1}^{n} E(\xi_k^{(n)})^2 \to 0 .$$

**Problem 4.3.20.** Let $N = (N_t)_{t \geq 0}$ be the renewal process from Example 4 in [P §4.3, 4], i.e., $N_t = \sum_{n=1}^{\infty} I(T_n \leq t)$, where $T_n = \sigma_1 + \ldots + \sigma_n$ and $(\sigma_n)_{n \geq 1}$ is some sequence of independent and identically distributed random variables, chosen so that $E\sigma_1 = \mu, 0 < \mu < \infty$. By the strong law of large numbers, one has $\frac{N_t}{t} \to \frac{1}{\mu}$ as $t \to \infty$ (P-a. e.). Prove that

$$E\left(\frac{N_t}{t}\right)^r \to \frac{1}{\mu^r} \qquad \text{as } t \to \infty, \quad \text{for every } r > 0,$$

and notice that these results remain valid even with $\mu = \infty$, in which case $1/\mu = 0$.

**Problem 4.3.21.** Let $\xi_1, \xi_2, \ldots$ be any sequence of independent and identically distributed random variables, set $S_n = \xi_1 + \ldots + \xi_n$, and let $\{N_t, t \geq 0\}$ be any family of random variables that take values in the set $\{1, 2, \ldots\}$ and are chosen so that $N_t \to \infty$ as $t \to \infty$, (P-a. e.).
    Prove that:
    (a) If $E|\xi_1|^r < \infty, r > 0$, then

$$\frac{\xi_{N_t}}{(N_t)^{1/r}} \to 0 \quad \text{as } t \to \infty \quad \text{(P-a. e.),}$$

and if, moreover, $N_t/t \to \lambda$ (P-a. e.), for some $0 < \lambda < \infty$, then

$$\frac{\xi_{N_t}}{t^{1/r}} \to 0 \quad \text{as } t \to \infty \quad \text{(P-a. e.).}$$

(b) If $E|\xi_1|^r < \infty$ for some $0 < r < 2$, with the understanding that $E\xi_1 = 0$ if $1 \leq r < 2$, then

$$\frac{S_{N_t}}{(N_t)^{1/r}} \to 0 \quad \text{as } t \to \infty \quad \text{(P-a. e.)}.$$

and if, in addition, $N_t/t \to \lambda$ (P-a. e.), for some $0 < \lambda < \infty$, then

$$\frac{S_{N_t}}{t^{1/r}} \to 0 \quad \text{as } t \to \infty \quad \text{(P-a. e.)}.$$

(c) If $E|\xi_1| < \infty$ and $E\xi_1 = \mu$, then

$$\frac{S_{N_t}}{N_t} \to \mu \quad \text{as } t \to \infty \quad \text{(P-a. e.)}.$$

and if, in addition, $N_t/t \to \lambda$ (P-a. e.), where $0 < \lambda < \infty$, then

$$\frac{S_{N_t}}{t} \to \mu\lambda \quad \text{as } t \to \infty \quad \text{(P-a. e.)}.$$

*Hint.* To prove (a), use the Borel–Cantelli lemma, in conjunction with the result established in Problem 4.2.25(a). To prove (b), use Marcinkiewicz-Zygmund's strong law of large numbers, established in Problem 4.3.2. To prove (c), use Kolmogorov's strong law of large numbers [P §4.3, Theorem 3] and recall the statement in Problem 4.2.25(a).

**Problem 4.3.22.** Let $f = f(x)$ be any bounded and continuous function defined on $(0, \infty)$. Prove that for every $a > 0$ and every $x > 0$ one must have

$$\lim_{n \to \infty} \sum_{k=1}^{\infty} f\left(x + \frac{k}{n}\right) e^{-an} \frac{(an)^k}{k!} = f(x + a).$$

**Problem 4.3.23.** Let $\xi_1, \xi_2, \ldots$ be independent and identically distributed random variables, chosen so that $E|\xi_1| < \infty$ and $E\xi_1 = \mu$. Prove that as $n \to \infty$ one has:

(a)  $\dfrac{\ln n}{n} \displaystyle\sum_{k=2}^{n} \frac{\xi_k}{\ln k} \to \mu$  (P-a. e.);

(b)  $n^{\alpha-1} \displaystyle\sum_{k=1}^{n} \frac{\xi_k}{k^\alpha} \to \mu$  (P-a. e.),  for any $0 < \alpha < 1$.

## 4.4 The Law of the Iterated Logarithm

**Problem 4.4.1.** Let $\xi_1, \xi_2, \ldots$ be any sequence of independent and identically distributed random variables with $\xi_n \sim \mathcal{N}(0, 1)$. Prove that:

(a) $P\left\{\overline{\lim} \dfrac{\xi_n}{\sqrt{2 \ln n}} = 1\right\} = 1;$

(b) $P\{\xi_n > a_n \text{ i.o.}\} = \begin{cases} 0, & \text{if } \sum_n P\{\xi_1 > a_n\} < \infty, \\ 1, & \text{if } \sum_n P\{\xi_1 > a_n\} = \infty. \end{cases}$

*Hint.* (a) Given some fixed $c > 0$ and setting $A_n = \{\xi_n > c\sqrt{2 \ln n}\}$, by using [$\underline{P}$ §4.4, (10)] one can show that

$$P(A_n) \sim \frac{n^{-c^2}}{c\sqrt{4\pi \ln n}}.$$

The required statement then follows from the Borel–Cantelli Lemma ($\sum P(A_n) < \infty$ for $c > 1$ and $\sum P(A_n) = \infty$ for $0 < c < 1$), in conjunction with the implications [$\underline{P}$ §4.4, (3) and (4)].

**Problem 4.4.2.** Let $\xi_1, \xi_2, \ldots$ be any sequence of independent random variables, which are identically distributed with Poisson law of parameter $\lambda > 0$. Prove that (independently of $\lambda$) one has

$$P\left\{\overline{\lim} \frac{\xi_n \ln \ln n}{\ln n} = 1\right\} = 1.$$

*Hint.* Consider the event $A_n = \{\xi_n > c\varphi_n\}$, where $c > 0$ and $\varphi_n = \frac{\ln n}{\ln \ln n}$, and notice that $\sum P(A_n) < \infty$ for $c > 1$, and $\sum P(A_n) = \infty$ for $0 < c < 1$. Then use the Borel–Cantelli Lemma and the implications [$\underline{P}$ §4.4, (3) and (4)].

**Problem 4.4.3.** Let $\xi_1, \xi_2, \ldots$ be a sequence of independent and identically distributed random variables with

$$Ee^{it\xi_1} = e^{-|t|^\alpha}, \quad 0 < \alpha < 2$$

(comp. with [$\underline{P}$ §3.6, **4**]). Prove that

$$P\left\{\overline{\lim} \left|\frac{S_n}{n^{1/\alpha}}\right|^{\frac{1}{\ln \ln n}} = e^{1/\alpha}\right\} = 1.$$

**Problem 4.4.4.** Let $\xi_1, \xi_2, \ldots$ be any sequence of Bernoulli random variables with $P\{\xi_n = \pm 1\} = 1/2$ and let $S_n = \xi_1 + \ldots + \xi_n$. Prove the following result, which is due to G. H. Hardy and J. E. Littlewood:

$$\lim_n \frac{|S_n|}{\sqrt{2n \ln n}} \le 1 \quad \text{with Probability } 1.$$

*Hint.* By showing that

$$P\{S_n \geq a\} \leq e^{-ha} \mathsf{E} e^{hS_n}, \qquad \text{for } a > 0, h > 0,$$

and that $\cosh h \leq \exp\{\frac{h^2}{2}\}$, conclude that

$$P\{S_n \geq a\} \leq \exp\left\{-\frac{a^2}{2n}\right\}.$$

Finally, set $a = 1 + \varepsilon$, $\varepsilon > 0$, and use the Borel–Cantelli Lemma. (See also the bibliographical notes for [P Chap. 4] and [P Chap. 7] at the end of the book "Probability-2").

**Problem 4.4.5.** Prove the following generalization of the inequality [P §4.4, (9)]. Let $\xi_1, \ldots, \xi_n$ be independent random variables and set $S_0 = 0$ and $S_k = \xi_1 + \ldots + \xi_k$, $k \leq n$. Then for every real $a$ one has (*Lévy's inequality*):

$$P\left\{ \max_{0 \leq k \leq n} [S_k + \mu(S_n - S_k)] > a \right\} \leq 2P\{S_n > a\},$$

where $\mu(\xi)$ stands for the median of the random variable $\xi$, i.e., the constant defined by the relation

$$\max(P\{\xi > \mu(\xi)\}, P\{\xi < \mu(\xi)\}) \leq \frac{1}{2}.$$

(For the various definitions of the notion of "median," see Problem 1.4.23.)
  *Hint.* Let

$$\tau = \inf\{0 \leq k \leq n : S_k + \mu(S_n - S_k) > a\},$$

with the understanding that $\inf \varnothing = n + 1$, and prove that

$$P\{S_n > a\} \geq \frac{1}{2} \sum_{k=0}^{n} P\{\tau = k\} = \frac{1}{2}P\left\{ \max_{0 \leq k \leq n} [S_k - \mu(S_n - S_k)] > a \right\}.$$

**Problem 4.4.6.** Let $\xi_1, \ldots, \xi_n$ be independent random variables with $\mathsf{E}\xi_i = 0$, $1 \leq i \leq n$ and let $S_k = \xi_1 + \ldots + \xi_k$. Prove that

$$P\left\{ \max_{1 \leq k \leq n} S_k > a \right\} \leq 2P\{S_n \geq \varepsilon - \mathsf{E}|S_n|\} \qquad \text{for all } a > 0.$$

**Problem 4.4.7.** Let $\xi_1, \ldots, \xi_n$ be independent and identically distributed random variables, such that $\mathsf{E}\xi_i = 0$, $\sigma^2 = \mathsf{E}\xi_i^2 < \infty$, and $|\xi_i| \leq c$ (P-a. e.), $i \leq n$. Setting $S_n = \xi_1 + \cdots + \xi_n$, prove that

$$\mathsf{E}e^{xS_n} \leq \exp\{2^{-1}nx^2\sigma^2(1 + xc)\} \qquad \text{for every } 0 \leq x \leq 2c^{-1}.$$

Prove under the same assumptions that if $(a_n)$ is some sequence of real numbers, chosen so that $a_n/\sqrt{n} \to \infty$ and $a_n = o(n)$ as $n \to \infty$, then for every $\varepsilon > 0$ and for all sufficiently large $n$ one has

$$P\{S_n > a_n\} > \exp\left\{ -\frac{a_n^2}{2n\sigma^2}(1+\varepsilon)\right\} .$$

**Problem 4.4.8.** Let $\xi_1, \ldots, \xi_n$ be independent and identically distributed random variables, such that $E\xi_i = 0$ and $|\xi_i| \leq c$ (P-a. e.), $i \leq n$. Setting $S_n = \xi_1 + \ldots + \xi_n$ and $D_n = \sum_{i=1}^{n} D\xi_i$, prove the *Prokhorov inequality*:

$$P\{S_n \geq a\} \leq \exp\left\{ -\frac{a}{2c}\arcsin\frac{ac}{2D_n}\right\}, \quad a \in R.$$

**Problem 4.4.9.** Let $\xi_1, \xi_2, \ldots$ be any sequence of independent and identically distributed random variables, such that $E|\xi_n|^\alpha = \infty$, for some $\alpha < 2$. Prove that

$$\overline{\lim} \frac{|S_n|}{n^{1/\alpha}} = \infty \quad \text{(P-a.s.)}$$

(and that, consequently, the law of the iterated logarithm does not hold for this particular sequence).

**Problem 4.4.10.** Let $\xi_1, \xi_2, \ldots$ be any sequence of independent and identically distributed random variables with $E\xi_n = 0$ and $E\xi_n^2 = 1$. Setting $S_n = \xi_1 + \ldots + \xi_n$, $n \geq 1$, prove that with Probability 1 the collection of all limiting points of the sequence $\left(\frac{S_n}{\sqrt{2n \ln \ln n}}\right)_{n \geq 1}$ coincides with the interval $[-1, 1]$.

**Problem 4.4.11.** Let $\xi_1, \xi_2, \ldots$ be any sequence of independent and identically distributed random variables, all having normal distribution $\mathcal{N}(m, \sigma^2)$. Setting

$$\overline{m}_n = \frac{1}{n}\sum_{i=1}^{n} \xi_i$$

and using the result in the previous problem, prove that with Probability 1 the collection of limiting points of the sequence $\left(\sqrt{n}\frac{\overline{m}_n - m}{\sqrt{2n \ln \ln n}}\right)_{n \geq 1}$ coincides with the interval $[-\sigma, \sigma]$.

**Problem 4.4.12.** Let $\xi_1, \xi_2, \ldots$ be any sequence of independent and identically distributed random variables that share one and the same continuous distribution function $F(x)$, $x \in R$, and let

$$F_n(x; \omega) = \frac{1}{n}\sum_{k=1}^{n} I(\xi_k(\omega) \leq x), \quad x \in R, \quad n \geq 1,$$

be the associated sequence of empirical distribution functions.

Prove that with Probability 1

$$\varlimsup_n \frac{\sqrt{n}\, \sup_x |F_n(x;\omega) - F(x)|}{\sqrt{2\ln\ln n}} = \sup_x \sqrt{F(x)(1 - F(x))}\,.$$

**Problem 4.4.13.** Let $\xi_1, \xi_2, \ldots$ be any sequence of independent and identically distributed random variables with exponential distribution, chosen so that $P\{\xi_i > x\} = e^{-x}$, $x \geq 0$. By using the argument of the Borel–Cantelli lemma (see also Problem 4.1.20), prove that with Probability 1

$$\varlimsup \frac{\xi_n}{\ln n} = 1, \quad \varlimsup \frac{\xi_n - \ln n}{\ln\ln n} = 1, \quad \text{and} \quad \varlimsup \frac{\xi_n - \ln n - \ln\ln n}{\ln\ln\ln n} = 1\,.$$

How will this result change if $P\{\xi_i > x\} = e^{-\lambda x}$, $x \geq 0$, for some $\lambda > 0$?

**Problem 4.4.14.** Let everything be as in the previous Problem (with $P\{\xi_i > x\} = e^{-\lambda x}$, $x \geq 0$, $\lambda > 0$). Setting $M_n = \max(\xi_1, \ldots, \xi_n)$, prove that

$$\varlimsup \frac{M_n}{\lambda \ln n} = \varlimsup \frac{\xi_n}{\lambda \ln n} \quad \text{(P-a. e.)}.$$

**Problem 4.4.15.** Let $\xi_1, \ldots, \xi_n$ be independent random variables and set $S_0 = 0$ and $S_k = \xi_1 + \cdots + \xi_k$, $k \leq n$. Prove that:
(a) (As a continuation of Problem 4.4.5)

$$P\left\{\max_{1\leq k\leq n} |S_k + \mu(S_n - S_k)| \geq a\right\} \leq 2P\{|S_n| \geq a\},$$

where $\mu(\xi)$ stands for the median of the random variable $\xi$.
(b) If $\xi_1, \ldots, \xi_n$ are identically distributed and symmetric, then

$$1 - e^{-nP\{|\xi_1|>x\}} \leq P\left\{\max_{1\leq k\leq n} |\xi_k| > x\right\} \leq 2P\{|S_n| > x\}.$$

**Problem 4.4.16.** Let $\xi_1, \ldots, \xi_n$ be independent random variables and set $S_k = \xi_1 + \cdots + \xi_k$, $1 \leq k \leq n$. Prove the *Skorokhod inequality*: for every $\varepsilon > 0$ one has

$$P\left\{\max_{1\leq k\leq n} |S_k| \geq 2\varepsilon\right\} \leq \inf_{1\leq k\leq n} P\{|S_n - S_k| < \varepsilon\} \cdot P\{|S_n| \geq \varepsilon\}.$$

*Hint.* Consider the stopping time $\tau = \inf\{1 \leq k \leq n : |S_k| \geq 2\varepsilon\}$ (with the understanding that $\inf \emptyset = n + 1$) and use the idea outlined in the hint for Problem 4.4.5.

**Problem 4.4.17.** Let $\xi_1, \ldots, \xi_n$ be some random variables and set $S_k = \xi_1 + \cdots + \xi_k$, $1 \leq k \leq n$. Prove that for every $\varepsilon \geq 0$ one has

$$\mathsf{P}\Big\{ \max_{1 \leq k \leq n} |\xi_k| \geq \varepsilon \Big\} \leq 2\mathsf{P}\Big\{ \max_{1 \leq k \leq n} |S_k| \geq \frac{\varepsilon}{2} \Big\},$$

and if, furthermore, the random variables $\xi_1, \ldots, \xi_n$ are independent and have symmetric distributions, then for every $\varepsilon \geq 0$ one has

$$\mathsf{P}\Big\{ \max_{1 \leq k \leq n} |\xi_k| \geq \varepsilon \Big\} \leq 2\mathsf{P}\Big\{ |S_n| \geq \frac{\varepsilon}{2} \Big\}.$$

## 4.5 Rate of Convergence in the Strong Law of Large Numbers

**Problem 4.5.1.** Prove the inequalities [P §4.5, (8) and (20)].
  *Hint.* Set $\widetilde{\xi} = -\xi$ and convince yourself that

$$\widetilde{H}(a) = \sup_{\lambda \in \mathbb{R}} [a\lambda - \psi(\lambda)] = H(-a).$$

In addition, use the inequality [P §4.5, (7)].

**Problem 4.5.2.** Consider the set $\Lambda$ defined in [P §4.5, (5)] and verify the claim that in the interior of the set $\Lambda$ the function $\psi(\lambda)$ is convex (in fact, *strictly* convex, if the random variable $\xi$ is non-degenerate) and infinitely differentiable.
  *Hint.* Setting $\lambda_* = \inf_{\lambda \in \Lambda} \lambda$ and $\lambda^* = \sup_{\lambda \in \Lambda} \lambda$, prove that (under the assumption [P §4.5, (3)])

$$-\infty \leq \lambda_* < 0 < \lambda^* \leq \infty.$$

Then prove that the function $\varphi(\lambda) = \mathsf{E}e^{\lambda \xi}$ is infinitely differentiable on the interval $(\lambda_*, \lambda^*)$. The convexity of the function $\psi(\lambda) = \ln \varphi(\lambda)$ follows from the Hölder inequality.

**Problem 4.5.3.** Assuming that the random variable $\xi$ is non-degenerate, prove that the function $H(a)$ is differentiable on the entire real line and is also convex.

**Problem 4.5.4.** Prove the following inversion formula for the Cramér transform:

$$\psi(\lambda) = \sup_a [\lambda a - H(a)],$$

for all $\lambda$, except, perhaps, at the endpoints of the the set $\Lambda = \{\lambda : \psi(\lambda) < \infty\}$.

**Problem 4.5.5.** Let $S_n = \xi_1 + \ldots + \xi_n$, where $\xi_1, \ldots, \xi_n$, $n \geq 1$, are assumed to be independent and identically distributed simple random variables with $\mathsf{E}\xi_1 < 0$ and $\mathsf{P}\{\xi_1 > 0\} > 0$. Let $\varphi(\lambda) = \mathsf{E}e^{\lambda \xi_1}$ and let $\inf_\lambda \varphi(\lambda) = \rho$ $(0 < \rho < 1)$.

Prove the *Chernoff theorem*:

$$\lim_n \frac{1}{n} \ln P\{S_n \geq 0\} = \ln \rho. \tag{$*$}$$

**Problem 4.5.6.** By using ($*$), prove that in the Bernoullian case (i.e., when $P\{\xi_1 = 1\} = p$ and $P\{\xi_1 = 0\} = q$), with $p < x < 1$, one has

$$\lim_n \frac{1}{n} \ln P\{S_n \geq nx\} = -H(x), \tag{$**$}$$

where (comp. with the notation in [P §1.6])

$$H(x) = x \ln \frac{x}{p} + (1 - x) \ln \frac{1 - x}{1 - p}.$$

**Problem 4.5.7.** Let $\xi_1, \xi_2, \ldots$ be independent and identically distributed random variables with $E\xi_1 = 0$ and $D\xi_1 = 1$ and let $S_n = \xi_1 + \ldots + \xi_n$, $n \geq 1$. Let $(x_n)_{n \geq 1}$ be some sequence of real numbers, chosen so that $x_n \to \infty$ and $\frac{x_n}{\sqrt{n}} \to 0$ as $n \to \infty$.

Prove that

$$P\{S_n \geq x_n \sqrt{n}\} = e^{-\frac{x_n^2}{2}(1 + y_n)},$$

where $y_n \to 0$, $n \to \infty$.

**Problem 4.5.8.** By using ($**$), conclude that in the Bernoullian case (i.e., when $P\{\xi_1 = 1\} = p$ and $P\{\xi_1 = 0\} = q$) one can claim that:
(a) For $p < x < 1$ and for $x_n = n(x - p)$ one has

$$P\{S_n \geq np + x_n\} = \exp\left\{-nH\left(p + \frac{x_n}{n}\right)(1 + o(1))\right\}. \tag{$***$}$$

(b) For $x_n = a_n \sqrt{npq}$, with $a_n \to \infty$ and $\frac{a_n}{\sqrt{n}} \to 0$, one has

$$P\{S_n \geq np + x_n\} = \exp\left\{-\frac{x_n^2}{2npq}(1 + o(1))\right\}. \tag{$****$}$$

Compare the relations ($***$) and ($****$) and then compare these two relations with the respective results in [P §1.6].

**Problem 4.5.9.** Let $\xi_1, \xi_2, \ldots$ be any sequence of independent random variables, all distributed according to the Cauchy law with density $f(x) = \frac{1}{\pi(1+x^2)}$, $x \in \mathbb{R}$. Prove that

$$\lim_n P\left\{\frac{1}{n} \max_{1 \leq k \leq n} \xi_k < x\right\} = e^{-\frac{1}{\pi x}}.$$

**Problem 4.5.10.** Let $\xi_1, \xi_2, \ldots$ be independent and identically distributed random variables with $E|\xi_1| < \infty$. Prove that

$$\lim_n \frac{1}{n} E\left( \max_{1 \le k \le n} |\xi_k| \right) = 0.$$

(Comp. with the statement in Problem 4.3.8.)

**Problem 4.5.11.** Suppose that $\xi$ is some random variable chosen so that $E\xi = 0$ and $a \le \xi \le b$, for some constants $a$ and $b$. Show that the moment generating function of $\xi$ satisfies the relation

$$E e^{h\xi} \le e^{\frac{1}{8} h^2 (b-a)^2} \quad \text{for all } h > 0.$$

*Hint.* Use the fact that the function $x \rightsquigarrow e^{hx}$ is convex.

**Problem 4.5.12.** Let $\xi_1, \ldots, \xi_n$ be independent and identically distributed Bernoulli random variables with $P\{\xi_i = 1\} = p$, $P\{\xi_i = 0\} = q$, $p + q = 1$, and let $S_n = \xi_1 + \ldots + \xi_n$. Prove *the Chernoff inequalities*: for any $x \ge 0$ one has

$$P\{S_n - np \ge nx\} \le e^{-2nx^2},$$

$$P\{|S_n - np| \ge nx\} \le 2e^{-2nx^2}.$$

*Hint.* Just as in many of the following problems, here one must use the relation

$$P\{S_n \ge y\} \le e^{-hy} E e^{hS_n}, \quad y \ge 0, \ h \ge 0,$$

which is often referred to as *the Bernstein inequality*.

**Problem 4.5.13.** Prove that, in the setting of the previous problem, the following stronger result, known as "*the maximal inequalities*" is in force:

$$P\left\{ \max_{1 \le k \le n} (S_k - kp) \ge nx \right\} \le e^{-2nx^2},$$

$$P\left\{ \max_{1 \le k \le n} |S_k - kp| \ge nx \right\} \le 2e^{-2nx^2}.$$

*Hint.* Use the exponential analog of the Kolmogorov inequality

$$P\left\{ \max_{1 \le k \le n} (S_k - kp) \ge \varepsilon \right\} \le e^{-h\varepsilon} E e^{h(S_n - np)}$$

(see Problem 4.2.23).

**Problem 4.5.14.** Let $\xi_1, \ldots, \xi_n$ be independent (but not necessarily identically distributed) random variables with values in the interval $[0, 1]$ and let $S_n = \xi_1 + \ldots + \xi_n$. Setting $p = \frac{ES_n}{n}$ and $q = 1 - p$, prove that, for every $0 \leq x < q$, the following inequality is in force

$$P\{S_n - ES_n \geq nx\} \leq e^{n\psi(x)},$$

where

$$\psi(x) = \ln\left[\left(\frac{p}{p+x}\right)^{p+x}\left(\frac{q}{q-x}\right)^{q-x}\right].$$

*Hint.* First, use the inequalities

$$e^{hy}P\{S_n \geq y\} \leq Ee^{hS_n} = Ee^{hS_{n-1}}Ee^{h\xi_n}$$

$$\leq Ee^{hS_{n-1}}(1 - p + pe^h) \leq \ldots \leq (1 - p + pe^h)^n,$$

and then choose $h > 0$ accordingly.

**Problem 4.5.15.** Let everything be as in the previous problem, prove *the Hoeffding inequality*, which is a generalization of the Chernoff inequality from Problem 4.5.12: for any $x \geq 0$ one has

$$P\{S_n - ES_n \geq nx\} \leq e^{-2nx^2},$$

$$P\{|S_n - ES_n| \geq nx\} \leq 2e^{-2nx^2}.$$

*Hint.* Use the result established in the previous problem and remark that $\psi(x) \leq -2x^2$.

**Problem 4.5.16.** Let $\xi_1, \ldots, \xi_n$ be independent random variables with values in the interval $[0, 1]$. Prove that for every $\varepsilon > 0$ the following inequalities are in force:

$$P\{S_n \leq (1 - \varepsilon)ES_n\} \leq \exp\left\{-\frac{1}{2}\varepsilon^2 ES_n\right\},$$

$$P\{S_n \geq (1 + \varepsilon)ES_n\} \leq \exp\left\{-[(1 + \varepsilon)\ln(1 + \varepsilon) - \varepsilon]ES_n\right\}\left(\leq e^{-\frac{\varepsilon^2 ES_n}{2(1+\varepsilon/3)}}\right).$$

*Hint.* For the proof of the first inequality use the result from Problem 4.5.14 and remark that $\psi(-xp) \leq -px^2/2, 0 \leq x < 1$. For the proof of the second inequality use the hint for Problem 4.5.14, which implies the relation

$$P\{S_n - ES_n \geq nx\} \leq \left[e^{-(p+x)h}(1 - p + pe^h)\right]^n.$$

**Problem 4.5.17.** Let $\xi_1, \ldots, \xi_n$ be independent random variables, chosen so that $a_i \leq \xi_i \leq b_i$, for some constants $a_i$ and $b_i$, $i = 1, \ldots, n$. As a generalization of the Hoeffding inequality from Problem 4.5.15, prove that, for any $x \geq 0$, one has

$$P\{S_n - \mathsf{E}S_n \geq x\} \leq e^{-2x^2 \sum_{k=1}^{n}(b_k - a_k)^2},$$

$$P\{|S_n - \mathsf{E}S_n| \geq x\} \leq 2e^{-2x^2 \sum_{k=1}^{n}(b_k - a_k)^2}.$$

*Hint.* First, use the inequality established in Problem 4.5.11 to derive the estimates

$$P\{S_n - \mathsf{E}S_n \geq x\} \leq e^{-hx}\mathsf{E}e^{h(S_n - \mathsf{E}S_n)} \leq e^{-hx + \frac{1}{8}h^2 \sum_{k=1}^{n}(b_k - a_k)^2},$$

and then choose $h$ accordingly.

**Problem 4.5.18.** (*"Large deviations."*) Let $(\xi_n)_{n \geq 1}$ be any sequence of independent standard Gaussian random variables (i.e., Law($\xi_n$) = $\mathscr{N}(0, 1)$) and let $S_n = \xi_1 + \ldots + \xi_n$. Prove that for any set $A \in \mathscr{B}(\mathbb{R})$

$$\lim_{n \to \infty} \frac{1}{n} \ln P\left\{\frac{S_n}{n} \in A\right\} = - \operatorname{ess\,inf}\left\{\frac{x^2}{2} : x \in A\right\}.$$

(Given any real Borel function $f(x)$ defined on $(\mathbb{R}, \mathscr{B}(\mathbb{R}))$, by definition, ess inf $\{f(x) : x \in A\}$ is understood as $\sup\{c \in \overline{\mathbb{R}} : \lambda\{x \in A : f(x) < c\} = 0\}$, where $\lambda$ is the Lebesgue measure—comp. with the definition of essential supremum in Remark 3 in [$\underline{P}$ §2.10].)

*Hint.* The following relation "nearly holds" for a "very large" $n$:

$$P\left\{\frac{S_n}{n} \in A\right\} = \sqrt{\frac{n}{2\pi}} \int_A e^{-nx^2/2}\, dx.$$

**Problem 4.5.19.** Let $\xi = (\xi_1, \ldots, \xi_n)$ be some Gaussian vector, such that $\mathsf{E}\xi_i = 0$, $i = 1, \ldots, n$. Prove that

$$\lim_{r \to \infty} \frac{1}{\sigma^2} \ln P\left\{\max_{1 \leq i \leq n} \xi_i \geq r\right\} = -\frac{1}{2\sigma^2}.$$

*Hint.* Setting $\sigma = \max_{1 \leq i \leq n}(\mathsf{E}\xi_i^2)^{1/2}$, show that, for every $r \geq 0$,

$$P\left\{\max_{1 \leq i \leq n} \xi_i \geq \mathsf{E}\max_{1 \leq i \leq n} \xi_i + \sigma r\right\} \leq e^{-r^2/2},$$

and then check that

$$P\left\{\max_{1 \leq i \leq n} \xi_i \geq r\right\} \geq 1 - \Phi\left(\frac{r}{\sigma_i}\right) \geq \frac{\exp\{-r^2/(2\sigma_i^2)\}}{\sqrt{2\pi}(1 + r/\sigma_i)},$$

for every $1 \leq i \leq n$.

# Chapter 5
# Stationary (in Strict Sense) Random Sequences and Ergodic Theory

## 5.1 Stationary (in Strict Sense) Random Sequences: Measure-Preserving Transformations

**Problem 5.1.1.** Let $T$ be any measure preserving transformation acting on the sample space $\Omega$ and let $\xi = \xi(\omega)$, $\omega \in \Omega$, be any random variable, chosen so that the expected value $\mathsf{E}\xi(\omega)$ exists. Prove that $\mathsf{E}\xi(\omega) = \mathsf{E}\xi(T\omega)$.

*Hint.* If $\xi = I_A$, $A \in \mathscr{F}$, then the identity $\mathsf{E}\xi(\omega) = \mathsf{E}\xi(T(\omega))$ follows from the definition of a "measure-preserving transformation." By linearity, this property extends for all random variables $\xi$ of the form $\sum_{k=1}^{n} \lambda_k I_{A_k}$, $A_k \in \mathscr{F}$. In addition, for $\xi \geq 0$, one has to use the construction of the expected value as an "integral," in conjunction with the monotone convergence theorem. For a general $\xi$, use the representation $\xi = \xi^+ - \xi^-$.

**Problem 5.1.2.** Prove that the transformation $T$, from [P §5.1, Examples 1 and 2] is measure-preserving.[1]

*Hint.* (Example 2) The identity $\mathsf{P}(A) = \mathsf{P}(T^{-1}(A))$ is trivial for sets $A$ of the form $A = [a, b) \subseteq [0, 1)$. For the general case, consider the system

$$\mathscr{M} = \{A \in \mathscr{B}([0, 1]) : \mathsf{P}(A) = \mathsf{P}(T^{-1}(A))\}$$

and, by using "the suitable sets method," prove that $\mathscr{M} = \mathscr{B}([0, 1))$.

**Problem 5.1.3.** Let $\Omega = [0, 1)$, let $\mathscr{F} = \mathscr{B}([0, 1))$, and let $\mathsf{P}$ be any probability measure on $(\Omega, \mathscr{F})$, chosen so that the associated distribution function on $[0, 1)$ is continuous. Prove that the transformations $Tx = \lambda x$, $0 < \lambda < 1$, and $Tx = x^2$ are *not* measure-preserving.

*Hint.* Due to the continuity assumption, it is possible to find some points $a, b \in (0, 1)$, such that

---

[1]It is assumed throughout the entire chapter that the probability space $(\Omega, \mathscr{F}, \mathsf{P})$ is complete.

A.N. Shiryaev, *Problems in Probability*, Problem Books in Mathematics, DOI 10.1007/978-1-4614-3688-1_5, © Springer Science+Business Media New York 2012

$$P([0, a)) = \frac{1}{3}, \quad P([0, b)) = \frac{2}{3}.$$

By using this property one can easily show that the transformations $Tx = \lambda x$, $0 < \lambda < 1$, and $Tx = x^2$ are not measure-preserving.

**Problem 5.1.4.** Let $\Omega$ denote the space of all real sequences of the form

$$\omega = (\dots, \omega_{-1}, \omega_0, \omega_1, \dots),$$

let $\mathscr{F}$ denote the $\sigma$-algebra generated by all cylinder sets

$$\{\omega: (\omega_k, \dots, \omega_{k+n-1}) \in B_n\},$$

for all possible choices of $n = 1, 2, \dots$, $k = 0, \pm 1, \pm 2, \dots$, and $B_n \in \mathscr{B}(\mathbb{R}^n)$. Given some probability measure $P$ on $(\Omega, \mathscr{F})$, prove that the double sided transformation $T$, defined by

$$T(\dots, \omega_{-1}, \omega_0, \omega_1, \dots) = (\dots, \omega_0, \omega_1, \omega_2, \dots),$$

is a measure-preserving if and only if

$$P\{\omega: (\omega_0, \dots, \omega_{n-1}) \in B_n\} = P\{\omega: (\omega_k, \dots, \omega_{k+n-1}) \in B_n\}$$

for all $n = 1, 2, \dots$, all $k = 0, \pm 1, \pm 2, \dots$, and all $B_n \in \mathscr{B}(\mathbb{R}^n)$.

**Problem 5.1.5.** Let $\xi_0, \xi_1, \dots$ be a stationary sequence of random elements with values in the *Borel* space $S$ (see [P §2.7, Definition 9]). Prove that one can construct (perhaps on some enlargement of the underlying probability space) random elements $\xi_{-1}, \xi_{-2}, \dots$, with values in $S$, so that the double-sided sequence $\dots, \xi_{-1}, \xi_0, \xi_1, \dots$ is stationary.

**Problem 5.1.6.** Let $(\Omega, \mathscr{F}, P)$ be any probability space, let $T$ be any measurable transformation of $\Omega$, and let $\mathscr{E}$ be any $\pi$-system of subsets of $\Omega$ that generates $\mathscr{F}$ (i.e., $\pi(\mathscr{E}) = \mathscr{F}$). Prove that if the identity $P(T^{-1}A) = P(A)$ holds for all $A \in \mathscr{E}$, then it must hold for all $A \in \mathscr{F}$.

**Problem 5.1.7.** Let $T$ be any measure-preserving transformation on $(\Omega, \mathscr{F}, P)$ and let $\mathscr{G}$ be any sub-$\sigma$-algebra of $\mathscr{F}$. Prove that the following relation must hold for every $A \in \mathscr{F}$:

$$P(A \mid \mathscr{G})(T\omega) = P(T^{-1}A \mid T^{-1}\mathscr{G})(\omega) \quad \text{(P-a.e.).} \qquad (*)$$

In particular, if $\Omega$ is taken to be the space $\mathbb{R}^\infty$ of all real sequences of the form $\omega = (\omega_0, \omega_1, \dots)$, if $\xi_k(\omega) = \omega_k, k \geq 0$, is the associated family of coordinate maps on $\mathbb{R}^\infty$, and if $T$ denotes the shift-transformation on $\mathbb{R}^\infty$, given by $T(\omega_0, \omega_1, \dots) = (\omega_1, \omega_2, \dots)$ (i.e., $\xi_k(T\omega) = \omega_{k+1}, k \geq 0$), then $(*)$ can be written as

$$P(A \mid \xi_n)(T\omega) = P(T^{-1}A \mid \xi_{n+1})(\omega) \quad \text{(P-a.e.).}$$

**Problem 5.1.8.** Let $T$ be any measurable transformation acting on $(\Omega, \mathscr{F})$ and let $\mathscr{P}$ stand for the collection of all probability measures, P, on $(\Omega, \mathscr{F})$ with the property that $T$ is P-measure preserving. Prove that:

(a) The set of measures $\mathscr{P}$ is convex.

(b) The transformation $T$ is an ergodic transformation of the measure P if and only if P is an *extremal* element of the set $\mathscr{P}$ (i.e., P cannot be written as P $= \lambda_1 P_1 + \lambda_2 P_2$, for some $\lambda_1 > 0$ and $\lambda_2 > 0$ with $\lambda_1 + \lambda_2 = 1$ and some $P_1, P_2 \in \mathscr{P}$ with $P_1 \neq P_2$).

**Problem 5.1.9.** Let $T$ be any measure preserving transformation acting on the probability space $(\Omega, \mathscr{F}, \mathsf{P})$ and let $\xi = \xi(\omega)$, $\omega \in \Omega$, be any random variable on that space. Prove that $\xi = \xi(\omega)$ is almost invariant under $T$ (i.e., $\xi(\omega) = \xi(T\omega)$ (P-a. e.)) if and only if for any bounded and $\mathscr{F} \otimes \mathscr{B}(\mathbb{R})$-measurable functions $G(\omega, x)$ one can write

$$\mathsf{E}G(\omega, \xi(\omega)) = \mathsf{E}G(T\omega, \xi(\omega)).$$

*Hint.* Consider first functions $G(\omega, x)$ of the form $G_1(\omega)G_2(x)$.

## 5.2  Ergodicity and Mixing

**Problem 5.2.1.** Prove that the random variable $\xi$ is *invariant* if and only if it is $\mathscr{J}$-measurable.

**Problem 5.2.2.** Prove that the set $A$ is *almost invariant* if and only if $\mathsf{P}(T^{-1}A \setminus A) = 0$. Show also that if the random variable $X$ is almost invariant (i.e., $X(\omega) = X(T\omega)$ (P-a. e.)), then one can find an (everywhere) invariant random variable $\widetilde{X} = \widetilde{X}(\omega)$ (i.e., $\widetilde{X}(\omega) = \widetilde{X}(T\omega)$ for all $\omega \in \Omega$) with the property $\mathsf{P}\{X(\omega) = \widetilde{X}(\omega)\} = 1$.

**Problem 5.2.3.** Prove that the transformation $T$ represents *mixing* if and only if for any two random variables $\xi$ and $\eta$, with $\mathsf{E}\xi^2 < \infty$ and $\mathsf{E}\eta^2 < \infty$, one has

$$\mathsf{E}\xi(T^n\omega)\eta(\omega) \to \mathsf{E}\xi(\omega)\,\mathsf{E}\eta(\omega) \quad \text{as } n \to \infty. \tag{$*$}$$

*Hint.* If $\eta = I_A$ and $\xi = I_B$, then $(*)$ is precisely the mixing property. Each of the variables $\xi$ and $\eta$ is in $L^2$ and can be approximated (in the metric of $L^2$) with any precision by linear combinations of indicator functions. The required convergence of the expected values then follows easily from the mixing property.

**Problem 5.2.4.** Give an example of a measure preserving *ergodic* transformation which is not *mixing*.

*Hint.* Take $\Omega = \{a, b\}$, set $\mathsf{P}(\{a\}) = \mathsf{P}(\{b\}) = 1/2$, and consider the transformation $T$, given by $Ta = b$ and $Tb = a$.

**Problem 5.2.5.** Let $T$ be a measure preserving transformation acting on $(\Omega, \mathscr{F}, \mathsf{P})$ and let $\mathscr{F} = \sigma(\mathscr{A})$, where $\mathscr{A}$ is some algebra of sub-sets of $\Omega$. Suppose that in [$\underline{\mathrm{P}}$ §5.2, Definition 4] one assumes that

$$\lim_{n \to \infty} \mathsf{P}(A \cap T^{-n} B) = \mathsf{P}(A)\mathsf{P}(B)$$

*only* for sets $A$ and $B$ that are chosen from $\mathscr{A}$. Prove that the above property will then be satisfied for all sets $A$ and $B$ that belong to $\mathscr{F} = \sigma(\mathscr{A})$ (and, as a result, one can claim that $T$ represents mixing).

Show also that this statement remains valid if $\mathscr{A}$ is required to be a $\pi$-system and $\mathscr{F} = \pi(\mathscr{A})$.

**Problem 5.2.6.** Let $(\Omega, \mathscr{F}) = (\mathbb{R}^{\infty}, \mathscr{B}(\mathbb{R}^{\infty}))$ and suppose that $T$ is the usual shift-transformation on $\Omega$, given by $T(x_1, x_2, \ldots) = (x_2, x_3, \ldots)$, for any $\omega = (x_1, x_2, \ldots)$. Prove that any event from $\mathscr{F} = \mathscr{B}(\mathbb{R}^{\infty})$ that is invariant under $T$ must be a "tail" event; in other words, the entire $\sigma$-algebra $\mathscr{J}$, which comprises all $T$-invariant sets, is included in the "tail" $\sigma$-algebra $\mathscr{X} = \bigcap \mathscr{F}_n^{\infty}$, where $\mathscr{F}_n^{\infty} = \sigma(\omega : x_n, x_{n+1}, \ldots)$. Give examples of tail events which are not $T$-invariant.

**Problem 5.2.7.** By providing appropriate examples of measure-preserving transformations $T$, acting on $(\Omega, \mathscr{F}, \mathsf{P})$, prove that: (a) $A \in \mathscr{F}$ does not entail $TA \in \mathscr{F}$; (b) one cannot conclude from $A \in \mathscr{F}$ and $TA \in \mathscr{F}$ that $\mathsf{P}(A) = \mathsf{P}(TA)$.

## 5.3   Ergodic Theorems

**Problem 5.3.1.** Let $\xi = (\xi_1, \xi_2, \ldots)$ be some stationary Gaussian sequence with $\mathsf{E}\xi_n = 0$ and with covariance function $\mathbb{R}(n) = \mathsf{E}\xi_{k+n}\xi_k$. Prove that the condition $\mathbb{R}(n) \to 0$ is sufficient for claiming that the measure preserving transformation, associated with the sequence $\xi$, represents *mixing* (and is therefore *ergodic*).

*Hint.* If $A = \{\omega : (\xi_1, \xi_2, \ldots) \in A_0\}$, $B = \{\omega : (\xi_1, \xi_2, \ldots) \in B_0\}$ and $B_n = \{\omega : (\xi_n, \xi_{n+1}, \ldots) \in B_0\}$, then one must show that

$$\mathsf{P}(A \cap B_n) \to \mathsf{P}(A)\mathsf{P}(B) \quad \text{as } n \to \infty.$$

The proof can then be established with the following line of reasoning:

1. Given any $\varepsilon > 0$, find a number $m \in \mathbb{N} = \{1, 2, \ldots\}$ and sets $\widetilde{A}_0 \in \mathscr{B}(\mathbb{R}^m)$ and $\widetilde{B}_0 \in \mathscr{B}(\mathbb{R}^m)$, such that $\mathsf{P}(A \triangle \widetilde{A}) < \varepsilon$ and $\mathsf{P}(B \triangle \widetilde{B}) < \varepsilon$, where $\widetilde{A} = \{\omega : (\xi_1, \ldots, \xi_m) \in \widetilde{A}_0\}$ and $\widetilde{B} = \{\omega : (\xi_1, \ldots, \xi_m) \in \widetilde{B}_0\}$.

2. Then choose some open sets $\overline{A}_0 \in \mathscr{B}(\mathbb{R}^m)$ and $\overline{B}_0 \in \mathscr{B}(\mathbb{R}^m)$, so that for the sets

$$\overline{A} = \left\{\omega : (\xi_1, \ldots, \xi_m) \in \overline{A}_0\right\} \quad \text{and} \quad \overline{B} = \left\{\omega : (\xi_1, \ldots, \xi_m) \in \overline{B}_0\right\}$$

one has

$$\mathsf{P}(\widetilde{A} \triangle \overline{A}) < \varepsilon \quad \text{and} \quad \mathsf{P}(\widetilde{B} \triangle \overline{B}) < \varepsilon.$$

This would then imply that $\mathsf{P}(\overline{A} \triangle A) < 2\varepsilon$ and $\mathsf{P}(\overline{B} \triangle B) < 2\varepsilon$.

3. The sets $\overline{B}_n = \{\omega : (\xi_n, \ldots, \xi_{n+m-1}) \in \overline{B}_0\}$ have the property $P(\overline{B}_n \triangle B_n) < 2\varepsilon$.

4. Let $P$ stand for the probability distribution of the vector $(\xi_1, \ldots, \xi_m)$ and let $Q_n$ be the distribution of the vector $(\xi_1, \ldots, \xi_m, \xi_n, \ldots, \xi_{n+m-1})$. Then

$$R(n) \to 0 \quad \Rightarrow \quad Q_n \overset{w}{\to} P \otimes P \text{ as } n \to \infty.$$

5. In conjunction with [$\underline{P}$ §3.1, Theorem 1], (iv) gives

$$\underline{\lim}_n \, P(\overline{A} \triangle \overline{B}_n) \geq P(\overline{A})P(\overline{B}),$$

which, taking into account the relations (see above) $P(\overline{A} \triangle A) < 2\varepsilon$ and $P(\overline{B}_n \triangle B_n) < 2\varepsilon$, gives

$$\underline{\lim}_n \, P(A \cap B_n) \geq (P(A) - 2\varepsilon)(P(B) - 2\varepsilon) - 4\varepsilon,$$

which, taking into account that $\varepsilon > 0$ is arbitrarily chosen, gives

$$\underline{\lim}_n \, P(A \cap B_n) \geq P(A)P(B).$$

6. In analogous fashion one can prove that

$$\overline{\lim}_n \, P(A \cap B_n) \leq P(A)P(B)$$

(instead of the open sets $\overline{A}_0$ and $\overline{B}_0$ one must choose closed sets).

**Problem 5.3.2.** Prove that for any sequence $\xi = (\xi_1, \xi_2, \ldots)$ that consists of independent and identically distributed random variables, one can claim that the associated measure preserving transformation represents mixing.

*Hint.* (Observe that the ergodicity of the sequence $\xi$ follows from the "zero-one law.") The proof of the *mixing*-property can be established with the following line of reasoning:

1. Define the sets

$$A = \{\omega : (\xi_1, \xi_2, \ldots) \in A_0\} \quad \text{and} \quad B = \{\omega : (\xi_1, \xi_2, \ldots) \in B_0\},$$

for some choice of $A_0, B_0 \in \mathscr{B}(\mathbb{R}^\infty)$. Given any $\varepsilon > 0$, it is possible to find an integer $m \in \mathbb{N} = \{1, 2, \ldots\}$ and a set $\widetilde{A}_0 \in \mathscr{B}(\mathbb{R}^m)$, so that $P(A \triangle \widetilde{A}) < \varepsilon$ for $\widetilde{A} = \{\omega : (\xi_1, \ldots, \xi_m) \in \widetilde{A}_0\}$.

2. Define the sets

$$B_n = \{\omega : (\xi_n, \xi_{n+1}, \ldots) \in B_0\}, \quad n \geq 1,$$

and observe that for any $n > m$

$$P(\widetilde{A} \cap B_n) = P(\widetilde{A})P(B_n) = P(\widetilde{A})P(B).$$

3. Finally, prove that $|P(A \cap B_n) - P(A)P(B_n)| \leq 2\varepsilon$ and that

$$P(A \cap B_n) \to P(A)P(B) \text{ as } n \to \infty.$$

**Problem 5.3.3.** Prove that the stationary sequence $\xi = (\xi_1, \xi_2, \ldots)$ is *ergodic* if and only if for every $k = 1, 2, \ldots$ and every $B \in \mathcal{B}(\mathbb{R}^k)$ one has

$$\frac{1}{n} \sum_{i=1}^{n} I_B(\xi_i, \ldots, \xi_{i+k-1}) \to P\{(\xi_1, \ldots, \xi_k) \in B\} \text{ as } n \to \infty \quad \text{(P-a. e.)}.$$

*Hint.* To prove the *necessity* part, let $Q$ denote the distribution (on $\mathbb{R}^\infty$) of the sequence $\xi = (\xi_1, \xi_2, \ldots)$ and let $T$ stand for the shift

$$\mathbb{R}^\infty \ni x = (x_1, x_2, \ldots) \rightsquigarrow T(x) = (x_2, x_3, \ldots) \in \mathbb{R}^\infty.$$

In addition, given any $k = 1, 2, \ldots$ and any $B \in \mathcal{B}(\mathbb{R}^k)$, define the function $\mathbb{R}^\infty \ni x = (x_1, x_2, \ldots) \rightsquigarrow f(x) = I((x_1, \ldots, x_k) \in B) \in \mathbb{R}$, and then apply to that function the Birkhoff–Khinchin ergodic theorem.

In order to establish the *sufficiency* part, one has to prove that the transformation $T$, introduced above, is ergodic; in other words, the measure of any set from the associated collection $\mathcal{J}$ (i.e., any invariant set) is either 0 or 1.

The property

$$\frac{1}{n} \sum_{i=1}^{n} I((\xi_1, \ldots, \xi_k) \in B) \to P\{(\xi_1, \ldots, \xi_k) \in B\} \text{ as } n \to \infty \quad \text{(P-a. e.)}$$

translates into the claim that for every set $A \in \mathcal{B}(\mathbb{R}^\infty)$, of the form $\{(x_1, \ldots, x_k) \in B\}$, for some $B \in \mathcal{B}(\mathbb{R}^k)$, one must have

$$\frac{1}{n} \sum_{i=1}^{n} I_A(T^i x) \to Q(A) \text{ as } n \to \infty \quad \text{(Q-a. e.)}.$$

In conjunction with the Birkhoff–Khinchin ergodic theorem, the last relation yields the identity $E_Q(I_A \mid \mathcal{J}) = E_Q I_A$ (Q-a. e.), which, in turn, implies that the sets $A$ of the form $\{(x_1, \ldots, x_k) \in B\}$, for some choice of $B \in \mathcal{B}(\mathbb{R}^k)$, do not depend on $\mathcal{J}$. By using the "suitable sets" method, one can then conclude that the collection of sets

$$\mathcal{M} = \{A \in \mathcal{B}(\mathbb{R}^\infty) : A \text{ is independent from } \mathcal{J}\}$$

coincides with $\mathcal{B}(\mathbb{R}^\infty)$. Finally, one can conclude that $\mathcal{J}$ does not depend on $\mathcal{J}$ and, therefore, the Q-measure of every invariant set is either 0 or 1. This proves the ergodicity of the transformation $T$, and, therefore, the ergodicity of the sequence $\xi$.

**Problem 5.3.4.** Suppose that $T$ is some measure-preserving transformation on $(\Omega, \mathscr{F})$, under two different measures, P and $\overline{\text{P}}$. Prove that if, in addition, $T$ happens to be ergodic relative to both P and $\overline{\text{P}}$, then either $\text{P} = \overline{\text{P}}$ or $\text{P} \perp \overline{\text{P}}$.

**Problem 5.3.5.** Let $T$ be any measure preserving transformation on the space $(\Omega, \mathscr{F}, \text{P})$, let $\mathscr{A}$ be any algebra of sub-sets of $\Omega$, chosen so that $\sigma(\mathscr{A}) = \mathscr{F}$, and let

$$I_A^{(n)} = \frac{1}{n} \sum_{k=0}^{n-1} I_A(T^k \omega), \quad A \in \mathscr{A}.$$

Prove that the transformation $T$ is ergodic if and only if at least one of the following conditions holds:

1. $I_A^{(n)} \xrightarrow{\text{P}} \text{P}(A)$ for every $A \in \mathscr{A}$;
2. $\lim_n \frac{1}{n} \sum_{k=0}^{n-1} \text{P}(A \cap T^{-k} B) = \text{P}(A) \text{P}(B)$ for all $A, B \in \mathscr{A}$;
3. $I_A^{(n)} \xrightarrow{\text{P}} \text{P}(A)$ for every $A \in \mathscr{F}$.

**Problem 5.3.6.** Suppose that $T$ is some measure-preserving transformation on $(\Omega, \mathscr{F}, \text{P})$. Prove that $T$ is ergodic (for the measure P) if and only if there is no measure $\overline{\text{P}} \neq \text{P}$, defined on $(\Omega, \mathscr{F})$, that has the property $\overline{\text{P}} \ll \text{P}$, and is such that the transformation $T$ is measure-preserving for $\overline{\text{P}}$.

**Problem 5.3.7.** (*Bernoulli shifts.*) Let $S$ be any *finite* set (say, $S = \{1, \ldots, N\}$), let $\Omega = S^\infty$ be the space of all sequences of the form $\omega = (\omega_0, \omega_1, \ldots)$, $\omega_i \in S$, and let $\xi_k$, $k \geq 0$, be the canonical coordinate maps on $S^\infty$, given by $\xi_k(\omega) = \omega_k$, $\omega \in \Omega = S^\infty$. Define the *shift transformation* $T(\omega_0, \omega_1, \ldots) = (\omega_1, \omega_2, \ldots)$. The same transformation can be defined in terms of the coordinate maps through the relations $\xi_k(T\omega) = \omega_{k+1}$, $k \geq 0$. Assume that to every $i \in \{1, 2, \ldots, N\}$ one can attach a non-negative number, $p_i$, so that $\sum_{i=1}^{N} p_i = 1$ (i.e., the list $(p_1, \ldots, p_N)$ represents a probability distribution on $\Omega$). With the help of this distribution it is possible to define a measure P on $(S^\infty, \mathscr{B}(S^\infty))$ (see [P §2.3]), so that

$$\text{P}\{\omega: (\omega_1, \ldots, \omega_k) = (u_1, \ldots, u_k)\} = p_{u_1} \ldots p_{u_k}.$$

In other words, the probability measure P can be defined in such a way that the random variables $\xi_0(\omega), \xi_1(\omega), \ldots$ become independent. It is common to refer to the shift transformation $T$ as the *Bernoulli shift* or the *Bernoulli transformation* relative to the measure P.

Prove that the Bernoulli transformation, as described above, has the mixing property.

**Problem 5.3.8.** Let $T$ be a some measure-preserving transformation on $(\Omega, \mathscr{F}, \text{P})$. Setting $T^{-n} \mathscr{F} = \{T^{-n} A: A \in \mathscr{F}\}$, we say that the $\sigma$-algebra

$$\mathscr{F}_{-\infty} = \bigcap_{n=1}^{\infty} T^{-n} \mathscr{F}$$

is *trivial* (or P-*trivial*), if the P-measure of every set from $\mathscr{F}_{-\infty}$ is either 0 or 1. If the transformation $T$ is such that the associated $\sigma$-algebra $\mathscr{F}_{-\infty}$ is trivial, then we say that $T$ is "*Kolmogorov transformation*." Prove that every Kolmogorov transformation is ergodic and, furthermore, has the mixing property.

**Problem 5.3.9.** Let $1 \leq p < \infty$, let $T$ be any measure-preserving transformation acting on $(\Omega, \mathscr{F}, \mathsf{P})$, and let $\xi(\omega)$ be any random variable from the space $L^p(\Omega, \mathscr{F}, \mathsf{P})$.

Prove the *von Neumann ergodic theorem* for $L^p(\Omega, \mathscr{F}, \mathsf{P})$: one can construct a random variable, $\eta(\omega)$, on $(\Omega, \mathscr{F}, \mathsf{P})$, for which

$$\mathsf{E} \left| \frac{1}{n} \sum_{k=0}^{n-1} \xi(T^k \omega) - \eta(\omega) \right|^p \to 0 \quad \text{as } n \to \infty.$$

**Problem 5.3.10.** The *Borel normal numbers theorem* claims that (see [P §4.3, Example 2]) the proportion of zeroes, or of ones, in the binary expansion of $\omega \in [0, 1)$ converges almost surely, relative to the Lebesgue measure on $[0, 1)$, to $1/2$. Prove this result by introducing the transformation $T : [0, 1) \to [0, 1)$, given by

$$T(\omega) = 2\omega \pmod{1},$$

and by using the ergodic theorem—[P §5.3, Theorem 1].

**Problem 5.3.11.** Let everything be as in Problem 5.3.10 and let $\omega \in [0, 1)$. Consider the transformation $T : [0, 1) \to [0, 1)$, given by

$$T(\omega) = \begin{cases} 0, & \text{if } \omega = 0, \\ \{\frac{1}{\omega}\}, & \text{if } \omega \neq 0, \end{cases}$$

where $\{x\}$ denotes the fractional part of the number $x$.

The so called *Gauss measure* on the interval $[0, 1)$ is defined as

$$P(A) = \frac{1}{\ln 2} \int_A \frac{dx}{1 + x}, \quad A \in \mathscr{B}([0, 1)).$$

Prove that the transformation $T$ preserves the Gauss measure $P$.

**Problem 5.3.12.** By providing appropriate examples, prove that the Poincaré "reversibility" theorem (see [P §5.1, **3**]) may not hold for measurable spaces with *infinite* measures.

# Chapter 6
# Stationary (in Broad Sense) Random Sequences: $L^2$-theory

## 6.1 Spectral Representation of Covariance Functions

**Problem 6.1.1.** By using [ P §6.1, (11)], prove the relation [ P §6.1, (12)].

*Hint.* The required statement can be established by using appropriate values for $t_i$ and $a_i$. For instance, with $m = 2$, $t_1 = 0$ and $t_2 = n$, it is easy to prove that

$$(|a_1|^2 + |a_2|^2)R(0) + a_1\bar{a}_2 R(-n) + \bar{a}_1 a_2 R(n) \geq 0.$$

Setting $a_1 = a_2 = 1$ and $a_1 = 1$, $a_2 = i$ above, and taking into account the property $R(0) \in \mathbb{R}$, one finds that $R(n) + R(-n) \in \mathbb{R}$ and $i(R(n) - R(-n)) \in \mathbb{R}$, and, therefore, $R(-n) = \overline{R}(n)$.

**Problem 6.1.2.** Prove that if all zeroes of the polynomial $Q(z)$, defined in [ P §6.1, (27)], happen to be *outside* of the unit disk, then the auto-regression equation [ P §6.1, (24)] admits unique stationary solution, which can be written in the form of one-sided moving average.

**Problem 6.1.3.** In the context of [ P §6.1], prove that the spectral functions for the sequences (22) and (24) have densities given by, respectively, (23) and (29).

*Hint.* The formula in (23) may be established as follows: prove first that $R(n) = \sum_{k=0}^{p} a_{n+k}\bar{a}_k$ and after that verify the relation $R(n) = \int_{-\pi}^{\pi} e^{i\lambda n} f(\lambda)\, d\lambda$, where $f(\lambda)$ is given by (23). (It is useful to keep in mind that $\int_{-\pi}^{\pi} e^{i\lambda n}\, d\lambda = 2\pi\delta_{n0}$, where $\delta_{n0}$ is the usual Kronecker symbol.)

**Problem 6.1.4.** Prove that if $\sum_{n=-\infty}^{+\infty} |R(n)|^2 < \infty$, then the spectral function $F(\lambda)$ has density $f(\lambda)$, given by

$$f(\lambda) = \frac{1}{2\pi} \sum_{n=-\infty}^{\infty} e^{-i\lambda n} R(n),$$

where the series converges in the complex space $L^2 = L^2([-\pi, \pi), \mathcal{B}([-\pi, \pi)), \lambda)$, $\lambda$ being the usual Lebesgue measure.

A.N. Shiryaev, *Problems in Probability*, Problem Books in Mathematics, DOI 10.1007/978-1-4614-3688-1_6, © Springer Science+Business Media New York 2012

*Hint.* Use the fact that $\left\{\frac{1}{\sqrt{2\pi}}e^{i\lambda n}, n = 0, \pm1, \pm2, \dots\right\}$ is an orthonormal system in the space $L^2([-\pi, \pi), \mathscr{B}([-\pi, \pi)), \lambda)$.

**Problem 6.1.5.** Let $(\xi_n)_{n\geq 0}$ be any stationary Gauss-Markov sequence with vanishing mean. Prove that the associated covariance function, $R(n)$, admits the representation

$$R(n) = \sigma^2 \lambda^n,$$

for some $0 < \lambda \leq 1$.

**Problem 6.1.6.** Let $N = (N_t)_{t\geq 0}$ be a Poisson process (see [P §7.10]) of parameter $\lambda > 0$. Define the (continuous-time) process $\xi_t = \xi \times (-1)^{N_t}$, where $\xi$ is some random variable, which is independent from $N$, and is chosen so that $P\{\xi = 1\} = P\{\xi = -1\} = \frac{1}{2}$. Prove that $E\xi_t = 0$ and $E\xi_s\xi_t = e^{-2\lambda|t-s|}$, $s, t \geq 0$.

**Problem 6.1.7.** Consider the sequence $(\xi_n)_{n\geq 0}$ defined as

$$\xi_n = \sum_{k=1}^{N} a_k \cos(b_k n - \eta_k),$$

where $a_k, b_k > 0$, for $k = 1, \dots, N$, are given constants, and $\eta_1, \dots, \eta_N$ are independent random variables that are uniformly distributed in $(0, 2\pi)$. Prove that $(\xi_n)_{n\geq 0}$ is a stationary sequence.

**Problem 6.1.8.** Let $\xi_n = \cos n\varphi$, $n \geq 1$, for some random variable $\varphi$, that is uniformly distributed on $[-\pi, \pi]$. Prove that the sequence $(\xi_n)_{n\geq 1}$ is stationary in broad sense, but is not stationary in strict sense.

**Problem 6.1.9.** Consider the one-sided moving average model of order $p$ (MA($p$)):

$$\xi_n = a_0\varepsilon_n + a_1\varepsilon_{n-1} + \dots + a_p\varepsilon_{n-p},$$

where $n = 0, \pm1, \dots$ and $\varepsilon = (\varepsilon_n)$ is a white noise sequence (see [P §6.1, Example 3]). Compute the dispersion $D\xi_n$ and the covariance $\text{cov}(\xi_n, \xi_{n+k})$.

**Problem 6.1.10.** Consider the auto-regression model of order 1 (AR(1))

$$\xi_n = \alpha_0 + \alpha_1\xi_{n-1} + \sigma\varepsilon_n, \qquad n \geq 1$$

(comp. with formula [P §6.1, (25)]) with white noise $\varepsilon = (\varepsilon_n)$ and suppose that $|\alpha_1| < 1$. Prove that if $E|\xi_0| < \infty$, then

$$E\xi_n = \alpha_1^n E\xi_0 + \frac{\alpha_0(1 - \alpha_1^n)}{1 - \alpha_1} \to \frac{\alpha_0}{1 - \alpha_1} \quad \text{as } n \to \infty;$$

if, furthermore, $D\xi_0 < \infty$, then

$$D\xi_n = \alpha_1^{2n} D\xi_0 + \frac{\sigma^2(1 - \alpha_1^{2n})}{1 - \alpha_1^2} \to \frac{\sigma^2}{1 - \alpha_1^2} \quad \text{as } n \to \infty,$$

and

$$\text{cov}(\xi_n, \xi_{n+k}) \to \frac{\sigma^2 \alpha_1^k}{1 - \alpha_1^2}.$$

**Problem 6.1.11.** In the setting of the previous problem, suppose that $\xi_0$ is normally distributed with law $\mathcal{N}\left(\frac{\alpha_0}{1-\alpha_1}, \frac{\sigma^2}{1-\alpha_1^2}\right)$. Prove that the Gaussian sequence $\xi = (\xi_n)_{n \geq 0}$ is both strictly and broadly stationary, with

$$\text{E}\xi_n = \frac{\alpha_0}{1 - \alpha_1}, \quad \text{D}\xi_n = \frac{\sigma^2}{1 - \alpha_1^2} \quad \text{and} \quad \text{cov}(\xi_n, \xi_{n+k}) = \frac{\sigma^2 \alpha_1^2}{1 - \alpha_1^2}.$$

## 6.2 Orthogonal Stochastic Measures and Stochastic Integrals

**Problem 6.2.1.** Prove the equivalence of conditions [P §6.2, (5) and (6)].
  *Hint.* To prove the implication (5) $\Rightarrow$ (6), take $\Delta_n \downarrow \varnothing$, $\Delta_n \in \mathcal{E}_0$, $D_n = E \setminus \Delta_n$, $D_0 = \varnothing$; then $E = \sum_{k=1}^{\infty}(D_k \setminus D_{k-1})$ and [P §6.2, (5)] implies that $Z(\Delta_n) = Z(E) - Z(D_n) \xrightarrow{H^2} 0$.

**Problem 6.2.2.** Consider the function $f \in L^2$. By using the results from [P Chap. 2] (specifically, [P §2.4, Theorem 1], the Corollary to [P §2.6, Theorem 3], and Problem 2.3.8), prove that there is a sequence, $(f_n)_{n \geq 1}$, that consists of functions of the form specified in [P §6.2, (10)], and is such that $\|f - f_n\| \to 0$ as $n \to \infty$.
  *Hint.* The proof may be established with the following argument. Given any $\varepsilon > 0$, one can construct the simple function $g(\lambda) = \sum_{k=1}^{p} f_k I_{B_k}(\lambda)$, where $B_k \in \mathcal{E}$ and $f_k \in C$, in such a way that $\|f - g\|_{L^2} < \varepsilon/2$. Then construct the sets $\Delta_k \in \mathcal{E}_0$, so that the quantities $m(\Delta_k \triangle B_k)$, $k = 1, \ldots, p$, are as small as needed. Finally, the function $h(\lambda) = \sum_{k=1}^{p} f_k I_{\Delta_k}(\lambda)$ has the form specified in [P §6.2, (10)] and, furthermore, can be chosen so that $\|f - h\|_{L^2} < \varepsilon$.

**Problem 6.2.3.** Assuming that $Z(\Delta)$ is some orthogonal stochastic measure, with structural function $m(\Delta)$, verify the following relations:

$$\text{E}|Z(\Delta_1) - Z(\Delta_2)|^2 = m(\Delta_1 \triangle \Delta_2),$$

$$Z(\Delta_1 \setminus \Delta_2) = Z(\Delta_1) - Z(\Delta_1 \cap \Delta_2) \quad \text{(P-a.e.)},$$

$$Z(\Delta_1 \triangle \Delta_2) = Z(\Delta_1) + Z(\Delta_2) - 2Z(\Delta_1 \cap \Delta_2) \quad \text{(P-a.e.)}.$$

**Problem 6.2.4.** Let $\xi = (\xi_n)$, with $\text{E}\xi_n = 0$, be any stationary sequence with correlation function $R(n)$, and with spectral measure $F(d\lambda)$. Setting $S_n = \xi_1 + \ldots + \xi_n$, prove that the dispersion $\text{D}S_n$ can be written in the form:

$$\text{D}S_n = \sum_{|k|<n}(n - |k|)R(k) \quad \text{or} \quad \text{D}S_n = \int_{-\pi}^{\pi}\left(\frac{\sin\frac{n\lambda}{2}}{\sin\frac{\lambda}{2}}\right)^2 F(d\lambda).$$

**Problem 6.2.5.** Suppose that $f(\lambda)$ is a spectral density (i.e., for some spectral measure $F$ one can write $F(d\lambda) = f(\lambda)\,d\lambda$), which is continuous at $\lambda = 0$. By using the second formula for the dispersion $DS_n$, established in the previous problem, prove that

$$DS_n = 2\pi f(0) \cdot n + o(n).$$

(The kernel $\left(\frac{\sin(n\lambda/2)}{\sin(\lambda/2)}\right)^2$ is known as the *Fejér's kernel*—see [P §6.4, 2].)

## 6.3 Spectral Representations of Stationary (in Broad Sense) Sequences

**Problem 6.3.1.** In the notation adopted in the proof of [P §6.3, Theorem 1], prove that $\overline{L_0^2(F)} = L^2(F)$.

*Hint.* According to Problem 6.2.2, every function $f(\lambda) \in L^2(F)$ can be approximated arbitrarily closely in the norm of $L^2(F)$ with functions of the form $g(\lambda) = \sum_{k=1}^{p} f_k I_{B_k}(\lambda)$, where $B_k \in \mathcal{A}$, $\mathcal{A}$ being the algebra comprised of all finite unions of intervals of the form $[a, b)$, $-\pi \le a < b < \pi$. Consequently, it is enough to prove only that every function $I_{[a,b)}(\lambda)$ can be approximated with linear combinations of functions of the form $e_n(\lambda) = e^{i\lambda n}$, $n = 0, \pm 1, \pm 2, \ldots$. However, a function of the form $I_{[a,b)}(\lambda)$ can be approximated with continuous functions, which, in turn, can be approximated with linear combinations of functions of the form $e_n(\lambda)$, $n = 0, \pm 1, \pm 2, \ldots$ (the Weierstrass–Stone theorem).

**Problem 6.3.2.** Let $\xi = (\xi_n)$ be any stationary sequence, such that, for some fixed $N$, one can claim that $\xi_{n+N}(\omega) = \xi_n(\omega)$, $\omega \in \Omega$, for all $n \in \mathbb{Z} = \{0, \pm 1, \pm 2, \ldots\}$. Prove that the spectral representation of any such sequence comes down to the representation [P §6.1, (13)].

*Hint.* Since $R(N) = R(0)$, one can claim that the spectral measure $F$ is piecewise constant on $[-\pi, \pi)$ and has jumps at the points

$$\lambda_k = \frac{2\pi k}{N} + 2\pi p_k, \quad k = 1, \ldots, N,$$

where the integers $p_k$ are chosen so that $\lambda_k \in [-\pi, \pi)$. As a result, the spectral representation of $\xi$ must have of the form:

$$\xi_n = \int_{[-\pi,\pi)} e^{i\lambda n}\, Z(d\lambda) = \sum_{k=1}^{N} e^{i\lambda_k n}\, Z(\{\lambda_k\}).$$

**Problem 6.3.3.** Let $\xi = (\xi_n)$ be any stationary sequence, chosen so that $E\xi_n = 0$ and

$$\frac{1}{N^2} \sum_{k=0}^{N-1}\sum_{l=0}^{N-1} R(k-l) = \frac{1}{N} \sum_{|k| \le N-1} R(k)\left[1 - \frac{|k|}{N}\right] \le CN^{-\alpha},$$

for some constants $C > 0$ and $\alpha > 0$. By using the Borel–Cantelli lemma, prove that

$$\frac{1}{N}\sum_{k=0}^{N}\xi_k \to 0 \quad \text{as } N \to \infty \quad \text{(P-a. e.)}.$$

**Problem 6.3.4.** Suppose that the spectral density $f_\xi(\lambda)$ of the sequence $\xi = (\xi_m)$ is rational, in that

$$f_\xi(\lambda) = \frac{1}{2\pi}\frac{|P_{n-1}(e^{-i\lambda})|}{|Q_n(e^{-i\lambda})|},$$

where $P_{n-1}(z) = a_0 + a_1 z + \cdots + a_{n-1}z^{n-1}$ and $Q_n(z) = 1 + b_1 z + \cdots + b_n z^n$ are given polynomials. In addition, suppose that $Q_n$ has no roots on the unit circle.

Prove that one can construct a white noise sequence $\varepsilon = (\varepsilon_m)$, $m \in \mathbb{Z}$, in such a way that the sequence $(\xi_m)$ is a component of the $n$-dimensional sequence $(\xi_m^1, \ldots, \xi_m^n)$ (i.e., $\xi_m^1 = \xi_m$), which is determined by the relations

$$\xi_{m+1}^i = \xi_m^{i+1} + \beta_i \varepsilon_{m+1}, \quad i = 1, \ldots, n-1,$$

$$\xi_{m+1}^n = -\sum_{j=0}^{n-1} b_{n-j}\xi_m^{j+1} + \beta_n \varepsilon_{m+1},$$

where $\beta_1 = a_0$ and $\beta_i = a_{i-1} - \sum_{k=1}^{i-1}\beta_k b_{i-k}$, $i > 1$.

**Problem 6.3.5.** One says that the stationary (in strict sense) sequence $\xi = (\xi_n)$ satisfies the *strong mixing condition* if

$$\alpha_n(\xi) = \sup_{A \in \mathscr{F}_{-\infty}^0(\xi), B \in \mathscr{F}_n^\infty(\xi)} |P(AB) - P(A)P(B)| \to 0 \quad \text{as } n \to \infty,$$

where $\mathscr{F}_{-\infty}^n(\xi) = \sigma(\ldots \xi_{-1}, \xi_0)$ and $\mathscr{F}_n^\infty(\xi) = \sigma(\xi_n, \xi_{n+1}, \ldots)$. (Comp. with Problem 2.8.7.)

Prove that if $X$ and $Y$ are two bounded ($|X| \le C_1$ and $|Y| \le C_2$) random variables, that are measurable, respectively, for $\mathscr{F}_{-\infty}^n(\xi)$ and $\mathscr{F}_n^\infty(\xi)$, then

$$|\mathsf{E}XY - \mathsf{E}X\,\mathsf{E}Y| \le 4C_1 C_2 \alpha_n(\xi).$$

**Problem 6.3.6.** Let $\xi = (\xi_m)_{-\infty < m < \infty}$ be any stationary Gaussian sequence, and let

$$\rho_n^*(\xi) = \sup_{X,Y} \mathsf{E}XY,$$

the supremum being taken over all random variables $X$ and $Y$, with $\mathsf{E}|X|^2 = \mathsf{E}|Y|^2 = 1$, chosen from the closed linear manifolds $L_{-\infty}^n(\xi)$ and $L_n^\infty(\xi)$, that are generated, respectively, by the families $(\xi_m)_{m \le 0}$ and $(\xi_m)_{m \ge n}$.

Prove the *Kolmogorov–Rozanov inequality*:

$$\alpha_n(\xi) \le \rho_n^*(\xi) \le 2\pi\alpha_n(\xi).$$

(Comp. with the inequalities in Problem 2.8.7.)

**Problem 6.3.7.** Suppose that $\xi = (\xi_n)$ is some stationary Gaussian sequence, that has a continuous spectral density, $f(\lambda)$, which is uniformly bounded from below by some positive constant, i.e., $f(\lambda) > C > 0$, $\lambda \in [-\pi, \pi]$. By using the inequalities established in the previous problem, prove that the sequence $\xi$ must have the strong mixing property.

**Problem 6.3.8.** By considering sequences $\xi = (\xi_n)$, of the form

$$\xi_n = A\cos(\lambda n + \theta),$$

for an appropriate choice of the constant $A \neq 0$ and the independent random variables $\lambda$ and $\theta$, prove that a stationary in broad sense sequence may have periodic sample paths and non-periodic covariance function.

# 6.4   Statistical Estimates of Covariance Functions and Spectral Densities

**Problem 6.4.1.** Consider the estimation scheme [P §6.4, (15)] and suppose that $\varepsilon_n \sim \mathcal{N}(0, 1)$. Prove that for every fixed $n$ one must have

$$(N - |n|)D\widehat{R}_N(n; \xi) \to 2\pi \int_{-\pi}^{\pi} (1 + e^{2in\lambda}) f^2(\lambda)\,d\lambda \quad \text{as } N \to \infty.$$

*Hint.* By using the assumption that $\varepsilon_n$ is Gaussian for every fixed $n \geq 0$, argue that

$$(N - n)D\widehat{R}_N(n; \xi) = 2\pi \int_{-\pi}^{\pi}\int_{-\pi}^{\pi} [1 + e^{in(\lambda + v)}]\Phi_{N-n}(\lambda - v) f(v) f(\lambda)\,dv\,d\lambda,$$

where $\Phi_{N-n}(\lambda)$ is the associated Fejér kernel. The required result then follows from the above relation.

**Problem 6.4.2.** Prove formula [P §6.4, (16)] and its generalization:

$$\lim_{N \to \infty} \text{cov}(\hat{f}_N(\lambda; \xi), \hat{f}_N(v; \xi)) = \begin{cases} 2f^2(0), & \lambda = v = 0, \pm\pi, \\ f^2(\lambda), & \lambda = v \neq 0, \pm\pi, \\ 0, & \lambda \neq v. \end{cases}$$

**Problem 6.4.3.** Consider the first-order autoregressive model AR(1)

$$\xi_n = \theta\xi_{n-1} + \sigma\varepsilon_n, \quad n \geq 1, \quad \xi_0 = 0,$$

in which $\varepsilon = (\varepsilon_n)$ is a Gaussian white noise sequence (comp. with [P §6.1, (25)] and with the model discussed in Problem 6.1.10). Suppose that in this model $\sigma > 0$ is a known parameter, while $\theta \in \mathbb{R}$ is some unknown parameter, that must be estimated from the observations $\xi_1, \xi_2, \ldots$.

Let $\hat{\theta}_n = \arg\max p_\theta(x_1, \ldots, x_n)$ be the maximum likelihood estimate of the parameter $\theta$, obtained from the joint probability density of $\xi_1, \ldots, \xi_n$, namely

$$p_\theta(x_1, \ldots, x_n) = \frac{1}{(\sqrt{2\pi}\,\sigma)^n} \exp\left\{ -\frac{1}{2\sigma^2} \sum_{k=1}^{n} (x_k - \theta x_{k-1})^2 \right\}.$$

Prove that

$$\hat{\theta}_n = \frac{\sum_{k=1}^{n} X_{k-1} X_k}{\sum_{k=1}^{n} X_{k-1}^2}.$$

**Problem 6.4.4.** Consider the *Fisher Information*

$$I_n(\theta) = \mathsf{E}_\theta\left\{ -\frac{\partial^2 \ln p_\theta(\xi_1, \ldots, \xi_n)}{\partial \theta^2} \right\}$$

for the AR(1) model from Problem 6.4.3, $\mathsf{E}_\theta$ stands for the averaging operation, under the distribution $\mathsf{P}_\theta$, of the sequence $\xi_1, \xi_2, \ldots$ .
   Prove that
(a) $I_n(\theta) = \mathsf{E}_\theta \sum_{k=1}^{n} \xi_{k-1}^2$;
(b) as $n \to \infty$, one has

$$I_n(\theta) \sim \begin{cases} \frac{n}{1-\theta^2}, & |\theta| < 1, \\ \frac{n^2}{2}, & |\theta| = 1, \\ \frac{\theta^{2n}}{(\theta^2-1)^2}, & |\theta| > 1. \end{cases}$$

**Problem 6.4.5.** In the context of the AR(1) model discussed in Problems 6.4.3 and 6.4.4, prove that the maximum likelihood estimate, $\hat{\theta}_n$, has the following asymptotic properties:

$$\lim_n \mathsf{P}_n\left\{ \sqrt{I_n(\theta)}\,(\hat{\theta}_n - \theta) \leq x \right\} = \begin{cases} \Phi(x), & |\theta| < 1, \\ H_\theta^{(1)}(x), & |\theta| = 1, \\ \mathrm{Ch}(x), & |\theta| > 1, \end{cases}$$

where $\Phi(x)$ is the distribution function of the standard normal law and $H_\theta^{(1)}(x)$ is the distribution function of the random variable

$$\theta \times \frac{B_1^2 - 1}{2\sqrt{2}\int_0^1 B_s^2\,ds},$$

where $B = (B_s)_{0 \le s \le 1}$ is a standard Brownian motion (see [P §2.13]) and Ch$(x)$ is the distribution function of the Cauchy distribution law with density $\frac{1}{\pi(1+x^2)}$ (see [P §2.3, Table 3]).

**Problem 6.4.6.** As a continuation of the previous problem, prove that

$$\lim_n \mathsf{P}_\theta \left\{ \sqrt{\sum_{k=1}^n \xi_{k-1}^2} \, (\hat\theta_n - \theta) \le x \right\} = \begin{cases} \Phi(x), & |\theta| \ne 1, \\ H_\theta^{(2)}(x), & |\theta| = 1, \end{cases}$$

where $H_\theta^{(2)}(x)$ denotes the distribution function of the random variable

$$\theta \times \frac{B_1^2 - 1}{2\sqrt{\int_0^1 B_s^2 \, ds}}.$$

Thus, if $(\hat\theta_n - \theta)$ is normalized not by the Fisher information, but by the random variable $\left(\sum_{k=1}^n \xi_{k-1}^2\right)^{1/2}$, then one would end-up with only *two* probability distributions instead of *three*.

**Problem 6.4.7.** Prove that the maximum likelihood estimate, $\hat\theta_n$, from Problem 6.4.3 is *uniformly asymptotically consistent on average*:

$$\sup_\theta \mathsf{E}_\theta |\hat\theta_n - \theta| \to 0 \quad \text{as } n \to \infty.$$

## 6.5   Wold Decomposition

**Problem 6.5.1.** Prove that any stationary sequence with discrete spectrum (i.e., with spectral function $F(\lambda)$ that is piece-wise constant) must be *singular*.

*Hint.* If $\xi_n$, $n \in \mathbb{Z} = \{0, \pm 1, \pm 2, \dots\}$, is one such sequence, then one can write

$$\xi_n = \sum_{k=-\infty}^{\infty} z_k e^{i\lambda_k n},$$

with $z_k = Z(\{\lambda_k\})$, $k \in \mathbb{Z}$, being orthogonal random variables with $\mathsf{E}z_k = 0$ and $\mathsf{E}|z_k|^2 = \sigma_k^2$. Consequently, the spectral function can be written in the form $F(\lambda) = \sum_{\{k : \lambda_k \le \lambda\}} \sigma_k^2$, where $\sum_{k=-\infty}^{\infty} \sigma_k^2 < \infty$. Thus, one must show that $H(\xi) = S(\xi)$, where $H(\xi)$ is the closed linear sub-space of $H^2$, generated by the random variables $\xi = (\dots, \xi_{n-1}, \xi_n, \dots)$, and $S(\xi) = \bigcap_{n=-\infty}^{\infty} H_n(\xi)$, where each $H_n(\xi)$ is generated by the family $\xi^n = (\dots, \xi_{n-1}, \xi_n)$.

In order to prove that $H(\xi) = S(\xi)$, it is enough to prove that $\xi_n \in S(\xi)$, for every $n \in \mathbb{Z}$. However, due to the stationarity, it is enough to prove only that $\xi_0 \in S(\xi)$, i.e., for every integer $N \in \mathbb{Z}$ and every $\delta > 0$, one can find some $\eta \in H_N(\xi)$ with

$\|\xi_0 - \eta\|_{H^2} < \delta$. Thus, it would be enough to show that one can take $\eta = \xi_n$, for some appropriate choice of $n \le N$. For that purpose, given an arbitrary $\delta > 0$, one can choose $M$ so that $\sum_{|k|>M} \sigma_k^2 < \delta/2$, then prove that

$$\|\xi_n - \xi_0\|_{H^2} \le \frac{2}{3}\delta + \sum_{|k|\le M} \sigma_k^2 |e^{i\lambda_k n} - 1|,$$

and, finally, prove that, for any $N \in \mathbb{Z}$ and any $\varepsilon > 0$, there is an integer $n \le N$, with $|e^{i\lambda_k n} - 1| < \varepsilon$, for $|k| \le M$.

**Problem 6.5.2.** Let $\sigma_n^2 = \mathsf{E}|\xi_n - \widehat{\xi}_n|^2$, where $\widehat{\xi}_n = \widehat{\mathsf{E}}(\xi_n \mid H_0(\xi))$. Prove that if $\sigma_n^2 = 0$, for some fixed $n \ge 1$, then the sequence $\xi$ must be singular. If, furthermore, $\sigma_n^2 \to R(0)$ as $n \to \infty$, then $\xi$ must be also regular.

**Problem 6.5.3.** Prove that the (automatically stationary) sequence $\xi = (\xi_n)$, of the form $\xi_n = e^{in\varphi}$, for some random variable $\varphi$, which is uniformly distributed on $[0, 2\pi]$, must be *regular*. Find the *linear* estimate, $\widehat{\xi}_n$, of the variable $\xi_n$ and prove that the *non-linear* estimate

$$\tilde{\xi}_n = \left(\frac{\xi_0}{\xi_{-1}}\right)^n$$

gives an *error-free* forecast for $\xi_n$, based on the "past history" $\xi^0 = (\ldots, \xi_{-1}, \xi_0)$, i.e.,

$$\mathsf{E}|\tilde{\xi}_n - \xi_n|^2 = 0, \quad n \ge 1.$$

*Hint.* To prove the regularity of the sequence $\xi = (\xi_n)$, convince yourself that $\varepsilon_n = \xi_k/\sqrt{2\pi}$ represents a white noise sequence, and, therefore the representation $\xi_n = \sqrt{2\pi}\varepsilon_n$ is of the same form as in [$\underline{P}$ §6.5, (3)].

**Problem 6.5.4.** Prove that the decomposition [$\underline{P}$ §6.5, (1)] into a regular and a singular components is unique.

## 6.6   Extrapolation, Interpolation and Filtartion

**Problem 6.6.1.** Prove that the assertion of [$\underline{P}$ §6.6, Theorem 1] remains valid even without the assumption that $\Phi(z)$ has radius of convergence $r > 1$, while all zeroes of $\Phi(z)$ are in the domain $|z| > 1$.

**Problem 6.6.2.** Prove that, for a regular process, the function $\Phi(z)$, which appears in [$\underline{P}$ §6.6, (4)], may be written in the form

$$\Phi(z) = \sqrt{2\pi}\exp\left\{\frac{1}{2}c_0 + \sum_{k=1}^{\infty} c_k z^k\right\}, \quad |z| < 1,$$

where

$$c_k = \frac{1}{2\pi} \int_{-\pi}^{\pi} e^{ik\lambda} \ln f(\lambda) \, d\lambda.$$

Conclude from the above relation that the error in the one-step forecast $\sigma_1^2 = E|\hat{\xi}_1 - \xi_1|^2$ is given by the *Szegö-Kolmogorov formula*:

$$\sigma_1^2 = 2\pi \exp\left\{ \frac{1}{2\pi} \int_{-\pi}^{\pi} \ln f(\lambda) \, d\lambda \right\}.$$

*Hint.* The Szegö-Kolmogorov formula may be established with the following line of reasoning:

(i) First, prove that

$$\sigma_1^2 = \|\xi_1 - \hat{\xi}_1\|_{H^2}^2 = \int_{-\infty}^{\infty} |e^{i\lambda} - \hat{\varphi}_1(\lambda)|^2 f(\lambda) \, d\lambda, \qquad (*)$$

where $\hat{\varphi}_1(\lambda)$ is given by [P §6.6, (7)]. In conjunction with the notation adopted in [P §6.6, Theorem 1], taking into account (*), and the fact that $f(\lambda) = \frac{1}{2\pi}|\Phi(e^{-i\lambda})|^2$, one can show that $\sigma_1^2 = |b_0|^2$.

(ii) From the first part of the problem,

$$\Phi(z) = \sum_{k=0}^{\infty} b_k z^k = \sqrt{2\pi} \exp\left\{ \frac{1}{2}c_0 + \sum_{k=1}^{\infty} c_k z^k \right\},$$

which shows that $b_0 = \sqrt{2\pi} \exp\{\frac{1}{2}c_0\}$, and, consequently, that

$$\sigma_1^2 = 2\pi \exp\left\{ \frac{1}{2}c_0 \right\} = 2\pi \exp\left\{ \frac{1}{2\pi} \int_{-\pi}^{\pi} \ln f(\lambda) \, d\lambda \right\}.$$

**Problem 6.6.3.** Prove [P §6.6, Theorem 2] without assuming that [P §6.6, (22)] is in force.

**Problem 6.6.4.** Suppose that the signal $\theta$ and the noise $\eta$ are uncorrelated and have spectral densities

$$f_\theta(\lambda) = \frac{1}{2\pi} \cdot \frac{1}{|1 + b_1 e^{-i\lambda}|^2} \quad \text{and} \quad f_\eta(\lambda) = \frac{1}{2\pi} \cdot \frac{1}{|1 + b_2 e^{-i\lambda}|^2}.$$

By using [P §6.6, Theorem 3], find the estimate, $\tilde{\theta}_{n+m}$, of the variable $\theta_{n+m}$, from the observations $\xi_k, k \le n$, where $\xi_k = \theta_k + \eta_k$. Solve the same problem for the spectral densities

$$f_\theta(\lambda) = \frac{1}{2\pi}|2 + e^{-i\lambda}|^2 \quad \text{and} \quad f_\eta(\lambda) = \frac{1}{2\pi}.$$

## 6.7 The Kalman-Bucy Filter and Its Generalizations

**Problem 6.7.1.** Prove that in the observation scheme [P §6.7, (1)] the vectors $m_n$ and $\theta_n - m_n$ are uncorrelated:

$$\mathsf{E}[m_n^*(\theta_n - m_n)] = 0.$$

**Problem 6.7.2.** Suppose that in the observation scheme [P §6.7, (1)–(2)] the variable $\gamma_0$ and all coefficients, except, perhaps, $a_0(n, \xi)$ and $A_0(n, \xi)$, are chosen to be "event independent," i.e., independent of $\xi$. Prove that the conditional covariance $\gamma_n$ is also "event-independent," in that $\gamma_n = \mathsf{E}\gamma_n$.

**Problem 6.7.3.** Prove that the solution to the system [P §6.7, (22)] is given by formula [P §6.7, (23)].

**Problem 6.7.4.** Let $(\theta, \xi) = (\theta_n, \xi_n)$ be a Gaussian sequence, which is subject to the following special case of the observation scheme [P §6.7, (1)]:

$$\theta_{n+1} = a\theta_n + b\varepsilon_1(n+1) \quad \text{and} \quad \xi_{n+1} = A\theta_n + B\varepsilon_2(n+1).$$

Prove that if $A \neq 0$, $b \neq 0$ and $B \neq 0$, then the limiting error of the filtration $\gamma = \lim_{n\to\infty} \gamma_n$ exists, and is given by the positive root of the equation

$$\gamma^2 + \left[\frac{B^2(1-a^2)}{A^2} - b^2\right]\gamma - \frac{b^2 B^2}{A^2} = 0.$$

*Hint.* By using formula [P §6.7, (8)], one can show that

$$\gamma_{n+1} = b^2 + a^2 c^2 - \frac{a^2 c^2}{c^2 + \gamma_n},$$

where $c^2 = (\frac{B}{A})^2$. In other words, $\gamma_{n+1} = f(\gamma_n)$, with $f(x) = b^2 + a^2 c^2 - \frac{a^2 c^2}{c^2 + x}$, $x \geq 0$. Furthermore, it is easy to see that $f(x)$ is non-decreasing and bounded. From this property one can conclude that $\lim \gamma_n \; (= \gamma)$ exists and satisfies the following equation

$$\gamma^2 + [c^2(1-a^2) - b^2]\gamma - b^2 c^2 = 0,$$

which, due to the Viète formula, can have only one positive root.

**Problem 6.7.5.** (*Interpolation; [80, 13.3].*) Let $(\theta, \xi)$ be a partially observable sequence, which is subject to the recursive relations [P §6.7, (1) and (2)]. Suppose that the conditional distribution of the vector $\theta_m$, namely

$$\pi_a(m, m) = \mathsf{P}\left(\theta_m \leq a \mid \mathscr{F}_m^\xi\right),$$

is normal.

(a) Prove that for any $n \geq m$ the conditional distribution

$$\pi_a(m,n) = \mathsf{P}(\theta_m \leq a \mid \mathscr{F}_n^\xi)$$

is also normal, i.e., $\pi_a(m,n) \sim \mathcal{N}(\mu(m,n), \gamma(m,n))$.

(b) Find the interpolation estimate, $\mu(m,n)$, of $\theta_m$ from $\mathscr{F}_n^\xi$. Find also the matrix $\gamma(m,n)$.

**Problem 6.7.6.** (*Extrapolation*; *[80, 13.4]*.) Suppose that in the relations [ $\underline{\mathrm{P}}$ §6.7, (1) and (2)] one has

$$a_0(n,\xi) = a_0(n) + a_2(n)\xi_n, \qquad a_1(n,\xi) = a_1(n),$$

$$A_0(n,\xi) = A_0(n) + A_2(n)\xi_n, \quad A_1(n,\xi) = A_1(n).$$

(a) Prove that, with the above choice, one can claim that the distribution $\pi_{a,b}(m,n) = \mathsf{P}(\theta_n \leq a, \xi_n \leq b \mid \mathscr{F}_m^\xi), n \geq m$, is normal.

(b) Find the extrapolation estimates

$$\mathsf{E}(\theta_n \mid \mathscr{F}_m^\xi) \quad \text{and} \quad \mathsf{E}(\xi_n \mid \mathscr{F}_m^\xi), \qquad n \geq m.$$

**Problem 6.7.7.** (*Optimal control*; *[80, 14.3]*.) Consider some "controlled" and partially observable system $(\theta_n, \xi_n)_{0 \leq n \leq N}$, where

$$\theta_{n+1} = u_n + \theta_n + b\varepsilon_1(n+1),$$

$$\xi_{n+1} = \theta_n + \varepsilon_2(n+1).$$

The "control" $u_n$ is $\mathscr{F}_n^\xi$-measurable and such that $\mathsf{E}u_n^2 < \infty$, for all $0 \leq n \leq N-1$. The variables $\varepsilon_1(n)$ and $\varepsilon_2(n)$, $n = 1, \ldots, N$, are chosen as in [ $\underline{\mathrm{P}}$ §6.7, (1) and (2)] and $\xi_0 = 0$, $\theta_0 \sim \mathcal{N}(m, \gamma)$.

We say that the "control" $u^* = (u_0^*, \ldots, u_{N-1}^*)$ is optimal if $V(u^*) = \sup_u V(u)$, where

$$V(u) = \mathsf{E}\left[\sum_{n=0}^{N-1}(\theta_n^2 + u_n^2) + \theta_N^2\right].$$

Prove that the optimal control exists and is given by

$$u_n^* = -[1 + P_{n+1}]^\oplus P_{n+1} m_n^*, \qquad n = 0, \ldots, N-1,$$

where

$$a^\oplus = \begin{cases} a^{-1}, & a \neq 0, \\ 0, & a = 0, \end{cases}$$

and the quantities $(P_n)_{0 \le n \le N}$ are defined recursively through the relation

$$P_n = 1 + P_{n+1} - P_{n+1}^2 [1 + P_{n+1}]^{\oplus}, \quad P_N = 1,$$

while $(m_n^*)$ is defined by the relation

$$m_{n+1}^* = u_n^* + \gamma_n^*(1 + \gamma_n^*)^+ (\xi_{n+1} - m_n^*), \quad 0 \le n \le N - 1,$$

with $m_0^* = m$ and with

$$\gamma_{n+1}^* = \gamma_n^* + 1 - (\gamma_n^*)^2 (1 + \gamma_n^*)^{\oplus}, \quad 0 \le n \le N - 1,$$

where we suppose that $\gamma_0^* = \gamma$.

**Problem 6.7.8.** (*Nonlinear filtering and the "change-point" detection problem—see [117].*) Typically, in statistical control—and especially in quality control—one encounters quantities whose probabilistic nature changes abruptly at some random moment $\theta$. This moment represents "the change-point," say, in a particular production process. In what follows we will describe the *Bayesian* formulation of the problem of early detection of "the change-point," and will address questions related to the construction of sufficient statistics for this quantity.

Let $(\Omega, \mathscr{F})$ be some measurable space, let $\{\mathsf{P}^\pi; \pi \in [0, 1]\}$ be some family of probability measures on $(\Omega, \mathscr{F})$, let $\theta$ be some random variable on $(\Omega, \mathscr{F})$, which takes values in the space of integers $N = \{0, 1, 2, \ldots\}$, and, finally, let $X_1, X_2, \ldots$ be some sequence of observable random variables, defined on $(\Omega, \mathscr{F})$. Next, suppose that the following relations are in force:

(i) $\mathsf{P}^\pi\{\theta = 0\} = \pi$, $\mathsf{P}^\pi\{\theta = k\} = (1 - \pi)p_k$, where $p_k \ge 0$, $\sum_{k=1}^\infty p_k = 1$.
(ii) For every $\pi \in [0, 1]$ and every $n \ge 1$ one has

$$\mathsf{P}^\pi\{X_1 \le x_1, \ldots, X_n \le x_n\} = \pi \mathsf{P}^1\{X_1 \le x_1, \ldots, X_n \le x_n\}$$

$$+ (1 - \pi) \sum_{k=0}^{n-1} p_{k+1} \mathsf{P}^0\{X_1 \le x_1, \ldots, X_k \le x_k\} \mathsf{P}^1\{X_{k+1} \le x_{k+1}, \ldots, X_n \le x_n\}$$

$$+ (1 - \pi)(p_{n+1} + p_{n+2} + \ldots)\mathsf{P}^0\{X_1 \le x_1, \ldots, X_n \le x_n\}, \quad x_k \in R.$$

(iii) $\mathsf{P}^j\{X_1 \le x_1, \ldots, X_n \le x_n\} = \prod_{k=1}^n \mathsf{P}^j\{X_k \le x_k\}, j = 0, 1.$

The practical meaning of the relations (i)–(iii) can be summarized as follows. If $\theta = 0$ or $\theta = 1$, then "the change-point" has taken place before the observation process has begun. In this case, the variables $X_1, X_2, \ldots$ are all associated with the already "changed" production process and are independent and identically distributed, with distribution function $F_1(x) = \mathsf{P}^1\{X_1 \le x\}$. If $\theta > n$, i.e., the "change-point" occurs after the $n$-th observation, then the random variables $X_1, \ldots, X_n$ are associated with the "normal" production process and are independent and identically distributed, with distribution function $F_0(x) = \mathsf{P}^0\{X_1 \le x\}$. If $\theta = k$, for some $1 < k \le$

$n$, then $X_1, \ldots, X_{k-1}$ are independent and identically distributed with distribution function $F_0(x)$, while $X_k, \ldots, X_n$ are also independent and identically distributed, but with distribution function $F_1(x)$. We suppose that $F_0(x) \not\equiv F_1(x)$.

Let $f_0(x)$ and $f_1(x)$ stand for the densities of the distributions $F_0(x)$ and $F_1(x)$, with respect to some distribution, say, $(F_0(x) + F_1(x))/2$, relative to which $F_0(x)$ and $F_1(x))$ are both absolutely continuous.

Let $\tau$ denote the moment at which the "change-point" is declared. We suppose that $\tau$ is a Markov moment relative to $(\mathscr{F}_n)_{n \geq 0}$, where $\mathscr{F}_0 = \{\varnothing, \Omega\}$ and $\mathscr{F}_n = \sigma(X_1, \ldots, X_n)$. Essentially, $\tau$ represents a guess and the quality of this guess will be measured in terms of the quantities: $\mathsf{P}^\pi\{\tau < \theta\}$, which is the probability for "false alarm," and $\mathsf{E}^\pi(\tau - \theta)^+$, which is the expected delay in detecting the "change-point," when the "alarm" is real, in that $\tau \geq \theta$.

One would like to construct a moment $\tau$ that minimizes simultaneously the probability for "false alarm" and the expected delay in detection. But since such a moment does not exist (except for some trivial situations), we introduce the "Bayesian" risk (below we suppose that $c > 0$ is some appropriately chosen constant)

$$R^\pi(\tau) = \mathsf{P}^\pi\{\tau < \theta\} + c\,\mathsf{E}^\pi(\tau - \theta)^+,$$

and say that the moment $\tau^*$ is *optimal*, if, for any $\pi \in [0, 1]$, one can claim that $\mathsf{P}^\pi\{\tau^* < \infty\} = 1$ and that $R^\pi(\tau^*) \leq R^\pi(\tau)$, for every $\mathsf{P}^\pi$-finite Markov moment $\tau$.

According to Problem 8.9.8, a moment $\tau^*$ with the above properties exists and, in the special case where $p_k = (1-p)^{k-1}p, 0 < p \leq 1, k \geq 1$, can be expressed as:

$$\tau^* = \inf\{n \geq 0 : \pi_n \geq A\},$$

where the constant $A$, which, in general, may depend on $c$ and $p$, is the "alarm-trigger" threshold, while $\pi_n$ is the posterior probability for the "change-point" to occur no later than the $n$-th observation:

$$\pi_n = \mathsf{P}^\pi(\theta \leq n \mid \mathscr{F}_n), \quad n \geq 0, \qquad \pi_0 = \pi.$$

(a) Prove that the posterior probabilities $\pi_n$, $n \geq 0$, are subject to the following recursive relations:

$$\pi_{n+1} = \frac{\pi_n f_1(X_{n+1}) + p(1 - \pi_n) f_1(X_{n+1})}{\pi_n f_1(X_{n+1}) + p(1 - \pi_n) f_1(X_{n+1}) + (1 - \pi)(1 - \pi_n) f_0(X_{n+1})}.$$

(b) Prove that if $\varphi_n = \pi_n/(1 - \pi_n)$, then

$$\varphi_{n+1} = (p + \varphi_n) \frac{f_1(X_{n+1})}{(1 - p) f_0(X_{n+1})}.$$

(c) Setting $\varphi = \varphi_n(p)$, $\pi = 0$ and $\psi = \lim_{p \downarrow 0} \varphi_n(p)/p$, prove that

$$\psi_{n+1} = (1 + \psi_n) \frac{f_1(X_{n+1})}{f_0(X_{n+1})}, \quad \psi_0 = 0.$$

*Remark.* If we set $\theta_n = I(\theta \leq n)$, then $\pi_n = \mathsf{E}^\pi (\theta_n \mid \mathscr{F}_n)$ is the mean-square optimal estimate of $\theta_n$ from the observations $X_1, \ldots, X_n$. From (a), (b) and (c) one can conclude that the statistics $\pi_n$, $\varphi_n$ and $\psi_n$ are governed by *nonlinear* recursive relations, which are said to define the *nonlinear filter* (for the problem of estimating the values $(\theta_n)_{n \geq 0}$ from the observations $X_1, \ldots, X_n$).

(d) Prove that each of the sequences $(\pi_n)_{n \geq 0}$, $(\varphi_n)_{n \geq 0}$ and $(\psi_n)_{n \geq 0}$ constitutes a Markov chain.

# Chapter 7
# Martingale Sequences

## 7.1 The Notion of Martingale and Related Concepts

**Problem 7.1.1.** Show the equivalence of conditions [P §7.1, (2) and (3)].
  *Hint.* The proof can be established by contradiction.

**Problem 7.1.2.** Let $\sigma$ and $\tau$ be two Markov times. Show that $\tau + \sigma$, $\tau \vee \sigma$ and $\tau \wedge \sigma$ are also Markov times, and, if $\sigma(\omega) \leq \tau(\omega)$ for all $\omega \in \Omega$, then $\mathscr{F}_\sigma \subseteq \mathscr{F}_\tau$. Does this property still hold if $\sigma \leq \tau$ only with Probability 1?
  *Hint.* If $\sigma(\omega) \leq \tau(\omega)$ for all $\omega \in \Omega$, then, for every $A \in \mathscr{F}_\sigma$, one has

$$A \cap \{\tau = n\} = A \cap \{\sigma \leq n\} \cap \{\tau = n\} \in \mathscr{F}_n,$$

and therefore $A \in \mathscr{F}_\tau$.

**Problem 7.1.3.** Prove that $\tau$ and $X_\tau$ are both $\mathscr{F}_\tau$-measurable.

**Problem 7.1.4.** Let $Y = (Y_n, \mathscr{F}_n)$ be a martingale (submartingale) and let $V = (V_n, \mathscr{F}_{n-1})$ be some predictable sequence, for which one can claim that all random variables $(VY)_n$, $n \geq 0$, are integrable. Prove that $VY$ is a martingale (submartingale).

**Problem 7.1.5.** Let $\mathscr{G}_1 \supseteq \mathscr{G}_2 \supseteq \ldots$ be any *non-increasing* sequence of $\sigma$-algebras, and suppose that $\xi$ is some integrable random variable. Setting $X_n = \mathsf{E}(\xi \mid \mathscr{G}_n)$, prove that the sequence $(X_n)_{n \geq 1}$, forms a *reverse martingale*, i.e.,

$$\mathsf{E}(X_n \mid X_{n+1}, X_{n+2}, \ldots) = X_{n+1} \quad \text{(P-a.e.)}, \quad \text{for every } n \geq 1.$$

**Problem 7.1.6.** Let $\xi_1, \xi_2, \ldots$ be any sequence of independent random variables, chosen so that $\mathsf{P}\{\xi_i = 0\} = \mathsf{P}\{\xi_i = 2\} = \frac{1}{2}$, and let $X_n = \prod_{i=1}^n \xi_i$. Prove that it is not possible to find an integrable random variable $\xi$, and a non-decreasing family of $\sigma$-algebras $(\mathscr{F}_n)$, so that one can write: $X_n = \mathsf{E}(\xi \mid \mathscr{F}_n)$ (P-a.e.), for every $n \geq 1$.

A.N. Shiryaev, *Problems in Probability*, Problem Books in Mathematics,
DOI 10.1007/978-1-4614-3688-1_7,
© Springer Science+Business Media New York 2012

Conclude that there are martingales that cannot be expressed as $(E(\xi \mid \mathscr{F}_n))_{n \geq 1}$, for some appropriate choice of $\xi$ and $(\mathscr{F}_n)_{n \geq 1}$ (comp. with [P §1.11, Example 3]).

*Hint.* The proof can be established by contradiction.

**Problem 7.1.7.** (a) Let $\xi_1, \xi_2, \ldots$ be any sequence of *independent* random variables with $E|\xi_n| < \infty$ and $E\xi_n = 0, n \geq 1$. Prove that, for every fixed $k \geq 1$, the sequence

$$X_n^{(k)} = \sum_{1 \leq i_1 < \cdots < i_k \leq n} \xi_{i_1} \ldots \xi_{i_k}, \quad n = k, k+1, \ldots$$

forms a martingale.

(b) Let $\xi_1, \xi_2, \ldots$ be any sequence of *integrable* random variables, for which

$$E(\xi_{n+1} \mid \xi_1, \ldots, \xi_n) = \frac{\xi_1 + \cdots + \xi_n}{n}, \quad n \geq 1.$$

Prove that the sequence $X_n = \frac{1}{n}(\xi_1 + \cdots + \xi_n), n \geq 1$, forms a martingale.

**Problem 7.1.8.** Give an example of a martingale $X = (X_n, \mathscr{F}_n)_{n \geq 1}$, for which the family $\{X_1, X_2, \ldots\}$ is *not* uniformly integrable.

**Problem 7.1.9.** Let $X = (X_n)_{n \geq 0}$ be a Markov chain ([P §8.1]) with countable state-space $E = \{i, j, \ldots\}$ and with transition probabilities $p_{ij}$. Let $\psi = \psi(x)$, $x \in E$, be any bounded function with the property that, for some $\lambda > 0$, one has

$$\sum_{j \in E} p_{ij} \psi(j) \leq \lambda \psi(i), \quad \text{for any } i \in E.$$

(A function $\psi$ with the above properties is said to be $\lambda$-*excessive*, or $\lambda$-*harmonic*.) Prove that the sequence $(\lambda^{-n} \psi(X_n))_{n \geq 0}$ forms a supermartingale.

**Problem 7.1.10.** Let $\tau_1, \tau_2, \ldots$ be any sequence of stopping times, chosen so that either $\tau_n \downarrow \tau$, or $\tau_n \uparrow \tau$, in point-wise sense. Prove that $\tau$ must be a stopping time in either case.

**Problem 7.1.11.** Prove that if $\sigma$ and $\tau$ are stopping times, then

$$\mathscr{F}_{\sigma \wedge \tau} = \mathscr{F}_\sigma \cap \mathscr{F}_\tau \quad \text{and} \quad \mathscr{F}_{\sigma \vee \tau} = \sigma(\mathscr{F}_\sigma \cup \mathscr{F}_\tau).$$

**Problem 7.1.12.** Let $\sigma$ be any (finite) stopping time, and let $\tau_1, \tau_2, \ldots$ be any sequence of stopping times. Prove that if $\tau_n \uparrow \infty$, then

$$\mathscr{F}_{\sigma \wedge \tau_n} \uparrow \mathscr{F}_\sigma,$$

and, if $\tau_n \downarrow \sigma$, then $\mathscr{F}_\sigma = \bigcap_n \mathscr{F}_{\tau_n}$.

**Problem 7.1.13.** Let $\xi_1, \xi_2, \ldots$ be any sequence of independent standard normal random variables ($\xi_n \sim \mathcal{N}(0, 1)$) and let $S_n = \xi_1 + \ldots + \xi_n$, $n \geq 1$. Prove that the sequence $(X_n)_{n \geq 1}$, given by

$$X_n = \frac{1}{\sqrt{n+1}} \exp\left\{ \frac{S_n^2}{2(n+1)} \right\},$$

is a martingale relative to the filtration $(\mathscr{F}_n^\xi)_{n \geq 1}$, with $\mathscr{F}_n^\xi = \sigma(\xi_1, \ldots, \xi_n)$.

**Problem 7.1.14.** Let $X = (X_n, \mathscr{F}_n)_{n \geq 0}$ be any stochastic sequence, set $\Delta X_n = X_n - X_{n-1}$, $n \geq 1$, and let $\nu(\omega; \{n\} \times dx) = \mathsf{P}(\Delta X_n \in dx \mid \mathscr{F}_{n-1})(\omega)$ be any regular version of the respective conditional expectation. Given any $u \in \mathbb{R}$, set

$$A(u)_0 = 0 \quad \text{and} \quad A(u)_n = \sum_{1 \leq k \leq n} (e^{iux} - 1)\nu(0; \{k\} \times dx), \quad n \geq 1.$$

Prove that the process $M(u) = (M_n(u), \mathscr{F}_n)$, $n \geq 1$, with

$$M_n(u) = e^{iuX_n} - \sum_{k=1}^{n} e^{iuX_{k-1}} \Delta A(u)_k,$$

is a martingale.

**Problem 7.1.15.** With the notation adopted in the previous problem, given any $u \in \mathbb{R}$, set

$$G(u)_0 = 0 \quad \text{and} \quad G(u)_n = \prod_{1 \leq k \leq n} \int e^{iux} \nu(\omega; \{k\} \times dx), \quad n \geq 1,$$

and suppose that $G(u)_n > 0$, $n \geq 1$. Prove that the (*complex-valued*) sequence

$$\left( \frac{e^{iuX_n}}{G(u)_n} \right)_{n \geq 0},$$

i.e., the sequence

$$\left( \frac{e^{iuX_n}}{\prod_{k=1}^{n} \mathsf{E}(e^{iu\Delta X_k} \mid \mathscr{F}_{k-1})} \right)_{n \geq 1},$$

is a martingale.

**Problem 7.1.16.** Let $X = (X_n, \mathscr{F}_n)_{n \geq 0}$ be any stochastic sequence, chosen so that $|\Delta X_n| \leq c$ (P-a. e.), for some constant $c > 0$ and for all $n \geq 1$, where $\Delta X_n = X_n - X_{n-1}$. Consider the (*real-valued*) sequence $Y = (Y_n, \mathscr{F}_n)_{n \geq 1}$, given by

$$Y_n = \frac{e^{X_n}}{\prod_{i=1}^{n} \mathsf{E}(e^{\Delta X_i} \mid \mathscr{F}_{i-1})},$$

and prove that $Y = (Y_n, \mathscr{F}_n)_{n \geq 1}$ is a martingale (comp. with Problem 7.1.15.)

Will this property hold without the requirement for the variables $\Delta X_n$ to be uniformly bounded?

**Problem 7.1.17.** Let $\xi_1, \ldots, \xi_n$ be independent and normally distributed ($\mathcal{N}(0,1)$) random variables, and let $S_0 = 0$ and $S_k = \xi_1 + \ldots + \xi_k$, $1 \leq k \leq n$. Let $\Phi(x) = \mathsf{P}\{\xi_1 \leq x\}$, let $\mathcal{F}_k = \sigma(\xi_1, \ldots, \xi_k)$, $1 \leq k \leq n$, and let $\mathcal{F}_0 = \{\varnothing, \Omega\}$. Prove that, for every $a \in \mathbb{R}$, the sequence $X = (X_k, \mathcal{F}_k)_{0 \leq k \leq n}$, given by

$$X_k = \Phi\left(\frac{a - S_k}{\sqrt{n - k}}\right),$$

is a martingale.

**Problem 7.1.18.** Let $\xi_1, \ldots, \xi_n$ be independent and identically distributed random variables, whose distribution is symmetric. Set $S_0 = 0$ and $S_k = \xi_1 + \ldots + \xi_k$, $1 \leq k \leq n$. Let $F(x; k) = \mathsf{P}\{S_k \leq x\}$. As a generalization of the result stated in the previous problem, prove that the sequence $X = (X_k, \mathcal{F}_k)_{0 \leq k \leq n}$, given by

$$X_k = F(a - S_k, n - k) \quad \text{and} \quad \mathcal{F}_k = \sigma(\xi_1, \ldots, \xi_k),$$

is a martingale. (An application of this property can be found in Problem 7.2.12.)

**Problem 7.1.19.** Let $\xi_1, \xi_2, \ldots$ be any sequence of independent and identically distributed random variables with (shared) distribution function $F = F(x)$, $x \in \mathbb{R}$, and let

$$F_n(x; \omega) = \frac{1}{n} \sum_{k=1}^{n} I(\xi_k(\omega) \leq x), \quad x \in \mathbb{R}, \quad n \geq 1,$$

be the associated sequence of empirical distribution functions (see [P §3.13]). By using the result in Problem 7.1.5, prove that, for every fixed $x \in \mathbb{R}$, the sequence $(Y_n(x), \mathcal{G}_n(x))_{n \geq 1}$, given by $Y_n(x) = F_n(x; \omega) - F(x)$, $\mathcal{G}_n(x) = \sigma(Y_n(x), Y_{n+1}(x), \ldots)$, is a martingale.

**Problem 7.1.20.** Suppose that $X = (X_n, \mathcal{F}_n)_{n \geq 0}$ and $Y = (Y_n, \mathcal{F}_n)_{n \geq 0}$ are two submartingales.

  (a) Prove that $X \vee Y = (X_n \vee Y_n, \mathcal{F}_n)_{n \geq 0}$ is also a submartingale.

  (b) Can one claim that the following two sequences are submartingales:

$$X + Y = (X_n + Y_n, \mathcal{F}_n)_{n \geq 0}, \qquad XY = (X_n Y_n, \mathcal{F}_n)_{n \geq 0}?$$

If yes, explain under what conditions, if not, explain why?

  (c) Answer the analogous questions in the case where $X$ and $Y$ are martingales and also in the case where $X$ and $Y$ are supermartingales.

**Problem 7.1.21.** Let $\xi_1, \xi_2, \ldots$ be any infinite sequence of *exchangeable* random variables (i.e., random variables with the property that, for every $n \geq 1$, the probability distribution of the vector $(\xi_1, \ldots, \xi_n)$ coincides with the probability distribution of the vector $(\xi_{\pi_1}, \ldots, \xi_{\pi_n})$, for any permutation, $(\pi_1, \ldots, \pi_n)$, of the set $(1, \ldots, n)$—for an equivalent definition see Problem 2.5.4). Suppose that $\mathsf{E}\xi_1 < \infty$, $S_n = \xi_1 + \ldots + \xi_n$ and let $\mathcal{G}_n = \sigma\left(\frac{S_n}{n}, \frac{S_{n+1}}{n+1}, \ldots\right)$.

As a generalization of [ P §1.11,  Example 4], prove that one has

$$E\left(\frac{S_n}{n} \mid \mathcal{G}_{n+1}\right) = \frac{S_{n+1}}{n+1} \quad (\text{P-a.e.}), \quad n \geq 1,$$

i.e., the sequence $\left(\frac{S_n}{n}, \mathcal{G}_n\right)_{n \geq 1}$ forms a *reverse martingale*.

**Problem 7.1.22.** Prove that any reverse martingale is automatically uniformly integrable.

**Problem 7.1.23.** The $\sigma$-algebra $\mathcal{F}_\tau$, associated with the Markov time $\tau$, is defined as the collection of sets

$$\{A \in \mathcal{F} : A \cap \{\tau = n\} \in \mathcal{F}_n \text{ for all } n \geq 0\}.$$

Why can't one define this $\sigma$-algebra as $\mathcal{F}_\tau \overset{\text{def}}{=} \sigma(\mathcal{F}_n : n \leq \tau)$?

**Problem 7.1.24.** If $X = (X_n, \mathcal{F}_n)_{n \geq 1}$ is a martingale, then, for every sub-sequence, $(n_k) \subseteq (n)$, one can claim that $(X_{n_k}, \mathcal{F}_{n_k})_{k \geq 1}$ is also a martingale. By providing appropriate examples, prove that, in general, this property may not hold for *local* martingales.

**Problem 7.1.25.** In martingale theory, a uniformly integrable supermartingale $\Pi = (\Pi_n, \mathcal{F}_n)_{n \geq 0}$, with the property $\Pi_n(\omega) \to 0$ as $n \to \infty$, for every $\omega \in \Omega$ (point-wise convergence to 0), is called *potential*.

Suppose that $\Pi = (\Pi_n, \mathcal{F}_n)_{n \geq 0}$ is a potential and let $\mathcal{F}_{-1} = \mathcal{F}_0$. Prove that there is a unique predictable and non-decreasing sequence $A = (A_n, \mathcal{F}_{n-1})_{n \geq 0}$, starting from 0, i.e., with $A_0 = 0$, for which one can write

$$\Pi_n = E(A_\infty - A_n \mid \mathcal{F}_n), \quad n \geq 0.$$

**Problem 7.1.26.** Let $X = (X_n, \mathcal{F}_n)_{n \geq 0}$ be a supermartingale. Prove that the following conditions are equivalent:

(i) There is a submartingale, $Y = (Y_n, \mathcal{F}_n)_{n \geq 0}$, for which one can claim that $X_n \geq Y_n$ (P-a.e.), for all $n \geq 0$;

(ii) There is a unique *Riesz decomposition* of the form:

$$X_n = M_n + \Pi_n, \quad n \geq 0,$$

in which $M = (M_n, \mathcal{F}_n)_{n \geq 0}$ is a martingale and $\Pi = (\Pi_n, \mathcal{F}_n)_{n \geq 0}$ is a potential.

**Problem 7.1.27.** Let $X = (X_n, \mathcal{F}_n)_{n \geq 0}$ be any submartingale. Prove that one can find a non-negative martingale, $M = (M_n, \mathcal{F}_n)_{n \geq 0}$, with the following properties:

$$X_n^+ \leq M_n, \quad n \geq 0, \quad \text{and} \quad \sup_n E X_n^+ = \sup_n E M_n.$$

*Hint.* Use the fact that $X^+ = (X_n^+, \mathcal{F}_n)_{n \geq 0}$ is also a submartingale and set $M_n = \lim_{m \to \infty} E(X_{n+m}^+ \mid \mathcal{F}_n)$.

**Problem 7.1.28.** Suppose that the probability space $(\Omega, \mathcal{F}, \mathsf{P})$ is endowed with the filtration $\mathcal{F} = (\mathcal{F}_n)_{n\geq 0}$, let $\sigma$ and $\tau$ be any two Markov times (for $\mathcal{F}$) with the property $\sigma(\omega) \leq \tau(\omega)$ for every $\omega \in \Omega$, and let $A_n = \{\omega : \sigma(\omega) < n \leq \tau(\omega)\}$, $n \geq 1$. Prove that $A_n \in \mathcal{F}_{n-1}$ for every $n \geq 1$. In other words, the sequence $(X_n)_{n\geq 1}$, given by

$$X_n = \begin{cases} 1, & \text{if } \sigma(\omega) < n \leq \tau(\omega), \\ 0 & \text{otherwise}, \end{cases}$$

is predictable, in that $X_n$ is $\mathcal{F}_{n-1}$-measurable for every $n \geq 1$.

**Problem 7.1.29.** (On [P §7.1, Theorem 2].) Let $X = (X_n, \mathcal{F}_n)_{n\geq 0}$ be any submartingale with Doob-decomposition $X_n = m_n + A_n$, $n \geq 0$, where $A_0 = 0$ and therefore $m_0 = X_0$. Prove that if $\{X_0, X_1, \ldots\}$ is a uniformly integrable family, then $\mathsf{E}A_\infty < \infty$ and the family $\{m_0, m_1, \ldots\}$ is also uniformly integrable.

**Problem 7.1.30.** Suppose that $M = (M_n, \mathcal{F}_n)_{n\geq 0}$ is a square integrable martingale. Prove that

$$\sup_n \mathsf{E}M_n^2 < \infty \iff \sum_{k\geq 1} \mathsf{E}(M_k - M_{k-1})^2 < \infty.$$

**Problem 7.1.31.** Let $\tau = \tau(\omega)$ be any Markov time for the filtration $(\mathcal{F}_n)_{n\geq 0}$ and suppose that $f = f(n)$ is a non-decreasing function of $n \in \mathbb{N} = \{0, 1, 2, \ldots\}$, chosen so that $f(n) \geq n$. Prove that $\widetilde{\tau}(\omega) = f(\tau(\omega))$ is also a Markov time.

**Problem 7.1.32.** Consider the sequence $X = (X_n, \mathcal{F}_n)$ and suppose that this sequence is a martingale with respect to the probability measure $\mathsf{P}$. Then suppose that $\mathsf{Q}$ is another probability measure that is equivalent to $\mathsf{P}$ ($\mathsf{Q} \sim \mathsf{P}$). Prove by way of example that the sequence $X = (X_n, \mathcal{F}_n)$ is not necessarily a martingale relative to the measure $\mathsf{Q}$.

**Problem 7.1.33.** According to [P §7.1, Example 5], if $X = (X_n, \mathcal{F}_n)$ is a submartingale and $g = g(x)$ is some convex and *non-decreasing* function with the property $\mathsf{E}|g(X_n)| < \infty$, $n \geq 0$, then the sequence $(g(X_n), \mathcal{F}_n)$ is also a submartingale. Give an example of a submartingale $(X_n)$ and a function $g = g(x)$, which is convex but fails to be non-decreasing, for which $(g(X_n), \mathcal{F}_n)$ is not a submartingale.

## 7.2  Invariance of the Martingale Property Under Random Time-Change

**Problem 7.2.1.** Prove that [P §7.2, Theorem 1] remains valid in the case of submartingales, provided that condition [P §7.2, (4)] is replaced with

$$\varliminf_{n\to\infty} \int_{\{\tau_2 > n\}} X_n^+ \, d\mathsf{P} = 0.$$

*Hint.* The proof is essentially the same as in [P §7.2, Theorem 1]. One only has to notice that the relation $X_m \le X_m^+$ implies the following chain of inequalities:

$$\int_{B \cap \{\tau_2 \ge n\}} X_{\tau_2} \, d\,P \ge \varlimsup_{m \to \infty} \left[ \int_{B \cap \{\tau_2 \ge n\}} X_n \, d\,P - \int_{B \cap \{\tau_2 > m\}} X_m \, d\,P \right]$$

$$\ge \int_{B \cap \{\tau_2 \ge n\}} X_n \, d\,P - \varlimsup_{m \to \infty} \int_{B \cap \{\tau_2 > m\}} X_m^+ \, d\,P.$$

**Problem 7.2.2.** Let $X = (X_n, \mathscr{F}_n)_{n \ge 0}$ be any square integrable martingale, with $EX_0 = 0$, let $\tau$ be a stopping time, and suppose that

$$\lim_{n \to \infty} \int_{\{\tau > n\}} X_n^2 \, d\,P = 0.$$

Prove that

$$EX_\tau^2 = E\langle X \rangle_\tau \left( = E \sum_{j=0}^{\tau} (\Delta X_j)^2 \right),$$

where $\Delta X_0 = X_0$, $\Delta X_j = X_j - X_{j-1}$, $j \ge 1$.

*Hint.* In order to prove the inequality

$$EX_\tau^2 \le E \sum_{j=0}^{\tau} (\Delta X_j)^2,$$

use [P §7.2, Theorem 1] and Fatou's lemma ($E \varliminf_N X_{\tau \wedge N}^2 \le \varliminf_N ES_{\tau \wedge N}^2$). To prove the inequality in the opposite direction, observe that

$$EX_\tau^2 \ge EX_{\tau \wedge N}^2 = E \sum_{j=0}^{\tau \wedge N} (\Delta X_j)^2,$$

and use Fatou's lemma again.

**Problem 7.2.3.** Prove that for every martingale, or, for every non-negative submartingale, $X = (X_n, \mathscr{F}_n)_{n \ge 0}$, and for every stopping time $\tau$, one has

$$E|X_\tau| \le \lim_{n \to \infty} E|X_n|.$$

*Hint.* Use the fact that $|X|$ is a submartingale and that, by [P §7.2, Theorem 1], $E|X_{\tau \wedge N}| \le E|X_N|$, for every $N \ge 1$. Consequently, $\varliminf_N E|X_{\tau \wedge N}| \le \varliminf_N E|X_N|$. The proof can now be completed by using Fatou's lemma.

**Problem 7.2.4.** Let $X = (X_n, \mathscr{F}_n)_{n \ge 0}$ be a supermartingale, and suppose that there is a random variable $\xi$, with $E|\xi| < \infty$, for which one can write $X_n \ge E(\xi \mid \mathscr{F}_n)$

(P-a. e.), for every $n \geq 0$. Prove that if $\tau_1$ and $\tau_2$ are two stopping times with $P\{\tau_1 \leq \tau_2\} = 1$, then

$$X_{\tau_1} \geq E(X_{\tau_2} \mid \mathscr{F}_{\tau_1}) \quad \text{(P-a. e.)}.$$

*Hint.* By using the result from [P §7.2, Theorem 1], verify the relations $E|X_{\tau_1}| < \infty$, $E|X_{\tau_2}| < \infty$ and

$$\lim_n \int_{\{\tau_2 > n\}} |X_n| \, dP = 0.$$

**Problem 7.2.5.** Let $\xi_1, \xi_2, \ldots$ be any sequence of independent random variables with $P\{\xi_i = 1\} = P\{\xi_i = -1\} = 1/2$, and let $a$ and $b$ be any two positive numbers with $b > a$. Given any $n \geq 1$, set

$$X_n = a \sum_{k=1}^{n} I(\xi_k = +1) - b \sum_{k=1}^{n} I(\xi_k = -1)$$

and let

$$\tau = \inf\{n \geq 1 : X_n \leq -r\}, \quad r > 0.$$

Prove that $Ee^{\lambda \tau} < \infty$, for $\lambda \leq \alpha_0$, and that $Ee^{\lambda \tau} = \infty$, for $\lambda > \alpha_0$, where

$$\alpha_0 = \frac{b}{a+b} \ln \frac{2b}{a+b} + \frac{a}{a+b} \ln \frac{2a}{a+b}.$$

**Problem 7.2.6.** Suppose that $\xi_1, \xi_2, \ldots$ is some sequence of independent random variables with $E\xi_i = 0$ and $D\xi_i = \sigma_i^2$, and let $S_n = \xi_1 + \cdots + \xi_n$ and $\mathscr{F}_n^\xi = \sigma\{\xi_1, \ldots, \xi_n\}$, for $n \geq 1$. Prove the following generalization of Wald's identities [P §7.2, (13) and (14)]: if $E\sum_{j=1}^{\tau} E|\xi_j| < \infty$, then $ES_\tau = 0$, and, if $E\sum_{j=1}^{\tau} E\xi_j^2 < \infty$, then

$$ES_\tau^2 = E\sum_{j=1}^{\tau} \xi_j^2 = E\sum_{j=1}^{\tau} \sigma_j^2.$$

**Problem 7.2.7.** Let $X = (X_n, \mathscr{F}_n)_{n \geq 1}$ be a square integrable martingale and let $\tau$ be any stopping time for $(\mathscr{F}_n)$. Prove that

$$EX_\tau^2 \leq E\sum_{n=1}^{\tau} (\Delta X_n)^2.$$

In addition, prove that if

$$\lim_{n \to \infty} E(X_n^2 I(\tau > n)) < \infty, \quad \text{or} \quad \lim_{n \to \infty} E(|X_n| I(\tau > n)) = 0,$$

then $E(\Delta X_\tau)^2 = E\sum_{n=1}^{\tau} X_n^2$.

**Problem 7.2.8.** Let $X = (X_n, \mathscr{F}_n)_{n \geq 1}$ be any submartingale and let $\tau_1 \leq \tau_2 \leq \ldots$ be stopping times for $(\mathscr{F}_n)$, such that the expectations $\mathsf{E}X_{\tau_m}$ are well defined and

$$\lim_{n \to \infty} \mathsf{E}(X_n^+ I(\tau_m > n)) = 0, \quad m \geq 1.$$

Prove that the sequence $(X_{\tau_m}, \mathscr{F}_{\tau_m})_{m \geq 1}$ is a submartingale. (As usual, we define $\mathscr{F}_{\tau_m} = \{A \in \mathscr{F} : A \cap \{\tau_m = j\} \in \mathscr{F}_j, j \geq 1\}$.)

**Problem 7.2.9.** Let $X = (X_n, \mathscr{F}_n)_{n \geq 0}$ be a non-negative supermartingale and let $\tau_0 \leq \tau_1 \leq \ldots$ be stopping times for $(\mathscr{F}_n)$. Show that the sequence $(X_{\tau_n}, \mathscr{F}_n)_{n \geq 0}$ is also a supermartingale.

**Problem 7.2.10.** As an extension of the *elementary theorem* in renewal theory— see [P §7.2, 4]—prove that (under the assumption $\mathsf{D}\sigma_1 < \infty$ and with the notation $a = (\mathsf{E}\sigma_1)^{-1}$) one must have

$$\frac{\mathsf{D}N_t}{t} \to a^3 \mathsf{D}\sigma_1 \quad \text{as } t \to \infty.$$

Furthermore, the central limit theorem holds:

$$\mathsf{P}\left\{ \frac{N_t - at}{\sqrt{a^3 \mathsf{D}\sigma_1 t}} \leq x \right\} \to \Phi(x) \quad \text{as } t \to \infty.$$

**Problem 7.2.11.** Let $\xi_1, \xi_2, \ldots$ be any sequence of independent and identically distributed random variables, let $S_n = \xi_1 + \ldots + \xi_n, n \geq 1$, and let

$$\tau = \inf\{n \geq 1 : S_n > 0\}$$

(as usual, we set $\tau = \infty$, if $S_n \leq 0$ for all $n \geq 1$).
    Prove that if $\mathsf{E}\xi_1 = 0$ then $\mathsf{E}\tau = \infty$.

**Problem 7.2.12.** By using the martingale property of the sequence $X = (X_k, \mathscr{F}_k)_{0 \leq k \leq n}$ from Problem 7.1.18, and also the property $\mathsf{E}X_0 = \mathsf{E}X_{\tau_a}$ (see [P §7.2, Corollary 1]), where $\tau_a = \min\{0 \leq k \leq n : S_k > a\}, a > 0$ (with the understanding that $\tau_a = n + 1$, if $S_k \leq a$ for all $0 \leq k \leq n$), prove the inequality (see [P §4.4, Lemma 1])

$$\mathsf{P}\left\{ \max_{0 \leq k \leq n} S_k > a \right\} \leq 2\,\mathsf{P}\{S_n > a\}.$$

**Problem 7.2.13.** As an extension of the statements in [P §7.2, Theorems 1 and 2], prove the following result: Consider the martingale $X = (X_n, \mathscr{F}_n)$ and let $\tau$ be any stopping time with $\mathsf{P}\{\tau < \infty\} = a$, for which $\mathsf{E}|X_\tau| < \infty$ and $\lim_{n \to \infty} \mathsf{E}[|X_n|I(\tau > n)] = 0$. Then:

$$\lim_{n \to \infty} \mathsf{E}[|X_\tau|I(\tau > n)] = 0;$$

$$\mathsf{E}|X_\tau - X_{\tau \wedge n}| \to 0 \quad \text{as } n \to \infty;$$

and $\mathsf{E}X_\tau = \mathsf{E}X_0$.

**Problem 7.2.14.** *(On [ P §7.2, Theorems 1 and 2].)* Suppose that $\tau_1$ and $\tau_2$ are two finite stopping times with $P\{\tau_1 \leq \tau_2\} = 1$, and let $X = (X_n)_{n \geq 0}$ be some martingale (all defined on the same probability space). Prove that if

$$\mathsf{E} \sup_{n \leq \tau_2} |X_n| < \infty, \tag{$*$}$$

then $\mathsf{E}(X_{\tau_2} \mid \mathscr{F}_{\tau_1}) = X_{\tau_1}$ (P-a. e.).

Hint. Use the fact that condition $(*)$ implies that the family of random variables $\{|X_{\tau_2 \wedge 0}|, |X_{\tau_2 \wedge 1}|, \ldots\}$ is uniformly integrable.

**Problem 7.2.15.** In the context of [ P §7.2, Example 1], consider the stopping time $\tau$ defined in [ P §7.2, (16)], and prove that $\mathsf{E}\tau^p < \infty$, for every $p \geq 1$.

**Problem 7.2.16.** Give an example of a martingale $X = (X_n, \mathscr{F}_n)_{n \geq 0}$, and a stopping time $\tau$, with the property that (see [ P §7.2, Theorem 1]) the condition

$$\lim_n \int_{\{\tau > n\}} |X_n| \, d\mathsf{P} = 0$$

holds, but the condition $\mathsf{E}|X_\tau| < \infty$ fails, i.e., $\mathsf{E}|X_\tau| = \infty$.

**Problem 7.2.17.** Let $M = (M_n, \mathscr{F}_n)_{n \geq 0}$ be any martingale and, given any $N \geq 1$, set $\tau_N = \inf\{m \geq 0 : |M_m| \geq N\}$, with the understanding that $\inf \emptyset = \infty$. Prove that the martingale $M$ is uniformly integrable if and only if

$$\lim_N \mathsf{E}|M_{\tau_N}| I(\tau_N < \infty) = 0.$$

**Problem 7.2.18.** *(On [ P §7.2, Examples 1 and 2].)* Let $\xi_1, \xi_2, \ldots$ be any sequence of independent and symmetric Bernoulli random variables ($P\{\xi_i = 1\} = P\{\xi_i = -1\} = 1/2$, for $i \geq 1$). Consider the stopping time

$$\tau = \inf\{n \geq 0 : S_n = 1\},$$

where $S_0 = 0$ and $S_n = \xi_1 + \ldots + \xi_n$ (as usual, we suppose that $\inf \emptyset = \infty$).
   (a) Prove that, for every $\lambda \in \mathbb{R}$, the sequence $(X_n^\lambda)_{n \geq 0}$, given by

$$X_n^\lambda = \frac{e^{\lambda S_n}}{(\cosh \lambda)^n},$$

forms a martingale. By using this property, prove that $P\{\tau < \infty\} = 1$, $\mathsf{E}\tau = \infty$ and

$$\mathsf{E}(\cosh \lambda)^{-\tau} = e^{-\lambda a}, \quad \text{for every } \lambda \geq 0$$

(comp. with [ P §7.2, (18)]).
   (b) With $\alpha = 1/\cosh \lambda$, the above formula implies that

$$\bullet \qquad \mathsf{E}\alpha^\tau = \sum_{n \geq 1} \alpha^n P\{\tau = n\} = \frac{1}{\alpha}\left[1 - \sqrt{1 - \alpha^2}\right]$$

(see also Problem 8.8.19). By using the last relation, prove that

$$P\{\tau = 2n - 1\} = (-1)^{n+1}C^n_{1/2} \, ,$$

where

$$C^n_X = \frac{X(X-1)\ldots(X-n+1)}{n!}$$

(see Problem 1.2.22).

(c) Let $I = \inf\{S_n : n \leq \tau\}$. Prove that for every $k \geq 0$ one has

$$P\{I \leq -k\} = \frac{1}{k+1}.$$

(d) Let $\tau_k = \inf\{n \geq 0 : S_n = 1 \text{ or } S_n = -k\}$. Show that $\tau_k \to \tau$ (P-a. e.) and $S_{\tau_k} \to S_\tau$ as $k \to \infty$ (P-a. e.), and yet $ES_{\tau_k} \not\to ES_\tau$ (in fact, $ES_{\tau_k} = 0$, while $ES_\tau = 1$). Explain why the convergence of the expected values does not hold ($ES_{\tau_k}$ does not converge to $ES_\tau$ as $k \to \infty$), in spite of the fact that $S_{\tau_k} \to S_\tau$ (P-a. e.).

**Problem 7.2.19.** The argument of [ P §7.2,  Theorem 2] is based on the assumption that the expectation of the stopping time $\tau$ is *finite* (i.e., $E\tau < \infty$). Prove that if, for some $0 < \varepsilon < 1$ and some integer $N$, one can write

$$P(\tau \leq n + N \mid \mathscr{F}_n) > \varepsilon \quad \text{(P-a. e.)} \quad \text{for every } n \geq 1 \, ,$$

then one can claim $E\tau < \infty$.

*Hint.* Show by induction that $P\{\tau \geq kN\} \leq (1-\varepsilon)^k, k \geq 1$.

**Problem 7.2.20.** Let $m(t)$ denote the renewal function, introduced in [ P §7.2,  4 ]. The elementary theorem of renewal theory says that $m(t)/t \to 1/\mu$ as $t \to \infty$. The next two statements refine this claim further.

(a) Suppose that the renewal process $N = (N_t)_{t \geq 0}$ lives on a lattice of size $d$, i.e., for some fixed $d > 0$ one can claim that the the distribution of the random variable $\sigma_1$ is supported by the set $\{0, d, 2d, \ldots\}$. Then (Kolmogorov, 1936)

$$\sum_{k=1}^{\infty} P\{T_k = nd\} \to \frac{d}{\mu} \quad \text{as } n \to \infty.$$

(b) If there is no $d > 0$, for which one can claim that the renewal process $N = (N_t)_{t \geq 0}$ lives on a lattice of size $d$, then (Blackwell, 1948)

$$\sum_{k=1}^{\infty} P\{t < T_k \leq t + h\} \to \frac{h}{\mu} \quad \text{as } t \to \infty, \qquad (*)$$

for every $h > 0$. (Note that the sum in $(*)$ gives $m(t+h) - m(t)$.)

Argue that the above two statements are plausible, or, better yet, just prove them.

*Hint.* With regard to (a), one must become familiar with the proof [ $\underline{P}$ §8.6, Theorem 2].

**Problem 7.2.21.** Let $\xi_1, \xi_2, \ldots$ be any sequence of independent Bernoulli random variables, with $P\{\xi_i = 1\} = p$, $P\{\xi_i = -1\} = q$, $p + q = 1$, $i \geq 1$. Given some integers $x$, $a$ and $b$, with $a \leq 0 \leq b$, set $S_n(x) = x + \xi_1 + \ldots + \xi_n$ and let

$$\tau_a(x) = \inf\{n \geq 0 : S_n(x) \leq a\},$$

$$\tau^b(x) = \inf\{n \geq 0 : S_n(x) \geq b\},$$

$$\tau_a^b(x) = \inf\{n \geq 0 : S_n(x) \leq a \text{ or } S_n(x) \geq b\}.$$

Prove that:

$$P\{\tau_a(x) < \infty\} = \begin{cases} 1, & \text{if } p \leq q \text{ and } x > a, \\ (q/p)^{x-a}, & \text{if } p > q \text{ and } x > a; \end{cases}$$

$$P\{\tau^b(x) < \infty\} = \begin{cases} 1, & \text{if } p \geq q \text{ and } x < b, \\ (p/q)^{b-x}, & \text{if } p < q \text{ and } x < b; \end{cases}$$

$$P\{\tau_a^b(x) < \infty\} = 1, \qquad a \leq x \leq b;$$

and that for $a \leq x \leq b$

$$\mathsf{E}\,\tau_a^b(x) = \frac{x-a}{q-p} - \frac{b-a}{q-p}\left[\frac{(q/p)^x - (q/p)^a}{(q/p)^b - (q/p)^a}\right], \quad \text{if} \quad p \neq q,$$

$$\mathsf{E}\,\tau_a^b(x) = (b-a)(x-a), \qquad\qquad\qquad \text{if} \quad p = q = 1/2.$$

**Problem 7.2.22.** Let $\xi_1, \xi_2, \ldots$ be any sequence of independent and identically distributed random variables, with values in the set $\{-1, 0, 1, \ldots\}$, and with expected value $\mu < 0$. Let $S_0 = 1$, $S_n = 1 + \xi_1 + \ldots + \xi_n$, $n \geq 1$, and let $\tau = \inf\{n \geq 1 : S_n = 0\}$. Prove that $\mathsf{E}\,\tau = \frac{1}{|\mu|}$.

## 7.3   Fundamental Inequalities

**Problem 7.3.1.** Let $X = (X_n, \mathscr{F}_n)_{n \geq 0}$ be any non-negative submartingale and let $V = (V_n, \mathscr{F}_{n-1})_{n \geq 0}$ be any predictable sequence (as usual, we set $\mathscr{F}_{-1} = \mathscr{F}_0$), with $0 \leq V_{n+1} \leq V_n \leq C$ (P-a. e.), where $C$ is some constant. Prove the following generalization of the inequality in [ $\underline{P}$ §7.3, (1)]: for every fixed $\lambda > 0$, one has

$$\lambda\,\mathsf{P}\left\{\max_{0 \leq k \leq n} V_k X_k \geq \lambda\right\} + \int_{\max_{0 \leq k \leq n} V_k X_k < \lambda} V_n X_n \, d\,\mathsf{P} \leq \sum_{k=0}^{n} \mathsf{E} V_k \Delta X_k,$$

with the understanding that $\Delta X_0 = X_0$.

**Problem 7.3.2.** Prove the following result, known as *Krickeberg's decomposition*: every martingale $X = (X_n, \mathscr{F}_n)_{n \geq 0}$, that has the property $\sup \mathsf{E}|X_n| < \infty$, can be written as the difference between two *non-negative* martingales.

**Problem 7.3.3.** Let $\xi_1, \xi_2, \ldots$ be any sequence of independent random variables, let $S_n = \xi_1 + \cdots + \xi_n$, and let $S_{m,n} = \sum_{j=m+1}^{n} \xi_j$. Prove the following relation, which is known as the *Ottaviani inequality*,

$$\mathsf{P}\left\{ \max_{1 \leq j \leq n} |S_j| > 2t \right\} \leq \frac{\mathsf{P}\{|S_n| > t\}}{\min_{1 \leq j \leq n} \mathsf{P}\{|S_{j,n}| \leq t\}}, \quad t > 0,$$

and conclude that (under the assumption $\mathsf{E}\xi_i = 0$, for $i \geq 1$) one must have

$$\int_0^\infty \mathsf{P}\left\{ \max_{1 \leq j \leq n} |S_j| > 2t \right\} dt \leq 2\mathsf{E}|S_n| + 2 \int_{2\mathsf{E}|S_n|}^\infty \mathsf{P}\{|S_n| > t\} dt. \quad (*)$$

*Hint.* To establish the Ottaviani inequality, let $A = \left\{ \max_{1 \leq k \leq n} |S_k| \geq 2t \right\}$, and let

$$A_k = \{|S_i| < 2t, i = 1, \ldots, k - 1; |S_k| \geq 2t\}, \quad \text{for } 1 \leq k \leq n.$$

Then $A = \sum_{k=1}^{n} A_k$, and one can show that for any $t > 0$

$$\mathsf{P}(A)\left[ \min_{1 \leq j \leq n} \mathsf{P}\{|S_{j,n}| \leq t\} \right] = (\mathsf{P}(A_1) + \cdots + \mathsf{P}(A_n))\left[ \min_{1 \leq j \leq n} \mathsf{P}\{|S_{j,n}| \leq t\} \right]$$

$$\leq \mathsf{P}(A_1 \cap \{|S_n| > t\}) + \cdots + \mathsf{P}(A_n \cap \{|S_n| > t\}) \leq \mathsf{P}\{|S_n| > t\}.$$

In order to establish $(*)$ (under the assumption $\mathsf{E}\xi_i = 0$, for $i \geq 1$), one only has to show that

$$\int_0^\infty \mathsf{P}\left\{ \max_{1 \leq j \leq n} |S_j| > 2t \right\} dt$$

$$\leq \int_0^{2\mathsf{E}|S_n|} dt + \int_{2\mathsf{E}|S_n|}^\infty \frac{\mathsf{P}\{|S_n| > t\}}{1 - \max_{1 \leq j \leq n} \mathsf{P}\{|S_{j,n}| > t\}} dt,$$

and that for $t \geq 2\mathsf{E}|S_n|$

$$1 - \max_{1 \leq j \leq n} \mathsf{P}\{|S_{j,n}| > t\} \geq 1 - \max_{1 \leq j \leq n} \mathsf{P}\{|S_{j,n} > 2\mathsf{E}|S_n|\}$$

$$\geq 1 - \max_{1 \leq j \leq n} \frac{\mathsf{E}|S_{j,n}|}{2\mathsf{E}|S_n|} \geq 1 - \frac{1}{2} = \frac{1}{2}.$$

**Problem 7.3.4.** Let $\xi_1, \xi_2, \ldots$ be any sequence of independent random variables, with $\mathsf{E}\xi_i = 0$, $i \geq 1$. By using $(*)$ in Problem 7.3.3, prove that following stronger version of the inequality in [P §7.3, (10)]:

$$\mathsf{E}S_n^* \leq 8\,\mathsf{E}|S_n|.$$

**Problem 7.3.5.** Prove the formula in [ $\underline{P}$ §7.3, (16)].

**Problem 7.3.6.** Prove the inequality in [ $\underline{P}$ §7.3, (19)].

**Problem 7.3.7.** Consider the $\sigma$-algebras $\mathscr{F}_0, \ldots, \mathscr{F}_n$ with $\mathscr{F}_0 \subseteq \mathscr{F}_1 \subseteq \cdots \subseteq \mathscr{F}_n$, and let the events $A_k \in \mathscr{F}_k, k = 1, \ldots, n$, be arbitrarily chosen. By using [ $\underline{P}$ §7.3, (22)], prove *Dvoretzky's inequality*:

$$\mathsf{P}\left\{ \bigcup_{k=1}^{n} A_k \right\} \leq \lambda + \mathsf{P}\left\{ \sum_{k=1}^{n} \mathsf{P}(A_k \mid \mathscr{F}_{k-1}) > \lambda \right\}, \quad \text{for every } \lambda > 0.$$

*Hint.* Define $X_k = I_{A_k}, k = 1, \ldots, n$, and notice that

$$X_n^* = \max_{1 \leq k \leq n} |I_{A_k}| = I_{\bigcup_{k=1}^{n} A_k}.$$

If $B_n = \sum_{k=1}^{n} \mathsf{P}(A_k \mid \mathscr{F}_{k-1})$, then [ $\underline{P}$ §7.3, (22)] implies that

$$\mathsf{P}\{X_N^* \geq 1\} \leq \mathsf{E}(B_n \wedge \varepsilon) + \mathsf{P}\{B_n \geq \varepsilon\},$$

from where the required inequality easily follows.

**Problem 7.3.8.** Let $X = (X_n)_{n \geq 1}$ be any square integrable martingale, and let $(b_n)_{n \geq 1}$ be any non-decreasing sequence of positive real numbers. Prove *Hájek–Rényi's inequalitiy*:

$$\mathsf{P}\left\{ \max_{1 \leq k \leq n} \left| \frac{X_k}{b_k} \right| \geq \lambda \right\} \leq \frac{1}{\lambda^2} \sum_{k=1}^{n} \frac{\mathsf{E}(\Delta X_k)^2}{b_n^2}, \quad \Delta X_k = X_k - X_{k-1}, \ X_0 = 0.$$

**Problem 7.3.9.** Let $X = (X_n)_{n \geq 1}$ be any submartingale and let $g(x)$ be any increasing function, which is non-negative and convex. Prove that, for every positive $h$, and for every real $x$, one has

$$\mathsf{P}\left\{ \max_{1 \leq k \leq n} X_k \geq x \right\} \leq \frac{\mathsf{E}g(hX_n)}{g(hx)}.$$

In particular, one has the following exponential analog of Doob's inequality:

$$\mathsf{P}\left\{ \max_{1 \leq k \leq n} X_k \geq x \right\} \leq e^{-hx} \, \mathsf{E}e^{hX_n}.$$

(Comp. this result with the exponential analog of Kolmogorov's inequality, established in Problem 4.2.23.)

**Problem 7.3.10.** Let $\xi_1, \xi_2, \ldots$ be independent random variables, with $\mathsf{E}\xi_n = 0$ and $\mathsf{E}\xi_n^2 = 1$, for $n \geq 1$, and let $\tau = \inf\{n \geq 1 : \sum_{i=1}^{n} \xi_i > 0\}$, with the understanding that $\inf \varnothing = \infty$. Prove, that $\mathsf{E}\tau^{1/2} < \infty$.

**Problem 7.3.11.** Let $\xi = (\xi_n)_{n\geq 1}$ be any martingale difference. Prove that, for every $1 < p \leq 2$, one can find a constant $C_p$, for which the following inequality is in force:

$$\mathsf{E}\sup_{n\geq 1}\left|\sum_{j=1}^{n}\xi_j\right|^p \leq C_p\sum_{j=1}^{\infty}\mathsf{E}|\xi_j|^p.$$

**Problem 7.3.12.** Let $X = (X_n)_{n\geq 1}$ be any martingale, with $\mathsf{E}X_n = 0$ and $\mathsf{E}X_n^2 < \infty = 1$, for any $n \geq 1$. As a generalization of the inequality established in Problem 4.2.5, prove that, for every fixed $n \geq 1$, one has

$$\mathsf{P}\left\{\max_{1\leq k\leq n} X_k \geq \varepsilon\right\} \leq \frac{\mathsf{E}X_n^2}{\varepsilon^2 + \mathsf{E}X_n^2}, \quad \text{for every } \varepsilon \geq 0.$$

**Problem 7.3.13.** Let $\xi_1, \xi_2, \ldots$ be any sequence of independent random variables, with $\mathsf{P}\{\xi_n = 1\} = p$ and $\mathsf{P}\{\xi_n = -1\} = q$, where $p + q = 1, 0 < p < 1$, and let $S_0 = 0, S_n = \xi_1 + \ldots + \xi_n$.

Prove that the sequence $((q/p)^{S_n})_{n\geq 0}$ is a martingale and, if $p < q$, then the following *maximal inequality* is in force:

$$\mathsf{P}\left\{\sup_{n\geq 0} S_n \geq k\right\} \leq \left(\frac{p}{q}\right)^k.$$

(Note that the above inequality is trivial if $p \geq q$.)
    In addition, prove that when $p < q$ one has

$$\mathsf{E}\sup_{n\geq 0} S_n \leq \frac{p}{q-p}.$$

In fact, the above relations are actually *identities*, which shows that, for $p < q$, the random variable $\sup_{n\geq 0} S_n$ has geometric distribution (see [P §2.3, Table 2]), i.e.,

$$\mathsf{P}\left\{\sup_{n\geq 0} S_n = k\right\} = \left(\frac{p}{q}\right)^k\left(1 - \frac{p}{q}\right), \quad k = 0, 1, 2, \ldots.$$

**Problem 7.3.14.** Let $M = (M_k, \mathscr{F}_k)_{0\leq k\leq n}$ be a martingale that starts from 0, i.e., $M_0 = 0$, and is such that $-a_k \leq \Delta M_k \leq 1 - a_k$, for $k = 1, \ldots, n$, where $\Delta M_k = M_k - M_{k-1}$ and $a_k \in [0, 1]$. As a generalization of the result established in Problem 4.5.14, prove that, for every $0 \leq x < q$, with $q = 1 - p$ and $p = \frac{1}{n}\sum_{k=1}^{n} a_k$, one has

$$\mathsf{P}\{M_n \geq nx\} \leq e^{n\psi(x)}$$

where $\psi(x) = \ln\left[\left(\frac{p}{p+x}\right)^{p+x}\left(\frac{q}{q-x}\right)^{q-x}\right]$.

*Hint.* Use the reasoning mentioned in the hint to Problem 4.5.14, and take into account the fact that

$$Ee^{hM_n} = E\big[e^{hM_{n-1}}E\big(e^{h\Delta M_n} \mid \mathscr{F}_{n-1}\big)\big]$$
$$\leq E\big[e^{hM_{n-1}}\big((1-a_n)e^{-ha_n} + a_n e^{h(1-a_n)}\big)\big].$$

**Problem 7.3.15.** Let $M = (M_k, \mathscr{F}_k)_{k\geq 0}$ be a martingale with $M_0 = 0$, chosen so that for some non-negative constants $a_k$ and $b_k$ one has

$$-a_k \leq \Delta M_k \leq b_k, \quad k \geq 1,$$

where $\Delta M_k = M_k - M_{k-1}$.

(a) Prove that, for every $x \geq 0$ and every $n \geq 1$, one has

$$P\{M_n \geq x\} \leq \exp\left\{ - \frac{2x^2}{\sum_{k=1}^{n}(a_k + b_k)^2} \right\},$$

$$P\{M_n \leq -x\} \leq \exp\left\{ - \frac{2x^2}{\sum_{k=1}^{n}(a_k + b_k)^2} \right\},$$

which, obviously, implies that

$$P\{|M_n| \geq x\} \leq 2\exp\left\{ - \frac{2x^2}{\sum_{k=1}^{n}(a_k + b_k)^2} \right\}.$$

(Comp. with the respective inequalities in [P §1.6] and [P §4.5].)

(b) Prove that if $a_k = a$ and $b_k = b$, for all $k \geq 1$ (and, therefore, $-a \leq \Delta M_k \leq b$, $k \geq 1$), then the following *maximal inequalities* is in force: for every $\beta > 0$ and every $x > 0$ one has

$$P\{M_n - \beta n \geq x \text{ for some } n\} \leq \exp\left\{ - \frac{8x\beta}{(a+b)^2} \right\}; \qquad (*)$$

furthermore, for every $\beta > 0$ and every integer $m \geq 1$, one has

$$P\{M_n \geq \beta n \text{ for some } n \geq m\} \leq \exp\left\{ - \frac{2m\beta^2}{(a+b)^2} \right\},$$

$$\qquad\qquad\qquad\qquad\qquad\qquad\qquad\qquad\qquad\qquad (**)$$

$$P\{M_n \leq -\beta n \text{ for some } n \geq m\} \leq \exp\left\{ - \frac{2m\beta^2}{(a+b)^2} \right\}.$$

(Comp. with the inequalities in [P §4.5].)

*Remark.* The inequalities in (a) are known as *Hoeffding–Azuma's inequalities.* The generalization given in (b) is due to S. M. Ross and can be found in the book [107].

*Hint.* (a) Given any $c > 0$, one can write

$$P\{M_n \geq x\} \leq e^{-cx} E e^{cM_n}.$$

Setting $V_n = e^{cM_n}$, we have $V_n = V_{n-1} e^{c \Delta M_n}$, so that

$$E(V_n \mid M_{n-1}) = V_{n-1} E(e^{c \Delta M_n} \mid M_{n-1}).$$

Iterating over $n$ and using the assumption $-a_k \leq \Delta M_k \leq b_k$, one can show that

$$P\{M_n \geq x\} \leq e^{-cx} \prod_{k=1}^{n} \frac{b_k e^{-ca_k} + a_k e^{-cb_k}}{a_k + b_k} \leq e^{-cx} \prod_{k=1}^{n} \exp\left\{ \frac{c^2}{8} (a_k + b_k)^2 \right\}.$$

Consequently,

$$P\{M_n \geq x\} \leq \exp\left\{ -cx + c^2 \sum_{k=1}^{n} \frac{(a_k + b_k)^2}{8} \right\},$$

and, since $c > 0$ is arbitrary, one can claim that

$$P\{M_n \geq x\} \leq \min_{c>0} \exp\left\{ -cx + c^2 \sum_{k=1}^{n} \frac{(a_k + b_k)^2}{8} \right\}$$

$$= \exp\left\{ -\frac{2x^2}{\sum_{k=1}^{n} (a_k + b_k)^2} \right\}.$$

(b) To prove (∗), introduce the variables

$$V_n = \exp\{c (M_n - x - \beta n)\}, \quad n \geq 0,$$

and notice that, with $c = 8\beta/(a + b)^2$, the sequence $(V_n)_{n \geq 0}$ is a non-negative supermartingale. Consequently, for every finite Markov time $\tau(K) (\leq K)$, one must have

$$E V_{\tau(K)} \leq E V_0 = e^{-8x\beta/(a+b)^2}.$$

With $\tau(K) = \min\{n : M_n \geq x + \beta n \ n = K\}$, this yields $P\{M_{\tau(K)} \geq x + \beta\tau(K) = P\{V_{\tau(K)} \geq 1\} \leq E V_{\tau(K)} \leq E V_0$, and, as a result,

$$P\{M_n \geq x + \beta n \text{ for some } n \leq K\} \leq \exp\left\{ \frac{-8x\beta}{(a + b)^2} \right\}.$$

Taking $K \to \infty$ gives the inequality in $(*)$, from which $(**)$ obtains with the following manipulation:

$$P\{M_n \geq \beta n \text{ for some } n \geq m\} \leq$$

$$\leq P\left\{M_n \geq \frac{m\beta}{2} + \frac{n\beta}{2} \text{ for some } n\right\}$$

$$\leq \exp\left\{-\frac{8(m\beta/2)(\beta/2)}{(a+b)^2}\right\} = \exp\left\{-\frac{2m\beta^2}{(a+b)^2}\right\}.$$

**Problem 7.3.16.** Let $M = (M_n, \mathscr{F}_n)_{n \geq 0}$ be any martingale and, given some $\lambda > 0$, let $\tau = \inf\{n \geq 0 : |M_n| > \lambda\}$, with the understanding $\inf \varnothing = \infty$. Prove that

$$P\{\tau < \infty\} \leq \lambda^{-1} \|M\|_1 ,$$

where $\|M\|_1 = \sup_n E|M_n|$.

**Problem 7.3.17.** With the notation adopted in the previous problem, prove that

$$\sum_{k=0}^{\infty} E|M_k - M_{k-1}|^2 I(\tau > k) \leq 2\lambda \|M\|_1 ,$$

where $M_{-1} = 0$.

**Problem 7.3.18.** Let $M = (M_n, \mathscr{F}_n)_{n \geq 0}$ be any martingale with $M_0 = 0$, and let $[M] = ([M]_n, \mathscr{F}_n)_{n \geq 1}$, stand for its quadratic variation, i.e., $[M]_n = \sum_{k=1}^{n} (\Delta M_k)^2$, where $\Delta M_k = M_k - M_{k-1}$. Prove that

$$E \sup_n |M_n| < \infty \iff E[M]_\infty^{1/2} < \infty . \tag{$*$}$$

*Remark.* The well known Burkholder–Davis–Gundi inequalities

$$A_p \|[M]_\infty^{1/2}\|_p \leq \|M_\infty^*\|_p \leq B_p \|[M]_\infty^{1/2}\|_p , \qquad p \geq 1,$$

in which $M_\infty^* = \sup_n |M_n|$ and $A_p$ and $B_p$ are universal constants (comp. with [$\underline{P}$ §7.3, (27), (30)]; see also [79]), can be viewed as an "$L^p$-refinement" of the property $(*)$.

**Problem 7.3.19.** Let $M = (M_k, \mathscr{F}_k)_{k \geq 1}$ be any martingale. Prove the *Burkholder's inequality*: for every $r \geq 2$ there is a universal constant $B_r$, such that

$$E|M_n|^r \leq B_r \left\{E\left[\sum_{k=1}^{n} E((\Delta M_k)^2 \mid \mathscr{F}_{k-1})\right]^{r/2} + E \sup_{1 \leq k \leq n} |\Delta M_k|^r\right\},$$

where $\Delta M_k = M_k - M_{k-1}, k \geq 1$, with $M_0 = 0$.

**Problem 7.3.20.** (*Moment inequalities I.*) Let $\xi_1, \xi_2, \ldots$ be any sequence of independent and identically distributed random variables with $E\xi_1 = 0$ and $E|\xi_1|^r < \infty$, for some $r \geq 1$, and let $S_n = \xi_1 + \ldots + \xi_n$, $n \geq 1$. Due to the second Marcinkiewicz-Zygmund inequality (see [P §7.3, (26)]),

$$E|S_n|^r \leq B_r E\left(\sum_{i=1}^{n} \xi_i^2\right)^{r/2},$$

for some universal constant $B_r$.

By using Minkowski's inequality (see [P §2.6]) with $r \geq 2$, and the $c_r$-inequality from Problem 2.6.72 with $r < 2$, prove that

$$E|S_n|^r \leq B_r \begin{cases} n E|\xi_1|^r, & 1 \leq r \leq 2, \\ n^{r/2} E|\xi_1|^r, & r > 2. \end{cases}$$

In particular, with $r \geq 2$ the last relation gives the inequality

$$E n^{-1/2}|S_n|^r \leq B_r E|\xi_1|^r.$$

In conjunction with the result from Problem 3.4.22, one must have $\lim_n E n^{-1/2}|S_n|^r \to E|Z|^r$, where $Z \sim \mathcal{N}(0, \sigma^2)$, $\sigma^2 = E\xi_1^2$.

**Problem 7.3.21.** (*Moments inequalities II.*) Let $\xi_1, \xi_2, \ldots$ be any sequence of independent and identically distributed random variables, and let $S_0 = 0$ and $S_n = \xi_1 + \ldots + \xi_n$, $n \geq 1$. Let $\tau$ be any Markov time, relative to the filtration $(\mathscr{F}_n^S)_{n \geq 0}$, defined by $\mathscr{F}_0^S = \{\varnothing, \Omega\}$ and $\mathscr{F}_n^S = \sigma(S_1, \ldots, S_n)$, $n \geq 1$. Prove that:
 (a) If $0 < r \leq 1$ and $E|\xi_1|^r < \infty$, then

$$E|S_\tau|^r \leq E|\xi_1|^r E\tau.$$

 (b) If $1 \leq r \leq 2$ and $E|\xi_1|^r < \infty$, $E\xi_1 = 0$, then

$$E|S_\tau|^r \leq B_r E|\xi_1|^r E\tau.$$

 (c) If $r > 2$ and $E|\xi_1|^r < \infty$, $E\xi_1 = 0$, then

$$E|S_\tau|^r \leq B_r\left[(E\xi_1^2)^{r/2} E\tau^{r/2} + E|\xi_1|^r E\tau\right] \leq 2 B_r E|\xi_1|^r E\tau^{r/2},$$

where $B_r$ is an universal constant, that depends only on $r$.
 *Hint.* In all cases one must prove the required inequalities first for the "cut-off" (finite) times $\tau \wedge n$, $n \geq 1$, and then pass to the limit as $n \to \infty$.

**Problem 7.3.22.** Let $\xi_1, \ldots, \xi_n$ be independent random variables. Prove the *Marcinkiewicz-Zygmund inequality*: for every $r > 0$ and every $n \geq 1$, one can find a constant, $B_r$, which is universal, in that it depends only on $r$, so that one can write (comp. with the second inequality in [P §7.3, (26)])

$$E\left|\sum_{j=1}^{n}\xi_j\right|^{2r} \le B_r\, n^{r-1}\sum_{j=1}^{n}E|\xi_j|^{2r}.$$

*Hint.* It is enough to consider only the (much simpler) case where $r \ge 1$ is an integer.

**Problem 7.3.23.** Let $(\xi_n)_{n\ge1}$ be any orthonormal sequence of random variables in $L^2$ (i.e., $E\xi_i\xi_j = 0$, for $i \ne j$ and $E\xi_i^2 = 1$ for all $i \ge 1$). Prove *Rademacher-Menshov's maximal inequality*: for any sequence of real numbers $(c_n)_{n\ge1}$ and for any integer $n \ge 1$, one has

$$E\max_{1\le k\le n}\left(\sum_{j=1}^{k}c_j\xi_j\right)^2 \le \ln^2(4n)\sum_{j=1}^{n}c_j^2 .$$

**Problem 7.3.24.** Let $(\xi_n)_{n\ge1}$ be any orthonormal sequence of random variables in $L^2$, and let $(c_n)_{n\ge1}$ be any sequence real numbers with

$$\sum_{k=1}^{\infty}c_k^2\ln^2 k < \infty.$$

Prove that the series $\sum_{k=1}^{\infty}c_k\xi_k$ converges with Probability 1.
    *Hint.* Use the result from the previous problem.

**Problem 7.3.25.** (*On the extremality of the class of Bernoulli random variables: Part I.*) Let $\xi_1,\dots,\xi_n$ be independent Bernoulli random variables with $P\{\xi_i = 1\} = P\{\xi_i = -1\} = 1/2$.

(a) Prove that, with $p = 2m$ and $m \ge 1$, the second Khinchin inequality in [P §7.3, (25)] can be written in the form: for every $n \ge 1$ and every family, $X_1,\dots,X_n$, of independent standard normal ($\mathcal{N}(0,1)$) random variables, one has

$$E\left|\sum_{k=1}^{n}a_k\xi_k\right|^{2m} \le E\left|\sum_{k=1}^{n}a_k X_k\right|^{2m} .$$

(b) Let $\Sigma_n$ denote the class of independent and identically distributed symmetric random variables $X_1,\dots,X_n$, with $DX_i = 1$, $i = 1,\dots,n$. Prove that, for every $n \ge 1$ and every $m \ge 1$, one has

$$E\left|\sum_{k=1}^{n}a_k\xi_k\right|^{2m} = \inf_{(X_1,\dots,X_n)\in\Sigma_n} E\left|\sum_{k=1}^{n}a_k X_k\right|^{2m} .$$

*Hint.* (a) It is enough to prove that

$$\mathsf{E}\left|\sum_{k=1}^{n} a_k \xi_k\right|^{2m} = \sum_{\substack{k_1+\ldots+k_n=m \\ k_i \geq 0}} \frac{(2m)!}{(2k_1)!\ldots(2k_n)!} |a_1|^{2k_1} \ldots |a_n|^{2k_n},$$

$$\mathsf{E}\left|\sum_{k=1}^{n} a_k X_k\right|^{2m} = \frac{(2m)!}{2^m m!} \sum_{\substack{k_1+\ldots+k_n=m \\ k_i \geq 0}} \frac{m!}{k_1!\ldots k_n!} |a_1|^{2k_1} \ldots |a_n|^{2k_n},$$

and that $2^m k_1! \ldots k_n! \leq (2k_1)! \ldots (2k_n)!$, if $k_1 + \ldots + k_n = m$ and $k_i \geq 0$. (Note that $\frac{(2m)!}{2^m m!} = (2m-1)!! = \mathsf{E} X_1^{2m}$—see Problem 2.8.9.)

(b) With $m = 1$, the required inequality is obvious. In the case $m \geq 2$, one must prove first that the function $\varphi(t) = \mathsf{E}|x + \sqrt{t}\xi_1|^{2m}$ is convex in the domain $t \geq 0$. Next, by using Jensen's inequality for the associated conditional expectations, prove that, if the sequences $(\xi_1, \ldots, \xi_n)$ and $(X_1, \ldots, X_n)$ are independent, then the following inequality must be in force

$$\mathsf{E}\left|\sum_{k=1}^{n} a_k \xi_k\right|^{2m} \leq \mathsf{E}\left|\sum_{k=1}^{n} a_k \xi_k |X_k|\right|^{2m}.$$

Finally, prove that

$$(\xi_1|X_1|, \ldots, \xi_n|X_n|) \overset{\text{law}}{=} (X_1, \ldots, X_n).$$

**Problem 7.3.26.** (*On the extremality of the class of Bernoulli random variables: Part II.*) Let $X_1, \ldots, X_n$ be independent random variables, such that $\mathsf{P}\{0 \leq X_i \leq 1\} = 1$ and $\mathsf{E} X_i = p_i$, $i = 1, \ldots, n$. In addition, let $\xi_1, \ldots, \xi_n$ be independent and identically distributed Bernoulli random variables with $\mathsf{P}\{\xi_i = 1\} = p$ and $\mathsf{P}\{\xi_i = 0\} = 1 - p$, where $p = (p_1 + \ldots + p_n)/n$. Prove the *Bentkus inequality*: for every $n \geq 1$ and every $x = 0, 1, 2, \ldots$, one has

$$\mathsf{P}\{X_1 + \ldots + X_n \geq x\} \leq e\,\mathsf{P}\{\xi_1 + \ldots + \xi_n \geq x\},$$

where $e = 2.718\ldots$.

## 7.4   Convergence Theorems for Submartingales and Martingales

**Problem 7.4.1.** Let $\{\mathcal{G}_n, n \geq 1\}$ be some *non-increasing* family of $\sigma$-algebras (i.e., $\mathcal{G}_1 \supseteq \mathcal{G}_2 \supseteq \dots$ ), let $\mathcal{G}_\infty = \bigcap \mathcal{G}_n$, and let $\eta$ be a some integrable random variable. Prove the following analog of [P §7.4, Theorem 3]:

$$\mathsf{E}(\eta \mid \mathcal{G}_n) \to \mathsf{E}(\eta \mid \mathcal{G}_\infty) \quad \text{as } n \to \infty \quad \text{(P-a.e. and in } L^1\text{-sense)}.$$

*Hint.* Let $\beta_n(a,b)$ denote the number downcrossings of the interval $(a,b)$ for the sequence $M = (M_k)_{1 \leq k \leq n}$, given by $M_k = \mathsf{E}(\eta \mid \mathcal{G}_k)$. Show first that

$$\mathsf{E}\beta_\infty(a,b) \leq \frac{\mathsf{E}|\eta| + |a|}{b-a} < \infty,$$

and conclude that $\beta_\infty(a,b) < \infty$ (P-a.e.). The rest of the proof is similar to the proofs of [P §7.4, Theorems 1 and 3].

**Problem 7.4.2.** Let $\xi_1, \xi_2, \dots$ be any sequence independent and identically distributed random variables with $\mathsf{E}|\xi_1| < \infty$ and $\mathsf{E}\xi_1 = m$ and let $S_n = \xi_1 + \dots + \xi_n$, $n \geq 1$. Prove that (see Problem 2.7.2)

$$\mathsf{E}(\xi_1 \mid S_n, S_{n+1}, \dots) = \mathsf{E}(\xi_1 \mid S_n) = \frac{S_n}{n} \quad \text{(P-a.e.)}, \quad n \geq 1.$$

By using the result from Problem 7.4.1, prove the strong law of large numbers: *as $n \to \infty$ one has*

$$\frac{S_n}{n} \to m \quad \text{(P-a.e. and in } L^1\text{-sense)}.$$

*Hint.* Given any $B \in \sigma(S_n, S_{n+1}, \dots)$, show that $\mathsf{E}I_B\xi_1 = \mathsf{E}I_B\xi_i, i \leq n$, and conclude that

$$\mathsf{E}(S_n \mid S_n, S_{n+1}, \dots) = n\,\mathsf{E}(\xi_1 \mid S_n, S_{n+1}, \dots);$$

in particular, $\mathsf{E}(\xi_1 \mid S_n, S_{n+1}, \dots) = \frac{S_n}{n}$ (P-a.e.). In order to prove that $\frac{S_n}{n} \to m$ (P-a.e. and in $L^1$-sense), consider the $\sigma$-algebra $\mathscr{X}(s) = \bigcap_{n=1}^\infty \sigma(S_n, S_{n+1}, \dots)$ and, using the result from Problem 7.4.1, conclude that $\frac{S_n}{n} \to \mathsf{E}(\xi_1 \mid \mathscr{X}(s))$ (P-a.e. and in $L^1$-sense). Finally, use the fact that the events $A \in \mathscr{X}(s)$ obey the Hewitt–Savage 0-1 law ([P §2.1, Theorem 3]).

**Problem 7.4.3.** Prove the following result, which combines H. Lebesgue's dominated convergence theorem and P. Lévy's theorem. Let $(\xi_n)_{n \geq 1}$ be any sequence of random variables, such that $\xi_n \to \xi$ (P-a.e.) and $|\xi_n| \leq \eta$, for some random variable $\eta$ with $\mathsf{E}\eta < \infty$. Let $(\mathscr{F}_m)_{m \geq 1}$ be any non-decreasing family of $\sigma$-algebras, and let $\mathscr{F}_\infty = \sigma(\bigcup \mathscr{F}_m)$. Then one has (P-a. e.)

$$\lim_{\substack{m\to\infty \\ n\to\infty}} \mathsf{E}(\xi_n \mid \mathscr{F}_m) = \mathsf{E}(\xi \mid \mathscr{F}_\infty).$$

*Hint.* Use Lebesgue's dominated convergence theorem ([P §2.6, Theorem 3]) and P. Lévy's theorem ([P §7.4, Theorem 3]) to estimate, for large $n$ and $m$, the terms in the right side of the representation:

$$\mathsf{E}(\xi_n \mid \mathscr{F}_m) - \mathsf{E}(\xi \mid \mathscr{F}_\infty)$$
$$= [\mathsf{E}(\xi_n \mid \mathscr{F}_m) - \mathsf{E}(\xi \mid \mathscr{F}_m)] + [\mathsf{E}(\xi \mid \mathscr{F}_m) - \mathsf{E}(\xi \mid \mathscr{F}_\infty)].$$

**Problem 7.4.4.** Prove formula [P §7.4, (12)].

*Hint.* Notice first that the system $\{H_1(x), \ldots, H_n(x)\}$ is a basis in the space of functions that are measurable for $\mathscr{F}_n = \sigma(H_1, \ldots, H_n)$. As $\mathscr{F}_n$ has finitely many elements, every function that is measurable for $\mathscr{F}_n$ is automatically simple (see [P §2.4, Lemma 3]). As a result, formula [P §7.4, (12)] must hold for some constants $a_1 \ldots, a_n$. The fact that $a_k = (f, H_k)$ follows from the orthonormality of the basis $\{H_1(x), \ldots, H_n(x)\}$.

**Problem 7.4.5.** Let $\Omega = [0, 1)$, $\mathscr{F} = \mathscr{B}([0, 1))$, let P stand for the Lebesgue measure, and suppose that the function $f = f(x)$ belongs to $L^1$. Prove that $f_n(x) \to f(x)$ (P-a. e.), for

$$f_n(x) = 2^n \int_{k2^{-n}}^{(k+1)2^{-n}} f(y)\, dy, \quad k\, 2^{-n} \le x < (k+1)2^{-n}.$$

*Hint.* The main step in the proof is to show that $(f_n(x), \mathscr{F}_n)_{n\geq 1}$, with $\mathscr{F}_n = \sigma([j2^{-n}, (j+1)2^{-n})]$, $j = 0, 1, \ldots, 2^n - 1)$, forms a martingale. The result from [P §7.4, Theorem 1] will then conclude the proof.

**Problem 7.4.6.** Let $\Omega = [0, 1)$, $\mathscr{F} = \mathscr{B}([0, 1))$, let P stand for the Lebesgue measure and suppose that the function $f = f(x)$ belongs to $L^1$. Assuming that the function $f = f(x)$ is extened to the interval $[0, 2)$ by periodicity in the obvious way, and setting

$$f_n(x) = \sum_{i=1}^{2^n} 2^{-n} f(x + i2^{-n}),$$

prove that

$$f_n(x) \to f(x) \quad \text{(P-a. e.)}.$$

*Hint.* Just as in the previous Problem, the key step is to show that the sequence $(f_n(x), \mathscr{F}_n)_{n\geq 1}$, with analogously defined $\sigma$-algebras $(\mathscr{F}_n)_{n\geq 1}$, forms a martingale.

**Problem 7.4.7.** Prove that [P §7.4, Theorem 1] remains valid for generalized submartingales $X = (X_n, \mathscr{F}_n)$, for which

$$\inf_m \sup_{n \geq m} E(X_n^+ \mid \mathscr{F}_m) < \infty \quad \text{(P-a.e.)}.$$

**Problem 7.4.8.** Let $(a_n)_{n \geq 1}$ be some sequence of real numbers with the property: for some $\delta > 0$, one can claim that the limit $\lim_n e^{ita_n}$ exists for any $t \in (-\delta, \delta)$. Prove that $\lim_n a_n$ also exists and is finite.

 Hint. The existence of $\lim_n e^{ita_n}$ for every $t \in (-\delta, \delta)$ is tantamount to the existence of $\lim_n e^{ita_n}$ for all $t \in R$. Thus, it suffices to prove that the function $f(t) = \lim_n e^{ita_n}$ can be written in the form $e^{itc}$, for some finite constant $c$. This last property can be derived from the following properties of the function $f(t)$:

   (i) $|f(t)| = 1, t \in R$;
   (ii) $f(t_1 + t_2) = f(t_1) f(t_2), t_1, t_2 \in R$;
   (iii) the set of continuity points for the function $f(t)$ is everywhere dense in $R$.

**Problem 7.4.9.** Let $F = F(x)$, $x \in R$, be some distribution function, and let $\alpha \in (0, 1)$ be chosen so that, for some $\theta \in R$, one can write $F(\theta) = \alpha$. Define the sequence of random variables $X_1, X_2, \ldots$ according to the following rule (known as the *Robbins–Monro procedure*):

$$X_{n+1} = X_n - n^{-1}(Y_n - \alpha),$$

where $Y_1, Y_2, \ldots$ are random variables, defined in such a way that

$$P(Y_n = y \mid X_1, \ldots, X_n; Y_1, \ldots, Y_{n-1}) = \begin{cases} F(X_n), & \text{if } y = 1, \\ 1 - F(X_n), & \text{if } y = 0, \end{cases}$$

with the understanding that, for $n = 1$, the conditional probability in the left side is to be replaced by $P(Y_1 = y)$.

 Prove the following result from stochastic approximation theory: in the Robbins–Monro procedure one has $E|X_n - \theta|^2 \to 0$ as $n \to \infty$.

**Problem 7.4.10.** Let $X = (X_n, \mathscr{F}_n)_{n \geq 1}$ be a submartingale, for which one can claim that

$$E(X_\tau I(\tau < \infty)) \neq \infty,$$

for every stopping time $\tau$. Prove that the limit $\lim_n X_n$ exists with probability 1.

**Problem 7.4.11.** Let $X = (X_n, \mathscr{F}_n)_{n \geq 1}$ be a martingale and let $\mathscr{F}_\infty = \sigma(\bigcup_{n=1}^\infty \mathscr{F}_n)$. Prove that if the sequence $(X_n)_{n \geq 1}$ is uniformly integrable, then the limit $X_\infty = \lim_n X_n$ exists (P-a.e.), and the "closed" sequence $\overline{X} = (X_n, \mathscr{F}_n)_{1 \leq n \leq \infty}$ is a martingale.

**Problem 7.4.12.** Suppose that $X = (X_n, \mathcal{F}_n)_{n \geq 1}$ is a submartingale and let $\mathcal{F}_\infty = \sigma\left(\bigcup_{n=1}^\infty \mathcal{F}_n\right)$. Prove that if the sequence $(X_n^+)_{n \geq 1}$ is uniformly integrable, then the limit $X_\infty = \lim_n X_n$ exists (P-a. e.), and the "closed" sequence $\overline{X} = (X_n, \mathcal{F}_n)_{1 \leq n \leq \infty}$ is a submartingale.

**Problem 7.4.13.** [P §7.4, Corollary 1 to Theorem 1] states that, for any non-negative supermartingale $X$, one can claim the limit $X_\infty = \lim X_n$ exists and is finite that with probability 1. Prove that the following properties are also in force:
 (a) $E(X_\infty \mid \mathcal{F}_n) \leq X_n$ (P-a. e.), $n \geq 1$;
 (b) $EX_\infty \leq \lim_n EX_n$;
 (c) $E(X_\tau \mid \mathcal{F}_\sigma) \leq X_{\tau \wedge \sigma}$ for *arbitrary* stopping times $\tau$ and $\sigma$;
 (d) $Eg(X_\infty) = \lim_n Eg(X_n)$, for any continuous function $g = g(x)$, $x \geq 0$, with $\frac{g(x)}{x} \to 0$ as $x \to \infty$;
 (e) if $g(x) > g(0) = 0$ for all $x > 0$, then

$$X_\infty = 0 \quad \Leftrightarrow \quad \lim_n Eg(X_n) = 0;$$

 (f) for every given $0 < p < 1$, one has

$$P\{X_\infty = 0\} = 1 \quad \Leftrightarrow \quad \lim_n EX_n^p = 0.$$

**Problem 7.4.14.** In P. Lévy's convergence theorem ([P §7.4, Theorem 3]) it is assumed that $E|\xi| < \infty$. Prove by way of example that the requirement for $E\xi$ to exist $(\min(E\xi^+, E\xi^-) < \infty)$ alone, in other words, without insisting that $E\xi^+ + E\xi^- < \infty$, cannot guarantee the convergence $E(\xi \mid \mathcal{F}_n) \to E(\xi \mid \mathcal{F})$ (P-a. e.).

**Problem 7.4.15.** If $X = (X_n, \mathcal{F}_n)_{n \geq 1}$ is a martingale with $\sup_n E|X_n| < \infty$, then, according to [P §7.4, Theorem 1], $\lim X_n$ must exist with Probability 1. Give an example of a martingale $X$, for which $\sup_n E|X_n| = \infty$ and $\lim X_n$ does not exist with Probability 1.

**Problem 7.4.16.** Give an example of a martingale, $(X_n)_{n \geq 0}$, for which one has $X_n \to -\infty$ as $n \to \infty$ with Probability 1.

**Problem 7.4.17.** According to [P §7.4, Theorem 2], given any uniformly integrable submartingale (supermartingale) $X = (X_n, \mathcal{F}_n)_{n \geq 1}$, one can find a "terminal" random variable $X_\infty$, such that $X_n \to X_\infty$ (P-a. e.). Give an example of a submartingale (supermartingale) for which the "terminal" variable $X_\infty$, with $X_n \to X_\infty$ (P-a. e.), exists, but the sequence $(X_n)_{n \geq 1}$ is not uniformly integrable.

**Problem 7.4.18.** Prove that any martingale, $X = (X_n)_{n \geq 0}$, that has the property

$$\sup_n E(|X_n| \ln^+ |X_n|) < \infty,$$

must be a Lévy martingale.

**Problem 7.4.19.** Give an example of a non-negative martingale,

$$X \equiv (X_n, \mathscr{F}_n)_{n \geq 1},$$

such that $\mathsf{E} X_n = 1$ for all $n \geq 1$, $X_n(\omega) \to 0$ as $n \to \infty$ for any $\omega$, and yet $\mathsf{E} \sup_n X_n = \infty$.

**Problem 7.4.20.** Assuming that $X = (X_n, \mathscr{F}_n)_{n \geq 1}$ is a uniformly integrable submartingale, prove that, for any Markov time $\tau$, one has

$$\mathsf{E}(X_\infty \mid \mathscr{F}_\tau) \geq X_\tau \quad \text{(P-a.e.)},$$

where $X_\infty$ stands for $\lim X_n$, which, according to Problem 7.4.12, exists with Probability 1.

**Problem 7.4.21.** (*On [ P §7.4, Theorem 1 ].*) Give an example of a supermartingale, $X = (X_n, \mathscr{F}_n)_{n \geq 1}$, which satisfies the condition $\sup_n \mathsf{E} |X_n| < \infty$, and, therefore, $\lim X_n$ ($= X_\infty$) exists with Probability 1, and yet $X_n \not\to X_\infty$ in $L^1$.

**Problem 7.4.22.** Argue that, given any square integrable martingale, $M = (M_n, \mathscr{F}_n)_{n \geq 1}$, the condition

$$\sum_{k \geq 1} \mathsf{E}(M_k - M_{k-1})^2 < \infty,$$

or, equivalently, $\mathsf{E}\langle M \rangle_\infty < \infty$, where $\langle M \rangle_\infty = \lim_n \langle M \rangle_n$, guarantees the convergence $M_n \to M_\infty$ (P-a.e.), and also the convergence $M_n \overset{L^1}{\to} M_\infty$, for some random variable $M_\infty$, with $\mathsf{E} M_\infty^2 < \infty$.

**Problem 7.4.23.** Let $X = (X_n, \mathscr{F}_n)_{n \geq 0}$ be any submartingale. By the very definition of submartingale, one must have $\mathsf{E} |X_n| < \infty$, for every $n \geq 0$. Sometimes this condition is relaxed, by requiring only that $\mathsf{E} X_n^- < \infty$, for $n \geq 0$. Which of the properties of the general class of submartingales, listed in [ P §7.4, **2–4** ], remain valid under this weaker notion of submartingale?

**Problem 7.4.24.** Suppose that $X = (X_n, \mathscr{F}_n)_{n \geq 0}$ is a supermartingale, i.e., $X_n$ is $\mathscr{F}_n$-measurable, $\mathsf{E} |X_n| < \infty$ and $\mathsf{E}(X_{n+1} \mid \mathscr{F}_n) \leq X_n$, for $n \geq 0$. According to [ P §7.4, Theorem 1], if $\sup_n \mathsf{E} |X_n| < \infty$, then one can claim that with Probability 1 the limit $\lim_n X_n = X_\infty$ exists and $\mathsf{E} |X_\infty| < \infty$.

Notice, however, that the condition $\mathsf{E}(X_{n+1} \mid \mathscr{F}_n) \leq X_n$ is meaningful even without the requirement $\mathsf{E} |X_{n+1}| < \infty$, as the conditional expectation $\mathsf{E}(X_{n+1} \mid \mathscr{F}_n)$ would be well defined if, for example, $X_{n+1} \geq 0$, although in this case $\mathsf{E}(X_{n+1} \mid \mathscr{F}_n)$ may take the value $+\infty$ on some non-negligible set.

In lieu with the last observation, we say that $X = (X_n, \mathscr{F}_n)_{n \geq 0}$ is a *non-negative supermartingale sequence*, if, for every $n \geq 0$, one can claim that $X_n$ is $\mathscr{F}_n$-measurable, $\mathsf{P}\{X_n \geq 0\} = 1$ and

$$\mathsf{E}(X_{n+1} \mid \mathscr{F}_n) \leq X_n \quad \text{(P-a.e.)}.$$

Prove that for any non-negative supermartingale sequence,

$$X = (X_n, \mathscr{F}_n)_{n \geq 0},$$

the limit $\lim_n X_n$ ( $= X_\infty$) exists with Probability 1 and, furthermore, if $P\{X_0 < \infty\} = 1$, then $P\{X_\infty < \infty\} = 1$.

*Hint.* The proof is analogous to the proof of [P §7.4, Theorem 1] and hinges on the estimate (37) from [P §7.3, Theorem 5] for the number of up-crossings of a given interval.

**Problem 7.4.25.** (*Continuation of Problem 2.2.15.*) As was shown in Problem 2.2.15, the following relation between $\sigma$-algebras does not hold in general:

$$\bigcap_n \sigma(\mathscr{G}, \mathscr{E}_n) = \sigma\left(\mathscr{G}, \bigcap_n \mathscr{E}_n\right).$$

Show, however, that the last relation is guaranteed by the following condition:

the $\sigma$-algebras $\mathscr{G}$ and $\mathscr{E}_1$ are *conditionally independent*, relative to the $\sigma$-algebra $\mathscr{E}_n$, for every $n > 1$, i.e., one has (P-a. e.)

$$P(A \cap B \mid \mathscr{E}_n) = P(A \mid \mathscr{E}_n)P(B \mid \mathscr{E}_n),$$

for any $A \in \mathscr{G}$ and $B \in \mathscr{E}_1$.

*Hint.* It is enough to show that, for every $\mathscr{G} \vee \mathscr{E}_1$-measurable and bounded random variable $X$, one has (P-a. e.)

$$E\left(X \mid \bigcap_n (\mathscr{G} \vee \mathscr{E}_n)\right) = E\left(X \mid \mathscr{G} \vee \bigcap_n \mathscr{E}_n\right).$$

Furthermore, it would be enough to consider only random variables $X$ of the form

$$X = X_1 X_2,$$

where the bounded variables $X_1$ and $X_2$ are such that $X_1$ is $\mathscr{E}_1$-measurable and $X_2$ is $\mathscr{E}_2$-measurable. Finally, use the $L^1$-convergence established in Problem 7.4.1, in conjunction with the conditional independence established above.

**Problem 7.4.26.** Let $\xi_1, \xi_2, \ldots$ be independent non-negative random variables, with $E\xi_1 \leq 1$ and $P\{\xi_1 = 1\} < 1$. For $M_n = \xi_1 \ldots \xi_n$, $n \geq 1$, prove that $M_n \to 0$ as $n \to \infty$ (P-a. e.).

*Hint.* Use the fact that the sequence $(M_n)_{n \geq 1}$ forms a non-negative supermartingale.

**Problem 7.4.27.** Let $(\Omega, (\mathscr{F}_i)_{i \geq 0}, P)$ be any filtered probability space with $\mathscr{F}_0 = \{\varnothing, \Omega\}$, and let $\xi_1, \xi_2, \ldots$ be any sequence of random variables, chosen so that each $\xi_i$ is $\mathscr{F}_i$-measurable. Assuming that $\sup_i E |\xi_i|^\alpha < \infty$, for some $\alpha \in (1, 2]$, prove that

$$\frac{1}{n}\sum_{i=1}^{n}(\xi_i - \mathsf{E}(\xi_i \mid \mathscr{F}_{i-1})) \overset{\text{a.s., } L^\alpha}{\longrightarrow} 0.$$

(Comp. with the law of large numbers, established in [$\underline{P}$ §4.3] and the ergodic theorems of [$\underline{P}$ §5.3].)

## 7.5   On the Sets of Convergence of Submartingales and Martingales

**Problem 7.5.1.** Prove that any submartingale, $X = (X_n, \mathscr{F}_n)$, that satisfies the condition $\mathsf{E}\sup_n |X_n| < \infty$, must belong to the class $\mathbf{C}^+$.

**Problem 7.5.2.** Prove that [$\underline{P}$ §7.5, Theorems 1 and 2] remain valid for generalized submartingales.

**Problem 7.5.3.** Prove that, for any generalized submartingale, $X = (X_n, \mathscr{F}_n)$, up to a P-negligible set, one has the inclusion:

$$\left\{\inf_m \sup_{n \geq m} \mathsf{E}(X_n^+ \mid \mathscr{F}_m) < \infty\right\} \subseteq \{X_n \text{ converges}\}.$$

**Problem 7.5.4.** Prove that the corollary to [$\underline{P}$ §7.5, Theorem 1] remains valid for generalized submartingales.

**Problem 7.5.5.** Prove that any generalized submartingale from the class $\mathbf{C}^+$ is automatically a local submartingale.

**Problem 7.5.6.** Consider the sequence $a_n > 0$, $n \geq 1$, and let $b_n = \sum_{k=1}^{n} a_k$. Prove that $\sum_{n=1}^{\infty} \frac{a_n}{b_n^2} < \infty$.
    *Hint.* Consider separately the cases: $\sum_{n=1}^{\infty} a_n < \infty$ and $\sum_{n=1}^{\infty} a_n = \infty$.

**Problem 7.5.7.** Let $\xi_0, \xi_1, \xi_2, \ldots$ be any sequence of uniformly bounded random variables, i.e., $|\xi_n| \leq c$, for $n \leq 1$. Prove that the series $\sum_{n \geq 0} \xi_n$ and $\sum_{n \geq 1} \mathsf{E}(\xi_n \mid \xi_1, \ldots, \xi_{n-1})$ either simultaneously converge or simultaneously diverge (P-a. e.).

**Problem 7.5.8.** Let $X = (X_n)_{n \geq 0}$ be any martingale, with the property $\Delta X_n = X_n - X_{n-1} \leq c$ (P-a. e.), for some constant $c < \infty$ ($\Delta X_0 = X_0$). Prove that the sets $\{X_n \text{ converges}\}$ and $\{\sup_n X_n < \infty\}$ can differ only with a P-negligible set.

**Problem 7.5.9.** Let $X = (X_n, \mathscr{F}_n)_{n \geq 0}$ be any martingale, with

$$\sup_{n \geq 0} \mathsf{E}|X_n| < \infty.$$

Prove that $\sum_{n \geq 1} (\Delta X_n)^2 < \infty$ (P-a. e.). (Comp. with Problem 7.3.18.)

**Problem 7.5.10.** Let $X = (X_n, \mathscr{F}_n)_{n \geq 0}$ be any martingale, with

$$\mathsf{E} \sup_{n \geq 1} |\Delta X_n| < \infty.$$

Prove that, up to a P-negligible set,

$$\left\{ \sum_{n \geq 1} (\Delta X_n)^2 < \infty \right\} \subseteq \{X_n \text{ converges}\}.$$

In particular, if $\mathsf{E}(\sum_{n \geq 1} (\Delta X_n)^2)^{1/2} < \infty$, then one can claim that the sequence $(X_n)_{n \geq 0}$ converges with Probability 1.

**Problem 7.5.11.** Let $X = (X_n, \mathscr{F}_n)_{n \geq 0}$ be any martingale with

$$\sup_{n \geq 0} \mathsf{E}|X_n| < \infty,$$

and let $Y = (Y_n, \mathscr{F}_{n-1})_{n \geq 1}$ be any predictable sequence with

$$\sup_{n \geq 1} |Y_n| < \infty \quad (\text{P-a.e.}).$$

Prove that the series $\sum_{n=1}^{\infty} Y_n \Delta X_n$ converges (P-a.e.).

**Problem 7.5.12.** Consider the martingale $X = (X_n, \mathscr{F}_n)_{n \geq 0}$, chosen so that $\sup_{\tau} \mathsf{E}(|\Delta X_\tau| I(\tau < \infty)) < \infty$, the sup being taken over all finite stopping times $\tau$.
Prove that, up to a P-negligible set, one has

$$\left\{ \sum_{n \geq 1} (\Delta X_n)^2 < \infty \right\} \subseteq \{X_n \to \infty\}.$$

**Problem 7.5.13.** Let $M = (M_n, \mathscr{F}_n)$ be any square-integrable martingale. Prove that, for almost every $\omega$ from the set $\{\langle M \rangle_\infty = \infty\}$, one has

$$\lim_{n \to \infty} \frac{M_n}{\langle M \rangle_n} = 0.$$

## 7.6 Absolute Continuity and Singularity of Probability Distributions on Measurable Spaces with Filtrations

**Problem 7.6.1.** Prove the inequality in [$\underline{\text{P}}$ §7.6, (6)].

**Problem 7.6.2.** Let $\widetilde{\mathsf{P}}_n \sim \mathsf{P}_n$, for $n \geq 1$. Prove that:

$$\widetilde{\mathsf{P}} \sim \mathsf{P} \iff \widetilde{\mathsf{P}}\{z_\infty < \infty\} = \mathsf{P}\{z_\infty > 0\} = 1;$$

$$\widetilde{\mathsf{P}} \perp \mathsf{P} \iff \widetilde{\mathsf{P}}\{z_\infty = \infty\} = 1 \text{ or } \mathsf{P}\{z_\infty = 0\} = 1.$$

**Problem 7.6.3.** Let $\widetilde{P}_n \ll P_n, n \geq 1$, suppose that $\tau$ is a stopping time (relative to the filtration $(\mathscr{F}_n)$), and let $\widetilde{P}_\tau = \widetilde{P} \mid \mathscr{F}_\tau$ and $P_\tau = P \mid \mathscr{F}_\tau$ denote, respectively, the restrictions of the measures $\widetilde{P}$ and P to the $\sigma$-algebra $\mathscr{F}_\tau$. Prove that $\widetilde{P}_\tau \ll P_\tau$ if and only if $\{\tau = \infty\} = \{z_\infty < \infty\}$ ($\widetilde{P}$-a.e.). In particular, this result implies that, if $P\{\tau < \infty\} = 1$, then $\widetilde{P}_\tau \ll P_\tau$.

**Problem 7.6.4.** Prove the "conversion formulas" [ P §7.6, (21) and (22)].
   *Hint.* Show directly that, for every $A \in \mathscr{F}_{n-1}$, one has

$$E\left[I_A \widetilde{E}(\eta \mid \mathscr{F}_{n-1}) z_{n-1}\right] = E[I_A \eta z_n].$$

As for the proof of the second formula, it is enough to notice that

$$\widetilde{P}\{z_{n-1} = 0\} = 0.$$

**Problem 7.6.5.** Prove the estimates in [ P §7.6, (28), (29) and (32)].

**Problem 7.6.6.** Prove the relation [ P §7.6, (34)].

**Problem 7.6.7.** Suppose that the sequences

$$\xi = (\xi_1, \xi_2, \dots) \quad \text{and} \quad \tilde{\xi} = (\tilde{\xi}_1, \tilde{\xi}_2, \dots),$$

introduced in [ P §7.6, **2** ], consist of *independent and identically distributed* random variables.
   (a) Prove that if $P_{\tilde{\xi}_1} \ll P_{\xi_1}$, then $\widetilde{P} \ll P$ if and only if the measures $P_{\tilde{\xi}_1}$ and $P_{\xi_1}$ coincide. Furthermore, if $P_{\tilde{\xi}_1} \ll P_{\xi_1}$ and $P_{\tilde{\xi}_1} \neq P_{\xi_1}$, then $\widetilde{P} \perp P$.
   (b) Prove that if $P_{\tilde{\xi}_1} \sim P_{\xi_1}$, then the following dichotomy is in force: one has either $\widetilde{P} = P$ or $\widetilde{P} \perp P$ (comp. with the Kakutani Dichotomy Theorem—[ P §6.7, Theorem 3].

**Problem 7.6.8.** Let P and $\widetilde{P}$ be any two probability measures on the filtered space $(\Omega, \mathscr{F}, (\mathscr{F}_n)_{n \geq 1})$. Let $\widetilde{P} \overset{\text{loc}}{\ll} P$ (i.e., $\widetilde{P}_n \ll P_n$, for all $n \geq 1$, where $\widetilde{P}_n = \widetilde{P} \mid \mathscr{F}_n$ and $P_n = P \mid \mathscr{F}_n$), and let $z_n = \frac{d\widetilde{P}_n}{dP_n}$, for $n \geq 1$.
   Prove that if $\tau$ is a Markov time, then, on the set $\{\tau < \infty\}$, one has (P-a.e.)

$$\widetilde{P}_\tau \ll P_\tau \quad \text{and} \quad \frac{d\widetilde{P}_\tau}{dP_\tau} = z_\tau.$$

**Problem 7.6.9.** Prove that $\widetilde{P} \overset{\text{loc}}{\ll} P$ if and only if one can find an increasing sequence of stopping times, $(\tau_n)_{n \geq 1}$, with $P\{\lim \tau_n = \infty\} = 1$, and with the property $\widetilde{P}_{\tau_n} \ll P_{\tau_n}$, for $n \geq 1$.

**Problem 7.6.10.** Let $\widetilde{P} \overset{\text{loc}}{\ll} P$ and let $z_n = \frac{d\widetilde{P}_n}{dP_n}$, for $n \geq 1$. Prove that

$$\widetilde{\mathsf{P}}\Big\{\inf_n z_n > 0\Big\} = 1.$$

**Problem 7.6.11.** Let $\widetilde{\mathsf{P}} \overset{\text{loc}}{\ll} \mathsf{P}$, $z_n = \frac{d\widetilde{\mathsf{P}}_n}{d\mathsf{P}_n}$, for $n \geq 1$, and let $\mathscr{F}_\infty = \sigma(\bigcup \mathscr{F}_n)$. Prove that the following conditions are equivalent:

(i) $\widetilde{\mathsf{P}}_\infty \ll \mathsf{P}_\infty$, where $\widetilde{\mathsf{P}}_\infty = \widetilde{\mathsf{P}} \mid \mathscr{F}_\infty$ and $\mathsf{P}_\infty = \mathsf{P} \mid \mathscr{F}_\infty$;
(ii) $\widetilde{\mathsf{P}}\{\sup_n z_n < \infty\} = 1$;
(iii) the martingale $(z_n, \mathscr{F}_n)_{n\geq 1}$ is uniformly integrable.

**Problem 7.6.12.** Let $(\Omega, \mathscr{F}, \mathsf{P})$ be any probability space and let $\mathscr{G}$ be any *separable* $\sigma$-sub-algebra inside $\mathscr{F}$, which is generated by the sets $\{G_n, n \geq 1\}$, all included in $\mathscr{F}$. Let $\mathscr{G}_n = \sigma(G_1, \ldots, G_n)$ and let $\mathscr{D}_n$ be the smallest partition of $\Omega$, which is generated by $\mathscr{G}_n$.

Let $\mathsf{Q}$ be any measure on $(\Omega, \mathscr{F})$ and set

$$X_n(\omega) = \sum_{A \in \mathscr{D}_n} \frac{\mathsf{Q}(A)}{\mathsf{P}(A)} I_A(\omega)$$

(with the understanding that $0/0 = 0$).
    Prove that:

(a) The sequence $(X_n, \mathscr{G}_n)_{n\geq 1}$ forms a supermartingale (relative to the measure $\mathsf{P}$).

(b) If $\mathsf{Q} \ll \mathsf{P}$, then the sequence $(X_n, \mathscr{G}_n)_{n\geq 1}$ must be a martingale.

**Problem 7.6.13.** As a continuation of the pervious problem, prove that, if $\mathsf{Q} \ll \mathsf{P}$, then one can find a $\mathscr{G}$-measurable random variable, $X_\infty = X_\infty(\omega)$, for which $X_n \overset{L^1}{\to} X_\infty$, $X_n = \mathsf{E}(X_\infty \mid \mathscr{G}_n)$ (P-a. e.) and, for every $A \in \mathscr{G}$, one can claim that

$$\mathsf{Q}(A) = \int X_\infty \, d\mathsf{P}.$$

    This is nothing but a special version of the Radon-Nikodym theorem from [P §2.6], stated for *separable* $\sigma$-sub-algebras $\mathscr{G} \subseteq \mathscr{F}$.

**Problem 7.6.14.** (*On the Kakutani dichotomy.*) Let $\alpha_1, \alpha_2, \ldots$ be any sequence of non-negative and independent random variables, with $\mathsf{E}\alpha_i = 1$, and let $z_n = \prod_{k=1}^n \alpha_k$, $z_0 = 1$.
    Prove that:

(a) The sequence $(z_n)_{n\geq 0}$ is a non-negative martingale.
(b) The limit $\lim_n z_n$ ($= z_\infty$) exists with probability 1.
(c) The following conditions are equivalent:

(i)   $\mathsf{E}z_\infty = 1$;        (ii)   $z_n \overset{L^1}{\to} z_\infty$;

(iii)   the family $(z_n)_n$ is uniformly integrable;

$$\text{(iv)} \quad \sum_{n=1}^{\infty} (1 - \mathsf{E}\sqrt{\alpha_n}) < \infty; \qquad \text{(v)} \quad \prod_{n=1}^{\infty} \mathsf{E}\sqrt{\alpha_n} > 0.$$

## 7.7   On the Asymptotics of the Probability for a Random Walk to Exit on a Curvilinear Boundary

**Problem 7.7.1.** Prove that the sequence defined in [P §7.7, (4)] is a martingale. Can one make this claim without the condition $|\alpha_n| \leq c$ (P-a. e.), for $n \geq 1$?

**Problem 7.7.2.** Prove the formula in [P §7.7, (13)].
   *Hint.* It is enough to write the expression $\mathsf{E} z_n^{\mathsf{P}}$ in the form

$$\mathsf{E} z_n^{\mathsf{P}} = \prod_{k=2}^{n} \mathsf{E}\left( p \exp\left\{ \alpha_k \xi_k - \frac{1}{2}\alpha_k^2 \right\} \right),$$

and use the fact that all $\xi_i$ are normally ($\mathscr{N}(0, 1)$) distributed.

**Problem 7.7.3.** Prove the formula in [P §7.7, (17)].

**Problem 7.7.4.** Let $\xi_1, \xi_2, \ldots$ be any sequence of independent and identically distributed random variables, and let $S_n = \xi_1 + \ldots + \xi_n$, for $n \geq 1$. Given any constant $c > 0$, set

$$\tau_{(\geq 0)} = \inf\{n \geq 1 : S_n \geq 0\} \quad \text{and} \quad \tau_{(>c)} = \inf\{n \geq 1 : S_n > c\},$$

with the understanding that $\inf \varnothing = \infty$. Prove that:

   (a) $\mathsf{P}\{\tau_{(\geq 0)} < \infty\} = 1 \Leftrightarrow \mathsf{P}\{\overline{\lim}\, S_n = \infty\} = 1$;
   (b) $(\mathsf{E}\tau_{(\geq 0)} < \infty) \Leftrightarrow (\mathsf{E}\tau_{(>c)} < \infty$ for all $c > 0)$.

**Problem 7.7.5.** Assume the notation introduced in previous problem, and set
$$\tau_{(>0)} = \inf\{n \geq 1 : S_n > 0\}, \quad \tau_{(\leq 0)} = \inf\{n \geq 1 : S_n \leq 0\},$$

and

$$\tau_{(<0)} = \inf\{n \geq 1 : S_n < 0\}.$$

   Prove that

$$\mathsf{E}\tau_{(\geq 0)} = \frac{1}{\mathsf{P}\{\tau_{(<0)} = \infty\}} \quad \text{and} \quad \mathsf{E}\tau_{(>0)} = \frac{1}{\mathsf{P}\{\tau_{(\leq 0)} = \infty\}}.$$

**Problem 7.7.6.** Let $\xi_1, \xi_2, \ldots$ be any sequence of independent and identically distributed random variables with $\mathsf{E}|\xi_1| > 0$, chosen so that $\frac{S_n}{n} \xrightarrow{\mathsf{P}} 0$, where $S_n = \xi_1 + \ldots + \xi_n$.

Prove that the Markov times $\tau_{(\geq 0)}$ and $\tau_{(\leq 0)}$, defined in the previous problem, are finite, and that $\overline{\lim}\, S_n = \infty$ and $\underline{\lim}\, S_n = -\infty$, both with Probability 1.

**Problem 7.7.7.** Let everything be as in the previous problem. Prove that $S_n \to \infty$ with Probability 1 if and only if one can find a stopping time $\tau$ (relative to the filtration $(\mathscr{F}_n^\xi)_{n\geq 1}$, with $\mathscr{F}_n^\xi = \sigma(\xi_1,\ldots,\xi_n)$), for which $\mathsf{E}\tau < \infty$ and $\mathsf{E}S_\tau > 0$.

**Problem 7.7.8.** Let $(\Omega, \mathscr{F}, (\mathscr{F}_n)_{n\geq 0}, \mathsf{P})$ be some filtered probability space, and let $h = (h_n)_{n\geq 1}$ be some sequence of the form

$$h_n = \mu_n + \sigma_n \xi_n, \quad n \geq 1,$$

where $\mu_n \in \mathbb{R}$ and $\sigma_n > 0$ are $\mathscr{F}_{n-1}$-measurable random variables and $\xi = (\xi_n, \mathscr{F}_n)_{n\geq 0}$ is some stochastic sequence of independent and normally distributed ($\mathscr{N}(0,1)$) random variables. Prove that the sequence $h = (h_n, \mathscr{F}_n)_{n\geq 1}$ is conditionally Gaussian, i.e.,

$$\mathrm{Law}(h_n \mid \mathscr{F}_{n-1}; \mathsf{P}) = \mathscr{N}(\mu_n, \sigma_n^2) \quad \text{(P-a.\,e.)}.$$

Setting

$$Z_n = \exp\left\{ -\sum_{k=1}^n \frac{\mu_k}{\sigma_k}\xi_k - \frac{1}{2}\sum_{k=1}^n \left(\frac{\mu_k}{\sigma_k}\right)^2 \right\}, \quad n \geq 1,$$

prove that the following properties hold:
(a) The sequence $Z_n = (Z_n)_{n\geq 1}$ is a martingale relative to the measure $\mathsf{P}$.
(b) If

$$\mathsf{E}\exp\left\{ \frac{1}{2}\sum_{k=1}^\infty \left(\frac{\mu_k}{\sigma_k}\right)^2 \right\} < \infty \quad \text{(Novikov's condition)}$$

and

$$Z_\infty = \exp\left\{ -\sum_{k=1}^\infty \frac{\mu_k}{\sigma_k}\xi_k - \frac{1}{2}\sum_{k=1}^\infty \left(\frac{\mu_k}{\sigma_k}\right)^2 \right\},$$

then $Z_n = (Z_n)_{n\geq 1}$ is a uniformly integrable martingale, $Z_\infty = \lim Z_n$ with probability 1, and $Z_n = \mathsf{E}(Z_\infty \mid \mathscr{F}_n)$ (P-a.\,e.), for any $n \geq 1$.

**Problem 7.7.9.** Adopting the notation introduced in the previous problem, let $\widetilde{\mathscr{F}} = \sigma(\bigcup \mathscr{F}_n)$, and let $\widetilde{\mathsf{P}}$ be the probability measure defined by

$$\widetilde{\mathsf{P}}(d\omega) = Z_\infty\, \mathsf{P}(d\omega).$$

Prove that if $\mathsf{E}Z_\infty = 1$, then one can claim that

$$\mathrm{Law}(h_n \mid \mathscr{F}_{n-1}; \widetilde{\mathsf{P}}) = \mathscr{N}(0, \sigma_n^2) \quad (\widetilde{\mathsf{P}}\text{-a.\,e.}).$$

If, furthermore, $\sigma_n^2 = \sigma_n^2(\omega)$ is independent from $\omega$, then

$$\mathrm{Law}(h_n \mid \widetilde{\mathsf{P}}) = \mathcal{N}(0, \sigma_n^2),$$

and the random variables $h_1, h_2, \ldots$ are independent, relative to the measure $\widetilde{\mathsf{P}}$.

**Problem 7.7.10.** Let $\mu_k$, $\sigma_k$, $\xi_k$ and $h_k$, for $k \geq 1$, be as in Problem 7.7.8, let $H_n = h_1 + \ldots + h_n$, $n \geq 1$, and let $X_n = e^{H_n}$.
  Prove that if

$$\mu_k + \frac{\sigma_k^2}{2} = 0 \quad \text{(P-a.e.)} \quad \text{for } k \geq 1,$$

then the sequence $X = (X_n, \mathscr{F}_n)_{n \geq 1}$ is a martingale.
  Now suppose that, for some $k \geq 1$, the above condition fails, and set

$$Z_\infty = \exp\left\{ -\sum_{k=1}^{\infty} \left( \frac{\mu_k}{\sigma_k} + \frac{\sigma_k}{2} \right) \xi_k - \frac{1}{2} \sum_{k=1}^{\infty} \left( \frac{\mu_k}{\sigma_k} + \frac{\sigma_k}{2} \right)^2 \right\}.$$

Assuming that $\mathsf{E} Z_\infty = 1$, define the measure

$$\widetilde{\mathsf{P}}(d\omega) = Z_\infty \, \mathsf{P}(d\omega),$$

and let $\mathscr{F} = \sigma(\bigcup \mathscr{F}_n)$.
  Prove that relative to the measure $\widetilde{\mathsf{P}}$ the sequence $(X_n, \mathscr{F}_n)_{n \geq 1}$, with $X_n = e^{H_n}$, is a martingale.


## 7.8   The Central Limit Theorem for Sums of Dependent Random Variables

**Problem 7.8.1.** Consider the random variables $\xi_n = \eta_n + \zeta_n$, $n \geq 1$, and suppose that $\eta_n \xrightarrow{d} \eta$ and $\zeta_n \xrightarrow{d} 0$. Prove that $\xi_n \xrightarrow{d} \eta$.

**Problem 7.8.2.** Let $(\xi_n(\varepsilon))$, $n \geq 1$, be some family of random variables, which is parameterized by $\varepsilon > 0$, and suppose that $\xi_n(\varepsilon) \xrightarrow{\mathsf{P}} 0$ as $n \to \infty$, for every $\varepsilon > 0$. By using, for example, the result in Problem 2.10.11, prove that one can construct the sequence $\varepsilon_n \downarrow 0$ in such a way that $\xi_n(\varepsilon_n) \xrightarrow{\mathsf{P}} 0$.
  *Hint.* Choose the sequence $\varepsilon_n \downarrow 0$ so that $\mathsf{P}\{|\xi_n(\varepsilon)| \geq 2^{-n}\} \leq 2^{-n}$, $n \geq 1$.

**Problem 7.8.3.** Consider the complex-valued random variables $(\alpha_k^n)$, $1 \leq k \leq n$, $n \geq 1$, chosen so that for some constant $C > 0$ and for some positive sequence $(a_n)_{n \geq 1}$, with $a_n \downarrow 0$, one has for every $n \geq 1$:

$$\sum_{k=1}^{n} |\alpha_k^n| \leq C \quad \text{and} \quad |\alpha_k^n| \leq a_n, \quad \text{for } 1 \leq k \leq n, \quad \text{(P-a.e.)}.$$

Prove that

$$\lim_{n\to\infty}\prod_{k=1}^{n}(1+\alpha_k^n)e^{-\alpha_k^n}=1 \quad \text{(P-a.e.)}.$$

**Problem 7.8.4.** Prove the statement formulated in Remark 2, following [P §7.8, Theorem 1].

**Problem 7.8.5.** Prove the statement formulated in the remark following the lemma in [P §7.8, 4].

**Problem 7.8.6.** Prove [P §7.8, Theorem 3].

**Problem 7.8.7.** Prove [P §7.8, Theorem 5].

**Problem 7.8.8.** Assuming that $\xi=(\xi_n)_{-\infty<n<\infty}$ is some sequence of independent and identically distributed random variables, with $E\xi_n=0$ and $D\xi_n<\infty$, consider the sequence $\eta=(\eta_n)_{n\geq1}$, given by

$$\eta_n=\sum_{j=-\infty}^{\infty}c_{n-j}\xi_j, \quad \text{with} \quad \sum_{j=-\infty}^{\infty}|c_j|^2<\infty,$$

and suppose that

$$D_n^2=E(\eta_1+\ldots+\eta_n)^2\to\infty.$$

Prove the following central limit theorem:

$$P\left\{\frac{\eta_1+\ldots+\eta_n}{D_n}\leq x\right\}\to\frac{1}{\sqrt{2\pi}}\int_{-\infty}^{x}e^{-t^2/2}\,dt.$$

**Problem 7.8.9.** Let $(\Omega^n,\mathscr{F}^n,(\mathscr{F}_k^n)_{0\leq k\leq n},P^n)$, $n\geq1$, be some sequence of filtered probability spaces and suppose that, given any $n\geq1$, the random variables $\xi^n=(\xi_k^n)_{1\leq k\leq n}$ are chosen so that each $\xi_k^n$ is $\mathscr{F}_k^n$-measurable.

Let $\mu$ be any infinitely divisible distribution on $(\mathbb{R},\mathscr{B}(\mathbb{R}))$, with characteristics $(b,c,F)$ (see Problem 3.6.17 and the continuous cutoff function $h=h(x)$ in that problem).

Consider the sequence of probability distributions associated with the random variables $Z^n=\sum_{k=1}^{n}\xi_k^n$, $n\geq1$, and prove that in order to guarantee the weak convergence of that sequence (of distributions) to some infinitely divisible distribution $\mu$, it is enought to require that the following conditions hold:

$$\sup_{1\leq k\leq n}P^n\left\{|\xi_k^n|>\varepsilon\mid\mathscr{F}_{k-1}^n\right\}\xrightarrow{P}0, \quad \varepsilon>0,$$

$$\sum_{1\leq k\leq n}E^n[h(\xi_k^n)\mid\mathscr{F}_{k-1}^n]\xrightarrow{P}b,$$

$$\sum_{1 \le k \le n} \left( \mathsf{E}^n\left[h^2(\xi_k^n) \mid \mathscr{F}_{k-1}^n\right] - \left(\mathsf{E}^n\left[h(\xi_k^n) \mid \mathscr{F}_{k-1}^n\right]\right)^2 \right) \xrightarrow{\mathsf{P}} \widetilde{c},$$

$$\sum_{1 \le k \le n} \mathsf{E}^n\left[g(\xi_k^n) \mid \mathscr{F}_{k-1}^n\right] \xrightarrow{\mathsf{P}} F(g), \quad g \in \mathfrak{G}_1,$$

where $\widetilde{c} = c + \int h^2(x) \, F(dx)$, $\mathfrak{G}_1 = \{g\}$ stands for the class of functions of the form $g_a(x) = (a|x| - 1)^+ \wedge 1$ for various choices of the rational number $a$, and $F(g) = \int g(x) \, F(dx)$.

**Problem 7.8.10.** Let $\xi_0, \xi_1, \xi_2, \ldots$ be some stationary in strict sense sequence with $\mathsf{E}\xi_0 = 0$. Let (comp. with Problem 6.3.5)

$$\alpha_k = \sup |\mathsf{P}(A \cap B) - \mathsf{P}(A)\mathsf{P}(B)|, \quad k \ge 1,$$

where the supremum is taken over all sets

$$A \in \mathscr{F}_0 = \sigma(\xi_0), \quad B \in \mathscr{F}_k^\infty = \sigma(\xi_k, \xi_{k+1}, \ldots).$$

Prove that if the strong mixing coefficients, $\alpha_k$, $k \ge 1$, are such that, for some $p > 2$, one has

$$\sum_{k \ge 1} \alpha_k^{\frac{p-2}{p}} < \infty \quad \text{and} \quad \mathsf{E}|\xi_0|^p < \infty,$$

then the joint distribution, $P_{t_1,\ldots,t_k}^n$, of the variables $X_{t_1}^n, \ldots, X_{t_k}^n$, given by

$$X_t^n = \frac{1}{\sqrt{n}} \sum_{k=1}^{\lfloor nt \rfloor} \xi_k, \quad t \ge 0,$$

converges weakly to the distribution, $P_{t_1,\ldots,t_k}$, of the variables $(\sqrt{c}\,B_{t_1}, \ldots, \sqrt{c}\,B_{t_k})$, where $B = (B_t)_{t \ge 0}$ is a Brownian motion process and the constant $c$ is given by

$$c = \mathsf{E}\xi_0^2 + 2 \sum_{k \ge 1} \mathsf{E}\xi_k^2.$$

## 7.9 Discrete Version of the Itô Formula

**Problem 7.9.1.** Prove the formula in $[\,\underline{P}\,\S7.9,\,(15)]$.

**Problem 7.9.2.** Based on the central limit theorem for the random walk $S = (S_n)_{n \ge 0}$, establish the following formula

$$\mathsf{E}|S_n| \sim \sqrt{\frac{2}{\pi} n}, \quad n \to \infty.$$

(Comp. with the hint to Problem 1.9.3.)

*Remark.* In formulas (17) and (18) in [P §7.9, Example 2] one can actually replace $2\pi$ in the denominator with $\pi/2$.

**Problem 7.9.3.** Prove the formula in [P §7.9, (22)].

**Problem 7.9.4.** Formula [P §7.9, (24)] remains valid for every function $F \in C^2$. Try to prove this claim.

**Problem 7.9.5.** Generalize formula [P §7.9, (11)] for the case where the function $F(X_k)$ is replaced by a non-homogeneous vector function of the form $F(k, X_k^1, \ldots, X_k^d)$.

**Problem 7.9.6.** Setting $f(x) = F'(x)$, consider the following trivial identity, which may be viewed as a *discrete version of the Itô formula*:

$$F(X_n) = F(X_0) + \sum_{k=1}^{n} f(X_{k-1})\Delta X_k + \sum_{k=1}^{n}[F(X_k) - F(X_{k-1}) - f(X_{k-1})\Delta X_k],$$

Outline the reasoning which, starting from the last relation, allows one to obtain the *discrete version of the Itô formula* (formula [P §7.9, (24)]), for twice continuously differentiable functions $F = F(x)$.

**Problem 7.9.7.** Generalize the identity in the previous problem for the case where the function $F(X_k)$ is replaced by a non-homogeneous vector function of the form $F(k, X_k^1, \ldots, X_k^d)$.

**Problem 7.9.8.** (*Discrete version of Tanaka formula; see Problem 1.9.3.*) Consider some symmetric Bernoulli scheme (i.e., a sequence of independent and identically distributed random variables), $\xi_1, \xi_2, \ldots$, with $\mathsf{P}\{\xi_n = +1\} = \mathsf{P}\{\xi_n = -1\} = 1/2$, $n \geq 1$, and let $S_0 = 0$ and $S_n = \xi_1 + \ldots + \xi_n$, for $n \geq 1$. Given any $x \in \mathbb{Z} = \{0, \pm 1, \pm 2, \ldots\}$, let

$$N_n(x) = \#\{k, 0 \leq k \leq n : S_k = x\}$$

be the number of the integers $0 \leq k \leq n$, for which $S_k = x$.
    Prove the following discrete analog of *Tanaka formula*:

$$|S_n - x| = |x| + \sum_{k=1}^{n} \text{sign}(S_{k-1} - x)\,\Delta S_k + N_{n-1}(x).$$

*Remark.* If $B = (B_t)_{t \geq 0}$ is a Brownian motion, then the renowned *Tanaka formula* gives

$$|B_t - x| = |x| + \int_0^t \text{sign}(B_s - x)\,dB_s + N_t(x),$$

where $N(x) = (N_t(x))_{t \geq 0}$ is the *local time* of the Brownian motion $B$ at level $x \in \mathbb{R}$. Recall that, originally, P. Lévy defined the local time $N_t(x)$ as (see, for example, [12] and [103]):

$$N_t(x) = \lim_{\varepsilon \downarrow 0} \frac{1}{2\varepsilon} \int_0^t I(x - \varepsilon < B_s < x + \varepsilon) \, ds.$$

## 7.10   The Probability for Ruin in Insurance. Martingale Approach

**Problem 7.10.1.** Prove that, under assumption **A** in [ $\underline{P}$ §7.10, **2** ], the process $N = (N_t)_{t \geq 0}$ has independent increments.

**Problem 7.10.2.** Prove that the process $X = (X_t)_{t \geq 0}$, defined [ $\underline{P}$ §7.10, **1** ], also has independent increments.

**Problem 7.10.3.** Consider the Cramér-Lundberg model, and formulate the analog of the Theorem in [ $\underline{P}$ §7.10, **3** ], for the case where the variables $\sigma_i$, $i = 1, 2, \ldots$, are independent and distributed with geometric law, i.e., $P\{\sigma_i = k\} = q^{k-1} p$, $k \geq 1$.

**Problem 7.10.4.** Let $N = (N_t)_{t \geq 0}$ be a *Poisson process* of parameter $\lambda$—see [ $\underline{P}$ §7.10, (3)]. Prove the following "Markov property:" for every choice of $0 = t_0 < t_1 < \ldots < t_n$ and $0 \leq k_1 \leq k_2 \leq \ldots \leq k_n$, one has

$$P(N_{t_n} = k_n \mid N_{t_1} = k_1, \ldots, N_{t_{n-1}} = k_{n-1}) = P(N_{t_n} = k_n \mid N_{t_{n-1}} = k_{n-1}).$$

**Problem 7.10.5.** Let $N = (N_t)_{t \geq 0}$ be a *standard* (i.e., of parameter $\lambda = 1$) Poisson process, and suppose that $\lambda(t)$ is some non-decreasing and continuous function, with $\lambda(0) = 0$. Then consider the process $N \circ \lambda = (N_{\lambda(t)})_{t \geq 0}$. Describe the properties of this process (finite dimensional distributions, moments, etc.).

**Problem 7.10.6.** Let $(T_1, \ldots, T_n)$ denote the times of the first $n$ jumps of a given Poisson process, let $(X_1, \ldots, X_n)$ be independent and identically distributed random variables, which are uniformly distributed on the interval $[0, t]$, and, finally, let $(X_{(1)}, \ldots, X_{(n)})$ denote the order statistics of the variables $(X_1, \ldots, X_n)$. Prove that

$$\text{Law}(T_1, \ldots, T_n \mid N_t = n) = \text{Law}(X_{(1)}, \ldots, X_{(n)}),$$

i.e., the conditional distribution of the vector $(T_1, \ldots, T_n)$, given the event $N_t = n$, coincides with distribution of the vector $(X_{(1)}, \ldots, X_{(n)})$.

**Problem 7.10.7.** Convince yourself that, if $(N_t)_{t \geq 0}$ is a Poisson process, then for any $s < t$ one can write

$$P(N_s = m \mid N_t = n) = \begin{cases} C_n^m \, (s/t)^m (1 - s/t)^{n-m}, & m \le n, \\ 0, & m > n. \end{cases}$$

**Problem 7.10.8.** It is an elementary matter to check that, if $X_1$ and $X_2$ are two independent random variables that have Poisson distribution with parameters, respectively, $\lambda_1$ and $\lambda_2$, then $X_1 + X_2$ also has Poisson distribution (and with parameters $\lambda_1 + \lambda_2$). Prove the converse statement (due to *D. Raikov*): if $X_1$ and $X_2$ are any two independent and non-degenerate random variables, for which one can claim that $X_1 + X_2$ is distributed with Poisson law, then $X_1$ and $X_2$ also must be distributed with Poisson law.

**Problem 7.10.9.** Suppose that $N = (N_t)_{t \ge 0}$ is a *standard* Poisson process, which is independent from the positive random variable $\theta$, and then consider the "hybrid" process $\widetilde{N} = (\widetilde{N}_t)_{t \ge 0}$, given by $\widetilde{N}_t = N_{t\theta}$. Prove the following properties:

(a) *Strong law of large numbers*:

$$\frac{N_t}{t} \to \theta \quad \text{as } t \to \infty \quad \text{(P-a.e.)}$$

(comp. with Example 4 in [ P §4.3, **4** ]).

(b) *Central limit theorem*:

$$P\left\{ \frac{N_t - \theta t}{\sqrt{\theta t}} \le x \right\} \to \Phi(x) \quad \text{as } t \to \infty.$$

(c) If $D\theta < \infty$, then

$$\frac{N_t - E N_t}{\sqrt{D N_t}} \to \frac{\theta - E\theta}{\sqrt{D\theta}}.$$

**Problem 7.10.10.** Prove that, for a given $u > 0$, the "ruin function"

$$\psi(u) = P\left\{ \inf_{t \ge 0} X_t \le 0 \right\} \quad (= P\{T < \infty\})$$

may be written in the form

$$\psi(u) = P\left\{ \sup_{n \ge 1} Y_n \ge u \right\},$$

where $Y_n = \sum_{i=1}^{n} (\xi_i - c\sigma_i)$.

In addition, prove that the estimate $\psi(u) \le e^{-Ru}$, which, under appropriate assumptions, was derived in [ P §7.10] by using "martingale" methods, can be established by using more elementary tools. Specifically, setting $\psi_n(u) = P\left\{ \max_{1 \le k \le n} Y_k \ge u \right\}$, $n \ge 1$, prove first that $\psi_1(u) \le e^{-Ru}$, and then prove by induction that $\psi_n(u) \le e^{-Ru}$, for any $n > 1$, so that $\psi(u) = \lim \psi_n(u) \le e^{-Ru}$.

**Problem 7.10.11.** The *time of ruin*, $T$, was defined by the formula $T = \inf\{t \geq 0 : X_t \leq 0\}$. Alternatively, the time of ruin may be defined as $\widetilde{T} = \inf\{t \geq 0 : X_t < 0\}$. Explain how the results established in [P §7.10] would change if the time $T$ is to be replaced by the time $\widetilde{T}$.

**Problem 7.10.12.** As a generalization of the (homogeneous) Poisson process, that was introduced in [P §7.10, 2], consider the *non-homogeneous Poisson process* $N = (N_t)_{t \geq 0}$, defined as:

$$N_t = \sum_{i \geq 1} I(T_i \leq t),$$

where $T_i = \sigma_1 + \ldots + \sigma_i$ and the random variables $\sigma_i$ are independent and identically distributed with

$$P\{\sigma_i \leq t\} = 1 - \exp\left\{ -\int_0^t \lambda(s)\, ds \right\}.$$

The function $\lambda(t)$ above, which is known as the *intensity function* of the process $N$, is assumed to satisfy: $\lambda(t) \geq 0$, $\int_0^t \lambda(s)\, ds < \infty$ and $\int_0^\infty \lambda(s)\, ds = \infty$. Prove that

$$P\{N_t < k\} = P\{T_k > t\} = \sum_{i=0}^{k-1} \exp\left\{ -\int_0^t \lambda(s)\, ds \right\} \frac{\left(\int_0^t \lambda(s)\, ds\right)^k}{k!}.$$

**Problem 7.10.13.** Let $N = (N_t)_{t \geq 0}$ be the non-homogeneous Poisson process defined in Problem 7.10.12 above, let $(\xi_n)_{n \geq 0}$ be a sequence of independent and identically distributed random variables, which are also independent from $N$, and, finally, let $g = g(t, x)$ be some non-negative function on $\mathbb{R} \times \mathbb{R}$. Prove the
   *Campbell formula*:

$$E \sum_{n=1}^{\infty} g(T_n, \xi_n) I(T_n \leq t) = \int_0^T E[g(s, \xi_1)]\lambda(s)\, ds.$$

**Problem 7.10.14.** Let $N = (N_t)_{t \geq 0}$ be a homogeneous Poisson process, defined by $N_0 = 0$ and $N_t = \sum_n I(T_n \leq t)$, for $t > 0$, the random variables $\sigma_{n+1} = T_{n+1} - T_n$ ($n \geq 1$, $T_0 = 0$) being independent and identically distributed, with law

$$P\{\sigma_{n+1} \geq x\} = e^{-\lambda x}, \quad x \geq 0.$$

Setting $U_t = t - T_{N_t}$ and $V_t = T_{N_t + 1}$, prove that

$$P\{U_t \leq u, V_t \leq v\} = \left[ I_{\{u \geq t\}} + I_{\{u < t\}}(1 - e^{-\lambda u}) \right](1 - e^{-\lambda v}).$$

(In particular, for any fixed $t > 0$, the variables $U_t$ and $V_t$ are independent, and $V_t$ is exponentially distributed with parameter $\lambda$.) Find the probability $P\{T_{N_t + 1} - T_{N_t} \geq$

$x\}$, and prove that $P\{T_{N_t+1} - T_{N_t} \geq x\} \neq e^{-\lambda x}$ $(= P\{T_{n+1} - T_n \geq x\})$. Prove that, as $t \to \infty$, the distribution of $T_{N_t+1} - T_{N_t}$ converges weakly to the distribution law of the sum of two independent exponentially distributed random variables of the same parameter $\lambda$.

## 7.11   On the Fundamental Theorem of Financial Mathematics: Martingale Characterization of the Absence of Arbitrage

**Problem 7.11.1.** Prove that with $N = 1$ the no-arbitrage condition is equivalent to the inequality [ $\underline{P}$ §7.11, (18)]. (It is assumed that $P\{\Delta S_1 = 0\} < 1$.)

**Problem 7.11.2.** Prove that in the proof of Lemma 1 in [ $\underline{P}$ §7.11, **4** ] condition (19) makes case (2) impossible.

**Problem 7.11.3.** Prove that the measure $\widetilde{P}$ from Example 1 in [ $\underline{P}$ §7.11, **5** ] is a *martingale measure* and that this measure is unique in the class $M(P)$.

**Problem 7.11.4.** Investigate the uniqueness of the martingale measure constructed in Example 2 in [ $\underline{P}$ §7.11, **5** ].

**Problem 7.11.5.** Prove that in the $(B, S)$-model the assumption $|M(P)| = 1$ implies that the variables $\frac{S_n}{B_n}$, $1 \leq n \leq N$, are "conditionally bi-valued."

**Problem 7.11.6.** According to Remark 1, following [ $\underline{P}$ §7.11, Theorem 1], the First Fundamental Theorem remains valid for any $N < \infty$ and any $d < \infty$. Prove by way of example that if $d = \infty$, then it could happen that the market is free of arbitrage and yet no martingale measure exists.

**Problem 7.11.7.** In addition to [ $\underline{P}$ §7.11, Definition 1], we say that the $(B, S)$-market is free of arbitrage in *weak sense*, if, for every self-financing portfolio $\pi = (\beta, \gamma)$, with $X_0^\pi = 0$ and $X_n^\pi \geq 0$ (P-a. e.), for $n \leq N$, one has $X_N^\pi = 0$ (P-a. e.). We say that the $(B, S)$-market is arbitrage-free in *strong sense* if, for every self-financing portfolio $\pi$, with $X_0^\pi = 0$ and $X_N^\pi \geq 0$ (P-a. e.), one has $X_n^\pi = 0$ (P-a. e.), for $0 \leq n \leq N$.

Assuming that all assumptions in [ $\underline{P}$ §7.11, Theorem 1] are in force, prove that the following conditions are equivalent:

   (i) The $(B, S)$-market is free of arbitrage.
  (ii) The $(B, S)$-market is free of arbitrage in weak sense.
 (iii) The $(B, S)$-market is free of arbitrage in strong sense.

**Problem 7.11.8.** Just as in [ $\underline{P}$ §7.11, Theorem 1], consider the family of all martingale measures:

$$M(P) = \{\widetilde{P} \sim P : S/B \text{ is a } \widetilde{P}\text{-martingale}\}$$

and let

$$M_{\text{loc}}(P) = \{\widetilde{P} \sim P : S/B \text{ is a } \widetilde{P}\text{-local martingale}\},$$

$$M_b(P) = \{\widetilde{P} \sim P : \widetilde{P} \in M(P) \text{ and } \frac{d\widetilde{P}}{dP}(\omega) \leq C(\widetilde{P}) \; (\widetilde{P}\text{-a.e.}) \text{ for some}$$
$$\text{constant } C(\widetilde{P})\}.$$

Prove that, in the setting of [P §7.11, Theorem 1], the following conditions are equivalent:

(i) $M(P) \neq \varnothing$;    (ii) $M_{\text{loc}}(P) \neq \varnothing$;    (iii) $M_b(P) \neq \varnothing$.

## 7.12  Hedging of Financial Contracts in Arbitrage-Free Markets

**Problem 7.12.1.** Find the price, $\mathbb{C}(f_N; P)$, of a standard call option with payoff $f_N = (S_N - K)^+$, in the $(B, S)$-market described in Example 2 in [P §7.11, 5].

**Problem 7.12.2.** Prove the inequality in [P §7.12, (10)] in the opposite direction.

**Problem 7.12.3.** Prove formulas [P §7.12, (12) and (13)].

**Problem 7.12.4.** Give a detailed derivation of formula [P §7.12, (23)].

**Problem 7.12.5.** Prove formulas [P §7.12, (25) and (28)].

**Problem 7.12.6.** Give a detailed derivation of formula [P §7.12, (32)].

**Problem 7.12.7.** Consider the one-period version of the CRR-model formulated in (17) in [P §7.12, 7]:

$$B_1 = B_0(1 + r), \quad S_1 = S_0(1 + \rho),$$

where we suppose that $\rho$ takes *two* values, $a$ and $b$, chosen so that $-1 < a < r < b$.

Now suppose that $\rho$ is *uniformly* distributed in the interval $[a, b]$ (with the same choice for $a$ and $b$) and consider the period 1 payoff $f(S_1) = f(S_0(1+\rho))$, for some convex-down and continuous payoff function $f = f(x), x \in [S_0(1+a), S_0(1+b)]$ (here we suppose that $S_0 = \text{const}$). Prove that the upper hedging price:

$$\widehat{\mathbb{C}}(f; P) = \inf\{x : \exists \pi, X_0^\pi = x \text{ and } X_1^\pi \geq f(S_0(1 + \rho)) \; \forall \rho \in [a, b]\},$$

coincides with the upper hedging price in [P §7.12, (19)], with $N = 1$ and with $P\{\rho = b\} = p$ and $P\{\rho = a\} = 1 - p, 0 < p < 1$, so that

$$\widehat{\mathbb{C}}(f; P) = \frac{r - a}{b - a} \cdot \frac{f(S_0(1 + b))}{1 + r} + \frac{b - r}{b - a} \cdot \frac{f(S_0(1 + a))}{1 + r}.$$

**Problem 7.12.8.** (*The Black–Scholes formula.*) As a generalization of the discrete-time $(B, S)$-market $B = (B_n)_{0 \le n \le N}$ and $S = (S_n)_{0 \le n \le N}$—see [P §7.12, 2]—consider the continuous-time $(B, S)$-market model

$$B_t = B_0 e^{-rt} \quad \text{and} \quad S_t = S_0 e^{\mu t + \sigma W_t}, \quad 0 \le t \le T, \tag{$*$}$$

in which $\mu, \sigma \in \mathbb{R}$ and $r > 0$ are exogenously specified constants and $W = (W_t)_{0 \le t \le T}$ is exogenously specified Brownian motion. Analogously to [P §7.12, (1)], for a given strike-price $K > 0$, consider an European-style call-option with termination payoff $f_T = (S_T - K)^+ \equiv \max[S_T - K, 0]$ and suppose that in $(*)$ the constant $\mu$ is chosen to be $\mu = r - \frac{1}{2}\sigma^2$. Under these conditions, prove that the following properties:
  (a) The process $(\frac{S_t}{B_t})_{0 \le t \le T}$ is a martingale.
  (b) The "fair" price of the call option, $\mathbb{C}(f_T; \mathsf{P})$, defined as

$$\mathbb{C}(f_T; \mathsf{P}) = B_0 \mathsf{E} \frac{f_T}{B_T},$$

can be computed according to the *Black–Scholes formula*:

$$\mathbb{C}(f_T; \mathsf{P}) = S_0 \, \Phi \left( \frac{\ln \frac{S_0}{k} + T \left( r + \frac{\sigma^2}{2} \right)}{\sigma \sqrt{T}} \right) - K e^{-rT} \Phi \left( \frac{\ln \frac{S_0}{k} + T \left( r - \frac{\sigma^2}{2} \right)}{\sigma \sqrt{T}} \right),$$

where $\Phi(x) = \frac{1}{\sqrt{2\pi}} \int_{-\infty}^{x} e^{-y^2/2} dy$, $x \in \mathbb{R}$.
   *Hint.* Prove that $\mathbb{C}(f_T; \mathsf{P}) = e^{-rT} \mathsf{E}(a \, e^{b\xi - b^2/2} - K)^+$, where $a = S_0 e^{rT}$, $b = \sigma \sqrt{T}$ and $\xi \in \mathcal{N}(0, 1)$. By using direct calculation prove that

$$\mathsf{E}(a \, e^{b\xi - b^2/2} - K)^+ = a \, \Phi \left( \frac{\ln \frac{a}{k} + \frac{1}{2}b^2}{b} \right) - K \, \Phi \left( \frac{\ln \frac{a}{k} - \frac{1}{2}b^2}{b} \right).$$

# 7.13   The Optimal Stopping Problem: Martingale Approach

**Problem 7.13.1.** Prove that the random variable $\xi(\omega) = \sup_{\alpha \in \mathfrak{A}_0} \xi_\alpha(\omega)$, introduced in the proof of the Lemma in [P §7.13, 3], satisfies conditions (a) and (b) in the definition of essential supremum (see [P §7.13, 3]).
   *Hint.* If $\alpha \notin \mathfrak{A}_0$, consider the expression $\mathsf{E} \max(\xi(\omega), \xi_\alpha(\omega))$.

**Problem 7.13.2.** Prove that the variable $\xi(\omega) = \tan \tilde{\xi}(\omega)$, introduced at the end of the proof of the Lemma in [P §7.13, 3], satisfies conditions (a) and (b) in the definition of essential supremum (see [P §7.13, 3]).

**Problem 7.13.3.** Let $\xi_1, \xi_2, \ldots$ be any sequence of independent and identically distributed random variables with $E|\xi_1| < \infty$. Consider the optimal stopping problem within the class $\mathfrak{M}_1^\infty = \{\tau : 1 \le \tau < \infty\}$:

$$V^* = \sup_{\tau \in \mathfrak{M}_1^\infty} E\left(\max_{i \le \tau} \xi_i - c\tau\right),$$

and let $\tau^* = \inf\{n \ge 1 : \xi_n \ge A^*\}$, where $A^*$ stands for the unique root of the equation $E(\xi_1 - A^*) = c$, with the understanding that $\inf \varnothing = \infty$. Prove that if $P\{\tau^* < \infty\} = 1$, then the time $\tau^*$ is optimal in the class of all finite stopping times $\tau$, for which $E\left(\max_{i \le \tau} \xi_i - c\tau\right)$ exists. Show also that $V^* = A^*$.

**Problem 7.13.4.** In addition to the notation introduced in [$\underline{P}$ §7.13, **1**] and [$\underline{P}$ §7.13, **2**], let

$$\mathfrak{M}_n^\infty = \{\tau : n \le \tau < \infty\},$$
$$V_n^\infty = \sup_{\tau \in \mathfrak{M}_n^\infty} E f_\tau,$$
$$v_n^\infty = \operatorname{ess\,sup}_{\tau \in \mathfrak{M}_n^\infty} E(f_\tau \mid \mathscr{F}_n),$$
$$\tau_n^\infty = \inf\{k \ge n : v_n^\infty = f_n\},$$

and assume that

$$E \sup f_n^- < \infty.$$

Prove that the following statements can be made for the limiting random variables $\tilde{v}_n = \lim_{N \to \infty} v_n^N$:

(a) For every $\tau \in \mathfrak{M}_n^\infty$, one has

$$\tilde{v}_n \ge E(f_\tau \mid \mathscr{F}_n).$$

(b) If $\tau_n^\infty \in \mathfrak{M}_n^\infty$, then

$$\tilde{v}_n = E(f_{\tau_n^\infty} \mid \mathscr{F}_n),$$
$$\tilde{v}_n = v_n^\infty \quad (= \operatorname{ess\,sup}_{\tau \in \mathfrak{M}_n^\infty} E(f_\tau \mid \mathscr{F}_n)).$$

**Problem 7.13.5.** Adopt the notation introduced in the previous problem and let $\tau_n^\infty \in \mathfrak{M}_n^\infty$. By using (a) and (b) in the previous problem, conclude that the time $\tau_n^\infty$ is optimal, in the sense that

$$\operatorname{ess\,sup}_{\tau \in \mathfrak{M}_n^\infty} E(f_\tau \mid \mathscr{F}_n) = E(f_{\tau_n^\infty} \mid \mathscr{F}_n) \quad \text{(P-a.e.)}$$

and

$$\sup_{\tau \in \mathfrak{M}_n^\infty} \mathsf{E} f_\tau = \mathsf{E} f_{\tau_n^\infty} \,,$$

i.e., $V_n^\infty = \mathsf{E} f_{\tau_n^\infty}$.

**Problem 7.13.6.** Suppose that the family of random variables $\Sigma = \{\xi_\alpha(\omega); \alpha \in \mathfrak{A}\}$, defined on some probability space $(\Omega, \mathscr{F}, \mathsf{P})$, is chosen so that, for some fixed constant $C$, one has $\mathsf{E}|\xi_\alpha| \leq C$ for all $\alpha \in \mathfrak{A}$. In addition, suppose that the family $\Sigma$ is "sufficiently rich," in the sense that: if $\xi_{\alpha_1} \in \Sigma$ and $\xi_{\alpha_2} \in \Sigma$, for some $\alpha_1, \alpha_2 \in \mathfrak{A}$, then

$$\xi = \xi_{\alpha_1} I_A + \xi_{\alpha_2} I_{\overline{A}} \in \Sigma$$

for every $A \in \mathscr{F}$. (A family $\Sigma$ with these properties is said to admit *needle variations*.) Setting

$$Q(A) = \sup_{\alpha \in \mathfrak{A}} \mathsf{E} \xi_\alpha I_A \,, \quad A \in \mathscr{F},$$

prove that:
   (a) The set function $Q = Q(\cdot)$ is $\sigma$-additive.
   (b) $Q \ll \mathsf{P}$.
   (c) The Radon-Nikodym derivative $\frac{dQ}{d\mathsf{P}}$ is given by

$$\frac{dQ}{d\mathsf{P}} = \operatorname*{ess\,sup}_{\alpha \in \mathfrak{A}} \xi_\alpha \quad (\mathsf{P}\text{-a.e.}).$$

(In particular, (c) above may be viewed as a proof of the fact that the essential supremum of a family of random variables that admits needle variations must be finite.)

   Prove that the statement (a), (b) and (c) above remain valid if the condition $\mathsf{E}|\xi_\alpha| \leq C, \alpha \in \mathfrak{A}$, is replaced with $\mathsf{E} \xi_\alpha^- < \infty, \alpha \in \mathfrak{A}$.

**Problem 7.13.7.** Let $\mathfrak{M}_n^\infty = \{\tau : n \leq \tau < \infty\}$. Prove that if $\tau_1, \tau_2 \in \mathfrak{M}_n^\infty$ and $A \in \mathscr{F}_n$, then the time $\tau = \tau_1 I_A + \tau_2 I_{\overline{A}}$ belongs to $\mathfrak{M}_n^\infty$.

**Problem 7.13.8.** Let $(\Omega, \mathscr{F}, (\mathscr{F}_n)_{n \geq 0}, \mathsf{P})$ be any filtered probability space, and let $f_n$ be any $\mathscr{F}_n$-measurable random variable with $\mathsf{E} f_n^- < \infty$, for $n \geq 0$. Prove that, for every fixed $n \geq 0$, the family of random variables $\{\mathsf{E}(f_\tau \mid \mathscr{F}_n); \tau \in \mathfrak{M}_n^\infty\}$ admits needle variations.

# Chapter 8
# Sequences of Random Variables that Form Markov Chains

## 8.1 Definitions and Basic Properties

**Problem 8.1.1.** Prove the statements formulated as Problems 1a, 1b and 1c in the proof of [ P §8.1, Theorem 1].

**Problem 8.1.2.** Prove that in [ P §8.1, Theorem 2] the function $\omega \to P_{n+1}(B - X_n(\omega))$ is $\mathscr{F}_n$-measurable.

**Problem 8.1.3.** By using [ P §2.2, Lemma 3], prove the relations [ P §8.1, (11) and (12)].

**Problem 8.1.4.** Prove the relations [ P §8.1, (20) and (27)].

**Problem 8.1.5.** Prove the identity in [ P §8.1, (33)].

**Problem 8.1.6.** Prove the relations (i), (ii) and (iii), formulated at the end of [ P §8.1, 8 ].

**Problem 8.1.7.** Can one conclude from the Markov property [ P §8.1, (3)] that for any choice of the sets $B_0, B_1, \ldots, B_n, B \in \mathscr{E}$, with $\mathsf{P}\{X_0 \in B_0, X_1 \in B_1, \ldots, X_n \in B_n\} > 0$, one must have:

$$\mathsf{P}(X_{n+1} \in B \mid X_0 \in B_0, X_1 \in B_1, \ldots, X_n \in B_n) = \mathsf{P}(X_{n+1} \in B \mid X_n \in B_n)?$$

**Problem 8.1.8.** Consider a cylindrical piece of chalk of length 1. Suppose that the piece is broken "randomly" into two pieces. Then the left piece is broken at "random" into two pieces—and so on. Let $X_n$ denote the length of the left piece after the $n^{\text{th}}$ breaking, with the understanding that $X_0 = 1$, and let $\mathscr{F}_n = \sigma(X_1, \ldots, X_n)$. Thus, the conditional distribution of $X_{n+1}$ given $X_n = x$ must be uniform on $[0, x]$.

Prove that the sequence $(X_n)_{n \geq 0}$ forms a homogeneous Markov chain. In addition, prove that for every $\alpha > -1$ the sequence

$$M_n = (1 + \alpha)^n X_n^\alpha, \quad n \geq 0,$$

A.N. Shiryaev, *Problems in Probability*, Problem Books in Mathematics,
DOI 10.1007/978-1-4614-3688-1_8,
© Springer Science+Business Media New York 2012

forms a non-negative martingale. Prove that with Probability 1 for every $0 < p < e$ one has

$$\lim_n p^n X_n = 0,$$

and for every $p > e$ one has

$$\lim_n p^n X_n = \infty.$$

(Given that "on average" every piece is broken in half, one may expect that $X_n$ would converge to 0 as $2^{-n}$. However, the property $\lim_n p^n X_n = 0$ (P-a.e.) implies that $X_n$ converges to 0 much faster—"almost" as $e^{-n}$.)

**Problem 8.1.9.** Let $\xi_1, \xi_2, \ldots$ be any sequence of independent and identically distributed random variables, which can be associated with a Bernoulli scheme, i.e., $P\{\xi_n = 1\} = P\{\xi_n = -1\} = \frac{1}{2}, n \geq 1$, and let $S_0 = 0, S_n = \xi_1 + \cdots + \xi_n$ and $M_n = \max\{S_k : 0 \leq k \leq n\}, n \geq 1$.

(a) Do the sequences $(|S_n|)_{n\geq 0}$, $(|M_n|)_{n\geq 0}$ and $(M_n - S_n)_{n\geq 0}$ represent Markov chains?

(b) Are these sequences going to be Markov chains if $S_0 = x \neq 0$ and $S_n = x + \xi_1 + \ldots + \xi_n$?

**Problem 8.1.10.** Consider the Markov chain $(X_n)_{n\geq 0}$, with state space $E = \{-1, 0, 1\}$, and suppose that $p_{ij} > 0$, for $i, j \in E$. Give necessary and sufficient conditions for the sequence $(|X_n|)_{n\geq 0}$ to be a Markov chain.

**Problem 8.1.11.** Give an example of a sequence of random variables $X = (X_n)_{n\geq 0}$, which is not a Markov chain, but for which the Chapman–Kolmogorov equation nevertheless holds.

**Problem 8.1.12.** Suppose that the sequence $X = (X_n)_{n\geq 0}$ forms a Markov chain in broad sense, and let $Y_n = X_{n+1} - X_n$, for $n \geq 0$. Prove that the sequence $(X, Y) = ((X_n)_{n\geq 0}, (Y_n)_{n\geq 0})$ is also a Markov chain. Does any of the following sequences represent a Markov chain: $(X_n, X_{n+1})_{n\geq 0}$, $(X_{2n})_{n\geq 0}$, $(X_{n+k})_{n\geq 0}$ for $k \geq 1$?

**Problem 8.1.13.** We say that a sequence of random variables $X = (X_n)_{n\geq 0}$, in which every $X_n$ takes values in some countable set $E$, forms a *Markov chain of order $r \geq 1$*, if

$$P(X_{n+1} = i_{n+1} \mid X_0 = i_0, \ldots, X_n = i_n) =$$
$$= P(X_{n+1} = i_{n+1} \mid X_{n-r+1} = i_{n-r+1}, \ldots, X_n = i_n),$$

for all $i_0, \ldots, i_{n+1}, n \geq r$.

Assuming that $X = (X_n)_{n\geq 0}$ is a Markov chain of order $r \geq 1$, let $\widetilde{X}_n = (X_n, X_{n+1}, \ldots, X_{n+r-1}), n \geq 0$. Prove that the sequence $\widetilde{X} = (\widetilde{X}_n)_{n\geq 0}$ represents a canonical (i.e., of order $r = 1$) Markov chain.

**Problem 8.1.14.** (*Random walk on groups.*) Let $G$ be some finite group, endowed with binary operation $\oplus$, so that the usual group properties hold:

(i) $x, y \in G$ implies $x \oplus y \in G$;
(ii) if $x, y, z \in G$, then $x \oplus (y \oplus z) = (x \oplus y) \oplus z$;
(iii) there is a unique $e \in G$, such that $x \oplus e = e \oplus x = x$, for all $x \in G$;
(iv) given any $x \in G$, there is an inverse $-x \in G$, which is characterized by the property $x \oplus (-x) = (-x) \oplus x = e$.

Let $\xi_0, \xi_1, \xi_2, \ldots$ be any sequence random elements in $G$, which are identically distributed with law $Q(g) = \mathsf{P}\{\xi_n = g\}$, $g \in G$, $n \geq 0$.
Prove that the random walk $X = (X_n)_{n \geq 0}$, given by $X_n = \xi_0 \oplus \xi_1 \oplus \ldots \oplus \xi_n$, forms a Markov chain and give the respective transition probability matrix.

**Problem 8.1.15.** (*Random walk on a circle.*) Let $\xi_1, \xi_2, \ldots$ be any sequence of independent random variables that are identically distributed in the interval $[0, 1]$, with a (common) continuous probability density $f(x)$. For a fixed $x \in [0, 1)$, consider the sequence $X = (X_n)_{n \geq 0}$, given by $X_0 = x$ and

$$X_n = x + \xi_1 + \ldots + \xi_n \qquad (\text{mod } 1).$$

Prove that $X = (X_n)_{n \geq 0}$ is a Markov chain with state space $E = [0, 1)$. Find the transition function for this Markov chain.

**Problem 8.1.16.** Suppose that $X = (X_n)_{n \geq 0}$ and $Y = (Y_n)_{n \geq 0}$ are two independent Markov chains, defined on the same probability space $(\Omega, \mathscr{F}, \mathsf{P})$, taking values in the same countable space $E = \{i, j, \ldots\}$, and sharing the same transition probability matrix. Prove that, for any choice of the initial values, $X_0 = x \in E$ and $Y_0 = y \in E$, the sequence $(X, Y) = (X_n, Y_n)_{n \geq 0}$ forms a Markov chain. Find the transition probability matrix for this Markov chain.

**Problem 8.1.17.** Let $X_1, X_2, \ldots$ be any sequence of independent and identically distributed non-negative random variables, that share a common continuous distribution function. Define the *record moments*:

$$\mathscr{R}_1 = 1, \quad \mathscr{R}_k = \inf\{n \geq \mathscr{R}_{k-1} : X_n \geq \max(X_1, \ldots, X_{n-1})\}, \quad k \geq 2,$$

and prove that $\mathscr{R} = (\mathscr{R}_k)_{k \geq 1}$ is a Markov chain. Find the transition probability matrix for this Markov chain.

**Problem 8.1.18.** Let $X_1, X_2, \ldots$ be some sequence of independent and identically distributed non-negative random variables that share the same *discrete* range of values. Assuming that the record times $\mathscr{R} = (\mathscr{R}_k)_{k \geq 1}$ are defined as in the previous problem, prove that the associated sequence of record values $V = (V_k)_{k \geq 1}$, with $V_k = X_{\mathscr{R}_k}$, forms a Markov chain. Find the transition probability matrix for this Markov chain.

**Problem 8.1.19.** (*Time reversibility for Markov chains.*) Suppose that $X = (X_n)_{0 \le n \le N}$ is some irreducible Markov chain with a countable state space $E$, with transition probability matrix $P = \| p_{ij} \|$, and with *invariant* initial distribution $q = (q_i)$, such that $q_i > 0$ for all $i \in E$ (for the definition of invariant distribution see [P §8.3, **1**]).

Next, consider the sequence $\widetilde{X}^{(N)} = (\widetilde{X}_n)_{0 \le n \le N}$, given by $\widetilde{X}_n = X_{N-n}$, which is nothing but the sequence $X$ in reverse time. Setting $\widetilde{P} = \| \widetilde{p}_{ij} \|$, where $\widetilde{p}_{ij} = p_{ji}$, prove that the matrix $\widetilde{P}$ is stochastic. In addition, prove that $\widetilde{X}^{(N)}$ is a Markov chain with transition matrix $\widetilde{P}$.

*Remark.* The Markov property comes down to saying that, conditioned to the "present", the "past" and the "future" are independent—see [P §8.1, (7)]. Because of this symmetry between past and future, one is lead to suspect that the Markov property of the sequence $X = (X_n)_{0 \le n \le N}$ may be preserved under time reversal, provided that in reverse time the initial distribution is chosen in a certain way. The statement in this problem makes this idea precise: the Markov property is preserved under time-reversal, possibly with a different transition probability matrix, provided that the initial distribution is chosen to be the invariant one.

**Problem 8.1.20.** (*Reversible Markov Chains.*) Let $X = (X_n)_{n \ge 0}$ be any Markov chain with countable state space $E$, with transition probability matrix $P = \| p_{ij} \|$, and with invariant distribution $q = (q_i)$. We say that the $(q, P)$-Markov chain $X = (X_n)_{n \ge 0}$ is *reversible* (see, for example, [22]) if, for *every* $N \ge 1$, the sequence $\widetilde{X}^{(N)} = (\widetilde{X}_n)_{0 \le n \le N}$, given by $\widetilde{X}_n = X_{N-n}$, is also a $(q, P)$-Markov chain.

Prove that an irreducible $(q, P)$-Markov chain is reversable if and only if the following condition holds:

$$q_i p_{ij} = q_j p_{ji}, \quad \text{for all } i, j \in E.$$

Convince yourself that, if the distribution $\lambda = (\lambda_i)$ ($\lambda_i \ge 0$, $\sum \lambda_i = 1$) and the matrix $P$ satisfy the *balance equation*

$$\lambda_i p_{ij} = \lambda_j p_{ji}, \quad i, j \in E,$$

then $\lambda = (\lambda_i)$ coincides with the invariant distribution $q = (q_i)$.

**Problem 8.1.21.** Consider the Ehrenfests' model (see [P §8.8, **3**]) with stationary distribution $q_i = C_N^i (1/2)^N$, $i = 0, 1, \ldots, N$, and prove that the following balance equation is satified:

$$q_i \, p_{i,i+1} = q_{i+1} p_{i+1,i} \, .$$

(Note that in this model $p_{ij} = 0$, if $|i - j| > 1$.)

**Problem 8.1.22.** Prove that a Markov chain with transition probability matrix

$$P = \begin{pmatrix} 0 & 2/3 & 1/3 \\ 1/3 & 0 & 2/3 \\ 2/3 & 1/3 & 0 \end{pmatrix}$$

has invariant distribution $q = (1/3, 1/3, 1/3)$. Convince yourself that, for any $N \geq 1$, the sequence $\widetilde{X}^{(N)} = (X_{N-n})_{0 \leq n \leq N}$ forms a Markov chain with transition probability matrix $\widetilde{P}$, which is simply the transpose of $P$. Argue that the chain $X = (X_n)_{n \geq 0}$ is not reversible and give the intuition behind this feature.

**Problem 8.1.23.** Let $X = (X_n)_{n \geq 0}$ be any stationary (in strict sense) and non-negative Gaussian sequence. Prove that this sequence has the Markov property if and only if the covariance $\mathrm{cov}(X_n, X_{n+m})$, $m, n \geq 0$, has the form:

$$\mathrm{cov}(X_n, X_{n+m}) = \sigma^2 \rho^m,$$

for some choice of $\sigma > 0$ and $-1 \leq \rho \leq 1$.

## 8.2    The Strong and the Generalized Markov Properties

**Problem 8.2.1.** Prove that the function $\psi(x) = \mathsf{E}_x H$, introduced in the Remark in [P §8.2, 1] is $\mathcal{E}$-measurable.

**Problem 8.2.2.** Prove the relation in [P §8.2, (12)].

**Problem 8.2.3.** Prove the relation in [P §8.2, (13)].

**Problem 8.2.4.** Are the random variables $X_n - X_{\tau \wedge n}$ and $X_{\tau \wedge n}$, from the Example in [P §8.2, 3], independent?

**Problem 8.2.5.** Prove the formula in [P §8.2, (23)]

**Problem 8.2.6.** Suppose that the space $E$ is at most countable, let $(\Omega, \mathscr{F}) = (E^\infty, \mathscr{E}^\infty)$, and let $\theta_n : \Omega \to \Omega, n \geq 1$, denote the usual shift operators

$$\theta_n(\omega) = (x_n, x_{n+1}, \ldots), \qquad \omega = (x_0, x_1, \ldots).$$

Let $X = (X_n(\omega))_{n \geq 0}$ be the canonical coordinate process on $\Omega$, defined as $X_n(\omega) = x_n, \omega = (x_0, x_1, \ldots)$, for $n \geq 0$.

Given any $\mathscr{F}$-measurable function $H = H(\omega)$, set (see [P §8.2, (1)])

$$(H \circ \theta_n)(\omega) = H(\theta_n(\omega)), \qquad n \geq 1,$$

and, given any $B \in \mathscr{F}$, set (comp. with [P §5.1, Definition 2])

$$\theta_n^{-1}(B) = \{\omega : \theta_n(\omega) \in B\}, \qquad n \geq 1.$$

With the above definitions in mind, prove the following properties:
(a) For any $m \geq 0$ and $n \geq 1$, one has

$$X_m \circ \theta_n = X_{m+n}$$

i.e., $(X_m \circ \theta_n)(\omega) = X_{m+n}(\omega)$.

(b) For any $m \geq 0$ and $n \geq 1$, one has

$$\theta_n^{-1}\{X_m \in A\} = \{X_m \circ \theta_n \in A\} = \{X_{m+n} \in A\}$$

i.e., for every $A \in \mathscr{E}$,

$$\theta_n^{-1}\{\omega : X_m(\omega) \in A\} = \{\omega : (X_m \circ \theta_n)(\omega) \in A\} = \{\omega : X_{m+n}(\omega) \in A\};$$

and, more generally,

$$\theta_n^{-1}\{X_0 \in A_0, \ldots, X_m \in A_m\} = \{X_0 \circ \theta_n \in A_0, \ldots, X_m \circ \theta_n \in A_m\}$$
$$= \{X_n \in A_0, \ldots, X_{m+n} \in A_m\}.$$

In addition, prove that

$$\theta_n^{-1}(\mathscr{F}_m) = \sigma(X_n, \ldots, X_{m+n}), \qquad\qquad (*)$$

with the obvious meaning of the symbols $\theta_n^{-1}(\mathscr{F}_m)$ and $\sigma(X_n, \ldots, X_{m+n})$ (explain).

**Problem 8.2.7.** Adopt the notation introduced in Problem 8.2.6, $H = H(\omega)$ be any $\mathscr{F}$-measurable function on $(\Omega, \mathscr{F})$, and let $A \in \mathscr{B}(\mathbb{R})$. Prove that

$$(H \circ \theta_n)^{-1}(A) = \theta_n^{-1}(H^{-1}(A)). \qquad\qquad (**)$$

**Problem 8.2.8.** Adopt the notation introduced in Problem 8.2.6 and let $\tau = \tau(\omega)$ be some stopping time (i.e., a finite Markov moment) on $(\Omega, \mathscr{F}, (\mathscr{F}_k)_{k \geq 0})$, where $\mathscr{F}_k = \sigma(X_0, X_1, \ldots, X_k)$, $k \geq 0$. Based on $(**)$ and $(*)$ in Problems 8.2.7 and 8.2.6, prove that, for any given $n \geq 0$, the moment $n + \tau \circ \theta_n$ is also a stopping time, i.e., $\{\omega : n + (\tau \circ \theta_n)(\omega) = m\} \in \mathscr{F}_m$, for every $m \geq n$.

**Warning:** Problems 8.2.9–8.2.21 below assume the notation and the terminology introduced in Problems 8.2.6 and 8.2.8.

**Problem 8.2.9.** Let $\sigma = \sigma(\omega)$ be any stopping time on $(\Omega, \mathscr{F}, (\mathscr{F}_k)_{k \geq 0})$ and let $H = H(\omega)$ be any $\mathscr{F}$-measurable function on $\Omega$. The symbol $(H \circ \theta_\sigma)(\omega)$ is understood as the function $H(\theta_{\sigma(\omega)}(\omega))$, i.e., $H(\theta_n(\omega))$, for $\omega \in \{\omega : \sigma(\omega) = n\}$. As a generalization of Problem 8.2.8, prove that $\sigma + \tau \circ \theta_\sigma$ is also a stopping time.

**Problem 8.2.10.** Given any two stopping times, $\tau$ and $\sigma$, on $(\Omega, \mathscr{F}, (\mathscr{F}_k)_{k \geq 0})$, the random variable $X_\tau \circ \theta_\sigma$ will be understood as $X_{\tau(\theta_\sigma(\omega))}(\theta_\sigma(\omega))$, i.e., as $X_{\tau(\theta_n(\omega))}(\theta_n(\omega))$, for $\omega \in \{\omega : \sigma(\omega) = n\}$, for any $n \geq 0$. As a generalization of the property $X_m \circ \theta_n = X_{m+n}$ from Problem 8.2.6, prove that

$$X_\tau \circ \theta_\sigma = X_{\tau \circ \theta_\sigma + \sigma}.$$

**Problem 8.2.11.** Given any set $B \in \mathscr{E}$, let

$$\tau_B(\omega) = \inf\{n \geq 0 : X_n(\omega) \in B\} \quad \text{and} \quad \sigma_B(\omega) = \inf\{n > 0 : X_n(\omega) \in B\}$$

denote, respectively, the time of *the first* and the time of *the first after time* 0 visit of the sequence $X$ to the set $B$. Suppose that the times $\tau_B(\omega)$ and $\sigma_B(\omega)$ are finite for all $\omega \in \Omega$, and let $\gamma = \gamma(\omega)$ be any stopping time on $(\Omega, \mathscr{F}, (\mathscr{F}_k)_{k \geq 0})$.
    Prove that $\tau_B$ and $\sigma_B$ are stopping times and, furthermore,

$$\gamma + \tau_B \circ \theta_\gamma = \inf\{n \geq \gamma : X_n \in B\}, \quad \gamma + \sigma_B \circ \theta_\gamma = \inf\{n > \sigma : X_n \in B\}.$$

Argue that, after appropriate change in the respective definitions, the above relations remain valid even in the case where the stopping times $\gamma$, $\tau_B$ and $\sigma_B$ may take infinite values and the sets in the right sides may be empty.

**Problem 8.2.12.** Let $\tau$ and $\sigma$ be any two Markov times. Prove that $\nu = \tau \circ \theta_\sigma + \sigma$, with the understanding that $\nu = \infty$ on the set $\{\sigma = \infty\}$, is also a Markov time.

**Problem 8.2.13.** Prove that the strong Markov property [P §8.2, (7)], from [P §8.2, Theorem 2], remains valid for *every* Markov time $\tau \leq \infty$, and can be expressed as

$$\mathsf{E}_\pi[I(\tau < \infty)(H \circ \theta_\tau) \mid \mathscr{F}_\tau^X] = I(\tau < \infty)\mathsf{E}_{X_\tau} H \quad (\mathsf{P}_\pi\text{-a. e.}).$$

(Recall that $H$ is a bounded and non-negative $\mathscr{F}$-measurable function and $\mathsf{E}_{X_\tau} H$ is a random variable of the form $\psi(X_\tau)$, where $\psi(x) = \mathsf{E}_x H$.)
    In addition, prove that, if $K = K(\omega)$ is some $\mathscr{F}_\tau$-measurable function and $H$ and $K$ are either bounded or non-negative, then, for every Markov time $\tau \leq \infty$, one has

$$\mathsf{E}_\pi[(I(\tau < \infty)K)(H \circ \theta_\tau)] = \mathsf{E}_\pi[(I(\tau < \infty)K)\mathsf{E}_{X_\tau} H].$$

**Problem 8.2.14.** Prove that the sequence $(X_{\tau \wedge n}, \mathsf{P}_x)_{n \geq 0}$, $x \in E$, introduced in [P §8.2, 3], is a Markov chain. Does this property hold for an *arbitrary* Markov chain (with countable state space) and for an *arbitrary* Markov time of the form $\tau = \inf\{n \geq 0 : X_n \in A\}$, for some choice of the set $A \subseteq E$ (comp. with [P §8.2, (15)])?

**Problem 8.2.15.** Let $h = h(x)$ be a non-negative function and let $H(x) = (Uh)(x)$ be the potential of $h$ (see Sect. A.7). Prove that $H(x)$ is the *minimal* solution of the equation $V(x) = h(x) + TV(x)$, within the class of non-negative functions $V = V(x)$.

**Problem 8.2.16.** Given any $y° \in E$, prove that the Green function $G(x, y°)$ is the minimal non-negative solution to the system

$$V(x) = \begin{cases} 1 + TV(x), & x = y°, \\ TV(x), & x \neq y°. \end{cases}$$

**Problem 8.2.17.** Prove that if $\tau$ and $\sigma$ are any two Markov times and $T_n$, $n \geq 0$, are the transition operators associated with $X = (X_n)_{n \geq 0}$, then:

$$T_\sigma T_\tau = T_{\sigma + \tau \circ \theta_\sigma}.$$

*Hint.* Use the strong Markov property and the identity $X_\tau \circ \theta_\sigma = X_{\tau \circ \theta_\sigma + \sigma}$, established in Problem 8.2.10.

**Problem 8.2.18.** Given any domain $D \in \mathscr{E}$, let

$$\tau(D) = \inf\{n \geq 0 : X_n \in D\} \quad \text{and} \quad \sigma(D) = \inf\{n > 0 : X_n \in D\}.$$

Prove that

$$X_{\sigma(D)} = X_{\tau(D) \circ \theta_1} \quad \text{on } \{\sigma(D) < \infty\},$$

$$T_{\sigma(D)} = T T_{\tau(D)}.$$

**Problem 8.2.19.** With the notation introduced in the previous two problems, let $g \geq 0$ and $V_D(x) = T_{\tau(D)} g(x)$. Prove that $V_D(x)$ is the smallest non-negative solution to the system

$$V(x) = \begin{cases} g(x), & x \in D, \\ T V(x), & x \notin D. \end{cases}$$

In particular, if $g \equiv 1$, then the function $V_D(x) = P_x\{\tau(D) < \infty\}$ is the smallest non-negative solution to the system

$$V(x) = \begin{cases} 1, & x \in D, \\ T V(x), & x \notin D. \end{cases}$$

**Problem 8.2.20.** By using the strong Markov property, prove that the function $m_D(x) = E_x \tau(D)$ solves the system:

$$V(x) = \begin{cases} 0, & x \in D, \\ 1 + T V(x), & x \notin D. \end{cases}$$

In addition, prove that $m_D(x)$ is the smallest non-negative solution to the above system.

**Problem 8.2.21.** Prove that any non-negative excessive function $f = f(x)$ admits the *Riesz decomposition*:

$$f(x) = \tilde{f}(x) + U\tilde{h}(x),$$

in which

$$\tilde{f}(x) = \lim_n (T_n f)(x),$$

is a harmonic function and $U\tilde{h}(x)$ is the potential of the function

$$\tilde{h}(x) = f(x) - Tf(x).$$

**Problem 8.2.22.** Let $X = (X_n)_{n\geq 1}$ and $Y = (Y_n)_{n\geq 1}$ be any two independent Markov chains, with the same state space $E = \{1, 2\}$ and the same transition probability matrix $\left(\begin{smallmatrix} \alpha & 1-\alpha \\ 1-\beta & \beta \end{smallmatrix}\right)$, for some choice of $\alpha, \beta \in (0, 1)$. Let $\tau = \inf\{n \geq 0 : X_n = Y_n\}$ (with $\inf \varnothing = \infty$) be the time of the first meeting between $X$ and $Y$. Find the probability distribution of the time $\tau$.

**Problem 8.2.23.** Let $X = (X_n, \mathscr{F}_n)_{n\geq 0}$ be any stochastic sequence and let $B \in \mathscr{B}(\mathbb{R})$. As was already established, the random variables $\tau_B = \inf\{n \geq 0 : X_n \in B\}$ and $\sigma_B = \inf\{n > 0 : X_n \in B\}$ (with $\inf \varnothing = \infty$) are Markov times. Prove that, for any fixed integer $N \geq 0$, the last visit of $B$ between times $0$ and $N$, i.e., the random variable

$$\gamma_B = \sup\{0 \leq n \leq N : X_n \in B\} \quad \text{with } (\sup \varnothing = 0)$$

is *not* a Markov time.

**Problem 8.2.24.** Prove that the statements in Theorems 1 and 2 in [ P §8.2] remain valid if the requirement for the function $H = H(\omega)$ to be *bounded* is replaced by the requirement that this function is *non-negative*.

**Problem 8.2.25.** Let $X = (X_n, \mathscr{F}_n)_{n\geq 0}$ be any Markov sequence and let $\tau$ be any Markov time. Prove that the random sequence $\overline{X} = (\overline{X}_n, \overline{\mathscr{F}}_n)_{n\geq 0}$, with

$$\overline{X}_n = X_{n+\tau} \quad \text{and} \quad \overline{\mathscr{F}}_n = \mathscr{F}_{n+\tau},$$

is also a Markov sequence, which, in fact, has the same transition function as the sequence $X$. (This fact may be seen as the *simplest form of the strong Markov property*.)

**Problem 8.2.26.** Let $(X_1, X_2, \ldots)$ be any sequence of independent and identically distributed random variables, with common distribution function $F = F(x)$. Set $\mathscr{F}_0 = \{\varnothing, \Omega\}$, $\mathscr{F}_n = \sigma(X_1, \ldots, X_n)$, $n \geq 1$, let $\tau$ be any Markov time for $(\mathscr{F}_n)_{n\geq 0}$, and let $A \in \mathscr{F}_\tau$.

Assuming that $\tau$ is globally bounded, i.e, $0 \leq \tau(\omega) \leq T < \infty$, for $\omega \in \Omega$, prove that:

(a) The variables $I_A, X_{1+\tau}, X_{2+\tau}, \ldots$ are independent.

(b) The variables $X_{n+\tau}$ share the same distribution function $F = F(x)$, i.e., $\mathrm{Law}(X_{n+\tau}) = \mathrm{Law}(X_1)$, $n \geq 1$.

(One consequence from (a) and (b) above is that the probabilistic structure of the sequence $(X_{1+\tau}, X_{2+\tau}, \ldots)$ is the same as the probabilistic structure of the sequence $(X_1, X_2, \ldots)$, i.e., $\text{Law}(X_{1+\tau}, X_{2+\tau}, \ldots) = \text{Law}(X_1, X_2, \ldots)$; plainly, the distribution of the sequence $(X_n)_{n \geq 1}$ is invariant under the random shift $n \rightsquigarrow n + \tau$).

Suppose now that $\tau$ is any, i.e., not necessarily bounded, Markov time, with $0 \leq \tau \leq \infty$. Prove that, in this case, property (a) can be written in the form:

$$P(A \cap \{\tau < \infty\}; X_{1+\tau} \leq x_1, \ldots, X_{n+\tau} \leq x_n)$$
$$= P(A \cap \{\tau < \infty\}) F(x_1) \ldots F(x_n),$$

which relation must hold for all $n \geq 1$ and $x_n \in \mathbb{R}$.

*Hint.* It is enough to notice that

$$P(A \cap \{\tau < \infty\}; X_{1+\tau} \leq x_1, \ldots, X_{n+\tau} \leq x_n) =$$

$$= \sum_{k=0}^{\infty} P(A \cap \{\tau = k\}; X_{1+k} \leq x_1, \ldots, X_{n+k} \leq x_n),$$

where the events $A \cap \{\tau = k\}$ and $\{X_{1+k} \leq x_1, \ldots, X_{n+k} \leq x_n\}$ are independent.


## 8.3   Limiting, Ergodic and Stationary Distributions of Markov Chains

**Problem 8.3.1.** Give examples of Markov chains for which the limit $\pi_j = \lim_n p_{ij}^{(n)}$ exists and
   (a) *Does not* depend on the initial state $i$.
   (b) *Does* depend on the initial state $i$.

**Problem 8.3.2.** Give examples of ergodic and non-ergodic chains.

**Problem 8.3.3.** Give an example of a Markov chain that has a non-ergodic stationary distribution.

**Problem 8.3.4.** Give an example of a transition probability matrix for which *any* probability distribution on the respective state space is a stationary distribution.


## 8.4   Markov Chain State Classification Based on the Transition Probability Matrix

**Problem 8.4.1.** Formulate the notions of "essential" and "inessential" states (see [P §8.4, 1]) in terms of the transition probabilities $p_{ij}^{(n)}$, $i, j \in E, n \geq 1$.

**Problem 8.4.2.** Let $\mathbb{P}$ be the transition probability matrix for some irreducible Markov chain, and suppose that $\mathbb{P}$ has the additional property that $\mathbb{P}^2 = \mathbb{P}$. Describe the structure of the matrix $\mathbb{P}$.

**Problem 8.4.3.** Let $\mathbb{P}$ denote the transition probability matrix for some finite Markov chain $X = (X_n)_{n \geq 0}$. Suppose that $\sigma_1, \sigma_2, \ldots$ is some sequence of independent and identically distributed non-negative integer-valued random variables, which are also independent from $X$. Let $\tau_0 = 0$ and $\tau_n = \sigma_1 + \cdots + \sigma_n$, $n \geq 1$. Prove that the sequence $\widetilde{X} = (\widetilde{X}_n)_{n \geq 0}$, given by $\widetilde{X}_n = X_{\tau_n}$ is a Markov chain and find the transition probability matrix $\widetilde{\mathbb{P}}$ for this chain. Prove that if the states $i$ and $j$ communicate for the chain $X$, then these two states must communicate also for the chain $\widetilde{X}$.

**Problem 8.4.4.** Consider the Markov chain $X = (X_n)_{n \geq 0}$, with state space $E = \{0, 1\}$, and suppose that its transition probability matrix is given by $\mathbb{P} = \left(\begin{smallmatrix} \alpha & 1-\alpha \\ 1-\beta & \beta \end{smallmatrix}\right)$, for some choice of $\alpha, \beta \in (0, 1)$. Then define the Markov moment

$$\nu = \inf\{n \geq 1 : X_{n-1} = X_n = 0\}$$

and prove that

$$\mathsf{E}_0 \nu = \frac{2 - (\alpha + \beta)}{\alpha(1 - \beta)}.$$

**Problem 8.4.5.** Consider the Markov chain with state-space $E = \{1, 2, 3\}$ and transition probability matrix

$$\mathbb{P} = \begin{pmatrix} \alpha & 1-\alpha & 0 \\ 0 & \beta & 1-\beta \\ 1-\gamma & 0 & \gamma \end{pmatrix},$$

where $\alpha, \beta, \gamma \in (0, 1)$. Prove that this Markov chain is irreducible. What can be said about the existence of a stationary distribution for this Markov chain?

**Problem 8.4.6.** Explain whether it may be possible for all states of a given Markov chain to be inessential in each of the following two cases:

1. The state space is finite.
2. The state space is countably infinite.

## 8.5   Markov Chain State Classification Based on the Asymptotics of the Transition Probabilities

**Problem 8.5.1.** Prove that an irreducible Markov chain, with state space $\{0, 1, 2, \ldots\}$ and with transition probabilities $p_{i,j}$, is *transient* if and only if the system of equations $u_j = \sum_i u_i p_{ij}$, $j = 0, 1, \ldots$, admits a bounded solution $u_j$,

$j = 0, 1, \ldots$, which is not constant (i.e., $\left| u_j \right| <$ const, for all $j$, and $u_i \neq u_j$, for at least one pair $(i, j)$).

**Problem 8.5.2.** Prove that, in order for an irreducible Markov chain, with state space $\{0, 1, 2, \ldots\}$ and with transition probabilities $p_{i,j}$, to be *recurrent*, it is enough to establish the existence of a sequence $(u_0, u_1, \ldots)$, with $\lim_{i \to \infty} u_i = \infty$, for which $u_j \geq \sum_i u_i p_{ij}$, for all $j \neq 0$.

**Problem 8.5.3.** Prove that an irreducible Markov chain, with state space $\{0, 1, 2, \ldots\}$ and with transition probabilities $p_{i,j}$, is *positive recurrent* if and only if the system of equations $u_j = \sum_i u_i p_{ij}$, $j = 0, 1, \ldots$, admits a solution $u_j$, $j = 0, 1, \ldots$, with $0 < \sum_j \left| u_j \right| < \infty$.

**Problem 8.5.4.** Consider a Markov chain with state space $\{0, 1, \ldots\}$ and with transition probabilities

$$p_{00} = r_0, \quad p_{01} = p_0 > 0,$$

$$p_{ij} = \begin{cases} p_i > 0, & j = i + 1, \\ r_i \geq 0, & j = i, \\ q_i > 0, & j = i - 1, \\ 0 & \text{in all other cases.} \end{cases}$$

Setting $\rho_0 = 1$ and $\rho_m = \frac{q_1 \ldots q_m}{p_1 \ldots p_m}$, prove the following statements:

$$\text{the chain is recurrent} \iff \sum \rho_m = \infty;$$

$$\text{the chain is transient} \iff \sum \rho_m < \infty;$$

$$\text{the chain is positive recurrent} \iff \sum \rho_m = \infty, \quad \sum \frac{1}{p_m \rho_m} < \infty;$$

$$\text{the chain is null recurrent} \iff \sum \rho_m = \infty, \quad \sum \frac{1}{p_m \rho_m} = \infty.$$

**Problem 8.5.5.** Prove that

$$f_{ik} \geq f_{ij} f_{jk},$$

$$\sup_n p_{ij}^{(n)} \leq f_{ij} \leq \sum_{n=1}^{\infty} p_{ij}^{(n)}.$$

**Problem 8.5.6.** Prove that, for any Markov chain with countable state space, the *Cesàro limits* of the $n$-step transition probabilities $p_{ij}^{(n)}$ always exist, and one has

$$\lim_n \frac{1}{n} \sum_{k=1}^{n} p_{ij}^{(k)} = \frac{f_{ij}}{\mu_j}.$$

**Problem 8.5.7.** Let $\eta_1, \eta_2, \ldots$ be any sequence of independent and identically distributed random variables, with $P\{\eta_k = j\} = p_j$, $j = 0, 1, \ldots$, and suppose that the Markov chain $\xi_0, \xi_1, \ldots$ is chosen so that $\xi_{k+1} = (\xi_k)^+ + \eta_{k+1}$, $k \geq 0$. Compute the transition probabilities for this Markov chain and prove that, if $p_0 > 0$ and $p_0 + p_1 < 1$, then the chain would be recursive if and only if $\sum_k k p_k \leq 1$.

**Problem 8.5.8.** Let $\sigma_i = \inf\{n > 0 : X_n = i\}$ (with $\inf \varnothing = \infty$) and then define $\sigma_i^n$ recursively through the relations:

$$\sigma_i^n = \begin{cases} \sigma_i^{n-1} + \sigma_i \circ \theta_{\sigma_i^{n-1}}, & \text{if } \sigma_i^{n-1} < \infty, \\ \infty, & \text{if } \sigma_i^{n-1} = \infty. \end{cases}$$

Prove that

$$P_i\{\sigma_i^n < \infty\} = (P_i\{\sigma_i < \infty\})^n \quad (= f_{ii}^n).$$

**Problem 8.5.9.** Let $N_{\{i\}}$ denote the number of visits of a particular Markov chain to the state $i$.

(a) Prove that

$$E_i N_{\{i\}} = \frac{1}{1 - P_i\{\sigma_i < \infty\}} \quad \left(= \frac{1}{1 - f_{ii}}\right).$$

(b) Reformulate the criteria for recurrence and transience of the state $i \in E$ from [P §8.5, Theorem 1] in terms of the average number of visits $E_i N_{\{i\}}$.

(c) Prove that

$$E_i N_{\{j\}} = P_i\{\sigma_j < \infty\} \cdot E_i N_{\{i\}}.$$

**Problem 8.5.10.** (*Necessary and sufficient condition for transience.*) Let $X = (X_n)_{n \geq 0}$ be some irreducible Markov chain with countable state space $E$ and transition probability matrix $\|p_{xy}\|$. Prove that the chain $X$ is transient if and only if there is a nontrivial and bounded functions $f = f(x)$ and a state $x^\circ \in E$, for which one can claim that

$$f(x) = \sum_{y \neq x^\circ} p_{xy} f(y), \quad x \neq x^\circ,$$

(harmonicity on the set $E \setminus \{x^\circ\}$).

**Problem 8.5.11.** (*Sufficient condition for transience.*) Let $X = (X_n)_{n \geq 0}$ be some irreducible Markov chain with countable state space $E$. Suppose that there is a *bounded* function $f = f(x)$, such that, for some set $B \subseteq \mathbb{R}$, one has

$$f(x^\circ) < h(x), \quad \text{for some } x^\circ \in B \text{ and all } x \in B,$$

and

$$\sum_{y \in E} p_{xy} f(y) \leq f(x), \quad x \in \overline{B} \quad (= E \setminus B)$$

(superharmonicity on the set $\overline{B}$).

Prove that if the above condition holds, then the chain must be transient.

**Problem 8.5.12.** (*Sufficient condition for recurrence.*) Let $X = (X_n)_{n \geq 0}$ be some irreducible Markov chain with countable state space $E$. Suppose that there is a function $h = h(x)$, $x \in E$, with the property that, for any constant $c$, one can claim that the set $B_c = \{x : h(x) < c\}$ is finite and, for some finite set $A \subseteq \mathbb{R}$, one has

$$\sum_{y \in E} p_{xy} h(y) \leq h(x), \quad x \in \overline{A}$$

(superharmonicity on the $\overline{A}$ ($= E \setminus A$)).
    Prove that the chain $X$ is recurrent.

**Problem 8.5.13.** Prove that the sufficient condition formulated in the previous problem is also necessary.

**Problem 8.5.14.** Let $(\xi_n)_{n \geq 1}$ be any sequence of independent and identically distributed random variables, and let $X = (X_n)_{n \geq 1}$ be the random walk defined as $X_0 = 0$ and $X_n = \xi_1 + \ldots + \xi_n$, for $n \geq 1$. Let $U(B) = \mathsf{E} N_B$ denote the *expected number* of the visits, $N_B = \sum_{n \geq 0} I_B(X_n)$, of the random walk $X$ to the set $B$. The set function $U(\cdot)$ is called *potential-measure* (in this case, for the starting point $x = 0$)—see Sect. A.7.
    Analogously to the definitions of transience and recurrence for Markov chains with *countable* state spaces (see Definitions 1 and 2 in [$\underline{P}$ §8.5, **2**]), we will say that the random walk $X$, which, in general, lives in the space $\mathbb{R}$, is *recurrent* if

$$U(I) = \infty,$$

and will say that it is *transient* if

$$U(I) < \infty,$$

for every *finite* interval $I \subseteq \mathbb{R}$.
    Assuming that the expectation $\mathsf{E}\xi_1$ is well defined, prove that one of the following three properties always holds:

1. $X_n \to \infty$ (P–a. e.) and the random walk $X$ is transient;

2. $X_n \to -\infty$ (P–a. e.) and the random walk $X$ is transient;

3. $\underline{\lim} \, X_n = -\infty$, $\overline{\lim} \, X_n = +\infty$, i.e., the random walk oscillates between $-\infty$ and $+\infty$, in which case transience and recurrence are both possible.

**Problem 8.5.15.** Let everything be as in Problem 8.5.14 and again suppose that the expectation $\mu = \mathsf{E}\xi_1$ is well defined. Prove that:

1. If $0 < \mu \leq \infty$, then $X_n \to \infty$ (P-a. e.).

2. If $-\infty \leq \mu < 0$, then $X_n \to -\infty$ (P-a. e.).

3. If $\mu = 0$, then $\underline{\lim} \, X_n = -\infty$, $\overline{\lim} \, X_n = +\infty$ and the random walk is recurrent.

**Problem 8.5.16.** Prove that a necessary and sufficient condition for the random walk $X = (X_n)_{n \geq 0}$ to be transient is that $|X_n| \to \infty$ as $n \to \infty$ with probability 1.

**Problem 8.5.17.** Consider the Markov chain $X = (X_n)_{n \geq 0}$ with transition probabilities $p_{ij}$, $i, j \in E = \{0, \pm 1, \pm 2, \ldots\}$, chosen so that $p_{ij} = 0$ if $|i - j| > 1$, i.e., for every $i \in E$ one has

$$p_{i,i-1} + p_{ii} + p_{i,i+1} = 1,$$

all probabilities in the left side being strictly positive.

Prove that any such chain must be irreducible and aperiodic. Under what conditions for the transition probabilities $(p_{ii}, p_{i,i-1}, p_{i,i+1}; i \in E)$, is the Markov chain $X$ transient, recurrent, positive recurrent and null-recurrent (comp. with Problem 8.5.4)?

*Hint.* Write down the recursive rule that governs the probabilities $V(i) = P_i\{\tau_{j^\circ} = \infty\}$, $i \in E$, for any fixed $j^\circ$.

**Problem 8.5.18.** (*On the probability for degeneracy in the Galton–Watson model.*) In their study of the extinction of family names in England, in the late nineteenth century F. Galton and H. W. Watson proposed the following model, which carries their names:

Let $\xi_0, \xi_1, \xi_2, \ldots$ be some sequence of random variables that take values in $N = \{0, 1, 2, \ldots\}$ and can be written as *random* sums of random variables:

$$\xi_{n+1} = \eta_1^{(n)} + \ldots + \eta_{\xi_n}^{(n)}, \quad n \geq 0.$$

(Comp. with [P §1.12, Example 4].) Suppose further, that the family $\{\eta_i^{(n)}, i \geq 1, n \geq 0\}$ is comprised of independent random variables, every one of which is distributed as the random variable $\eta$, chosen so that $P\{\eta = k\} = p_k, k \geq 0$, and $\sum_{k=0}^{\infty} p_k = 1$. In this model, each $\xi_n$ represents "the number of parents" in the $n^{\text{th}}$-generation on the family tree, while each $\eta_i^{(n)}$ represents the "the number of offsprings," produced by the $i^{\text{th}}$ parent. Thus, $\xi_{n+1}$ is exactly the number of offsprings that comprise the $(n + 1)^{\text{st}}$ generation, with the understanding that if $\xi_n = 0$, then $\xi_k = 0$ for all $k > n$.

Let $\tau = \inf\{n \geq 0 : \xi_n = 0\}$ denote the time of extinction for the family, with the understanding $\tau = \infty$, if $\xi_n > 0$ for all $n \geq 0$. The main question is how to calculate the probability for extinction in finite time, namely the probability

$$q = P\{\tau < \infty\}.$$

It turns out that the most efficient method for calculating the above probability is the method of generating functions (see Problem 2.6.28). Consider the generating functions $g(s) = E s^\eta \equiv \sum_{k=0}^{\infty} p_k s^k$, $|s| \leq 1$, and $f_n(s) = E s^{\xi_n}$, $n \geq 1$, and prove the following properties of the Galton–Watson model:

(a) $f_n(s) = f_{n-1}(g(s)) = f_{n-2}(g(g(s))) = \ldots = f_0(g^{(n)}(s))$, where $g^{(n)}(s) = (g \circ \ldots \circ g)(s)$ ($n$ times);

(b) if $\xi_0 = 1$, then $f_0(s) = s$ and $f_n(s) = g^{(n)}(s) = g(f_{n-1}(s))$;

(c) $f_n(0) = \mathsf{P}\{\xi_n = 0\}$;

(d) $\{\xi_n = 0\} \subseteq \{\xi_{n+1} = 0\}$;

(e) $\mathsf{P}\{\tau < \infty\} = \mathsf{P}\left(\bigcup_{n=1}^{\infty}\{\xi_n = 0\}\right) = \lim_{N \to \infty} \mathsf{P}\{\xi_N = 0\} = \lim_{N \to \infty} f_N(0)$;

(f) if $q = \mathsf{P}\{\tau < \infty\}$, then $q$ is one of the roots of equation $x = g(x)$, $0 \le x \le 1$.

**Problem 8.5.19.** (*Continuation of Problem 8.5.18.*) Let $g(s) = \mathsf{E}s^{\eta}$ be the generating function of the random variable $\eta$, which takes values in the set $\{0, 1, 2, \ldots\}$. Prove that:

(a) The function $g = g(s)$ is non-decreasing and convex on $[0, 1]$.

(b) If $\mathsf{P}\{\eta = 0\} < 1$, then the function $g = g(s)$ is strictly increasing.

(c) If $\mathsf{P}\{\eta \le 1\} < 1$, then the function $g = g(s)$ is strictly convex.

(d) If $\mathsf{P}\{\eta \le 1\} < 1$ and $\mathsf{E}\eta \le 1$, then the equation $x = g(x)$, $0 \le x \le 1$, has unique solution $q \in [0, 1]$.

(e) If $\mathsf{P}\{\eta \le 1\} < 1$ and $\mathsf{E}\eta > 1$, then the equation $x = g(x)$, $0 \le x \le 1$, has two solutions: $x = 1$ and $x = q \in (0, 1)$.

*Hint.* Show that $g'(x) \ge 0$ and $g''(x) \ge 0$, $x \in [0, 1]$, and consider separately the graphs of the function $g = g(x)$ in the case $\mathsf{E}\eta \le 1$ and in the case $\mathsf{E}\eta > 1$.

**Problem 8.5.20.** (*Continuation of Problems 8.5.18 and 8.5.19.*) Consider the Galton–Watson model with $\mathsf{E}\eta > 1$ and prove that the probability for extinction $q = \mathsf{P}\{\tau < \infty\}$ can be identified with the only root of the equation $x = g(x)$ that is located strictly between 0 and 1, i.e.,

$$\mathsf{E}\eta > 1 \;\Rightarrow\; 0 < \mathsf{P}\{\tau < \infty\} < 1.$$

If $\mathsf{E}\eta \le 1$ and $p_1 \ne 1$, then the probability for extinction occurs with probability 1, i.e.,

$$\mathsf{E}\eta \le 1 \;\Rightarrow\; \mathsf{P}\{\tau < \infty\} = 1.$$

**Problem 8.5.21.** Consider the Galton–Watson model with $p_1 < 1$. Prove that for every fixed $k \ge 1$ one has $\mathsf{P}\{\xi_n = k \text{ i.o.}\} = 0$. Conclude that $\mathsf{P}\{\lim_n \xi_n \in \{0, \infty\}\} = 1$.

**Problem 8.5.22.** Consider the Markov chain $X = (X_n)_{n \ge 0}$, with countably infinite state space $E = \{1, 2, \ldots\}$, and suppose that all states are inessential. Prove that each of the following conditions is necessary and sufficient for the chain to be irreducible and recurrent:

(a) $f_{ij} = 1$ for all $i, j \in E$ (i.e., $\mathsf{P}_i\{\sigma(j) < \infty\} = 1$, where $\sigma(j) = \inf\{n > 0 : X_n = j\}$).

(b) Every finite and non-negative function $h = h(i)$, $i \in E$, which is excessive for the chain $X$ (i.e., $h(i) \geq \sum_{j \in E} p_{ij} h(j)$, $i \in E$, where $p_{ij}$ are the transition probabilities for $X$), must be a constant.

*Hint.* The necessity of (b) is established in Sect. A.7 in the Appendix. To establish the sufficiency, prove that for any $i, j \in E$ one must have

$$f_{ij} = p_{ij} + \sum_{k \neq j} p_{ik} f_{kj},$$

and conclude that if all excessive functions are constants, then $f_{ij} = 1$ for all $i, j \in E$, which, according to (a), is equivalent to the claim that the chain is irreducible and recurrent.

## 8.6-7   On the Limiting, Stationary, and Ergodic Distributions of Markov Chains with at Most Countable State Space

**Problem 8.6-7.1.** Describe the limiting, stationary and ergodic distributions of the Markov chain with transition probability matrix

$$\mathbb{P} = \begin{pmatrix} 1/2 & 0 & 1/2 & 0 \\ 0 & 0 & 0 & 1 \\ 1/4 & 1/2 & 1/4 & 0 \\ 0 & 1/2 & 1/2 & 0 \end{pmatrix}.$$

**Problem 8.6-7.2.** Let $\mathbb{P} = \| p_{ij} \|$ be some $m \times m$-matrix ($m < \infty$), which is *doubly-stochastic* (i.e., $\sum_{j=1}^{m} p_{ij} = 1$, for $i = 1, \ldots, m$, and $\sum_{i=1}^{m} p_{ij} = 1$, for $j = 1, \ldots, m$). Prove that the uniform distribution $\mathbb{Q} = (1/m, \ldots, 1/m)$ is stationary for the associated Markov chain.

**Problem 8.6-7.3.** Let $X = (X_n)_{n \geq 0}$ be some Markov chain with state space $E = \{0, 1\}$ and with transition probability matrix $\mathbb{P} = \begin{pmatrix} \alpha & 1-\alpha \\ 1-\beta & \beta \end{pmatrix}$, for some choice of $0 < \alpha < 1$ and $0 < \beta < 1$.

Prove that:

(a)

$$\mathbb{P}^n = \frac{1}{2 - (\alpha + \beta)} \begin{pmatrix} 1 - \beta & 1 - \alpha \\ 1 - \beta & 1 - \alpha \end{pmatrix} + \frac{(\alpha + \beta - 1)^n}{2 - (\alpha + \beta)} \begin{pmatrix} 1 - \alpha & -(1 - \alpha) \\ -(1 - \beta) & 1 - \beta \end{pmatrix};$$

(b) if the initial distribution is $\pi = (\pi(0), \pi(1))$, then

$$P_x\{X_n = 0\} = \frac{1 - \beta}{2 - (\alpha + \beta)} + (\alpha + \beta - 1)^n \left[ \pi(0) - \frac{1 - \beta}{2 - (\alpha + \beta)} \right].$$

**Problem 8.6-7.4.** (*Continuation of Problem 8.6-7.3.*) Find the stationary distribution, $\pi°$, of the Markov chain $X$ and calculate the covariance

$$\text{cov}_{\pi°}(X_n, X_{n+1}) = \mathsf{E}_{\pi°} X_n X_{n+1} - \mathsf{E}_{\pi°} X_n \, \mathsf{E}_{\pi°} X_{n+1}.$$

Setting $S_n = X_1 + \cdots + X_n$, prove that

$$\mathsf{E}_{\pi°} S_n = \frac{n(1 - \alpha)}{2 - (\alpha + \beta)} \quad \text{and} \quad \mathsf{D}_{\pi°} S_n \le c\, n,$$

where $c$ is some constant.

Finally, prove that almost surely (with respect to any of the measures $\mathsf{P}_0$, $\mathsf{P}_1$ and $\mathsf{P}_{\pi°}$) one has

$$\frac{S_n}{n} \to \frac{1 - \alpha}{2 - (\alpha + \beta)} \quad \text{as } n \to \infty.$$

**Problem 8.6-7.5.** Let $\mathbb{P} = \| p_{ij} \|$ be a transition probability matrix ($i, j \in E = \{0, 1, 2, \ldots\}$), chosen so that for any $i \in E \setminus \{0\}$ one has $p_{i,i+1} = p_i$ and $p_{i,0} = 1 - p_i$, for some $0 < p_i < 1$, and for $i = 0 \in E$ one has $p_{i,0} = 1$.

Prove that all states of the associated Markov chain would be recurrent if and only if $\lim_n \prod_{j=1}^{n} p_j = 0$ (or, equivalently, $\sum_{j=1}^{\infty}(1 - p_j) = \infty$).

Show also, that, if all states are recurrent, then all states can be claimed to be positive recurrent if and only if

$$\sum_{k=1}^{\infty} \prod_{j=1}^{k} p_j < \infty.$$

**Problem 8.6-7.6.** Prove that if $X = (X_k)_{k \ge 0}$ is some irreducible and positive recurrent Markov chain, with invariant distribution $\pi°$, then, for every fixed $x \in E$, one has ($\mathsf{P}_\pi$-a. e. for *every* initial distribution $\pi$)

$$\frac{1}{n} \sum_{k=0}^{n-1} I_{\{x\}}(X_k) \to \pi°(x) \quad \text{as } n \to \infty,$$

and

$$\frac{1}{n} \sum_{k=0}^{n-1} p_{yx}^{(k)} \to \pi°(x) \quad \text{as } n \to \infty, \quad \text{for every } y \in E.$$

(comp. with the law of large numbers from [ $\underline{\mathsf{P}}$ §1.12].)

In addition, prove that if the chain is irreducible and null recurrent, then one has ($\mathsf{P}_\pi$-a. e., for *every* initial distribution $\pi$)

$$\frac{1}{n} \sum_{k=0}^{n-1} I_{\{x\}}(X_k) \to 0 \quad \text{as } n \to \infty,$$

and

$$\frac{1}{n} \sum_{k=0}^{n-1} p_{yx}^{(k)} \to 0 \quad \text{as } n \to \infty, \quad \text{for every } y \in E.$$

**Problem 8.6-7.7.** Consider the Markov chain $X = (X_n)_{n \geq 0}$, with finite state space $E = \{0, 1, \ldots, N\}$, and suppose that this chain is also a martingale. Prove that:

(a) The states $\{0\}$ and $\{N\}$ must be absorbing (i.e., $p_{0,0} = p_{N,N} = 1$).

(b) If $\tau(x) = \inf\{n \geq 0 : X_n = x\}$, then $P_x\{\tau(N) < \tau(0)\} = x/N$.

## 8.8 Simple Random Walks as Markov Chains

**Problem 8.8.1.** Prove Stirling's formula ($n! \sim \sqrt{2\pi} \, n^{n+1/2} e^{-n}$) by using the following argument ([9, Problem 27.18]). Let $S_n = X_1 + \ldots + X_n$, $n \geq 1$, where $X_1, X_2, \ldots$ are independent random variables, all distributed with Poisson law of parameter $\lambda = 1$. Then:

(a) $\quad E\left(\dfrac{S_n - n}{\sqrt{n}}\right)^- = e^{-n} \sum_{k=0}^{n} \left(\dfrac{n-k}{\sqrt{n}}\right) \dfrac{n^k}{k!} = \dfrac{n^{n+1/2} e^{-n}}{n!};$

(b) $\quad \text{Law}\left[\left(\dfrac{S_n - n}{\sqrt{n}}\right)^-\right] \to \text{Law}[N^-],$

where $N$ is some standard normal random variable;

(c) $\quad E\left[\left(\dfrac{S_n - n}{\sqrt{n}}\right)^-\right] \to E N^- = \dfrac{1}{\sqrt{2\pi}};$

(d) $\quad n! \sim \sqrt{2\pi} \, n^{n+1/2} e^{-n}.$

**Problem 8.8.2.** Prove the Markov property in [P §8.8, (28)].

**Problem 8.8.3.** Prove the formula in [P §8.8, (30)].

**Problem 8.8.4.** Consider the Markov chain in the Ehrenfests' model and prove that all states in that chain are recurrent.

**Problem 8.8.5.** Verify the formulas [P §8.8, (31) and (32)].

**Problem 8.8.6.** Consider the simple random walk on $\mathbb{Z} = \{0, \pm 1, \pm 2, \ldots\}$, with $p_{x,x+1} = p$, $p_{x,x-1} = 1 - p$, and prove that the function $f(x) = \left(\dfrac{1-p}{p}\right)^x$, $x \in \mathbb{Z}$, is harmonic.

**Problem 8.8.7.** Let $\xi_1, \ldots, \xi_n$ be independent and identically distributed random variables and let $S_k = \xi_1 + \ldots + \xi_k$, $k \leq n$. Prove that

$$\sum_{k \leq n} I(S_k > 0) \overset{d}{=} \min\left\{1 \leq k \leq n : S_k = \max_{j \leq n} S_j\right\},$$

where $\overset{d}{=}$ stands for "identity in distribution." (This result, which is due to E. Sparre Andersen, clarifies why, in the Bernoulli scheme, the law of the time spent on the positive axis and the law of the location of the maximum are asymptotically the same as the arc-sine law—see [P §1.10] and Problems 1.10.4 and 1.10.5.)

**Problem 8.8.8.** Let $\xi_1, \xi_2, \ldots$ be a sequence independent Bernoulli random variables with $P\{\xi_n = 1\} = P\{\xi_m = -1\} = 1/2, n \geq 1$. Setting $S_0 = 0$ and $S_n = \xi_1 + \ldots + \xi_n$, prove that if $\tau$ is any stopping time and

$$\overline{S}_n = S_{n \wedge \tau} - (S_n - S_{n \wedge \tau}) = \begin{cases} S_n, & n \geq \tau, \\ 2S_\tau - S_n, & n > \tau, \end{cases}$$

then $(\overline{S}_n)_{n \geq 0} \overset{d}{=} (S_n)_{n \geq 0}$, i.e., the distribution laws of the sequences $(\overline{S}_n)_{n \geq 0}$ and $(S_n)_{n \geq 0}$ coincide. (This result is known as *André's reflection principle* for the symmetric random walk $(S_n)_{n \geq 0}$—comp. with other versions of the reflection principle described in [P §1.10].)

**Problem 8.8.9.** Suppose that $X = (X_n)_{n \geq 0}$ is a random walk on the lattice $\mathbb{Z}^d$, defined by: $X_0 = 0$ and

$$X_n = \xi_1 + \ldots + \xi_n, \quad \text{for } n \geq 1, \quad \text{and } P\{\xi_i = e\} = \frac{1}{2d},$$

where the vector $e = (e_1, \ldots, e_d) \in \mathbb{R}^d$ is chosen so that $e_i = 0, -1, +1$ and $|e| \equiv |e_1| + \ldots + |e_d| = 1$.

Prove the following multivariate analog of the Central Limit Theorem, in which $A$ stands for any open ball in $\mathbb{R}^d$ centered at the origin $\mathbf{0} = (0, \ldots, 0)$:

$$\lim_n P\left\{ \frac{X_n}{\sqrt{n}} \in A \right\} = \int_A \left( \frac{d}{2\pi} \right)^d e^{-\frac{d|x|^2}{2}} dx_1 \ldots dx_d.$$

*Hint.* Prove first that the characteristic function $\varphi(t) = Ee^{i(t, \xi_1)}, t = (t_1, \ldots, t_d)$, is given by the formula $\varphi(t) = d^{-1} \sum_{j=1}^d \cos(t_j)$, and then use the multivariate version of the continuity theorem (see [P §3.3, Theorem 1]) and Problem 3.3.5).

**Problem 8.8.10.** Let $X = (X_n)_{n \geq 0}$ be the random walk introduced in Problem 8.8.9 and let $N_n = \sum_{k=0}^{n-1} I(X_k = 0)$ be the number of moments $k \in \{0, 1, \ldots, n-1\}$ at which $X_k = 0$. It is shown in [P §7.9] that, for $d = 1$, one has $EN_n \sim \sqrt{\frac{2}{\pi}n}$ as $n \to \infty$. (In formulas [P §7.9, (17) and (18)], one must replace $\frac{1}{2\pi}$ with $\frac{2}{\pi}$.)

(a) Prove that for $d \geq 2$ one has:

$$EN_n \sim \begin{cases} \frac{1}{\pi} \ln n, & d = 2, \\ c_d, & d \geq 3, \end{cases}$$

where $c_d = 1/P\{\sigma_d = \infty\}$, with $\sigma_d = \inf\{k > 0 : X_k = 0\}$ ($\sigma_d = \infty$ when the infimum is taken over the empty set). Calculate the values of the constant $c_d$.

(b) Prove that for $d = 2$ one has:

$$\lim_n P\left\{\frac{N_n}{\ln n} \geq x\right\} = e^{-\pi x}, \quad x > 0,$$

and

$$P\{\sigma_1 > n\} = P\{N_n = 0\} \sim \frac{\pi}{\ln n} \quad \text{as } n \to \infty.$$

(c) Prove that for $d \geq 3$ one has, as $n \to \infty$:

$$P\{\sigma_1 = 2n\} \sim \frac{P\{X_{2n} = 0\}}{\left[1 + \sum_{k=1}^{\infty} P\{X_{2k} = 0\}\right]^2}.$$

(d) Prove that for $d \geq 1$ one has:

$$P\{\sigma_1 = \infty\} = \frac{1}{1 + \sum_{k=1}^{\infty} P\{X_{2k} = 0\}}.$$

*Remark.* Property (d) is essentially established in the proof of [P §8.5, Theorem 1]. It is also useful to notice that *Pólya's Theorem*:

$P\{\sigma_1 < \infty\} = 1$    for $d = 1$ and $d = 2$ (*recurrence* with probability 1);

$P\{\sigma_1 < \infty\} < 1$    for $d \geq 3$ (*transience* with positive probability),

obtains directly from property (d), in conjunction with the asymptotic property $P\{X_{2k} = 0\} \sim \frac{c(d)}{n^{d/2}}$, for $d \geq 1$, and with $c(d) > 0$.

**Problem 8.8.11.** Consider the Dirichlet problem for the Poisson equation in the domain $C \subset E$, where $E$ is an at most countable set, namely: find a non-negative function $V = V(x)$, such that

$$LV(x) = -h(x), \quad x \in C,$$
$$V(x) = g(x), \quad x \in D = E \setminus C,$$

where $h(x)$ and $g(x)$ are given non-negative functions.

Prove that *the smallest non-negative solution $V_D(x)$ for this problem is given by*:

$$V_D(x) = E_x\left[I(\tau(D) < \infty)g(X_{\tau(D)})\right] + I_C(x)E_x\left[\sum_{k=0}^{\tau(D)-1} h(X_k)\right],$$

where $\tau(D) = \inf\{n \geq 0 : X_n \in D\}$ (as usual, we suppose that $\tau(D) = \infty$, if the infiumum is taken over the empty set).

*Hint.* Write the function $V_D(x)$ in the form $V_D(x) = \varphi_D(x) + \psi_D(x)$, where $(\overline{D} = C)$

$$\varphi_D(x) = \mathsf{E}_x\big[I(\tau(D) < \infty)g(X_{\tau(D)})\big],$$

$$\psi_D(x) = I_{\overline{D}}(x)\mathsf{E}_x\left[\sum_{k=0}^{\tau(D)-1} h(X_k)\right].$$

Then observe that the functions $\varphi_D(x)$ and $\psi_D(x)$ can be written in the form

$$\varphi_D(x) = I_D(x)g(x) + I_{\overline{D}}(x)T\varphi_D(x),$$
$$\psi_D(x) = I_{\overline{D}}(x)h(x) + I_{\overline{D}}(x)T\psi_D(x),$$

and conclude from the last relations that

$$V_D(x) = I_D(x)g(x) + I_{\overline{D}}(x)[h(x) + TV_D(x)],$$

which implies that the above function gives a non-negative solution to the system: $LV(x) = -h(x)$ in the domain $C$ and $V(x) = g(x)$, for $x \in D$.

To prove that $V(x) \geq V_D(x)$, for every non-negative solution $V(x)$ to this system, it is enough to notice that $V(x) = I_D(x)g(x) + I_{\overline{D}}(x)[h(x) + TV(x)]$, from where one finds that

$$V(x) \geq I_D(x)g(x) + I_{\overline{D}}(x)h(x),$$

and conclude by induction that

$$V(x) \geq \sum_{k=0}^{n}(I_{\overline{D}}T^k)[I_D g + I_{\overline{D}}h](x),$$

for every $n \geq 0$. This implies that

$$V(x) \geq \sum_{k\geq 0}(I_{\overline{D}}T^k)[I_D g + I_{\overline{D}}h](x) = \varphi_D(x) + \psi_D(x) = V_D(x).$$

**Problem 8.8.12.** Let $X = (X_n)_{n\geq 0}$ be a simple symmetric random walk on the lattice $\mathbb{Z}^d$ and let

$$\sigma(D) = \inf\{n > 0 : X_n \in D\}, \quad D \subset \mathbb{Z}^d,$$

assuming that the set $\overline{D}$ is finite. Prove that one can find positive constants $c = c(D)$ and $\varepsilon = \varepsilon(D) < 1$, such that

$$\mathsf{P}_x\{\sigma(D) \geq n\} \leq c\,\varepsilon^n,$$

for all $x \in \overline{D}$. (Comp. with the inequality in [ P §1.9, (20)].)

**Problem 8.8.13.** Consider two independent simple symmetric random walks, $X^1 = (X_n^1)_{n \geq 0}$ and $X^2 = (X_n^2)_{n \geq 0}$, that start, respectively, from $x \in \mathbb{Z}$ and $y \in \mathbb{Z}$, and are defined on the same probability space $(\Omega, \mathscr{F}, \mathsf{P})$. Let $\tau^1(x) = \inf\{n \geq 0 : X_n^1 = 0\}$ and $\tau^2(y) = \inf\{n \geq 0 : X_n^2 = 0\}$. Find the probability $\mathsf{P}\{\tau^1(x) < \tau^2(y)\}$.

**Problem 8.8.14.** Prove that for a simple symmetric random walk, $X = (X_n)_{n \geq 0}$, on the lattice $\mathbb{Z} = \{0, \pm 1, \pm 2, \ldots\}$ that starts from $0 \in \mathbb{Z}$, one must have

$$\mathsf{P}_0\{\tau(y) = N\} \sim \frac{|y|}{\sqrt{2\pi}} N^{-3/2} \quad \text{as } N \to \infty,$$

where $\tau(y) = \inf\{n \geq 0 : X_n = y\}$, $y \neq 0$. (Comp. with the results in Problem 2.4.16.)

**Problem 8.8.15.** Consider the simple random walk $X = (X_n)_{n \geq 0}$ with $X_n = x + \xi_1 + \ldots + \xi_n$, where $\xi_1, \xi_2, \ldots$ are independent and identically distributed random variables, with $\mathsf{P}\{\xi_1 = 1\} = p$, $\mathsf{P}\{\xi_1 = -1\} = q$, $p + q = 1$, and $x \in \mathbb{Z}$. Setting $\sigma(x) = \inf\{n > 0 : X_n = x\}$, prove that

$$\mathsf{P}_x\{\sigma(x) < \infty\} = 2\min(p, q).$$

**Problem 8.8.16.** Consider the random walk introduced in the previous problem in the special case $x = 0$, and let $\mathscr{R}_n$ denote the total number of (different) integer values that appear in the set $\{X_0, X_1, \ldots, X_n\}$ (note that $X_0 = 0$). Prove that

$$\mathsf{E}_0 \frac{\mathscr{R}_n}{n} \to |p - q| \quad \text{as } n \to \infty.$$

**Problem 8.8.17.** Let $X = (X_n)_{n \geq 0}$ be the simple random walk on $\mathbb{Z} = \{0, \pm 1, \pm 2, \ldots\}$, given by $X_0 = 0$ and $X_n = \xi_1 + \ldots + \xi_n$, $n \geq 1$, for some sequence, $\xi_1, \xi_2, \ldots$, of independent and identically distributed random variables with $\mathsf{P}\{\xi_1 = 1\} = p$ and $\mathsf{P}\{\xi_1 = -1\} = q \, (= 1 - p)$, for some fixed $0 < p < 1$. Prove that the sequence $|X| = (|X_n|)_{n \geq 0}$ is a Markov chain with state space $E = \{0, 1, 2, \ldots\}$ and with transition probabilities

$$p_{i,i+1} = \frac{p^{i+1} + q^{i+1}}{p^i + q^i} = 1 - p_{i,i-1}, \quad i > 0, \quad p_{0,1} = 1.$$

In addition, prove that

$$\mathsf{P}(X_n = i \mid |X_n| = i, |X_{n-1}| = i_{n-1}, \ldots, |X_1| = i_1) = \frac{p^i}{p^i + q^i}, \quad \text{for } n \geq 1.$$

**Problem 8.8.18.** Let $\xi = (\xi_0, \xi_1, \xi_2, \ldots)$ be any sequence of independent and identically distributed random variables with $\mathsf{P}\{\xi_1 = 1\} = p$ and $\mathsf{P}\{\xi_1 = -1\} = q$,

$p + q = 1$, and suppose that the sequence $X = (X_n)_{n \geq 0}$ is defined as $X_n = \xi_n \xi_{n+1}$. Is this sequence a Markov chain? Is the sequence $Y = (Y_n)_{n \geq 1}$, defined as $Y_n = \frac{1}{2}(\xi_{n-1} + \xi_n)$, $n \geq 1$, a Markov chain?

**Problem 8.8.19.** Suppose that $X = (X_n)_{n \geq 0}$ is a simple symmetric random walk on the lattice $\mathbb{Z} = \{0, \pm 1, \pm 2 \dots\}$, which starts from 0, and let $\sigma_1, \sigma_2, \dots$ be the moments of return to 0, i.e., $\sigma_1 = \inf\{n > 0 : X_n = 0\}$, $\sigma_2 = \inf\{n > \sigma_1 : X_n = 0\}$, etc.

Prove that:

(a) $\mathsf{P}_0\{\sigma_1 < \infty\} = 1$;

(b) $\mathsf{P}_0\{X_{2n} = 0\} = \sum_{k \geq 1} \mathsf{P}_0\{\sigma_k = 2n\}$;

(c) $\mathsf{E}_0 z^{\sigma_k} = (\mathsf{E}_0 z^{\sigma_1})^k$, $|z| < 1$;

(d) $\sum_{n \geq 0} \mathsf{P}_0\{X_{2n} = 0\} z^{2n} = \frac{1}{1 - \mathsf{E}_0 z^{\sigma_1}}$;

(e) $\mathsf{E}_0 z^{\sigma_1} = 1 - \sqrt{1 - z^2}$, so that $\sum_{n \geq 0} \mathsf{P}_0\{X_{2n} = 0\} z^{2n} = \frac{1}{\sqrt{1 - z^2}}$;

(f) if $N(k)$ denotes the number of visits of state $k$ before the first return to 0, i.e., before time $\sigma_1$, then $\mathsf{E}N(k) = 1$, for any $z \in \mathbb{Z} \setminus \{0\}$, $k \neq 0$.

**Problem 8.8.20.** Let $X = (X_n)_{n \geq 0}$ and $Y = (Y_n)_{n \geq 0}$ be any two simple symmetric random walks on $\mathbb{Z}^d$, $d \geq 1$, and let

$$R_n = \sum_{i=0}^{n} \sum_{j=0}^{n} I(X_i = Y_j).$$

Prove that when $d = 1$ the expectation $\mathsf{E}R_n$, i.e., the expected number of periods during which the two random walks meet before time $n$ (taking into account multiple visits of the same state), behaves, asymptotically, for large values of $n$, as $c\, n^{3/2}$, for some constant $c > 0$. It is well known—see [75], for example—that in dimensions $d > 1$ one has (for large $n$)

$$\mathsf{E}R_n \sim \begin{cases} c\, n, & d = 2; \\ c\, \sqrt{n}, & d = 3; \\ c\, \ln n, & d = 4; \\ c, & d \geq 5, \end{cases}$$

where the constant $c = c_d$ depends on the dimension $d$. Verify the above asymptotics for $\mathsf{E}R_n$ and compute the constants $c = c_d$.

**Problem 8.8.21.** Suppose that $B$ is some finite set inside $\mathbb{Z}^d$ and the function $f = f(x)$ is defined for $x \in B \cup \partial B$, where $\partial B = \{x \notin B : \|x - y\| = 1 \text{ for some } y \in B\}$. Then suppose that the function $f = f(x)$ is subharmonic in $B$, i.e., $Tf(x) \geq f(x)$, $x \in B$, where $T$ is the one-step transition operator. Prove the following *maximum principle*:

$$\sup_{x \in B \cup \partial B} f(x) = \sup_{x \in \partial B} f(x).$$

**Problem 8.8.22.** Prove that every bounded harmonic function on $\mathbb{Z}^d$ must be constant.

**Problem 8.8.23.** Prove that all *bounded* solutions $V = V(x)$ to the problem:

$$\Delta V(x) = 0 \quad \text{for } x \in C, \quad \text{and} \quad V(x) = g(x) \quad \text{for } x \in \partial C,$$

where $C \subset \mathbb{Z}^d$ is a given domain, and $g = g(x)$ is some (also given) bounded function on $\partial C$, can be written as

$$V(x) = \mathsf{E}_x\big[g(X_{\tau(\partial C)})I(\tau(\partial C) < \infty)\big] + \alpha\,\mathsf{P}_x\{\tau(\partial C) = \infty\},$$

for some $\alpha \in \mathbb{R}$, where $\tau(\partial C) = \inf\{n \geq 0 : X_n \in \partial C\}$.

**Problem 8.8.24.** Prove the following results about the solution to the homogeneous *Dirichlet problem*: find a function $V = V(x)$ that is harmonic in the domain $C \subseteq \mathbb{Z}^d$ (i.e., $\Delta V(x) = 0$, $x \in C$) and satisfies the boundary condition $V(x) = g(x)$, $x \in \partial C$, for a given function $g = g(x)$ that is defined on $\partial C$ as follows:

(a) if $d \leq 2$ and the function $g = g(x)$ is bounded, then, in the class of bounded functions, the solution is unique and is given by the formula $V_{\partial C}(x) = \mathsf{E}_x g(X_{\tau(\partial C)})$;

(b) if $d \geq 3$, $g = g(x)$ is bounded and $\mathsf{P}_x\{\tau(\partial C) < \infty\} = 1$, for all $x \in C$, then, in the class of bounded functions, the solution is again unique and is given by the formula in (a).

**Problem 8.8.25.** Let $X = (X_n)_{n\geq 0}$ be a simple symmetric random walk on $\mathbb{Z}^d$, $d \geq 1$, and suppose that the domain $C \subset \mathbb{Z}^d$ is bounded and its boundary $\partial C$ is defined as $\{x \in \mathbb{Z}^d : x \notin C \text{ and } \|x - y\| = 1 \text{ for some } y \in C\}$. Prove that the Dirichlet problem:

find a function $V = V(x)$ on $C \cup \partial C$, such that

$$\Delta V(x) = -h(x) \quad \text{for } x \in C, \quad \text{and} \quad V(x) = g(x) \quad \text{for } x \in \partial C,$$

where $h = h(x)$, $x \in C$, and $g = g(x)$, $x \in \partial C$, are given functions,

has a unique solution, given by the formula

$$V_{\partial C}(x) = \mathsf{E}_x g(X_{\tau(\partial C)}) + \mathsf{E}_x\left[\sum_{k=0}^{\tau(\partial C)-1} h(X_k)\right], \qquad \mathsf{E}_x\left[\sum_{k=0}^{\tau(\partial C)-1} |h(X_k)|\right] < \infty.$$

*Hint.* Use the method described in Sect. A.7 in the Appendix and the fact that, because of the finiteness of $C$, one must have $\mathsf{P}_x\{\tau(\partial C) < \infty\} = 1$, $x \in C$ (comp. with Problem 8.8.11).

**Problem 8.8.26.** Consider the simple random walk $S = (S_n)_{n\geq 0}$, defined on $\mathbb{Z} = \{0, \pm 1, \pm 2, \ldots\}$ by $S_0 = 0$ and $S_n = \xi_1 + \ldots + \xi_n$, $n \geq 1$, where $\xi_1, \xi_2, \ldots$ are independent Bernoulli random variables, with $\mathsf{P}\{\xi_i = 1\} = \mathsf{P}\{\xi_i = -1\} = 1/2$, and let $\tau = \inf\{n \geq 0 : S_n = 1\}$. In Problem 7.2.18 it was proposed to show (by using martingale methods) that, given any $|\alpha| \leq 1$, one must have

$$\mathsf{E}\alpha^\tau = \alpha^{-1}[1 - \sqrt{1 - \alpha^2}].$$

Derive the last relation from the strong Markov property, by showing first that the function $\varphi(\alpha) = \mathsf{E}\alpha^\tau$ satisfies the relation $\varphi(\alpha) = \frac{1}{2}\alpha + \frac{1}{2}\alpha\varphi^2(\alpha)$.

**Problem 8.8.27.** In this problem it is proposed to carry out certain calculations in the model developed by T. and P. Ehrenfest, which was meant to explain the absence of (the seemingly existent) contradiction between "irreversibility" and "recurrence" in Boltzmann's *kinetic theory* of heat propagation.

As is well known, this theory stems from the representation of the molecular structure of the matter and the consequent treatment of the heat exchange as a diffusion process. It was developed by Boltzmann for the purpose of explaining the (mostly phenomenological) theoretical conclusions of thermodynamics, based on the hypothesis that the distribution of heat is irreversible and moves toward a thermal equilibrium. Although Boltzmann also believed that thermal equilibrium in the system prevails and leads to a state that maximizes the entropy, the "stochastic" theory that he proposed did not exclude—in theory at least—the possibility that over time the system may return to its original thermodynamic disequilibrium, which was the basis for criticism of the kinetic theory. (Poincaré noted the possibility for "recurrence" in the case of dynamical systems described in terms of measure-preserving transformations—see [P §5.1].)

Boltzmann himself claimed that there was no contradiction between "irreversibility" and the physically unobservable "recurrence," since in a stochastic system the return to states of macroscopic non-equilibrium is possible, but occurs after such a long period of time that it is practically unobservable.

From a physical point of view, the model developed by the Ehrenfests', which was formulated in terms of the theory of Markov chains, was quite adequate, as it was able to describe the exchange of heat between two bodies that are in contact with each other, but are otherwise isolated from their environment. This aspect of the model allows for an interesting *quantitative* analysis of the average time for transition from one state to another.

Let $E = \{0, 1, \ldots, 2k\}$, where "state $i$" means "there are $i$ molecules in camera $A$" (a detailed description of the model proposed by the Ehrenfests can be found in [P §8.8, 3]). Denote by

$$\tau(i) = \inf\{n \geq 0 : X_n = i\} \quad \text{and} \quad \sigma(i) = \inf\{n > 0 : X_n = i\},$$

respectively, the time of the first visit and the time of the first return to state $i$, with the usual understanding that $\inf \varnothing = \infty$.

Prove that:

(a) $\mathsf{E}_i\sigma(i) = 2^{2k}\frac{i!(2k-i)!}{(2k)!}$ and, in particular, the average recurrence time to the null state is given by $\mathsf{E}_0\sigma(0) = 2^{2k}$;

(b) $\mathsf{E}_k \tau(0) = \frac{1}{2k} 2^{2k} (1 + O(k))$;
(c) $\mathsf{E}_0 \tau(k) = k \ln k + k + O(1)$.

(In [8] one can find the following numerical results: if $k = 10{,}000$ and the exchange of molecules occurs once per second, then the expected time $\mathsf{E}_0 \tau(k)$ is less than 29 h, whereas $\mathsf{E}_k \tau(0)$ is astronomically large: $10^{6.000}$ years (!)).

## 8.9   The Optimal Stopping Problem for Markov Chains

**Problem 8.9.1.** Prove by way of example that for Markov chains with *countable* state space the optimal stopping time may not exist (within the class $\mathfrak{M}_0^\infty$).

**Problem 8.9.2.** Verify that the time $\tau_y$, introduced in the proof of [P §8.9, Theorem 2], is a Markov time.

**Problem 8.9.3.** Prove that the sequence $X = (X_1, X_2, \ldots)$, which was defined in in [P §8.9, 7] in the description of "the marriage problem," forms a Markov chain.

**Problem 8.9.4.** Let $X = (X_n)_{n \geq 0}$ be some homogeneous Markov chain with values in $\mathbb{R}$ and with transition function $P = P(x; B)$, $x \in \mathbb{R}$, $B \in \mathcal{B}(\mathbb{R})$. We say that the $\mathbb{R}$-valued function $f = f(x)$, $x \in \mathbb{R}$, is *harmonic* (or $P$-harmonic, or harmonic relative to the transition function $P$), if

$$\mathsf{E}_x |f(X_1)| = \int_R |f(y)| \, P(x; dy) < \infty, \quad x \in R,$$

and

$$f(x) = \int_R f(y) \, P(x; dy), \quad x \in R. \tag{$*$}$$

If the identity "$=$" in $(*)$ is replaced by the inequality "$\geq$" we say that the function $f$ is *superharmonic*—see also Sect. A.7.

Prove that if $f$ is a superharmonic function, then, for every $x \in \mathbb{R}$, the sequence $(f(X_n))_{n \geq 0}$, with $X_0 = x$, is a supermartingale (relative to the measure $\mathsf{P}_x$).

**Problem 8.9.5.** Prove that the time $\bar\tau$, which appears in [P §8.9, (40)], belongs to the class $\mathfrak{M}_1^\infty$.

**Problem 8.9.6.** Analogously to Example 1 in [P §8.9, 6], consider the optimal stopping problem

$$s_N(x) = \sup_{\tau \in \mathfrak{M}_0^N} \mathsf{E}_x g(X_\tau)$$

and

$$s(x) = \sup_{\tau \in \mathfrak{M}_0^\infty} \mathsf{E}_x g(X_\tau),$$

for all simple random walks from the Examples in [P §8.8].

**Problem 8.9.7.** (*Controlled Markov chains and optimization.*) Let $\{\mathbb{P}(a), a \in A\}$ be some *family* of transition probability matrices $\mathbb{P}(a) = \|p_{ij}(a)\|$, parametrized by the collection $A$ of all possible choices for the "control." The associated phase space $E = \{i, j, \ldots\}$ is assumed to be either finite or countably infinite and any function $u = u(i), i \in E$, which takes values in the space $A$, will be treated as a "possible control strategy," i.e., a prescription for the value of the control parameter in every state $i \in E$.

Given a particular choice for the control $u = u(i), i \in E$, we denote by $\mathbb{P}^u$ the associated transition probability matrix $\|p_{ij}(u(i))\|$, from which one can obtain (see, for example, the Ionescu Tulcea Theorem in [P §2.9]) the respective probability distribution, $\mathsf{P}_i^u, i \in E$, in the space $E^\infty$—this is nothing but the probability distribution of the Markov chain $X = (X_n)_{n \geq 0}$ that starts from state $X_0 = i$ and is being "steered" by the control $u$.

Let $C$ be some domain inside the phase space $E$, set $D = E \setminus C$, and consider the functions $h = h(i, a), i \in C, a \in A$, and $g = g(i, A), i \in D, a \in A$, which, for now, will be assumed non-negative. For every (fixed) choice of the control $u = u(i)$, $i \in E$, we write $h^u(i) = h(i, u(i))$ and $g^u(i) = g(i, u(i))$. The "gain" associated with the control $u = u(i), i \in E$, when the chain is in state $j \in E$ is given by

$$V^u(j) = \mathsf{E}_j^u \left[ g^u(X_{\tau(D)}) I(\tau(D) < \infty) + \sum_{k=0}^{\tau(D)-1} h^u(X_k) \right],$$

where $\tau(D) = \inf\{n \geq 0 : X_n \in D\}$. The meaning of the quantity $V^u(j)$ should be clear: it represents the expected aggregate gains, including the cumulative gains $h^u$ and the termination gain $g^u$, collected while the chain remains in the domain $C$, assuming that the initial state is $X_0 = j$ and the chain is subjected to the control $u = u(i), i \in E$.

The *optimal control problem* associated with the gain function $V^u(j)$ comes down to computing the value function

$$V^*(j) = \sup_u V^u(j), \quad j \in E,$$

and the optimal control $u^* = u^*(i), i \in E$, if one exist, with $V^*(j) = \sup_u V^{u^*}(j)$. Prove the following statement, which is known as the "*verification theorem:*"

Suppose that

(i) There is a function $V = V(j), j \in E$, such that

$$V(j) = \sup_{a \in A} \left\{ \sum_{j \in E} p_{ji}(a) V(i) + h(j, a) \right\}, \quad j \in C,$$

and

$$V(j) = \sup_{a \in A} g(j, a), \quad j \in D;$$

(ii) In the class of admissible controls, one can find a control $u^* = u^*(i)$, $i \in E$, such that for any fixed $j$ both supremums above are achieved with $a = u^*(j)$.

Then the control $u^* = u^*(i)$, $i \in E$, is *optimal*: for every admissible control $u$ one has

$$V^{u^*}(j) \geq V^u(j) \quad \text{and} \quad V^{u^*}(j) = V(j), \quad j \in E.$$

*Hint.* Use the fact that for *every* admissible control $u$ one must have

$$V(j) \geq T^u V(j) + h^u(j), \quad j \in C, \quad \text{where} \quad T^u V(j) = E_j^u V(X_1),$$

and

$$V(j) \geq g^u(j), \quad j \in D.$$

Then use the fact that with $u \equiv u^*$ the above inequalities turn into equalities.

**Problem 8.9.8.** (*The "disorder" problem.*) Consider the *Bayesian risk*

$$R^\pi(\tau) = P^\pi\{\tau < \theta\} + cE^\pi(\tau - \theta)^+,$$

which was introduced in Problem 6.7.8. According to that problem, the infimum of the quantity $R^\pi(\tau)$, taken over the class $\mathfrak{M}_0^\infty$ of all $P^\pi$-finite ($\pi \in [0, 1]$) Markov times $\tau$, is attained at the Markov time

$$\tau^* = \inf\{n \geq 0 : \pi_n \geq A\}, \tag{*}$$

where $A$ is some constant that may depend on $c$ and $p$.

Prove that

(a) The Bayesian risk $R^\pi(\tau)$ can be written in the form

$$R^\pi(\tau) = E^\pi\left\{(1 - \pi_\tau) + cI(\tau \geq 1)\sum_{k=0}^{\tau-1}\pi_k\right\};$$

(b) In the optimal stopping problem for the Markov chain $(\pi_n)_{n \geq 0}$

$$R^\pi = \inf_{\tau \in \mathfrak{M}_0^\infty} E^\pi\left\{(1 - \pi_\tau) + cI(\tau \geq 1)\sum_{k=0}^{\tau-1}\pi_k\right\},$$

the infimum is achieved with the stopping time $\tau^*$ defined in $(*)$.

# Appendix A
# Review of Some Fundamental Concepts and Results from Probability Theory and Combinatorics

## A.1 Elements of Combinatorics

In its early stages, the "calculus of probability" was comprised mostly of combinatorial methods for counting the (usually finite) number of configurations that lead to the realization of certain random events. Even today, these counting techniques remain indispensable for the theory of probability—especially for "the elementary theory of probability," which deals with finite spaces of elementary outcomes. In fact, combinatorial methods play a crucial role in many domains of discrete mathematics, including graph theory and the theory of algorithms.

What follows is a brief summary of some basic notions and result from combinatorics that are used in the books "Probability" and also in the present collection of problems.

• Let $\mathbf{A}$ be some *collection* of $N < \infty$ elements $a_1, \ldots, a_N$ (so that $|\mathbf{A}| = N$). If all of these elements are *distinct*, then the collection $\mathbf{A}$ can be referred to as a *set* and may be expressed as

$$\mathbf{A} = \{a_1, \ldots, a_N\}.$$

In the above notation the *order* in which the elements $a_1, \ldots, a_N$ are written is irrelevant. For example, $\{1, 2, 3\}$ and $\{2, 3, 1\}$ refer to one and the same set comprised of the elements $\{1\}$, $\{2\}$ and $\{3\}$.

With each set $\mathbf{A} = \{a_1, \ldots, a_N\}$ one can associate two different types of *samples* (sometimes called *sequences*) of size $n$:

$$(a_{i_1}, \ldots, a_{i_n}) \quad \text{and} \quad [a_{i_1}, \ldots, a_{i_n}],$$

where $i_1, \ldots, i_n \in \{1, \ldots, N\}$ and the symbols $a_{i_j}$ stand for elements of the set $\mathbf{A}$, which may or may not coincide for different values of $j$.

The token $(a_{i_1}, \ldots, a_{i_n})$ is used to denote *ordered* samples, i.e., samples identified not only by the collection of its members, but also by the *order* in which those members are listed.

A.N. Shiryaev, *Problems in Probability*, Problem Books in Mathematics,
DOI 10.1007/978-1-4614-3688-1, © Springer Science+Business Media New York 2012

The token $[a_{i_1}, \ldots, a_{i_n}]$ is used to denote *unordered* samples, i.e., samples identified only by the collection of its members, but *not* by the order in which those members are listed.

For example, the samples $(a_4, a_1, a_3, a_1)$ and $(a_1, a_1, a_4, a_3)$, represent one and the same collection of elements, but are nevertheless *different*, because these are ordered samples that differ in the order in which their (identical) members are listed. At the same time, the samples $[a_4, a_1, a_3, a_1]$ and $[a_1, a_1, a_4, a_3]$ are two identical unordered samples.

If samples of the form $(a_{i_1}, \ldots, a_{i_n})$, or of the form $[a_{i_1}, \ldots, a_{i_n}]$, are taken from the set $\mathbf{A}$ by way of "sampling without replacement," obviously, all elements in the sample must be different and, of course, one must have $n \leq N$. If samples of the form $(a_{i_1}, \ldots, a_{i_n})$, or of the form $[a_{i_1}, \ldots, a_{i_n}]$, are taken from the set $\mathbf{A}$ by way of "sampling with replacement," obviously, the sample may contain identical elements. Furthermore, in this case the size of the sample, $n$, could be arbitrarily large.

A *partition* of the set $\mathbf{A}$, with $|\mathbf{A}| = N$, is any collection, $\mathscr{D} = \{D_1, \ldots, D_n\}$, $n \leq N$, of subsets $D_i \subseteq \mathbf{A}$, $1 \leq i \leq n$, with $D_i \neq \varnothing$, $D_i \cap D_j = \varnothing$ for $i \neq j$, and $D_1 + \ldots + D_n = \mathbf{A}$. The sets $D_i$ are the *atoms* of the partition $\mathscr{D}$.

Counting Various Samples
from a generic set $\mathbf{A} = \{a_1, \ldots, a_N\}$;
Combinatorial Numbers and Their Interpretation.

(a)  $(N)_n \equiv N(N-1) \ldots (N-n+1)$ ("number of placements" $N$ to $n$, $1 \leq n \leq N$) — The number of *ordered* samples $(\ldots)$ of size $n$, comprised of elements of the set $\mathbf{A}$ with $|\mathbf{A}| = N$, by way of "sampling without replacement;"

(b)  $C_N^n \equiv \frac{(N)_n}{n!} \left(= \frac{N!}{n!(N-n)!}\right)$ ("number of combinations" $n$ of $N$, binomial coefficients) — The number of *unordered* samples $[\ldots]$ of size $n$, comprised of elements of the set $\mathbf{A}$, with $|\mathbf{A}| = N$, by way of "sampling without replacement;"

(c)  $N^n$ — The number of *ordered* samples $(\ldots)$ of size $n \geq 1$, comprised of elements of the set $\mathbf{A}$, with $|\mathbf{A}| = N$, by way of "sampling with repetition;"

(d)  $C_{N+n-1}^n$ — The number of *unordered* samples $[\ldots]$ of size $n \geq 1$, comprised of elements of the set $\mathbf{A}$, with $|\mathbf{A}| = N$, by way of "sampling with repetition."

For various combinatorial interpretations of the above numbers, see the problems from Sects. 1.1 and 1.2. In particular, according to Problem 1.1.3 the number of ordered sequences $(\ldots)$ of length $N$, that consist of $n$ "ones" and $N - n$ "zeroes", equals $C_N^n$. This result is particularly important in the elementary theory of probability, in connection with the binomial distribution.

*Example A.1.1.* Consider the set $\mathbf{A} = \{a_1, a_2, a_3, a_4\}$, in which $|\mathbf{A}| = 4$, and let $n = 2$. Then one has

(a) $(4)_2 = 4(4-1) = 12$. There are 12 ordered samples of size $n$:

$$(a_1, a_2), \quad (a_1, a_3), \quad (a_1, a_4), \quad (a_2, a_1), \quad (a_2, a_3), \quad (a_2, a_4),$$
$$(a_3, a_1), \quad (a_3, a_2), \quad (a_3, a_4), \quad (a_4, a_1), \quad (a_4, a_2), \quad (a_4, a_3).$$

(b) $C_4^2 = \frac{4!}{2!\,2!} = 6$. There are 6 unordered samples of size $n$:

$$[a_1, a_2], \quad [a_1, a_3], \quad [a_1, a_4], \quad [a_2, a_3], \quad [a_2, a_4], \quad [a_3, a_4].$$

(c) $4^2 = 16$. In addition to the 12 samples listed in (a) one must include the samples $(a_1, a_1), (a_2, a_2), (a_3, a_3)$ and $(a_4, a_4)$.

(d) $C_{4+2-1}^2 = C_5^2 = \frac{5!}{2!\,3!} = 10$. In addition to the 6 samples listed in (b) one must include also the samples $[a_1, a_1], [a_2, a_2], [a_3, a_3]$ and $[a_4, a_4]$.

Counting Subsets and Partitions
of a generic set $\mathbf{A} = \{a_1, \ldots, a_N\}$;

Combinatorial Numbers and Their Interpretation

(e) $2^N$ — The number of all subsets of $\mathbf{A}$ (including the empty set $\varnothing$ and the set $\mathbf{A}$ with $|\mathbf{A}| = N$).

(f) $C_N^n = \frac{N!}{n!\,(N-n)!}$ — The number of subsets $D \subseteq \mathbf{A}$, of size $0 \le n \le N$ ($|D| = n$, $|\mathbf{A}| = N$, with the understanding that $D = \{\varnothing\}$ when $n = 0$ and $C_N^0 = 1$).

(g) $C_N(n_1, \ldots, n_r) = \frac{N!}{n_1!\ldots n_r!}$ (the "multinomial," or "polynomial" coefficients, $n_1 + \ldots + n_r = N$) — The number of partitions $\mathscr{D} = \{D_1, \ldots, D_r\}$ of the set $\mathbf{A}$ with $|\mathbf{A}| = N$ into $r$ disjoint sets $D_1, \ldots, D_r$, $r \le n$, with $|D_1| = n_1, \ldots, |D_r| = n_r$, $n_1 + \ldots + n_r = N$.

(h) $D_N(\lambda_1, \ldots, \lambda_N)$
$= \frac{N!}{(1!)^{\lambda_1}\ldots(N!)^{\lambda_N}\,(\lambda_1)!\ldots(\lambda_N)!}$
($\lambda_i \ge 0$ for all $i$ and $\sum_{i=1}^N i\lambda_i = N$) — The number of "block" partitions of the set $\mathbf{A}$ with $|\mathbf{A}| = N$, of the form

$$\mathscr{D} = \{D_{11}, \ldots, D_{1\lambda_1}; \ldots$$
$$\ldots; D_{N1}, \ldots, D_{N\lambda_N}\},$$

where the "block" $[\![D_{i1}, \ldots, D_{i\lambda_i}]\!]$ consists of $\lambda_i$ sets, every one of which has $i$ elements ($|D_{ik}| = i$, $1 \le k \le \lambda_i$); if $\lambda_i = 0$, then the respective block is undefined and is not included in the partition $\mathscr{D}$.

(i) $S_N^n = \sum D_N(\lambda_1, \ldots, \lambda_N)$ (the summation is taken over all choices of $\lambda_1, \ldots, \lambda_N$, for which $\sum_{i=1}^{N} \lambda_i = n$, $\sum_{i=1}^{N} i \lambda_i = N$, and $\lambda_i \geq 0$ for all $i$) — The number of partitions $\mathscr{D}$ of the set $\mathbf{A}$, with $|\mathbf{A}| = N$, that consist of exactly $n$ classes.

The numbers $S_N^n$, $1 \leq n \leq N$, are known as *Stirling numbers of the second kind.*[1]

(j) $B_N = \sum_{n=1}^{N} S_N^n$ — The number of partitions of the set $\mathbf{A}$, with $|\mathbf{A}| = N$.

The numbers $B_N$ are known as *Bell numbers.*

(Some additional properties of the numbers introduced in (f)–(j) above can be found in the problems from Sect. 1.2.)

*Example A.1.2.* Consider again the set $\mathbf{A} = \{a_1, a_2, a_3, a_4\}$ and let $N = |\mathbf{A}| = 4$ and $n = 2$.

(e) $2^4 = 16$. The 16 sets are given by:

$$\varnothing, \ \{a_1\}, \ \{a_2\}, \ \{a_3\}, \ \{a_4\},$$

$$\{a_1, a_2\}, \ \{a_1, a_3\}, \ \{a_1, a_4\}, \ \{a_2, a_3\}, \ \{a_2, a_4\}, \ \{a_3, a_4\},$$

$$\{a_1, a_2, a_3\}, \ \{a_1, a_2, a_4\}, \ \{a_1, a_3, a_4\}, \ \{a_2, a_3, a_4\}, \ \{a_1, a_2, a_3, a_4\}.$$

(f) $C_4^2 = 6$. The 6 sets are given by:

$$\{a_1, a_2\}, \ \{a_1, a_3\}, \ \{a_1, a_4\}, \ \{a_2, a_3\}, \ \{a_2, a_4\}, \ \{a_3, a_4\}.$$

(g) If $r = 2$, $n_1 = 1$ and $n_2 = 3$, then $C_4(1, 3) = \frac{4!}{1! \, 3!} = 4$. The 4 partitions are given by:

$$\{a_1\} \text{ and } \{a_2, a_3, a_4\}, \quad \{a_2\} \text{ and } \{a_1, a_3, a_4\},$$

$$\{a_3\} \text{ and } \{a_1, a_2, a_4\}, \quad \{a_4\} \text{ and } \{a_1, a_2, a_3\}.$$

(h) $\lambda_1 = 2$, $\lambda_2 = 1$, $\lambda_3 = 0$, $\lambda_4 = 0$; $\sum_{i=1}^{4} i \lambda_i = 4$,

---

[1] For the Stirling numbers of the first kind, see p. 377 below.

$$D_4(2,1,0,0) = \frac{4!}{(1!)^2(2!)^1(3!)^0(4!)^0 2!\,1!\,0!\,0!} = 6.$$

The 6 "block" partitions are given by:

$$[\{a_1\},\{a_2\}] \text{ and } [\{a_3\},\{a_4\}], \quad [\{a_1\},\{a_3\}] \text{ and } [\{a_2\},\{a_4\}],$$

$$[\{a_1\},\{a_4\}] \text{ and } [\{a_2\},\{a_3\}], \quad [\{a_2\},\{a_3\}] \text{ and } [\{a_1\},\{a_4\}],$$

$$[\{a_2\},\{a_4\}] \text{ and } [\{a_1\},\{a_3\}], \quad [\{a_3\},\{a_4\}] \text{ and } [\{a_1\},\{a_2\}].$$

(i) $S_4^2 = D_4(0,2,0,0) + D_4(1,0,1,0) = 3 + 4 = 7$. The 7 partitions are: $\{a_1\}$ and $\{a_2,a_3,a_4\}$,

$$\{a_2\} \text{ and } \{a_1,a_3,a_4\}, \quad \{a_3\} \text{ and } \{a_1,a_2,a_4\}, \quad \{a_4\} \text{ and } \{a_1,a_2,a_3\},$$

$$\{a_1,a_2\} \text{ and } \{a_3,a_4\}, \quad \{a_1,a_3\} \text{ and } \{a_2,a_4\}, \quad \{a_1,a_4\} \text{ and } \{a_2,a_3\}.$$

For example, analogous calculation shows that:

$$S_4^1 = 1, \quad S_4^3 = 6, \quad S_4^4 = 1,$$

$$S_5^1 = 1, \quad S_5^2 = 15, \quad S_5^3 = 25, \quad S_5^4 = 10, \quad S_5^5 = 1,$$

$$S_6^1 = 1, \quad S_6^2 = 31, \quad S_6^3 = 90, \quad S_6^4 = 65, \quad S_6^5 = 15, \quad S_6^6 = 1.$$

(j) With $N = 4$ property (i) implies that

$$B_4 = S_4^1 + S_4^2 + S_4^3 + S_4^4 = 1 + 7 + 6 + 1 = 15,$$

$$B_5 = 1 + 15 + 25 + 10 + 1 = 52,$$

$$B_6 = 1 + 31 + 90 + 65 + 15 + 1 = 203.$$

The respective 15 partitions (for $N = 4$) are:

$\{a_1,a_2,a_3,a_4\};$     $\{a_1\}, \{a_2\}, \{a_3\}, \{a_4\};$

$\{a_1\}$ and $\{a_2,a_3,a_4\};$     $\{a_2\}$ and $\{a_1,a_3,a_4\};$

$\{a_3\}$ and $\{a_1,a_2,a_4\};$     $\{a_4\}$ and $\{a_1,a_2,a_3\};$

$\{a_1,a_2\}$ and $\{a_3,a_4\};$     $\{a_1,a_3\}$ and $\{a_2,a_4\};$     $\{a_1,a_4\}\ \{a_2,a_3\};$

$\{a_1\}, \{a_2\}$ and $\{a_3,a_4\};$     $\{a_1\}, \{a_3\}$ and $\{a_2,a_4\};$     $\{a_1\}, \{a_4\}$ and $\{a_2,a_3\};$

$\{a_2\}, \{a_3\}$ and $\{a_1, a_4\}$;     $\{a_2\}, \{a_4\}$ and $\{a_1, a_3\}$;     $\{a_3\}, \{a_4\}$ and $\{a_2, a_3\}$.

• There is more to combinatorics then the mere counting of all favorable configurations for various events encountered in the elementary theory of probability. For example, combinatorial reasoning is often used to establish identities like the following one:

$$n^N = \sum_{k=1}^{n} S_N^k \times (n)_k , \quad 1 \le n \le N , \tag{$*$}$$

where $S_N^k$ are the Stirling numbers of the second kind. (Note that $S_N^1 = S_N^N = 1$ and, by the usual convention, $S_N^0 = 0$ and $S_N^n = 0$, for $n > N$.)

The combinatorial proof of the above identity is based on the idea that both sides represent one and the same number of configurations, except that these configurations are counted in two different ways. More specifically, let $A$ and $B$ denote any two finite sets with $|A| = N$ and $|B| = n$. Consider a generic function of the form $y = f(x)$, which is defined for $x \in A$ and takes values in the set $B$. How many such functions can one find? Since one can assign to each of the $N$ possible values of $x$ any one of the $n$ possible values $y$, it is clear that the total number of functions from $A$ to $B$ must be $n^N$.

The total number of functions from $A$ to $B$ can be counted also by considering the pre-image $f^{-1}(y) = \{x : f(x) = y\}$ of any given $y \in B$. With this construction in mind, given any subset $C \subseteq B$ with $|C| = k$, for some $1 \le k \le n$, consider the collection of all functions $y = f(x)$, for which one can claim that $\text{Range}(f) = C$. Since $|C| = k$, then any function $f$ from this collection defines a partition of $A$ that consists $k$ disjoint classes, characterized by the property that $f$ takes one and the same value on each class and different values on different classes, i.e., the classes in the partition are simply the pre-images under $f$ of the elements of $C$. As the total number of such partitions is $S_N^k$, and for each partition there are $(k)_k = k!$ functions that assign *different* values from the set $C$ to the classes in the partition, the total number of functions from $A$ to $B$ whose range is precisely $C$, must be $S_N^k \times k!$.

As there are $C_n^k$ possible selections of the set $C$ with $|C| = k$, the total number of functions $y = f(x)$, defined for $x \in A$ and taking values in $B$, with $|A| = N$ and $|B| = n$, must be

$$\sum_{k=1}^{n} C_n^k \times S_N^k \times k! = \sum_{k=1}^{n} S_N^k \times (n)_k .$$

But the number of the elements in the same collection of functions was also found to be $n^N$, so that the identity ($*$) is now established. (Many problems and examples of the use of "double counting" and other combinatorial techniques can be found in the books [20, 27, 46, 110, 111].)

**Table A.1**  Factorials and their logarithms

| $n$ | $n!$ | $\ln n!$ |
|---|---|---|
| 1 | ........................................1 | 0 |
| 2 | ........................................2 | 0,3010300 · |
| 3 | ........................................6 | 0,7781513 |
| 5 | .................................... 120 | 2,0791812 |
| 10 | ............................. 3 628 800 | 6,5597630 |
| 15 | ...................... 1 307 674 368 000 | 12,1164996 |
| 20 | ............... 2 432 902 008 176 640 000 | 18,3861246 |
| 25 | ........ 15 511 210 043 330 985 984 000 000 | 25,1906457 |
| 30 | 265 252 859 812 191 058 636 308 480 000 000 | 32,4236601 |

**Table A.2**  Binomial coefficients

| 1 | 1 | 1 | | | | | | | | | | | | | | |
|---|---|---|---|---|---|---|---|---|---|---|---|---|---|---|---|---|
| 2 | 1 | 2 | 1 | | | | | | | | | | | | | |
| 3 | 1 | 3 | 3 | 1 | | | | | | | | | | | | |
| 4 | 1 | 4 | 6 | 4 | 1 | | | | | | | | | | | |
| 5 | 1 | 5 | 10 | 10 | 5 | 1 | | | | | | | | | | |
| 6 | 1 | 6 | 15 | 20 | 15 | 6 | 1 | | | | | | | | | |
| 7 | 1 | 7 | 21 | 35 | 35 | 21 | 7 | 1 | | | | | | | | |
| 8 | 1 | 8 | 28 | 56 | 70 | 56 | 28 | 8 | 1 | | | | | | | |
| 9 | 1 | 9 | 36 | 84 | 126 | 126 | 84 | 36 | 9 | 1 | | | | | | |
| 10 | 1 | 10 | 45 | 120 | 210 | 252 | 210 | 120 | 45 | 10 | 1 | | | | | |
| 11 | 1 | 11 | 55 | 165 | 330 | 462 | 330 | 165 | 165 | 55 | 11 | 1 | | | | |
| 12 | 1 | 12 | 66 | 220 | 495 | 792 | 924 | 792 | 495 | 220 | 66 | 12 | 1 | | | |
| 13 | 1 | 13 | 78 | 286 | 715 | 1287 | 1716 | 1716 | 1287 | 715 | 286 | 78 | 13 | 1 | | |
| 14 | 1 | 14 | 91 | 364 | 1001 | 2002 | 3003 | 3432 | 3003 | 2002 | 1001 | 364 | 91 | 14 | 1 | |
| 15 | 1 | 15 | 105 | 455 | 1365 | 3003 | 5005 | 6435 | 6435 | 5005 | 3003 | 1365 | 455 | 105 | 15 | 1 |
| $n$ $k$ | 0 | 1 | 2 | 3 | 4 | 5 | 6 | 7 | 8 | 9 | 10 | 11 | 12 | 13 | 14 | 15 |

The first several values of the quantities $n!$ and $C_n^k$ are given in Tables A.1 and A.2.

# A.2  Basic Probabilistic Structures and Concepts

The most basic structure, on which essentially any probabilistic or statistical analysis is usually carried out is that of a *probability space*, or, a *probabilistic model*, which as a triplet of the form (see [P §2.1])

$$(\Omega, \mathscr{F}, P),$$

where:

$\Omega$ is the space of elementary outcomes $\omega$;

$\mathscr{F}$ is a $\sigma$-algebra of subsets of $\Omega$;

P is a probability measure on $\mathscr{F}$, i.e., $\sigma$-additive function of the set
$A \in \mathscr{F}$, such that $0 \le P(A) \le 1$, $P(\varnothing) = 0$, $P(\Omega) = 1$.

• In addition to the notion of $\sigma$-algebra, which is an essential ingredient in the structure of any probability space, sometimes one must work with other *systems of subsets*: algebras, separable $\sigma$-algebras, monotone classes, $\pi$-systems, $\lambda$-systems, $\pi$-$\lambda$-systems, families of cylindrical sets, etc.—see [ P §2.2].

• The events (or the sets) $A$ and $B$ are said to be *independent*, if $P(A \cap B) = P(A) \times P(B)$.

*Two systems*, $\mathscr{G}_1$ and $\mathscr{G}_2$, of subsets of $\mathscr{F}$ are said to be *independent* if for any set $A \in \mathscr{G}_1$ and any set $B \in \mathscr{G}_2$ one can claim that $A$ and $B$ are independent.

The sets $A_1, \ldots, A_n \in \mathscr{F}$ are said to be independent, if, for every $k = 1, \ldots, n$ and $1 \le i_1 < i_2 < \ldots < i_k \le n$, one has

$$P(A_{i_1} \cap \ldots \cap A_{i_k}) = P(A_{i_1}) \ldots P(A_{i_k}).$$

The independence of the systems $\mathscr{G}_1, \ldots, \mathscr{G}_n$, all comprised of sets from $\mathscr{F}$, can be defined in a similar fashion.

• A *measurable space* is a pair of the form $(E, \mathscr{E})$, where $E$ is a set and $\mathscr{E}$ is a $\sigma$-algebra comprised of subsets of $E$.

The most common measurable spaces are (see [ P §2.2]):

$(\mathbb{R}, \mathscr{B}(\mathbb{R}))$ — the real line $\mathbb{R}$ endowed with the Borel $\sigma$-algebra $\mathscr{B}(\mathbb{R})$ (often denoted simply by $\mathscr{B}$);

$(\mathbb{R}^n, \mathscr{B}(\mathbb{R}^n))$ — the space $\mathbb{R}^n = \mathbb{R} \times \ldots \times \mathbb{R}$ endowed with $\sigma$-algebra $\mathscr{B}(\mathbb{R}^n) = \mathscr{B}(\mathbb{R}) \otimes \ldots \otimes \mathscr{B}(\mathbb{R})$;

$(\mathbb{R}^\infty, \mathscr{B}(\mathbb{R}^\infty))$ — the space $\mathbb{R}^\infty = \mathbb{R} \times \ldots \times \mathbb{R} \times \ldots$ endowed with the $\sigma$-algebra $\mathscr{B}(\mathbb{R}^\infty)$, generated by all cylinder sets;

$(\mathbb{R}^T, \mathscr{B}(\mathbb{R}^T))$ — the space $\mathbb{R}^T$ of all functions that map the (generic) set $T$ into the real line $\mathbb{R}$, endowed with the $\sigma$-algebra $\mathscr{B}(\mathbb{R}^T)$, generated by all cylinder sets in the space $\mathbb{R}^T$;

$(\mathscr{C}, \mathscr{B}(\mathscr{C}))$ — the space $\mathscr{C}$ of continuous functions (e.g., continuous functions on $[0, 1]$ or $[0, \infty)$) endowed with the $\sigma$-algebra $\mathscr{B}(\mathscr{C})$, generated by all open sets for the usual topology of convergence on compacts (or, which amounts to the same, generated by all cylinder sets);

$(D, \mathscr{B}(D))$ — the space $D$ of right-continuous functions with left limits (e.g, function on $[0, 1]$ that are right-continuous at any $t < 1$ and have left limits at any $t > 0$), endowed with the $\sigma$-algebra $\mathscr{B}(D)$, generated by all open sets for the Skorokhod's metric (or, which amounts to the same, generated by all cylinder sets).

• A *random variable* is any function $X = X(\omega)$ which is defined on some measurable space $(\Omega, \mathscr{F})$, takes values in $(\mathbb{R}, \mathscr{B}(\mathbb{R}))$, and is $\mathscr{F}$-measurable, in the sense that

$$\{\omega : X(\omega) \in B\} \in \mathscr{F}$$

for any Borel set $B \in \mathscr{B}(\mathbb{R})$.

The simplest, and at the same time very important, example of a random variable is the *indicator*, $X(\omega) = I_A(\omega)$, of a generic set $A \in \mathscr{F}$, which is given by

$$I_A(\omega) = \begin{cases} 1, & \omega \in A \\ 0, & \omega \notin A \end{cases}.$$

A *random element* is any $\mathscr{F}/\mathscr{E}$-measurable map $X = X(\omega)$ from $\Omega$ into $E$ (i.e., $\{\omega : X(\omega) \in B\} \in \mathscr{F}$ for any $B \in \mathscr{E}$), where $(\Omega, \mathscr{F})$ and $(E, \mathscr{E})$ are two measurable spaces.

A *n-dimensional random vector* $(X_1(\omega), \ldots, X_n(\omega))$ is simply an ordered list of random variables $X_1(\omega), \ldots, X_n(\omega)$.

A *random sequence*, or, equivalently, a *random process in discrete time*, $X = (X_n(\omega))_{n \geq 1}$, is simply a sequence of random variables $X_1(\omega), X_2(\omega), \ldots$

A *random process*, $X = (X_t(\omega))_{t \in T}$, on the time interval $T \subseteq \mathbb{R}$ is simply a collection of random variables parameterized by the set $T$: $X_t(\omega), t \in T$.

• A *distribution function*, $F = F(x)$, on $(\mathbb{R}, \mathscr{B}(\mathbb{R}))$, is any $\mathscr{B}(\mathbb{R})$-measurable function on $\mathbb{R}$ which has the following properties:

1.  $F(x)$ is non-decreasing;
2.  $\lim_{x \to -\infty} F(x) = 0$ and $\lim_{x \to +\infty} F(x) = 1$;
3.  $F(x)$ is right-continuous and admits left limits at any point $x \in \mathbb{R}$.

If $X = X(\omega)$ is a random variable defined on the probability space $(\Omega, \mathscr{F}, \mathsf{P})$, then the probability measure $P_X$ on $(\mathbb{R}, \mathscr{B}(\mathbb{R}))$, given by

$$P_X(B) = \mathsf{P}\{\omega : X(\omega) \in B\},$$

is known as the *probability distribution of the random variable* $X = X(\omega)$. It is easy to see that the function $F_X(x) = P_X((-\infty, x])$ is a distribution function on $(\mathbb{R}, \mathscr{B}(\mathbb{R}))$. This function is known as the *distribution function of the random variable* $X = X(\omega)$.

If $X = (X_t(\omega))_{t \in T}$ is a random process then the probability distributions on $\mathbb{R}^n$, for various choices of $n \geq 1$ and $t_1 < \ldots < t_n, t_i \in T$, given by

$$P_{t_1,\ldots,t_n}(B) = \mathsf{P}\{\omega : (X_{t_1}(\omega),\ldots,X_{t_n}(\omega)) \in B\}, \quad B \in \mathscr{B}(\mathbb{R}^n),$$

are known as the *finite-dimensional distributions of the random process* $X$. The associated distribution functions

$$F_{t_1,\ldots,t_n}(x_1,\ldots,x_n) = \mathsf{P}\{\omega : X_{t_1}(\omega) \le x_1,\ldots,X_{t_n}(\omega) \le x_n\}$$

are known as *finite-dimensional distribution functions of the random process* $X$.

• If the *reference measure* of choice on $(\mathbb{R},\mathscr{B}(\mathbb{R}))$ is the Lebesgue measure $\lambda = \lambda(dx)$, then the "Lebesgue decomposition" (see [P §3.9, (29)] or [P §7.6, (3)]) leads to the following result: any distribution function, $F = F(x)$, on $(\mathbb{R},\mathscr{B}(\mathbb{R}))$ can be decomposed into the sum

$$F(x) = a\,F_{\text{abc}}(x) + b\,F_{\text{sing}}(x),$$

where the constants $a \ge 0$ and $b \ge 0$ are chosen so that $a + b = 1$ and

$F_{\text{abc}}(x)$ is some *absolutely continuous* distribution function on $\mathbb{R}$ with (Borel-measurable) density $f = f(y)$, i.e., $f(y) \ge 0$, $\int_{-\infty}^{\infty} f(y)\,\lambda(dy) = 1$ and $F_{\text{abc}}(x) = \int_{-\infty}^{x} f(y)\,\lambda(dy)$, $x \in \mathbb{R}$;

$F_{\text{sing}}(x)$ is some *singular* distribution function on $(\mathbb{R},\mathscr{B}(\mathbb{R}))$, in the sense that the respective probability law, $P_{\text{sing}}$, on $(\mathbb{R},\mathscr{B}(\mathbb{R}))$ is singular with respect to the Lebesgue measure $\lambda$ ($P_{\text{sing}} \perp \lambda$).

The singular function $F_{\text{sing}}(x)$ can be further decomposed into the sum

$$F_{\text{sing}}(x) = d \cdot F_{\text{d-sing}}(x) + c \cdot F_{\text{c-sing}}(x),$$

in which the constants $d \ge 0$ and $c \ge 0$ are chosen so that $d + c = 1$, $F_{\text{d-sing}}(x)$ is a discrete distribution function with the property that the support of the associated probability measure, $P_{\text{d-sing}}$, is some set inside $\mathbb{R}$ that is at most countable, and $F_{\text{c-sing}}(x)$ is a continuous distribution function characterized by the property that the support of the associated probability measure, $P_{\text{c-sing}}$, is some uncountable set inside $\mathbb{R}$ which is negligible for the Lebesgue measure $\lambda$. (The canonical example of such a function is the Cantor function $F_{\text{c-sing}}(x)$—see [P §2.3, 1].)

Recall that the *support* of any measure $\mu$ on $(\mathbb{R},\mathscr{B}(\mathbb{R}))$ is defined as the set

$$\text{supp}(\mu) = \{x \in R : \mu\{y : |y - x| \le r\} > 0, \forall r > 0\}.$$

As a direct application of the above decompositions, one arrives at the *canonical decomposition* (see Problem 2.3.18) of a generic distribution function $F = F(x)$ on $(\mathbb{R},\mathscr{B}(\mathbb{R}))$:

$$F = \alpha_1 F_{\text{d}} + \alpha_2 F_{\text{abc}} + \alpha_3 F_{\text{sc}},$$

where the constants $\alpha_1 \geq 0$, $\alpha_2 \geq 0$ and $\alpha_3 \geq 0$ are such that $\alpha_1 + \alpha_2 + \alpha_3 = 1$, and $F_d$ ($= F_{d\text{-sing}}$), $F_{abc}$ and $F_{sc}$ ($= F_{c\text{-sing}}$) are distribution functions on $(\mathbb{R}, \mathscr{B}(\mathbb{R}))$, which are, respectively, discrete, absolutely continuous, and continuous and singular.

## A.3   Analytical Methods and Tools of Probability Theory

• An important characteristic of any random variable $X = X(\omega)$, defined on some probability space $(\Omega, \mathscr{F}, \mathsf{P})$, is its expected value (or simply "expectation") $\mathsf{E}X$.

If $X = X(\omega)$ is a non-negative random variable, then its *expected value* $\mathsf{E}X$ is defined as the Lebesgue integral of the function $\omega \rightsquigarrow X(\omega)$ with respect to the measure $\mathsf{P}$:

$$\mathsf{E}X = \int_\Omega X(\omega)\,\mathsf{P}(d\omega)\,.$$

If $X = X(\omega)$ is an arbitrary (i.e., not necessarily non-negative) random variable, then one can write $X = X^+ - X^-$, where $X^+ = \max(X, 0)$ and $X^- = -\min(X, 0)$, and the expected value $\mathsf{E}X$ is said to *exist*, or to be *well defined*, if at least one of the expectations $\mathsf{E}X^+$ and $\mathsf{E}X^-$ is finite (i.e., $\min(\mathsf{E}X^+, \mathsf{E}X^-) < \infty$), in which case $\mathsf{E}X$ is defined as

$$\mathsf{E}X = \mathsf{E}X^+ - \mathsf{E}X^-\,.$$

The expectation $\mathsf{E}X$ is said to be *finite* (equivalently, $X$ is said to be *integrable*), if $\mathsf{E}X^+ < \infty$ and $\mathsf{E}X^- < \infty$, which is equivalent to the requirement $\mathsf{E}|X| < \infty$, since $|X| = X^+ + X^-$ (see [P §2.6]).

• An important analytical "trick" in probability theory is the passage to the limit under the Lebesgue integral. This operation is justified by the monotone convergence theorem, Fatou's lemma, and Lebesgue's dominated convergence theorem. The following tools are fundamental in probability theory: the concept of uniform integrability, the fundamental inequalities (Chebyshev, Cauchy-Schwarz, Jensen, Lyapunov, Hölder, Minkowski and others), the Radon-Nikodym theorem, Fubini's theorem and the "change of variables" theorem for the Lebesgue integral (see [P §2.6]).

• The *dispersion* of the random variable $X = X(\omega)$ is defined as

$$\mathsf{D}X = \mathsf{E}(X - \mathsf{E}X)^2\,.$$

The quantity $\sigma = +\sqrt{\mathsf{D}X}$ is known as *standard (linear) deviation* of the random variable $X$ (from the mean value $\mathsf{E}X$).

If it exists, the *covariance* of any given pair of random variables, $(X, Y)$, is defined as

$$\mathrm{cov}(X, Y) = \mathsf{E}(X - \mathsf{E}Y)(Y - \mathsf{E}Y)\,.$$

If $X$ and $Y$ are random variables with $0 < DX < \infty$ and $0 < DY < \infty$, then the quantity

$$\rho(X, Y) = \frac{\text{cov}(X, Y)}{\sqrt{DX\, DY}}$$

is known as the *correlation coefficient* of $X$ and $Y$.

For a given random variable $X$ and an integer $n$, if it exists, the expectation $EX^n$ is called the *moment of order n*, or the $n^{th}$ *moment*, of $X$. The quantity $E(X)_n = EX(X - 1) \ldots (X - n + 1)$ is called *factorial moment of order n*.

• If $F = F(x)$ is any distribution function, then the function

$$\varphi(t) = \int_{\mathbb{R}} e^{itx}\, dF(x) \quad \left( = \int_{\mathbb{R}} \cos tx\, dF(x) + i \int_{\mathbb{R}} \sin tx\, dF(x) \right), \quad t \in \mathbb{R},$$

is the *characteristic function* of $F$. In particular, if $X$ is a random variable and $F_X$ is its distribution function, then the characteristic function of $F_X$, namely,

$$\varphi_X(t) = \int_{\mathbb{R}} e^{itx}\, dF_X(x) = Ee^{itX(\omega)}, \quad t \in \mathbb{R},$$

is also called *characteristic function of the random variable* $X = X(\omega)$ (see [P §2.12]).

• Given any non-negative random variable $X$ with distribution function $F_X = F_X(x)$, the *Laplace transform* of $X$—or, equivalently, of $F_X$—is defined as the function

$$\widehat{F}_X(\lambda) = \int_0^\infty e^{-\lambda x}\, dF_X(x) = Ee^{-\lambda X}, \quad \lambda > 0.$$

Tables of the most commonly used discrete probability distributions and distributions with densities can be found in [P §2.3].

• The method of *generating functions* is particularly useful in the study of discrete random variables. This method is widely used also in other areas of mathematics as a convenient tool for studying some special numerical sequences, whose structure is not immediately obvious.

In probability theory, the generating function, $G(s)$, of the discrete random variable $X$, which takes the values $0, 1, 2, \ldots$ with probabilities $p_0, p_1, p_2, \ldots$ ($p_k \geq 0$, $\sum_{k=0}^\infty p_k = 1$), is given by

$$G(s) = Es^X \quad \left( = \sum_{k=0}^\infty p_k s^k \right), \quad |s| \leq 1.$$

The distribution of the random variable $X$, namely, $(p_k)_{k \geq 0}$, is uniquely determined from the generating function $G(s)$ through the formula

$$p_k = \mathsf{P}\{X = k\} = \frac{G^{(k)}(0)}{k!} .$$

If the components of the discrete *vector*-valued random variable $X = (X_1, \ldots, X_d)$, take values in the set $\mathbb{N} = \{0, 1, 2, \ldots\}$, then its generating function, $G(s)$, with $s = (s_1, \ldots, s_d)$, is defined by:

$$G(s_1, \ldots, s_d) = \mathsf{E} s_1^{X_1} \ldots s_d^{X_d} = \sum_{k_1, \ldots, k_d = 0}^{\infty} p_{k_1, \ldots, k_d} s_1^{k_1} \ldots s_d^{k_d} ,$$

where $p_{k_1, \ldots, k_d} = \mathsf{P}\{X_1 = k_1, \ldots, X_d = k_d\}, |s_k| \leq 1, k = 1, \ldots, d$.
If the variables $X_1, \ldots, X_d$ are independent, then

$$G(s_1, \ldots, s_d) = G_1(s_1) \ldots G_d(s_d) ,$$

where $G_k(s_k) = \mathsf{E} s_k^{X_k}, k = 1, \ldots, d$.

The above definition of the generating function $G(s)$ assumes that the random variable $X$ is non-negative and takes values in the set $\mathbb{N} = \{0, 1, 2, \ldots\}$. For various reasons it is useful to expand this construction also for the case where $X$ takes positive and negative values, i.e., $\mathsf{P}\{X = k\} = p_k$, for $k = 0, \pm 1, \pm 2, \ldots$, and $\sum_{k=-\infty}^{\infty} p_k = 1$, without supposing that all $p_{-1}, p_{-2}, \ldots$ must vanish. The generating function, $G(s)$, of any such random variable $X$ is given by the formula

$$G(s) = \mathsf{E} s^X = \sum_{k=-\infty}^{\infty} p_k s^k ,$$

for those $s$ for which $\mathsf{E}|s^X| < \infty$.

Typically, generating functions of the above type are used when working with the *difference*, $X = X_1 - X_2$, of two random variables, $X_1$ and $X_2$, that take values in the set $\mathbb{N} = \{0, 1, 2, \ldots\}$. For example, if $X_1$ and $X_2$ are independent and have generating functions, respectively, $G_{X_1}(s)$ and $G_{X_2}(s)$, then

$$G_X(s) \equiv G_{X_1 - X_2}(s) = G_{X_1}(s) \, G_{X_2}\left(\frac{1}{s}\right) .$$

In particular, if $X_i$ is distributed with Poisson law of parameter $\lambda_i$, $i = 1, 2$, then $G_{X_i}(s) = e^{-\lambda_i (1-s)}$ and the generating function $G_X(s)$ of $X = X_1 - X_2$ is given by

$$G_X(s) = e^{-(\lambda_1 + \lambda_2) + s\lambda_1 + \frac{1}{s}\lambda_2} = e^{-(\lambda_1 + \lambda_2)} \times e^{\sqrt{\lambda_1 \lambda_2}(t + \frac{1}{t})} ,$$

where $t = s\sqrt{\lambda_1 / \lambda_2}$.

It is well known from analysis that, for every fixed $x \in \mathbb{R}$, one can write

$$e^{x(t+\frac{1}{t})} = \sum_{k=-\infty}^{\infty} t^k I_k(2x),$$

where $I_k(2x)$ is the modified Bessel function of the first kind of order $k$, namely

$$I_k(2x) = x^k \sum_{r=0}^{\infty} \frac{x^{2r}}{r!\,\Gamma(k+r+1)}, \qquad k = 0, \pm 1, \pm 2, \ldots$$

(Alternatively, one can say that for every fixed $x \in \mathbb{R}$, the generating function of the sequence $(I_k(2x))_{k=0,\pm 1,\ldots}$ is $e^{x(t+\frac{1}{t})}$.)

Thus, we find that the probability distribution of the random variable $X = X_1 - X_2$ can be expressed as

$$P\{X = k\} = e^{-(\lambda_1+\lambda_2)} \left(\frac{\lambda_1}{\lambda_2}\right)^{k/2} I_n(2\sqrt{\lambda_1\lambda_2}),$$

where $k = 0, \pm 1, \pm 2, \ldots$.

For other examples related to the calculation of some concrete generating functions, and for various applications of the method of generating functions, see, Problems 2.6.28, 2.6.32, 7.2.18, and 8.8.19.

• As was noted earlier, the method of generating functions plays a significant role in several important domains of mathematics; in particular, in discrete mathematics and combinatorics.

In fact, it was the method of generating functions that brought to light the *algebraic* methods for solving various combinatorial problems, thus giving rise to a new direction in combinatorics, called *algebraic combinatorics*.

In general, many important combinatorial properties, operations and relations can be interpreted in such a way that they become algebraic in nature.

As an illustration of the use of the algebraic properties of certain generating functions for the purpose of a concrete combinatorial calculation, consider the following lottery-problem. The tickets in a particular lottery are identified by the six-digit numbers from 000000 to 999999. Suppose that one must compute the probability that a randomly chosen ticket has a number in which the sum of the first three digits equals the sum of the last three digits. Clearly, this is a combinatorial problem, which comes down to computing the respective number of favorable configurations. One may try to compute this number by brute force, i.e., by counting those configurations one-by-one. However, as we are about to see, this number is quite large $(55, 252$, to be precise) and straight counting would be rather impractical.

In contrast, the method of generating functions allows one to solve fairly quickly a more general problem: calculate the probability for a randomly selected ticket in a lottery with $10^{2n}$ tickets, identified with the $2n$-digit numbers from 0 to $10^{2n} - 1$, to have a number in which the sum of the first $n$ digits equals the sum of the last $n$ digits. Assuming that $n = 3$, let $X = (X_1, \ldots, X_6)$ be a vector of independent random variables, chosen so that $p_k = P\{X_i = k\} = 1/10$, for $k = 0, 1, \ldots, 9$, and consider the generating function

$$G_{X_i}(s) = \sum_{k=0}^{9} p_k s^k = \frac{1}{10}(1 + s + \ldots + s^9) = \frac{1}{10}\frac{1 - s^{10}}{1 - s}.$$

Because of the independence of the random variables $X_i$, $1 \le i \le 6$, one can write

$$G_{X_1+X_2+X_3}(s) = G_{X_1}(s) \times G_{X_2}(s) \times G_{X_3}(s) = \frac{1}{10^3}\left(\frac{1 - s^{10}}{1 - s}\right)^3.$$

Analogous expression can be written also for $G_{X_4+X_5+X_6}(s)$.

Now consider the random variable $Y = (X_1 + X_2 + X_3) - (X_4 + X_5 + X_6)$. Clearly, due to the independence, one must have

$$G_Y(s) = G_{X_1+X_2+X_3}(s)\, G_{X_4+X_5+X_6}\left(\frac{1}{s}\right) = \frac{1}{10^6}\frac{1}{s^{27}}\left(\frac{1 - s^{10}}{1 - s}\right)^6.$$

In addition, the coefficient $q_0$ (for the term $s^0$) in the expansion

$$G_Y(s) = \sum_k q_k s^k$$

is nothing but the probability $P\{Y = 0\}$, which is precisely the probability that the sum of the first three digits on the randomly selected ticket equals the sum of the last three digits. After a somewhat involved but otherwise straight-forward calculation, from the expansions of $(1 - s^{10})^6$, $(1 - s)^{-6}$ and $\frac{1}{s^{27}}\left(\frac{1-s^{10}}{1-s}\right)^6$ into power series (for the powers $s^k$, $k = 0, \pm 1, \pm 2, \ldots$) one finds that

$$q_0 = \frac{55252}{10^6} = 0.05525$$

(see also Problem 2.6.79).

In general, the *generating function* associated with an *arbitrary* numerical sequence $a = (a_n)_{n \ge 0}$ is defined as the (formal) power series

$$G_a(x) = a_0 + a_1 x + a_2 x^2 + \ldots, \qquad x \in \mathbb{R}.$$

If the above series has a non-trivial radius of convergence, then it would define a true (i.e., not just formal) *function* (on the respective interval of convergence). According to the general theory of generating functions, the function $G_a(x)$ is nothing but a special "encryption" of the sequence $a = (a_n)_{n \geq 0}$, in that there is a bijective correspondence

$$(a_n) \leftrightarrow G_a(x).$$

It is a trivial matter to check that if $(b_n) \leftrightarrow G_b(x)$ and $c$ is some constant, then

$$(a_n + cb_n) \leftrightarrow G_a(x) + cG_b(x).$$

Perhaps the most important features of the bijection "$\leftrightarrow$" is the relation

$$\left( \sum_{i=0}^{n} a_i b_{n-i} \right)_{n \geq 0} \leftrightarrow G_a(x) \times G_b(x),$$

which simply says that under the bijection "$\leftrightarrow$" the convolution of the sequences $a = (a_n)_{n \geq 0}$ and $b = (b_n)_{n \geq 0}$ corresponds to the multiplication of their respective generating functions. It is not hard to see that the formal operations introduced above (addition, multiplication by scalars and multiplication between formal series) posses the associativity, commutativity and distributivity properties, so that the space of all formal series can be treated as an *algebraic* structure—for more details, see [27, 46, 110, 111].

In addition to the power series $G_a(x)$, constructed from the sequence $(a_n)$, one can define the *exponential generating function*

$$E_a(x) = \sum_{n \geq 0} a_n \frac{x^n}{n!},$$

the series again being understood as a formal series. Just as in the case of generating functions, one has the one-to-one correspondence

$$(a_n) \leftrightarrow E_a(x)$$

and the following properties hold

$$(a_n + cb_n) \leftrightarrow E_a(x)\big(cE_b(x)\big),$$

$$\left( \sum_{i=0}^{n} C_n^i a_i b_{n-i} \right) \leftrightarrow E_a(x)E_b(x).$$

Now we turn to some examples. If the sequence $(a_n)_{n \geq 0}$ is chosen so that $a_n = 1$, $n \geq 0$, then

$$G_a(x) = \sum_{n=0}^{\infty} x^n \quad \left( = \frac{1}{1-x}, \quad |x| < 1 \right)$$

and one has the formula

$$[G_a(x)]^N = \left[ \sum_{n=0}^{\infty} x^n \right]^N = \sum_{n=0}^{\infty} C_{N+n-1}^n x^n ,$$

which comes as a result of the following argument.

What is the coefficient for $x^n$ in the formal expansion of $(1 + x + x^2 + \ldots)^N$?
Since

$$(1+x+x^2+\ldots)^N = (1+x+x^2+\ldots)(1+x+x^2+\ldots)\ldots(1+x+x^2+\ldots),$$

it is clear that if one extracts from the first factor $x^{n_1}$, extracts from the second factor
$x^{n_2}$, $\ldots$, and, finally, extracts from the $N^{\text{th}}$ factor the term $x^{n_N}$, then one would
end up with the term $x^{n_1} x^{n_2} \ldots x^{n_N} = x^n$. The total number of all such choices
$(n_1, n_2, \ldots, n_N)$, with $n_1 + n_2 + \ldots + n_N = n$ and $n_i \geq 0$, is simply the number of
all non-negative integer-valued solutions to the equation $n_1 + n_2 + \ldots + n_N = n$,
which, according to Problem 1.1.6, is precisely $C_{N+n-1}^n$.

This shows that the generating function of the sequence $(C_{N+n-1}^n)_{n\geq 0}$ is simply
$(1-x)^{-N}$, $|x| < 1$. In particular, the following identity must hold:

$$\sum_{n=0}^{\infty} 2^{-n} C_{N+n-1}^n = 2^N .$$

Furthermore, the generating function of the sequence $(C_N^n)_{n\geq 0}$, with the understanding that $C_N^n = 0$, $n > N$, is nothing but $(1 + x)^N$; in other words,

$$(1 + x)^N = \sum_{n=0}^{N} C_N^n x^n .$$

The proof of the last relation is analogous to the proof of the formula for the
generating function of the sequence $(C_{N+n-1}^n)_{n\geq 0}$.

Consider the identity

$$(1 + x)^N (1 + x)^M = (1 + x)^{N+M},$$

and observe that after expanding both sides in the powers $x^k$ and comparing the
respective terms one finds that

$$\sum_{j=0}^{N} C_N^j C_M^{k-j} = C_{N+M}^k .$$

The last identity is known as "Vandermonde's convolution," or the hypergeometric identity—see also Problem 1.2.2. Its derivation is a good illustration of the use of the method of generating functions for deriving various combinatorial identities.

To conclude the discussion of the topic "generating functions," we now turn to the Stirling numbers of the second kind, $S_N^n$, and the Bell numbers, $B_N$, which were introduced earlier. Recall that $S_N^n$ gives the number of all partitions $\mathcal{D}$ of a set $\mathbf{A}$ with $N$ elements, such that $\mathcal{D}$ consists of exactly $n$ classes. Recall also that $B_N = \sum_{n=1}^{N} S_N^n$ gives the number of all possible partitions of the set $\mathbf{A}$ with $|\mathbf{A}| = N$.

In Sect. A.1 we established the formula $n^N = \sum_{k=1}^{n} S_N^k (n)_k$ by using combinatorial considerations. (Recall that $S_N^1 = S_N^N = 1$, $S_N^0 = 0$ and $S_N^n = 0$ for $n > N$.) It is easy to see from this formula that for any $N \geq 1$ the polynomial

$$P_N(x) = x^N - \sum_{n=1}^{N} S_N^n (x)_n, \quad x \in \mathbb{R},$$

has roots $x = 1, \ldots, N$. Since $x = 0$ is also a root, it follows that $P_N(x) \equiv 0$. Consequently, for any $N \geq 1$ and $x \in \mathbb{R}$ one has

$$x^N = \sum_{n=1}^{N} S_N^n (x)_n.$$

If we set $S_0^0 = 1$, $(x)_0 = 1$ and $S_N^0 = 0$ for $N \geq 1$, one finds that

$$x^N = \sum_{n=0}^{N} S_N^n (x)_n,$$

for all $N = 0, 1, 2, \ldots$ and all $x \in \mathbb{R}$.

With the above relations in mind, one can write

$$\sum_{n \geq 0} \left( \sum_{N \geq 0} S_N^n \frac{y^N}{N!} \right) (x)_n = \sum_{N \geq 0} \frac{y^N}{N!} \left( \sum_{n \geq 0} S_N^n (x)_n \right)$$

$$= \sum_{N \geq 0} \frac{(yx)^N}{N!} = e^{yx} = (e^y)^x = (1 + (e^y - 1))^x = \sum_{n \geq 0} \frac{1}{n!} (e^y - 1)^n (x)_n,$$

due to the Taylor expansion $(1 + z)^x = \sum_{n \geq 0} \frac{z^n}{n!} (x)_n$. By comparing the left and right sides of the above chain of identities, one finds that, for every $n \geq 0$, the exponential generating function for the Stirling numbers of the second kind, $S_N^n$, $N \geq 0$, is given by the formula

$$\sum_{N \geq 0} S_N^n \frac{y^N}{N!} = \frac{1}{n!} (e^y - 1)^n , \quad n \geq 0 .$$

(with the convention $S_0^0 = 1$, $S_N^0 = 0$ for $N \geq 1$ and $S_N^n = 0$ for $N < n$).

In much the same way one can obtain the generating function of the sequence $(S_N^n)_{n \geq 0}$:

$$\sum_{n \geq 0} S_N^n x^n = e^{-x} \sum_{m \geq 0} \frac{m^N x^m}{m!} . \tag{**}$$

Indeed, taking into account that $(m)_n = 0$ for $m \leq n - 1$, one can write

$$x^n e^x = \sum_{i \geq 0} \frac{x^{i+n}}{i!} = \sum_{m \geq 0} (m)_n \frac{x^m}{m!} .$$

Furthermore, the formula $m^N = \sum_{n \geq 0} S_N^n (m)_n$ yields

$$e^x \sum_{n \geq 0} S_N^n x^n = \sum_{n \geq 0} S_N^n x^n e^x = \sum_{n \geq 0} S_N^n \left( \sum_{m \geq 0} (m)_n \frac{x^m}{m!} \right)$$

$$= \sum_{m \geq 0} \frac{x^m}{m!} \left( \sum_{n \geq 0} S_N^n (m)_n \right) = \sum_{m \geq 0} \frac{m^N x^m}{m!} ,$$

which gives $(**)$.

With $x = 1$ $(**)$ gives *Dobinski's formula* for the Bell numbers:

$$B_N = e^{-1} \sum_{m \geq 0} \frac{m^N}{m!} .$$

The definition of the Stirling numbers of the *second* kind, $S_N^n$, was based on the combinatorial interpretation of these quantities as the total number of partitions of a set $\mathbf{A}$ that has $N$ elements into $n$ disjoint classes. Then we showed that $x^N = \sum_{n=0}^{N} S_N^n (x)_n$, for every $N \geq 0$.

The algebraic Stirling numbers of the *first* kind, $s_N^n$, $0 \leq n \leq N$, can be defined by the relation

$$(x)_N = \sum_{n=0}^{N} s_N^n x^n . \tag{*}$$

The combinatorial interpretation of the numbers $s_N^n$ can be explained as follows. Let $\pi = (\pi_1, \ldots, \pi_N)$ be any permutation of the numbers $(1, \ldots, N)$ and let $c_N^n$ denote the number of permutations with exactly $n$ cycles. (For example, the permutation $\begin{pmatrix} 1, 2, 3, 4, 5 \\ 2, 1, 4, 5, 3 \end{pmatrix}$ has two cycles.) One can then show that

$$c_N^n = c_{N-1}^{n-1} + (N-1)c_{N-1}^n$$

(with $c_0^0 = 1$), and conclude that

$$\sum_{n=0}^{N} c_N^n x^n = x(x+1)\ldots(x+N-1).$$

By comparing the above generating function for the sequence $c_N^0, c_N^1, \ldots, c_N^N$ with the generating function in $(*)$, for the sequence of Stirling numbers of the first kind, $s_N^0, s_N^1, \ldots, s_N^N$, one finds that

$$c_N^n = (-1)^{N-n} s_N^n .$$

This shows that the Stirling numbers of the first kind $s_N^n$ coincide up to their sign with the number of permutations of the set $(1, \ldots, N)$, that have precisely $n$ cycles.

The *generating function* $G(s) = \mathsf{E}s^X$, $|s| \leq 1$, associated with a discrete random variable $X$ that takes values in the set $N = \{0, 1, 2, \ldots\}$ with probabilities $p_k = \mathsf{P}\{X = k\}, k \in N$, can be written as

$$G(s) = \sum_{k=0}^{\infty} p_k s^k ,$$

and therefore can be identified with the generating function of the *numerical sequence* $(p_k)_{k \geq 0}$.

Closely related to the notion of generating function is the notion of *moment generating function* (see Problem 2.6.32). The moment generating function of the random variable $X$ is defined as

$$M(s) = \mathsf{E}e^{sX}.$$

Notice that if $X \geq 0$ (a. e.) the expectation $\mathsf{E}e^{sX}$ would be well defined for $-1 < s < 0$. Assuming that all moments $m^{(k)} = \mathsf{E}X^k, k \geq 1$, are finite, the moment generating function $M(s)$ can be expanded into the (formal) series

$$M(s) = \sum_{k=0}^{\infty} m^{(k)} \frac{s^k}{k!} ,$$

which is nothing but the exponential generating function for the sequence $(m^{(k)})_{k \geq 0}$. As was noted earlier, in addition to the usual moments $m^{(k)} = \mathsf{E}X^k$, in probability theory it is often useful to work with the *factorial moments*

$$(m)_k = \mathsf{E}(X)_k \equiv \mathsf{E}X(X-1)\ldots(X-k+1)$$

and the *binomial moments*

$$b_{(k)} = \mathsf{E}\,\frac{(X)_k}{k!} = \frac{(m)_k}{k!}$$

(the term "binomial" is justified by the relation $b_{(k)} = \mathsf{E}C_X^k$, where $C_X^k = (X)_k/k!$).

The sequence of factorial moments $((m)_k)_{k\geq 0}$ and the sequence of binomial moments $(b_{(k)})_{k\geq 0}$ give rise, respectively, to the exponential generating function

$$(M)(s) = \sum_{k=0}^{\infty} (m)_k \frac{s^k}{k!}$$

and the generating function

$$B(s) = \sum_{k=0}^{\infty} b_{(k)} s^k \,.$$

Clearly, one must have

$$M(s) = G(e^s) \quad \text{and} \quad (M)(s) = B(s) = G(s+1)\,.$$

It is useful to point out that the following two identities, established earlier, in which $S_N^n$ and $s_N^n$ stand for the Stirling numbers, respectively, of the second and the first kind,

$$x^N = \sum_{n=0}^{\infty} S_N^n\,(x)_n, \quad (x)_N = \sum_{n=0}^{\infty} s_N^n x^n \,,$$

entail the following connection between the moments $m^{(n)} = \mathsf{E}X^n$ and the factorial moments $(m)_n = \mathsf{E}(X)_n, n \geq 0$:

$$m^{(N)} = \sum_{n=0}^{N} S_N^n\,(m)_n \,, \quad (m)_N = \sum_{n=0}^{N} s_N^n m^{(n)} \,.$$

• It is useful to notice that many special sequences in mathematics (e.g., Bernoulli, Euler, etc.) and special polynomials (Bernoulli, Euler, Hermite, Appell, etc.) are defined in terms of generating functions.

(a) The *Bernoulli numbers*, $b_0, b_1, b_2, \ldots,$ and the *Bernoulli polynomials*, $B_0(x), B_1(x), B_2(x), \ldots,$ are defined through the respective exponential generating functions as:

$$\frac{s}{e^s - 1} = \sum_{n=0}^{\infty} b_n \frac{s^n}{n!} \quad \text{and} \quad \frac{s e^{xs}}{e^s - 1} = \sum_{n=0}^{\infty} B_n(x) \frac{s^n}{n!}\,.$$

(The first several Bernoulli numbers are: $b_0 = 1$, $b_1 = -\frac{1}{2}$, $b_2 = \frac{1}{6}$, $b_4 = -\frac{1}{30}$, $b_6 = \frac{1}{42}$, $b_8 = -\frac{1}{30}$, $b_0(x) = 1$, $B_1(x) = x - \frac{1}{2}$, $B_2(x) = x^2 - x + \frac{1}{6}$, $B_3(x) = x^3 - \frac{3}{2}x^2 + \frac{1}{2}x$—see [109], for example.) All odd-numbered Bernoulli numbers (except for $b_1 = -\frac{1}{2}$) are equal to zero. What follows is a list of some key properties and relations:

1. $b_N = \sum_{n=0}^{N} C_N^n b_{N-n}$, $N = 2, 3, \ldots$;
2. all numbers $b_N$ are rational;
3. $B_N(0) = b_N$, $B_N(1) = (-1)^N b_N$, $N \geq 0$;
4. $B_N(x) = \sum_{n=0}^{N} C_N^n b_n x^{N-n}$, $N \geq 1$;
5. $B_N'(x) = N B_{N-1}(x)$, $N \geq 1$.

(b) The *Euler numbers*, $e_0, e_1, e_2, \ldots$, and the *Euler polynomials* $E_0(x)$, $E_1(x)$, $E_2(x), \ldots$, are also defined through the exponential generating functions as:

$$\frac{2e^s}{e^{2s} + 1} = \sum_{n=0}^{\infty} e_n \frac{s^n}{n!} \quad \text{and} \quad \frac{2e^{xs}}{e^s + 1} = \sum_{n=0}^{\infty} E_n(x) \frac{s^n}{n!} .$$

Since $\frac{2e^s}{e^{2s}+1} = \frac{1}{\cosh s}$, the exponential generating function for the sequence of Euler numbers $e_0, e_1, e_2, \ldots$ is simply $\frac{1}{\cosh s}$.

The above definitions imply that:

1. $e_N = 2^N E_N(\frac{1}{2})$, $N \geq 0$;
2. $E_N(x) = \sum_{n=0}^{N} C_N^n E_n \frac{1}{2^n} \left(x - \frac{1}{2}\right)^{N-n}$, $N \geq 0$;
3. $E_N'(x) = N E_{N-1}(x)$, $N \geq 1$;
4. all odd-numbered Euler numbers are equal to zero, while even-numbered Euler numbers are integers.

The first several Euler numbers can be computed as: $e_0 = 1$, $e_2 = -1$, $e_4 = 5$, $e_6 = -61$, $e_8 = 1,385$—see [109].

(c) The *Hermite polynomials* are defined somewhat differently in analysis and probability theory.

The Hermite polynomials, $H_n(x)$, $n \geq 0$, of the type commonly used in analysis, are defined as

$$H_n(x) = (-1)^n \frac{D^n \psi(x)}{\psi(x)},$$

where $\psi(x) = \frac{1}{\sqrt{2\pi}} e^{-x^2}$. The respective exponential generating function is given by

$$\sum_{n=0}^{\infty} H_n(x) \frac{s^n}{n!} = e^{2xs-s^2}, \quad s \in \mathbb{R}, \ x \in \mathbb{R}.$$

The Hermite polynomials, $He_n(x)$, $n \geq 0$, of the type commonly used in probability theory, are defined as:

$$He_n(x) = (-1)^n \frac{D^n \varphi(x)}{\varphi(x)}, \quad n \geq 0,$$

where $\varphi(x) = \frac{1}{\sqrt{2\pi}} e^{-x^2/2}$ is the density of the standard distribution $\mathcal{N}(0, 1)$. (Note that in [P §2.11]—and, commonly, in probability theory—the above polynomials $\mathrm{He}_n(x)$ are denoted by $H_n(x)$.) The exponential generating function associated with the sequence $\mathrm{He}_n(x)$, $n \geq 0$, is given by:

$$\sum_{n=0}^{\infty} \mathrm{He}_n(x) \frac{s^n}{n!} = e^{xs-s^2/2}, \quad s \in \mathbb{R}, \ x \in \mathbb{R}.$$

One can easily verify the relation

$$\mathrm{He}_n(x) = 2^{-n/2} H_n(2^{-1/2}x).$$

The first several Hermite polynomials can be computed as:

$$
\begin{aligned}
H_0(x) &= 1, & \mathrm{He}_0(x) &= 1, \\
H_1(x) &= 2x, & \mathrm{He}_1(x) &= x, \\
H_2(x) &= 4x^2 - 2, & \mathrm{He}_2(x) &= x^2 - 1, \\
H_3(x) &= 8x^3 - 12x, & \mathrm{He}_3(x) &= x^3 - 3x.
\end{aligned}
$$

A more general version of the Hermite polynomials, written as $\mathrm{He}_n(x, t)$, $n \geq 0$, $x \in \mathbb{R}$, $t \in \mathbb{R}_+$, can be defined through the relation

$$\sum_{n=0}^{\infty} \mathrm{He}_n(x, t) \frac{s^n}{n!} = e^{xs - \frac{s^2}{2}t}, \quad s \in \mathbb{R}, \ x \in \mathbb{R}.$$

The polynomials $\mathrm{He}_n(x, t)$, $n \geq 0$, play an important role in the study of the Brownian motion, due to the following property: if $B = (B_t)_{t\geq0}$ is any standard Brownian motion, then the following processes can be claimed to be martingales relative to the filtration of $B = (B_t)_{t\geq0}$:

$$(\mathrm{He}_n(B_t, t))_{t\geq0}, \ \text{for any} \quad n \geq 0, \quad \text{and} \quad \left(e^{sB_t - \frac{s^2}{2}t}\right)_{t\geq0}, \quad \text{for any} \quad s \in \mathbb{R}.$$

Note that in the literature the polynomials $\mathrm{He}_n(x, t)$ are usually written as $H_n(x, t)$, the exact meaning being made clear from the context.

(d) Suppose that $X$ is some random variable and the associated generating function,

$$G(s) = \mathrm{E}e^{sX},$$

is finite for all $|s| < \lambda$, for some $\lambda > 0$.

We now define the function

$$A(s, x) = \frac{e^{sx}}{G(s)}, \quad x \in \mathbb{R}, \quad |s| < \lambda.$$

In actuarial and financial mathematics the map $x \leadsto \frac{e^{sx}}{G(s)}$ is called Escher's transform (of the random variable $X$)—see [P §7.11]. The function $A(s, x)$ gives rise to the *Appell polynomials* (also known as *Sheffer polynomials*) $Q_0(x), Q_1(x), \ldots$ through the expansion

$$A(s, x) = \sum_{k=0}^{\infty} Q_k(x) \frac{s^k}{k!}.$$

In other words, $A(s, x) = \frac{e^{sx}}{\mathsf{E}e^{sX}}$ is simply written as the generating function of the sequence of polynomials $(Q_k(x))_{k \geq 0}$.

The generating function of a random variables $X$ that is uniformly distributed in the interval $[0, 1]$ is

$$G(s) = \mathsf{E}e^{sX} = \frac{e^s - 1}{s}.$$

Consequently, in this special case one has

$$A(s, x) = \frac{se^{sx}}{e^s - 1}$$

and the Appell polynomials $Q_k(x)$ are nothing but the Bernoulli polynomials $B_k(x)$, considered earlier.

If $X$ is a Bernoulli random variable with $\mathsf{P}\{X = 1\} = \mathsf{P}\{X = 0\} = 1/2$, then its generating function is

$$G(s) = \mathsf{E}e^{sX} = \frac{e^s + 1}{2}$$

and, consequently,

$$A(s, x) = \frac{2e^{sx}}{e^s + 1},$$

and that in this case the Appell polynomials coincide with the Euler polynomials.

A standard normal ($\mathcal{N}(0, 1)$) random variable $X$ has generating function

$$G(s) = e^{s^2/2},$$

and it is easy to check that in this case one has

$$A(s, x) = e^{xs - s^2/2},$$

and that the Appell polynomials $Q_k(x)$ coincide with the Hermite polynomials $\mathrm{He}_k(x)$.

Next, let $\varkappa_1, \varkappa_2, \ldots$ denote the cumulants (a.k.a. semi-invariants) of the random variable $X$. The following relations are easy to verify:

$$Q_0(x) = 1,$$
$$Q_1(x) = x - \varkappa_1,$$
$$Q_2(x) = (x - \varkappa_1)^2 - \varkappa_2,$$
$$Q_3(x) = (x - \varkappa_1)^3 - 3\varkappa_2(x - \varkappa_1) - \varkappa_3.$$

In the special case where $X \sim \mathcal{N}(0, 1)$, the cumulants are $\varkappa_1 = 0$, $\varkappa_2 = 1$, and $\varkappa_3 = \varkappa_4 = \ldots = 0$. As a result, one can write:

$$Q_0(x) = He_0(x) = 1,$$
$$Q_1(x) = He_1(x) = x,$$
$$Q_2(x) = He_2(x) = x^2 - 1,$$
$$Q_3(x) = He_3(x) = x^3 - 3x.$$

Notice that in order to claim that the polynomials $Q_k(x)$, $k = 1, \ldots, n$, are uniquely defined it is enough to require that $E|X|^n < \infty$. Furthermore, one has (with the understanding that $Q_0(x) \equiv 1$):

$$Q'_k(x) = kQ_{k-1}(x), \qquad 1 \le k \le n.$$

The above identities are known as the *Appell relations*.

• Given any non-negative random variable $X$, defined on $(\Omega, \mathcal{F}, P)$, and given any $\sigma$-sub-algebra $\mathcal{G} \subseteq \mathcal{F}$, the conditional expectation of $X$ relative to $\mathcal{G}$ is any non-negative (not necessarily finite, i.e., with values in the extended real line $\overline{\mathbb{R}}$) random variable $E(X \mid \mathcal{G}) = E(X \mid \mathcal{G})(\omega)$ that shares the following two properties
1. $E(X \mid \mathcal{G})$ is $\mathcal{G}$-measurable,
2. For every set $A \in \mathcal{G}$ one has:

$$E[XI_A] = E[E(X \mid \mathcal{G})I_A].$$

For a general (i.e., not necessarily non-negative) random variable $X$ ($= X^+ - X^-$) the conditional expectation of $X$ relative to the $\sigma$-sub-algebra $\mathcal{G} \subseteq \mathcal{F}$ is considered to be well defined if one has (P-a. e.)

$$\min[E(X^+ \mid \mathcal{G})(\omega), E(X^- \mid \mathcal{G})(\omega)] < \infty,$$

in which case one can write

$$E(X \mid \mathscr{G})(\omega) = E(X^+ \mid \mathscr{G})(\omega) - E(X^- \mid \mathscr{G})(\omega).$$

If $X(\omega) = I_A(\omega)$, i.e., if $X$ is the indicator of the set $A \in \mathscr{F}$, the conditional expectation $E(I_A \mid \mathscr{G}) = E(I_A \mid \mathscr{G})(\omega)$ is usually written as $P(A \mid \mathscr{G})$, or as $P(A \mid \mathscr{G})(\omega)$, and is called the conditional probability of $A$ relative to the $\sigma$-algebra $\mathscr{G} \subseteq \mathscr{F}$.

If the $\sigma$-algebra $\mathscr{G}$ is generated by the random element $Y = Y(\omega)$ (i.e., $\mathscr{G} = \mathscr{G}_Y = \sigma(Y)$), the quantities $E(X \mid \mathscr{G}_Y)$ and $P(A \mid \mathscr{G}_Y)$ are usually written as $E(X \mid Y)$ and $P(A \mid Y)$ and are referred to, respectively, as *the conditional expectation of $X$ given $Y$* and *the conditional probability of the event $A$ given $Y$*. (See [P §2.7].)

• Just as in mathematical analysis one deals with various types of convergence, in probability theory, too, one deals with various types of convergence for sequence of random variables: convergence in probability $(X_n \xrightarrow{P} X)$; convergence almost surely or almost everywhere $(X_n \to X$ (P-a. e.)); convergence in distribution $(X_n \xrightarrow{d} X$, or $X_n \xrightarrow{\text{law}} X$, or $\mathrm{Law}(X_n) \to \mathrm{Law}(X)$, or $\mathrm{Law}(X_n) \xrightarrow{w} \mathrm{Law}(X))$; $L^p$-convergence, $p > 0$, $(X_n \xrightarrow{L^p} X)$; point-wise convergence $(X_n(\omega) \to X(\omega), \omega \in \Omega)$. (See [P §2.10].)

• In addition to the various types of convergence of sequences of random variables, in probability theory one also deals with convergence of probability measures and convergence of probability distributions and their characteristics.

One of the most important types of convergence of probability measures is the *weak* convergence $P_n \xrightarrow{w} P$, for a given sequence of probability measures $P_n$, $n \geq 1$, and a probability measure $P$, defined on various metric spaces, including the spaces $R^n$, $R^\infty$, $C$ and $D$ that were introduced earlier.

Many classical results from probability theory (e.g., the central limit theorem, Poisson theorem, convergence to infinitely divisible distributions, etc.), are essentially statements about weak convergence of certain sequences of probability measures—see [P Chap. 3].

• Most fundamental results in probability theory—e.g., the zero-one law, the strong law of large numbers, the law of the iterated logarithm—are concerned exclusively with properties that hold "with Probability 1" (or "almost surely"). A particularly interesting and useful result is contained in the *Borel–Cantelli lemma*:

Let $A_1, A_2, \ldots$ be any sequence of events and let $\{A_n \text{ i. o}\}$ ( $\equiv \overline{\lim}_n A_n \equiv \bigcap_{n=1}^{\infty} \bigcup_{k=n}^{\infty} A_k$) stand for the set of those $\omega \in \Omega$ which belong to infinitely many events from the sequence $A_1, A_2, \ldots$ Then
  (a) $\sum_{n=1}^{\infty} P(A_n) < \infty$ implies that $P\{A_n \text{ i. o.}\} = 0$;
  (b) If the events $A_1, A_2, \ldots$ are independent, then $\sum_{n=1}^{\infty} P(A_n) = \infty$ implies that $P\{A_n \text{ i. o.}\} = 1$.

(See [P Chap. 4] for more details.)

## A.4   Stationary (in Strict Sense) Random Sequences

• The random sequence $X = (X_1, X_2, \ldots)$, defined on the probability space $(\Omega, \mathscr{F}, \mathsf{P})$ is said to be *stationary in strict sense*, if its distribution law, $\mathrm{Law}(X)$, (or, equivalently, its probability distribution, $P_X$) coincides with the distribution law, $\mathrm{Law}(\theta_k X)$, of the "shifted" sequence $\theta_k X = (X_{k+1}, X_{k+2}, \ldots)$, for any $k \geq 1$.

It is convenient to study the probabilistic properties of such sequences (as is done [P Chap. 5]) by using the notions, ideas and methods of *the theory of dynamical systems*.

• The main object of study in dynamical systems theory are the (measurable) measure-preserving transformations of a given configuration space.

The map $T: \Omega \rightarrow \Omega$ is said to be *measurable* if, for any given $A \in \mathscr{F}$, the set $T^{-1}A = \{\omega : T\omega \in A\}$ belongs to $\mathscr{F}$, or $T^{-1}A \in \mathscr{F}$ for short. The map $T: \Omega \rightarrow \Omega$ is said to be a *measure preserving transformation* (of the configuration space $\Omega$) if it is measurable (for $\mathscr{F}$) and

$$\mathsf{P}(T^{-1}A) = \mathsf{P}(A), \quad \text{for every } A \in \mathscr{F}.$$

The intrinsic connection between "stationary in strict sense random sequences" and "measure-preserving transformations" can be explained as follows.

Let $T$ be any measure-preserving transformation and let $X_1 = X_1(\omega)$ be any random variable on $\Omega$ (automatically measurable for $\mathscr{F}$). Given any $n \geq 1$, define $X_n(\omega) = X_1(T^{n-1}\omega)$, where $T^{n-1} = T \circ T \circ \cdots \circ T$ ($(n-1)$-times) is the $(n-1)^{\text{st}}$ power of $T$ (as a transformation of $\Omega$). The sequence $X = (X_1, X_2, \ldots)$ is easily seen to be stationary in strict sense.

The converse statement can also be made, if one is allowed to reconstruct the probability space. Specifically, if $X = (X_1, X_2, \ldots)$ is any stationary in strict sense random sequence (defined on some probability space $(\Omega, \mathscr{F}, \mathsf{P})$), then it is possible to produce a probability space $(\widetilde{\Omega}, \widetilde{\mathscr{F}}, \widetilde{\mathsf{P}})$, on which one can construct a measure-preserving (for $\widetilde{\mathsf{P}}$) transformation $\widetilde{T}: \widetilde{\Omega} \rightarrow \widetilde{\Omega}$ and a random variable $\widetilde{X}_1 = \widetilde{X}_1(\widetilde{\omega})$, so that $\mathrm{Law}(X) = \mathrm{Law}(\widetilde{X})$, where $\widetilde{X} = (\widetilde{X}_1(\widetilde{\omega}), \widetilde{X}_1(\widetilde{T}\widetilde{\omega}), \ldots)$.

The main results of [P Chap. 5] are concerned with the fundamental properties of certain measure-preserving transformations, such as recurrence ("Poincaré recurrence theorem"), ergodicity and mixing. The key result in that chapter is the Birkhoff–Khinchin theorem, one invariant of which (that covers both measure-preserving transformations and stationary in strict sense random sequences) can be stated as follows:

(a) Let $T$ be some measure-preserving ergodic transformation on $(\Omega, \mathscr{F}, \mathsf{P})$ and let $\xi = \xi(\omega)$ be any random variable on $\Omega$ with $\mathsf{E}|\xi| < \infty$. Then

$$\lim_n \frac{1}{n} \sum_{k=0}^{n-1} \xi(T^k \omega) = \mathsf{E}\xi \quad (\mathsf{P}\text{-a.\,e.}).$$

(b) Let $X = (X_1, X_2, \ldots)$ be any stationary in strict sense ergodic sequence of random variables on $(\Omega, \mathscr{F}, \mathsf{P})$, for which $\mathsf{E}|X_1| < \infty$. Then

$$\lim_n \frac{1}{n} \sum_{k=0}^{n-1} X_k(\omega) = \mathsf{E}X_1.$$

## A.5   Stationary (in Broad Sense) Random Sequences

From the point of view of both theory and practice, in the study of random sequences of the form $X = (X_n)$, it is important to allow the random variables $X_n$ to take complex values and to be defined for all $n \in \mathbb{Z} = \{0, \pm 1, \pm 2, \ldots\}$. We will then write $X = (\ldots, X_{-1}, X_0, X_1, \ldots)$ and will suppose that each $X_n$ is a complex random variable of the form $(a_n + ib_n)$ with $\mathsf{E}|X_n|^2 = \mathsf{E}(a_n^2 + b_n^2) < \infty$ for all $n \in \mathbb{Z}$—see [P §6.1].

Our main assumption "stationarity in broad sense" comes down to $\mathsf{E}X_n = \mathsf{E}X_0$ and $\mathsf{cov}(X_{n+m}, X_m) = \mathsf{cov}(X_n, X_0)$, for all $n, m \in \mathbb{Z}$.

Without any loss of generality we may and do suppose that $\mathsf{E}X_0 = 0$, so that $\mathsf{cov}(X_n, X_0) = \mathsf{E}X_n X_0$. The function $R(n) = \mathsf{E}X_n X_0$, $n \in \mathbb{Z}$, is called the *covariance function* of the sequence $X$.

• The following two results (the Herglotz theorem and the spectral representation theorem) demonstrate that, by nature, a stationary in broad sense random sequence is nothing but an infinite sum of harmonics with random amplitudes, the summation being taken (with an appropriate limiting procedure) over the entire range of frequencies of the harmonics.

The first result (see [P §6.1]) states that every covariance function $R(n)$, $n \in \mathbb{Z}$, admits the spectral representation:

$$R(n) = \int_{-\pi}^{\pi} e^{i\lambda n} F(d\lambda), \quad \text{for all } n \in \mathbb{Z},$$

where $F = F(B)$, $B \in \mathscr{B}([\pi, \pi))$, is some finite real-valued measure, and the integral is understood in the sense of Lebesgue–Stiltjes.

The second result (see [P §6.3]) gives the spectral representation of the random sequence $X = (X_n)_{n \in \mathbb{Z}}$:

$$X_n = \int_{-\pi}^{\pi} e^{i\lambda n} Z(d\lambda) \quad (\text{P-a.e.}), \quad \text{for all } n \in \mathbb{Z},$$

where $Z = Z(\Delta)$, $\Delta \in \mathcal{B}([-\pi, \pi))$, is some orthogonal (generally, complex-valued) random measure with $\mathsf{E}Z(\Delta) = 0$ and $\mathsf{E}|Z(\Delta)|^2 = F(\Delta)$ (recall that in our setting $\mathsf{E}X_0 = 0$).

If they exist, the spectral function $F = F(d\lambda)$ and the spectral density $f = f(\lambda)$ (related by $F(B) = \int_B f(\lambda)\,d\lambda$, $B \in \mathcal{B}([-\pi, \pi))$), play a fundamental role in the spectral and correlation analysis of the random sequence $X$, providing a description of the "spectral composition" of the covariance function.

At the same time, the relation $\mathsf{E}|Z(\Delta)|^2 = F(\Delta)$ reveals the connection between the spectral function and the "stochastic spectral component" in the representation $X_n = \int_{-\pi}^{\pi} e^{i\lambda n} Z(d\lambda)$, $n \in \mathbb{Z}$.

• Given the intrinsic nature of the spectral properties outlined above, it is easy to understand why results of this type are so important in the statistics of stationary sequences and the statistics of random processes in continuous time. More specifically, these features allow one to construct "reasonably good" *estimates* of the covariance function, the spectral density and their characteristics (see [P §5.4]). All of this is instrumental for building probabilistic models of observable phenomena, which are consistent with the data derived from experiments.

Finally, we note that the pioneering work of A. N. Kolmogorov and N. Wiener on the theory of filtering, extrapolation and interpolation of random sequences and processes, was developed almost entirely in the context of stationary in the broad sense random sequences and processes (see [P §6.6]).

## A.6 Martingales

In the very early stages of the development of the general theory of martingales it was recognized that it would be extremely useful to amend the underlying probability space $(\Omega, \mathcal{F}, \mathsf{P})$ with a flow of $\sigma$-algebras, i.e., a filtration, of the form $(\mathcal{F}_n)_{n\geq 0}$, where $\mathcal{F}_n \subseteq \mathcal{F}$. The filtration has the meaning of "flow of information," i.e., each $\mathcal{F}_n$ comprises all "pieces of information" that an observer may be able to receive by time $n$. The structure $(\Omega, \mathcal{F}, (\mathcal{F}_n)_{n\geq 0}, \mathsf{P})$ is called *filtered probability space*. With any such structure one can associate the notions "adapted" (to the filtration $(\mathcal{F}_n)_{n\geq 0}$), "predictable," "stochastic sequence," "martingale," "Markov times", "stopping times," etc.

• The sequence of random variables $X = (X_n)_{n\geq 0}$, defined on the structure $(\Omega, \mathcal{F}, (\mathcal{F}_n)_{n\geq 0}, \mathsf{P})$, is said to be adapted to the filtration $(\mathcal{F}_n)_{n\geq 0}$ if $X_n$ is $\mathcal{F}_n$-measurable for every $n \geq 0$. The same sequence is said to be a *martingale* on $(\Omega, \mathcal{F}, (\mathcal{F}_n)_{n\geq 0}, \mathsf{P})$ if, in addition to being adapted to $(\mathcal{F}_n)_{n\geq 0}$, it is integrable, in that $\mathsf{E}|X_n| < \infty$, $n \geq 0$, and has the property

$$\mathsf{E}(X_n \mid \mathcal{F}_{n-1}) = X_{n-1}, \quad \text{for all } n \geq 1.$$

If the equality in the above relation is replaced by the inequality $E(X_n \mid \mathcal{F}_{n-1}) \geq X_{n-1}$, or the inequality $E(X_n \mid \mathcal{F}_{n-1}) \leq X_{n-1}$, then the sequence $X = (X_n)_{n \geq 0}$ is said to be, respectively, *submartingale* and *supermartingale*.

   • The class of martingales includes many special sequences of random variables, encountered in many important practical applications (see [P §7.1]). More importantly, the general theory of martingales provides methods, insights and computational tools that are indispensable for certain aspects of probability theory and mathematical statistics—especially in connection with some important practical applications. The key insights from the martingale theory are: the invariance of the martingale property under random time-change (see [P §7.2]), the fundamental inequalities for martingales and submartingales (see [P §7.3]) and the convergence theorems for martingales and submartingales (see [P §7.4]).

   Some of the most important practical application of martingale theory, namely: the probability for ruin in insurance, the martingale characterization of the absence of arbitrage in financial markets, the construction of hedging strategies in complete financial markets and the optimal stopping problem, are discussed in [P §7.10] through [P §7.13].

## A.7   Markov Chains

In what follows we will expand and reformulate some of the main results from the general theory of Markov chains that was developed in [P Chap. 8]. The notation and the terminology introduced in [P Chap. 8] will be assumed, but will be modified and expanded, in connection with some new topics that were not included in [P Chap. 8].

   • Similarly to martingales, a generic Markov chain (in broad sense), $X = (X_n)_{n \geq 0}$, can be treated as a sequence of random variables that are defined on some filtered probability space $(\Omega, \mathcal{F}, (\mathcal{F}_n)_{n \geq 0}, P)$ and take values in some set $E$, called the "state space" of the Markov chain $X$. The state space $E$ will be endowed with the structure of a measurable space and will be denoted by $(E, \mathcal{E})$. As a sequence of random variables, the Markov chain $X = (X_n)_{n \geq 0}$ will always be assumed to be adapted to the filtration $(\mathcal{F}_n)_{n \geq 0}$, in the sense that $X_n$ is $\mathcal{F}_n/\mathcal{E}$-measurable for every $n \geq 0$. The fundamental property that characterizes $X = (X_n)_{n \geq 0}$ as a *Markov chain in broad sense* can be stated as follows for every $n \geq 0$ and every $B \in \mathcal{E}$ one has

$$P(X_{n+1} \in B \mid \mathcal{F}_n)(\omega) = P(X_{n+1} \in B \mid X_n)(\omega) \quad \text{(P-a.e.)}.$$

(With a slight abuse of the notation, we will write $P(X_{n+1} \in B \mid X_n(\omega))$ instead of $P(X_{n+1} \in B \mid X_n)(\omega)$.)

   If the filtration $(\mathcal{F}_n)_{n \geq 0}$ happens to be the *natural filtration* of the sequence $X = (X_n)_{n \geq 0}$, i.e., $\mathcal{F}_n = \mathcal{F}_n^X \equiv \sigma(X_0, X_1, \ldots, X_n)$ for every $n \geq 0$, then the Markov property in broad sense becomes *Markov property in strict sense*, and, if this property holds, the sequence $X = (X_n)_{n \geq 0}$ is said to be a *Markov chain*.

In the special case where $(E, \mathcal{E})$ is a Borel space, [P §2.7, Theorem 5] guarantees that for every fixed $n \geq 0$ there is a *regular conditional probability* $P_n(x; B)$, $x \in E$, $B \in \mathcal{E}$, with the property that for every $B \in \mathcal{E}$ one can write

$$P(X_n \in B \mid X_{n-1}(\omega)) = P_n(X_{n-1}(\omega); B), \quad \text{for P-a. e.} \omega \in \Omega.$$

In the theory of Markov processes, the regular conditional probabilities $P_n(x; B)$, $n \geq 0$, are called *transition functions* (from $E$ to $\mathcal{E}$), or *Markov kernels*. In the special case where the transition functions do not depend on $n$, i.e., one can write $P_n(x; B) = P(x; B)$, the associated Markov chain (in broad sense or in strict sense) is said to be *homogeneous*.

Another important element of the construction of any Markov chain, in addition to the transition functions $P_n(x; B)$, $n \geq 0$, is the *initial distribution* $\pi = \pi(B)$, $B \in \mathcal{E}$, which is simply the probability distribution of the random variable $X_0$, i.e.,

$$\pi(B) = P\{X_0 \in B\}, \quad B \in \mathcal{E}.$$

The initial distribution and the transition functions, i.e., the entire collection $(\pi, P_1, P_2, \ldots)$, which in the homogeneous case comes down to the pair $(\pi, P)$, uniquely determines the probability distribution (as a random sequence) of the Markov chain $X = (X_0, X_1, \ldots)$.

• Following the modern treatment of the subject, [P Chap. 8] adopts the view that the main building blocks in the general theory of Markov chains are the state space $(E, \mathcal{E})$ and the collection of transition functions $P_n(x; B)$, $x \in E$, $B \in \mathcal{E}$, $n \geq 0$ from $E$ to $\mathcal{E}$ (which reduces to a single transition function $P(x; B)$, $x \in E$, $B \in \mathcal{E}$ in the homogeneous case). This was a departure from the classical framework, in which the starting point is the filtered probability space $(\Omega, \mathcal{F}, (\mathcal{F}_n)_{n \geq 0}, P)$, the state space $(E, \mathcal{E})$, and the sequence $X = (X_0, X_1, \ldots)$ of $E$-valued random variables, chosen so that each $X_n$ is $\mathcal{F}_n / \mathcal{E}$-measurable. According to the Ionescu Tulcea Theorem (see [P §2.9]), for any given state space $(E, \mathcal{E})$ and any given family of transition functions from $E$ to $\mathcal{E}$, one can take $(\Omega, \mathcal{F})$ to be the canonical coordinate space $(E^\infty, \mathcal{E}^\infty)$ and then construct a family of probability measures, $\{P_x, x \in E\}$, on $(\Omega, \mathcal{F})$, in such a way that the sequence of coordinate maps, $X = (X_0, X_1, \ldots)$, given by $X_n(\omega) = x_n$ for $\omega = (x_0, x_1, \ldots)$, $n \geq 0$, forms a Markov chain under the probability measure $P_x$, with $P_x\{X_0 = x\} = 1$, for every $x \in E$, i.e., under the probability measure $P_x$ (on $(E^\infty, \mathcal{E}^\infty)$) the sequence of coordinate maps $X = (X_0, X_1, \ldots)$ (from $\Omega$ into $E$) behaves as a Markov chain that starts from $x \in E$ with probability 1.

Given any probability law $\pi = \pi(B)$, $B \in \mathcal{E}$ (think of this law as the "initial" distribution of some Markov chain), we denote by $P_\pi$ the probability measure on $(E^\infty, \mathcal{E}^\infty)$ given by $P_\pi(A) = \int_E P_x(A) \pi(dx)$, $A \in \mathcal{E}^\infty$. It is not very difficult to check that under the probability measure $P_\pi$ the sequence of coordinate maps $X$ behaves as a Markov chain with initial distribution $\pi$, i.e., $P_\pi\{X_0 \in B\} = \pi(B)$, for every $B \in \mathcal{E}$.

• In order to formulate two new variants of the Markov property—the so called *generalized Markov property* and *the strong Markov property*—we must introduce the shift operator $\theta$, its "powers" $\theta_n$, and the "random power" $\theta_\tau$, for any given Markov time $\tau$. The *shift operator* $\theta: \Omega \to \Omega$ is defined as

$$\theta(\omega) = (x_1, x_2, \ldots), \quad \text{for } \omega = (x_0, x_1, \ldots).$$

In other words, the operator $\theta$ "shifts" the time-scale one period forward (period 1 becomes period 0, period 2 becomes period 1, and so on), as a result of which the trajectory $(x_0, x_1, \ldots)$ turns into $(x_1, x_2, \ldots)$. (Recall that in [ P Chap. 5], which deals with stationary in strict sense random sequences and the related dynamical systems, we also had to introduce certain transformations of $\Omega$ into itself, which were denoted by $T$.)

If $\theta_0 = I$ stands for the identity map $\theta_0(\omega) = \omega$, the $n$-th power, $\theta_n$, of the operator $\theta$, is defined for $n \geq 1$ as $\theta_n = \theta_{n-1} \circ \theta \ (= \theta \circ \theta_{n-1})$, i.e., $\theta_n(\omega) = \theta_{n-1}(\theta(\omega))$.

Given any Markov time $\tau = \tau(\omega)$ with $\tau \leq \infty$, we denote by $\theta_\tau$ the operator that acts only on the set $\Omega_\tau = \{\omega : \tau(\omega) < \infty\}$ in such a way that $\theta_\tau = \theta_n$ on the set $\{\tau = n\}$, i.e., if $\omega \in \Omega$ is such that $\tau(\omega) = n$, then

$$\theta_\tau(\omega) = \theta_n(\omega).$$

If $H = H(\omega)$ is any $\mathscr{F}$-measurable function of $\omega \in \Omega$ (such as, for example, $\tau = \tau(\omega)$, or $X_m = X_m(\omega)$), then the function $H \circ \theta_n$ is defined as $(H \circ \theta_n)(\omega) \equiv H(\theta_n(\omega))$, $\omega \in \Omega$.

If $\sigma$ is a Markov time, then the function $H \circ \theta_\sigma$ is defined on the $\Omega_\sigma = \{\omega : \sigma(\omega) < \infty\}$ so that for every fixed $n \in \{0, 1, \ldots\}$ one has $H \circ \theta_\sigma = H \circ \theta_n$ everywhere in the subset $\{\sigma = n\} \subseteq \Omega_\sigma$, i.e., $(H \circ \theta_\sigma)(\omega) = (H \circ \theta_n)(\omega) = H(\theta_n(\omega))$, for every $\omega \in \{\sigma = n\}$, and every $n = 0, 1, \ldots$.

In particular, the above relations imply that, for any $m, n = 0, 1, \ldots$ and for any Markov time $\sigma$, one has

$$X_m \circ \theta_n = X_{m+n},$$

$$X_m \circ \theta_\sigma = X_{m+\sigma} \quad \text{on the set } \Omega_\sigma.$$

Furthermore, for every two finite Markov times, $\tau$ and $\sigma$, one has

$$X_\tau \circ \theta_\sigma = X_{\tau \circ \theta_\sigma + \sigma}.$$

The operators $\theta_n: \Omega \to \Omega$ give rise to the inverse operators $\theta_n^{-1}: \mathscr{F} \to \mathscr{F}$, defined in the obvious way as

$$\theta_n^{-1}(A) = \{\omega : \theta_n(\omega) \in A\}, \quad A \in \mathscr{F}.$$

If in the last relation the set $A$ is replaced by the set $\{\omega : X_m(\omega) \in B\}$, for some $B \in \mathcal{E}$, then one can write

$$\theta_n^{-1}(A) = \{\omega : X_{m+n}(\omega) \in B\},$$

which is the same as

$$\theta_n^{-1}(X_m^{-1}(B)) = X_{m+n}^{-1}(B).$$

(Additional properties of the operators $\theta_n$, $\theta_\sigma$, $\theta_n^{-1}$, etc., can be found in some of the problems included in Sect. 8.2 in the present book)

• With the help of the operators $\theta_n$ one can establish (see [P §8.2, Theorem 1]) the so called *generalized Markov property*: if $H = H(\omega)$ is any bounded (or non-negative) and $\mathcal{F}$-measurable function, then for every choice of the initial distribution $\pi$ and for every integer $n \geq 0$ one has

$$E_\pi(H \circ \theta_n \mid \mathcal{F}_n^X)(\omega) = E_{X_n(\omega)}H \quad (P_\pi\text{-a.e.}).$$

In the above relation $E_\pi$ denotes the averaging over the measure $P_\pi$ and the expression $E_{X_n(\omega)}H$ is understood as $\psi(X_n(\omega))$, where $\psi(x) = E_x H$.

In fact, the generalized Markov property can be generalized (i.e., weaken) even further, in that one can replace the deterministic time $n$ in the above relation with some finite Markov time $\tau$. To be precise, one can claim the following: if $(H_n)_{n \geq 0}$ is any family of bounded (or non-negative) and $\mathcal{F}$-measurable functions and if $\tau$ is any finite Markov time, then the Markov property implies the so called *strong Markov property*, according to which for any initial distribution $\pi$ one has

$$E_\pi(H_\tau \circ \theta_\tau \mid \mathcal{F}_\tau^X)(\omega) = \psi(\tau(\omega), X_{\tau(\omega)}(\omega)) \quad (P_\pi\text{-a.e.}),$$

where $\psi(n, x) = E_x H_n$.

Note that the expression $H_\tau \circ \theta_\tau = (H_\tau \circ \theta_\tau)(\omega)$ is understood as $(H_n \circ \theta_n)(\omega)$ for $\omega \in \{\tau = n\}$.

• As was pointed out earlier, the distribution (as a random sequence) of any homogeneous Markov chain $X = (X_n)_{n \geq 0}$ with state space $(E, \mathcal{E})$, is completely determined by the initial distribution $\pi = \pi(dx)$ and the transition function $P = P(x; B)$, $x \in E$, $B \in \mathcal{E}$. Furthermore, the distributions $P_x$, $x \in E$, which are defined on $(E^\infty, \mathcal{E}^\infty)$ are determined only by the transition function $P = P(x; B)$.

It is interesting that the concept of transition functions (or Markov kernels) also lies in the core of the (entirely deterministic) domain of mathematical analysis, which is known as *potential theory*. In fact, there is an intrinsic connection between potential theory and the theory of homogeneous Markov chains. This connection has been extremely beneficial for both fields.

We will now introduce some important notions in both potential theory and the Markovian theory, which will be needed later in this section.

The transition function $P = P(x; B)$, $x \in E$, $B \in \mathcal{E}$, gives rise to the linear (one step) *transition operator* $\mathbb{P}g$, which acts on functions $g = g(x)$ according to the formula

$$\mathbb{P}g(x) = \int_E g(y)\, P(x; dy)\,.$$

(It is quite common to also write $(\mathbb{P}g)(x)$.) The domain of the operator $\mathbb{P}$ consists of all $g \in \mathscr{L}^0(E, \mathscr{E}; \mathbb{R})$ ($\equiv$ the space of all $\mathscr{E}$-measurable functions on $E$ with values in $\mathbb{R}$), for which the integral $\int_E g(y)\, P(x; dy)$ is well defined for all $x \in E$. Clearly, this integral is well defined also on the class of all non-negative and $\mathscr{E}$-measurable functions on $E$, which class we denote by $\mathscr{L}^0(E, \mathscr{E}; \mathbb{R}_+)$, or on the class of all bounded functions $\mathscr{L}_b^0(E, \mathscr{E}; \mathbb{R})$.

Letting $\mathbb{I}$ denote the identity operator $\mathbb{I}g(x) = g(x)$, one can define the $n$-step transition operator $\mathbb{P}_n$, as $\mathbb{P}_n = \mathbb{P}(\mathbb{P}_{n-1})$ for $n \geq 1$, or, equivalently, $\mathbb{P}_n = \mathbb{P}_{n-1}(\mathbb{P})$ for $n \geq 1$, with the understanding that $\mathbb{P}_0 = \mathbb{I}$.

It is clear that one has

$$\mathbb{P}_n g(x) = \mathsf{E}_x g(X_n)$$

for every $g \in \mathscr{L}^0(E, \mathscr{E}; \mathbb{R})$, for which the integral $\int_E g(y)\, P^n(x; dy)$ is well defined, where $P^n = P^n(x; dy)$ is the $n$-step transition function (see [$\underline{P}$ §8.1]).

Given any Markov time $\tau$ for the filtration $(\mathscr{F}_n^X)_{n \geq 0}$ ($\mathscr{F}_n^X = \sigma(X_0, X_1, \ldots, X_n)$), let $\mathbb{P}_\tau$ denote the operator that acts on functions $g = g(x)$ according to the formula

$$\mathbb{P}_\tau g(x) = \mathsf{E}_x\big[I(\tau < \infty)g(X_\tau)\big]\,.$$

Notice that if $g(x) \equiv 1$, then

$$\mathbb{P}_\tau 1(x) = \mathsf{P}_x\{\tau < \infty\}\,.$$

The operators $\mathbb{P}_n$, $n \geq 0$, give rise to the (generally, unbounded) operator

$$\mathbb{U} = \sum_{n \geq 0} \mathbb{P}_n\,,$$

which is called *potential of the operator* $\mathbb{P}$ (or potential of the associated Markov chain).

For any $g \in \mathscr{L}^0(E, \mathscr{E}; \mathbb{R}_+)$ one has

$$\mathbb{U}g = \sum_{n \geq 0} \mathbb{P}_n g = (\mathbb{I} + \mathbb{P}\mathbb{U})g\,,$$

which may be abbreviated as

$$\mathbb{U} = \mathbb{I} + \mathbb{P}\mathbb{U}\,.$$

The function $\mathbb{U}g$ is usually called the *potential of the function g*.

If the function $g(x)$ is taken to be the indicator of the set $B \in \mathscr{E}$, i.e., $g(x) = I_B(x)$, then $N_B = \sum_{n \geq 0} I_B(X_n)$ is simply the number of visits of the chain $X$ to the set $B$, and one can write:

$$\mathbb{U} I_B(x) = \sum_{n \geq 0} \mathsf{E}_x I_B(X_n) = \mathsf{E}_x N_B \,.$$

For any fixed $x \in E$, treated as a function of $B \in \mathscr{E}$, the quantity $U(x, B) = \mathbb{U} I_B(x)$ gives a measure on $\mathscr{E}$, which is sometimes called *potential measure*. Choosing $B$ to be a singleton, namely $B = \{y\}$, for $y \in E$, turns $U(x, \{y\})$ into a function of $x, y \in E$, which is usually denoted by $G(x, y)$ and is called the *Green function* (of the operator $\mathbb{P}$, or, of the associated Markov chain). The meaning of the Green function should be clear: $G(x, y) = \mathsf{E}_x N_{\{y\}}$ is nothing but the *average number of visits* to state $y \in E$, starting from state $X_0 = x \in E$.

Analogously to the potential $\mathbb{U}$ of the operator $\mathbb{P}$, one can define the kernel $Q = Q(x; B)$ of the transition function $P = P(x; B)$ by the formula

$$Q(x; B) = \sum_{n \geq 0} P^n(x; B) \quad \left( = I_B(x) + PQ(x; B) \right) .$$

Since $\mathbb{P}_n I_B(x) = P^n(x; B)$, it is clear that $U(x; B) = Q(x; B)$.

• The operator $\mathbb{P}$ gives rise to another important operator, namely

$$\mathbb{L} = \mathbb{P} - \mathbb{I} \,,$$

where, as usual, $\mathbb{I}$ denotes the identity operator. In Markovian theory the operator $\mathbb{L}$ is called the *generating operator* (a.k.a. the *discrete generator*) of the homogeneous Markov chain with transition function $P = P(x; B)$. The domain, $\mathscr{D}_{\mathbb{L}}$, of the operator $\mathbb{L}$ is the space of all $g \in \mathscr{L}^0(E, \mathscr{E}; \mathbb{R})$ for which the expression $\mathbb{P} g - g$ is well defined.

If $h \in \mathscr{L}^0(E, \mathscr{E}; \overline{\mathbb{R}}_+)$ (i.e., $h$ takes values in $\overline{\mathbb{R}}_+$ and is $\mathscr{E}$-measurable), then, since $\mathbb{U} = \mathbb{I} + \mathbb{P}\mathbb{U}$, its potential $H = \mathbb{U} h$ satisfies the relation

$$H = h + \mathbb{P} H \,.$$

Consequently, if $H \in \mathscr{D}_{\mathbb{L}}$, then $H$ solves the (non-homogeneous) *Poisson equation*

$$\mathbb{L} V = -h \,, \quad V \in \mathscr{D}_{\mathbb{L}} \,.$$

If one can find a solution, $W \in \mathscr{L}^0(E, \mathscr{E}; \overline{\mathbb{R}}_+)$, of the equation $W = h + \mathbb{P} W$ (or to the equation $\mathbb{L} W = -h$, when $W \in \mathscr{D}_{\mathbb{L}}$), then, since $W = h + \mathbb{P} W \geq h$, one can show by induction that $W \geq \sum_{k=0}^n \mathbb{P}_k h$ for any $n \geq 1$, so that $W \geq H$. As a result, the potential $H = \mathbb{U} h$ is the *smallest* solution to the system $V = h + \mathbb{P} V$ within the class $\mathscr{L}^0(E, \mathscr{E}; \overline{\mathbb{R}}_+)$ (remind that $\mathbb{U} h(x) = \mathsf{E}_x \sum_{k=0}^\infty h(X_k)$).

• A function $f = f(x)$, $x \in E$, that belongs to the class $\mathscr{L}^0(E, \mathscr{E}; \overline{\mathbb{R}}_+)$, is said to be *excessive* for the operator $\mathbb{P}$ (or, for the associated Markov chain with transition function $P = P(x; B)$), if

$$\mathbb{P} f \leq f \,,$$

or, which amounts to the same, $E_x f(X_1) \leq f(x)$, for all $x \in E$. In particular, the potential $H = \mathbb{U}h$ of any function $h \in \mathcal{L}^0(E, \mathcal{E}; \overline{\mathbb{R}}_+)$ is an excessive function.

The function $f \in \mathcal{L}^0(E, \mathcal{E}; \overline{\mathbb{R}}_+)$ is said to be *harmonic* (or *invariant*), if

$$\mathbb{P}f = f,$$

i.e., $E_x f(X_1) = f(x)$, for all $x \in E$.

The connection between potential theory (to which the notion of *excessivity* belongs) and probability theory (specifically, the martingale theory) becomes evident from the following statement: if $X = (X_n)_{n \geq 0}$ is any homogeneous Markov chain with initial distribution $\pi$ and with transition function $P = P(x; B)$, if the associated distribution in the space $(E^\infty, \mathcal{E}^\infty)$ is $P_\pi$, and if $f = f(x)$ is any $\mathbb{P}$-excessive function, then one can claim that $Y = (Y_n, \mathcal{F}_n^X, P_\pi)_{n \geq 0}$, with $Y_n = f(X_n)$, is a non-negative supermartingale sequence, in that:

$$Y_n \text{ is } \mathcal{F}_n^X\text{-measurable,} \quad \text{for all } n \geq 0;$$

$$E_\pi(Y_{n+1} \mid \mathcal{F}_n^X) \leq Y_n \quad (P_\pi\text{-a.e.}), \quad \text{for all } n \geq 0.$$

If, in addition, one can claim that $EY_n < \infty$, for all $n \geq 0$, then $Y = (Y_n, \mathcal{F}_n^X, P_\pi)_{n \geq 0}$ is simply a supermartingale.

It is interesting to point out that some of the main properties of non-negative supermartingales (see [P §7.4, Theorem 1]) continue to hold also for non-negative supermartingale sequences of the type described above: the limit $\lim_n Y_n (= Y_\infty)$ exists with $P_\pi$-probability 1; furthermore, if $P_\pi\{Y_0 < \infty\} = 1$, then $P_\pi\{Y_\infty < \infty\} = 1$. The proof of this claim is delegated to Problem 7.4.24.

• Given any $h \in \mathcal{L}^0(E, \mathcal{E}; \mathbb{R}_+)$, or $h \in \mathcal{L}^0(E, \mathcal{E}; \overline{\mathbb{R}}_+)$, the potential, $H(x) = \mathbb{U}h(x)$, satisfies the relation $H(x) = h(x) + \mathbb{P}H(x)$, which, in turn, gives

$$H(x) \geq \max(h(x), \mathbb{P}H(x)), \quad x \in E.$$

Consequently, the potential $H(x) = \mathbb{U}h(x)$ does both: dominates the function $h(x)$ (i.e., $H(x) \geq h(x)$, $x \in E$) and belongs to the class of excessive functions (one usually says that the potential of given function is an example of an excessive *majorant* of that function).

In fact, many practical problems—the optimal stopping problem from [P §8.9] being a typical example—can be formulated as problems for computing the *smallest* excessive majorant of a given $\mathcal{E}$-measurable non-negative function $g = g(x)$. Potential theory provides a special technique for solving such problems, which is described next.

Let $\mathbb{Q}$ denote the operator that acts on all $\mathcal{E}$-measurable non-negative functions $g = g(x)$ according to the formula

$$\mathbb{Q}g(x) = \max\left(g(x), \mathbb{P}g(x)\right).$$

Next, notice that the *smallest excessive majorant*, $s(x)$, of any such function $g(x)$ is given by the formula

$$s(x) = \lim_n Q^n g(x)$$

and satisfies the equation

$$s(x) = \max\left(g(x), \mathbb{P}s(x)\right), \quad x \in E.$$

In particular, the last equation implies that for every $s \in \mathcal{D}_\mathbb{L}$ one must have

$$\mathbb{L}s(x) = 0, \qquad x \in C_g;$$

$$s(x) = g(x), \qquad x \in D_g,$$

where $C_g = \{x : s(x) > g(x)\}$ and $D_g = E \setminus C_g$. (The proof of this claim can be found in [P §8.9], where the token $\mathbb{P}$ is replaced by $T$, and the token $Q$ is replaced by $Q$.)

 • One of the central issues in potential theory is the description of the class of solutions to the *Dirichlet problem* for the operator $\mathbb{P}$: for a given domain $C \subseteq E$ and two $\mathcal{E}$-measurable non-negative functions $h$ and $g$, defined, respectively, on $C$ and $D = E \setminus C$, one must find a non-negative function $V = V(x)$, $x \in E$, chosen from one of the classes $\mathcal{L}^0(E, \mathcal{E}; \overline{\mathbb{R}}_+)$, $\mathcal{L}^0(E, \mathcal{E}; \mathbb{R}_+)$, $\mathcal{L}^0_b(E, \mathcal{E}; \mathbb{R})$, etc., which satisfies the equation

$$V(x) = \begin{cases} \mathbb{P}V(x) + h(x), & x \in C; \\ g(x), & x \in D. \end{cases}$$

If one looks for solutions $V$ only in the class $\mathcal{D}_\mathbb{L}$, then the above system is equivalent to the following one:

$$\mathbb{L}V(x) = -h(x), \qquad x \in C;$$

$$V(x) = g(x), \qquad x \in D.$$

The first equation above is commonly referred to as "the Poisson equation for the domain $C$" and, usually, the Dirichlet problem is understood as the problem for solving the Poisson equation in some domain $C$, with the requirement that the solution is defined everywhere in $E$ and coincides on the complement $D = E \setminus C$ with a given function $g$.

It is quite remarkable that the solution to the Dirichlet problem—which is entirely non-probabilistic—can be expressed in terms of the homogeneous Markov chain with transition function $P = P(x; B)$, which gives rise to the operator $\mathbb{P}$. To make this claim precise, let $X = (X_n)_{n \geq 0}$ be one such Markov chain and let $\tau(D) = \inf\{n \geq 0 : X_n \in D\}$ (with the usual convention $\inf\{\emptyset\} = \infty$). One can then claim that for every two functions, $h$ and $g$, from the class $\mathcal{L}^0(E, \mathcal{E}; \overline{\mathbb{R}}_+)$, a solution to the Dirichlet problem exists and the smallest non-negative solution, $V_D(x)$, is given by the formula:

$$V_D(x) = \mathsf{E}_x\big[I(\tau(D) < \infty)g(X_{\tau(D)})\big] + I_C(x)\mathsf{E}_x\left[\sum_{k=0}^{\tau(D)-1} h(X_k)\right].$$

(For the proof of the last statement see the hint to Problem 8.8.11.)

Some special choices for $h$ and $g$ are considered next.

(a) If $h = 0$, i.e., one is looking for a function $V = V(x)$ which is harmonic in the domain $C$ and coincides with the function $g$ on $D = E \setminus C$, then the smallest non-negative solution $V_D(x)$ is given by the formula

$$V_D(x) = \mathsf{E}_x\big[I(\tau(D) < \infty)g(X_{\tau(D)})\big].$$

In particular, if $g(x) \equiv 1, x \in D$, then

$$V_D(x) = \mathsf{P}_x\{\tau(D) < \infty\}.$$

At the same time, *the probability*, $\mathsf{P}_x\{\tau(D) < \infty\}$, *that the Markov chain will eventually reach $D$, starting from $X_0 = x$*, treated as function of $x \in E$, can be claimed to be harmonic in the domain $C$. It is clear that if $x \in D$, then $\mathsf{P}_x\{\tau(D) < \infty\} = 1$, since in this case $\tau(D) = 0$.

(b) With $g(x) = 0, x \in D$, and $h(x) = 1, x \in C$, the system becomes

$$V(x) = \begin{cases} \mathbb{P}V(x) + 1, & x \in C; \\ 0, & x \in D. \end{cases} \tag{$*$}$$

In this case the smallest non-negative solution is given by the formula

$$V_D(x) = I_C(x)\mathsf{E}_x\left[\sum_{k=0}^{\tau(D)-1} 1\right] = \begin{cases} \mathsf{E}_x\tau(D), & x \in C; \\ 0, & x \in D. \end{cases}$$

In particular, treated as a function of $x \in E$, *the expected time*, $\mathsf{E}_x\tau(D)$, *until the first visit to $D$* gives the smallest non-negative solution to the system $(*)$.

• A particularly important class of Markov sequences, associated with random walks on some state space $(E, \mathscr{E})$, is the class of *simple symmetric random walks* on the lattice

$$E = \mathbb{Z}^d = \{0 \pm 1, \pm 2, \ldots\}^d,$$

where $d$ is a finite integer chosen from the set $\{1, 2, \ldots\}$ (see [P §8.8]). Random walks in the "entire" space $E = \mathbb{Z}^d$, of the form $X = (X_n)_{n \geq 0}$, can be defined *constructively*, by setting

$$X_n = x + \xi_1 + \ldots + \xi_n,$$

where $\xi_1, \xi_2, \ldots$ is a sequence of independent $\mathbb{R}^d$-valued random variables, which are defined on some probability space $(\Omega, \mathscr{F}, \mathsf{P})$, and are distributed uniformly in the set of all basis vectors $e = (e_1, \ldots, e_d) \in \mathbb{R}^d$, defined by $e_i = 0, +1$ or $-1$ and $\|e\| \equiv |e_1| + \ldots + |e_d| = 1$; in particular

$$\mathsf{P}\{\xi_1 = e\} = (2d)^{-1}.$$

Such a random walk describes the movement of a "particle" which, starting from some point $x \in \mathbb{Z}^d$, during every period moves arbitrarily to one of the $2d$ neighboring points on the lattice, and in such a way that each neighboring point is equally likely to get selected.

The operator $\mathbb{P}$, associated with such a random walk, has a particularly simple form:

$$\mathbb{P} f(x) = \mathsf{E}_x f(x + \xi_1) = \frac{1}{2d} \sum_{|e|=1} f(x + e).$$

Consequently, the generating operator (or, the discrete generator) $\mathbb{L} = \mathbb{P} - \mathbb{I}$, which in this case is referred to as *the discrete Laplacian* and is commonly denoted by $\Delta$, has the following form

$$\Delta f(x) = \frac{1}{2d} \sum_{|e|=1} (f(x + e) - f(x)).$$

It is natural to reformulate the Dirichlet problem for the *simple* random walk by taking into account the fact that exit from $C \subseteq \mathbb{Z}^d$ can happen only on the "boundary"

$$\partial C = \{x : x \in \mathbb{Z}^d, x \notin C \text{ and } \|x - y\| = 1 \text{ for some } y \in C\}.$$

This observation leads to the following standard formulation of the (non-homogeneous) *Dirichlet problem* on the lattice: given some domain $C \subseteq \mathbb{Z}^d$ and functions $h = h(x), x \in C$, and $g = g(x), x \in \partial C$, find a function $V = V(x)$, $x \in C \cup \partial C$, which satisfies the equations

$$\Delta V(x) = -h(x), \qquad x \in C;$$
$$V(x) = g(x), \qquad x \in \partial C.$$

If the domain $C$ consists of *finitely many* points, then $\mathsf{P}_x\{\tau(\partial C) < \infty\} = 1$ for all $x \in C$, where $\tau(\partial C) = \inf\{n \geq 0 : X_n \in \partial C\}$ (see Problem 8.8.12). By using the method described earlier, one can show that the solution in the domain $C \cup \partial C$ is unique and is given by the formula:

$$V_{\partial C}(x) = \mathsf{E}_x \left[ g(X_{\tau(\partial C)}) \right] + \mathsf{E}_x \left[ \sum_{k=0}^{\tau(\partial C)-1} h(x) \right], \qquad x \in C \cup \partial C.$$

Since in this case the domain $C$ is finite, there is actually no need to suppose that the functions $h(x)$ and $g(x)$ are non-negative. In particular, setting $h = 0$, one finds that the only function on $C \cup \partial C$, which is harmonic in $C$ and equals $g$ on $\partial C$, is the function

$$V_{\partial C}(x) = \mathsf{E}_x g(X_{\tau(\partial C)}).$$

We now turn to the homogeneous Dirichlet problem:

$$\begin{aligned} \Delta V(x) &= 0, & x \in C, \\ V(x) &= g(x), & x \in \partial C. \end{aligned} \qquad (**)$$

treated on some *unbounded* domain $C$.

If $d \leq 2$, by Pólya's theorem (see [P §8.8]) one must have $\mathsf{P}_x\{\tau(\partial C) < \infty\} = 1$, which, by using the same reasoning as in the case of *finite domains*, leads to the following result: if the function $g = g(x)$ is *bounded*, then, in the class of *bounded* functions on $C \cup \partial C$, the solution to $(**)$ is unique and is given by

$$V_{\partial C}(x) = \mathsf{E}_x g(X_{\tau(\partial C)}).$$

One must realize that even with bounded $g = g(x)$ there could be multiple solutions in the class of *unbounded* functions on $C \cup \partial C$. A classical example of such situation is the following. In dimension $d = 1$ consider the domain $C = \mathbb{Z} \setminus \{0\}$, for which $\partial C = \{0\}$. Setting $g(0) = 0$, it is easy to see that every (automatically *unbounded*) function $V(x) = \alpha x, \alpha \in \mathbb{R}$, is a solution to the Dirichlet problem, i.e., one has $\Delta V(x) = 0, x \in \mathbb{Z} \setminus \{0\}$, and $V(0) = g(0)$.

In dimension $d \geq 3$ the question of existence and uniqueness of the solution to the Dirichlet problem $\Delta V(x) = 0, x \in C$, and $V(x) = g(x), x \in \partial C$, even in the class of bounded functions $V(x), x \in C \cup \partial C$, depends on the condition $\mathsf{P}_x\{\tau(\partial C) < \infty\} = 1$, for all $x \in C$. If this condition holds and $g = g(x)$ is bounded, then one can claim that there is precisely one solution in the class of bounded functions on $C \cup \partial C$, which is given by $V_{\partial C}(x) = \mathsf{E}_x g(X_{\tau(\partial C)})$, for all $x \in C \cup \partial C$.

However, if the condition $\mathsf{P}_x\{\tau(\partial C) < \infty\} = 1, x \in C$, does not hold, then, assuming that $g = g(x), x \in \partial C$, is bounded, every (automatically bounded) function of the form

$$V_{\partial C}^{(\alpha)}(x) = \mathsf{E}_x\big[I(\tau(\partial C) < \infty)g(X_{\tau(\partial C)})\big] + \alpha \mathsf{P}_x\{\tau(\partial C) = \infty\},$$

for all choices of $\alpha \in \mathbb{R}$, is a solution to the Dirichlet problem $\Delta V(x) = 0, x \in C$, and $V(x) = g(x), x \in \partial C$—see, for example, [75, Theorem 1.4.9].

• The discussion in [P Chap. 8] of the various aspects of the *classification* of Markov chains with countable state space follows the tradition established during the 1930s in the works of Kolmogorov, Fréchet, Döblin and others, which is based on the idea that any classification must reflect, on the one hand, the algebraic properties of the one-step transition probability matrix, and, on the other hand, the asymptotic properties of the transition probabilities as the time grows to $\infty$. Since then notions like

*essential and inessential states,*

*reachable and communicating states,*

*irreducibility and periodicity,*

which are determined from the properties of the one-step transition matrix, and notions like

*transience and recurrence,*

*positive recurrent and null recurrent states,*

*invariant (stationary) distributions,*

*ergodic distributions and ergodic theorems,*

which are determined from the limiting behavior of the transition probabilities, have become central in the theory of Markov chains.

Gradually, it became clear that it is more convenient to study the asymptotic properties of Markov chains by utilizing the tools of potential theory, the basic ingredients of which (e.g., the notion of potential, the notions of harmonic and excessive functions, and some basic results involving those notions) were introduced above.

The exposition in [$\underline{P}$ Chap. 8] makes it clear that the primary tool for studying the limiting behavior of Markov chains is a method that would be rather natural to call "the method of regenerating cycles," as is explained next.

Let $x \in E$ be any state and let $(\sigma_x^k)_{k \geq 0}$ be the sequence of "regenerating Markov times," which is constructed as follows: first, define $\sigma_x^0 = 0$ and $\sigma_x^1 = \sigma_x$, where

$$\sigma_x = \inf\{n > 0 : X_n = x\};$$

then define by induction, for any $k \geq 2$,

$$\sigma_x^k = \inf\{n > \sigma_x^{k-1} : X_n = x\} \quad \text{on the set } \{\sigma_x^{k-1} < \infty\}.$$

Equivalently, one can write

$$\sigma_x^k = \begin{cases} \sigma_x^{k-1} + \sigma_x \circ \theta_{\sigma_x^{k-1}}, & \text{if } \sigma_x^{k-1} < \infty; \\ \infty, & \text{if } \sigma_x^{k-1} = \infty. \end{cases}$$

The following properties explain the term "regenerating times" and its connection with the "regenerating cycles":

1. On the set $\{\sigma_x^k < \infty\}$ one has $X_{\sigma_x^k} = x$.

2. On the set $\{\sigma_x^k < \infty\}$ the sequence $(X_{\sigma_x^k + n})_{n \geq 0}$ is independent from random vector $(X_0, X_1, \ldots, X_{\sigma_x^k - 1})$, relative to the measure $P_x$.

3. If $\sigma_x^k(\omega) < \infty$ for all $\omega \in E^\infty$, then, relative to $P_x$, the distribution of the sequence $(X_{\sigma_x^k + n})_{n \geq 0}$ is the same as the distribution of the sequence $(X_n)_{n \geq 0}$.

4. If $\sigma_x^k(\omega) < \infty$ for all $\omega \in E^\infty$, then, relative to $P_x$, the "regenerating cycles"

$$(X_0, X_1, \ldots, X_{\sigma_x^1 - 1}), \ldots, (X_{\sigma_x^{k-1}}, X_{\sigma_x^{k-1} + 1}, \ldots, X_{\sigma_x^k - 1})$$

are independent.

5. $P_x\{\sigma_x^k < \infty\} = P_x\{\sigma_x^{k-1} < \infty\} P_x\{\sigma_x < \infty\}$ and, therefore, $P_x\{\sigma_x^n < \infty\} = [P_x\{\sigma_x < \infty\}]^n$.

6. Setting $N_x = \sum_{n \geq 0} I_{\{x\}}(X_n)$ (in the notation introduced previously, this is nothing but $N_{\{x\}}$, the number of visits to state $x$), then the expected time $E_x N_x$ (which is $E_x N_{\{x\}} = G(x, x)$) is given by

$$E_x N_x = 1 + \sum_{n \geq 1} P_x\{\sigma_x^n < \infty\} = 1 + \sum_{n \geq 1} [P_x\{\sigma_x < \infty\}]^n .$$

7. The above relations entail

$$P_x\{\sigma_x < \infty\} = 1 \iff E_x N_x = \infty \iff P_x\{N_x = \infty\} = 1,$$

$$P_x\{\sigma_x < \infty\} < 1 \iff E_x N_x < \infty \iff P_x\{N_x < \infty\} = 1.$$

8. For any $y \neq x$ one has

$$G(x, y) = P_x\{\sigma_y < \infty\} G(y, y),$$

or, equivalently,

$$E_x N_y = P_x\{\sigma_y < \infty\} E_y N_y .$$

9. If $P_x\{\sigma_x^k < \infty\} = 1$ for all $k \geq 1$, then the sequence of "regenerating periods", $(\sigma_x^k - \sigma_x^{k-1})_{k \geq 0}$, is a sequence of independent and identically distributed random variables.

Recall that according to the definitions in [P §8.5] the state $x \in E$ is called

$$\text{recurrent, if } P_x\{\sigma_x < \infty\} = 1,$$

$$\text{transient, if } P_x\{\sigma_x < \infty\} < 1.$$

Since (see [ P §8.5, Theorem 1])

$$P_x\{\sigma_x < \infty\} = 1 \iff P_x\{X_n = x \text{ i.o.}\} = 1,$$
$$P_x\{\sigma_x < \infty\} < 1 \iff P_x\{X_n = x \text{ i.o.}\} = 0,$$

the state $x \in E$ is (or may be called that by definition)

*recurrent*, if $P_x\{X_n = x \text{ i.o.}\} = 1$,

*transient*, if $P_x\{X_n = x \text{ i.o.}\} = 0$.

In fact, the intrinsic meaning of the terms "recurrent" and "transient" is better reflected in the relations "$P_x\{X_n = x \text{ i.o.}\} = 1$" and "$P_x\{X_n = x \text{ i.o.}\} = 0$," as opposed to the equivalent relations "$P_x\{\sigma_x < \infty\} = 1$" and "$P_x\{\sigma_x < \infty\} < 1$". Indeed, "recurrence of $x$" is to be understood as "eventual return to $x$ after *every* visit to $x$" and "transience of $x$" is to be understood as "non-recurrence of $x$," i.e., as "non-return after *some* visit to $x$".

Thus, the *recurrence* of the state $x$ is equivalent to each of the following properties

$$P_x\{X_n = x \text{ i.o.}\} = 1, \text{ or } P_x\{N_x = \infty\} = 1, \text{ or } E_x N_x = \infty,$$

while the *transience* is equivalent to each of the properties:

$$P_x\{X_n = x \text{ i.o.}\} = 0, \text{ or } P_x\{N_x < \infty\} = 1, \text{ or } E_x N_x < \infty.$$

• The use of potential theory and, in particular, the technique of "regenerating cycles" allows one to develop a more or less complete understanding of the structure of the *invariant* measures and distributions (i.e., probability distributions). The exposition below follows [85].

Recall that any (one-step) transition probability matrix $P = \|p_{xy}\|$, $x, y \in E$, gives rise to the linear operator $\mathbb{P}f$, which acts on functions $f \in \mathscr{L}^0(E, \mathscr{E}; \mathbb{R}_+)$ according to the rule

$$(\mathbb{P}f)(x) = \sum_{y \in E} p_{xy} f(y), \quad x \in E,$$

understood as

$$(\text{matrix } P) \otimes (\text{vector-column } f) = (\text{vector-column } Pf).$$

Let $q = q(A)$, $A \subseteq E$, be any non-trivial (i.e., not identically 0 or $\infty$) measure defined on the subsets of some countable set $E$. Such a measure is completely determined by its values, $q(\{x\})$, on the singleton sets $\{x\}$, $x \in E$ (for the sake of simplicity we will write $q(x)$ instead of $q(\{x\})$).

Let $\mathcal{M}_+$ denote the space of such measures $q$ and let $\mathbb{P}$ stand for the linear operator that transforms measures from $\mathcal{M}_+$ into measures from $\mathcal{M}_+$ according to the rule $\mathcal{M}_+ \ni q \rightsquigarrow q\mathbb{P} \in \mathcal{M}_+$, where $q\mathbb{P}$ is the measure

$$q\mathbb{P}(y) = \sum_{x \in E} q(x) p_{xy}, \quad y \in E,$$

i.e., $q\mathbb{P} \in \mathcal{M}_+$ is understood as the vector

$$(\text{vector-column } q\mathbb{P}) = (\text{vector-column } q) \otimes (\text{matrix } P).$$

The measure $q \in \mathcal{M}_+$ is said to be *invariant* or *stationary* for the Markov chain with operator $\mathbb{P}$ if $q\mathbb{P} = q$. The measure $q \in \mathcal{M}_+$ is said to be *excessive*, or $\mathbb{P}$-*excessive*, if $q\mathbb{P} \le q$.

Next, consider the *bi-linear form*

$$\langle q, f \rangle = \sum_x q(x) f(x), \quad f \in \mathscr{L}^0(E, \mathscr{E}; \mathbb{R}_+), \ q \in \mathcal{M}_+.$$

The following *duality* relation is easy to verify:

$$\langle q, \mathbb{P}f \rangle = \langle q\mathbb{P}, f \rangle, \quad f \in \mathscr{L}^0(E, \mathscr{E}; \mathbb{R}_+), \ q \in \mathcal{M}_+.$$

Essentially, the above relation says that the action of the operator $\mathbb{P}$ on functions and the action of the operator $\mathbb{P}$ on measures can be interchanged.

[P §8.6, Theorem 2] shows that, in the case of *irreducible* (there is only one class) and *positive recurrent* Markov chains with countable state space, an invariant distribution exists, it is unique, and is given by

$$q(x) = [\mathsf{E}_x \sigma_x]^{-1}, \quad x \in E,$$

where $\sigma_x = \inf\{n \ge 1 : X_n = x\}$ is the time of the first *recurrence* to $x$. (Note that $1 \le \mathsf{E}_x \sigma_x < \infty, x \in E$.)

As we are about to show, by using the characteristics of the first "regenerating cycle" the result about the existence and the structure of the invariant sets can be established for arbitrary *irreducible and recurrent* Markov chains, without the requirement for positive recurrence.

More specifically, one can claim the following:

Any irreducible and recurrent Markov chain $X = (X_n)_{n \ge 0}$, which has a countable state space $E$, admits an invariant measure $q = q(A)$, $A \subseteq E$, which is non-trivial, in that $0 < q(E), q(x) \ne \infty$ and $0 < q(x) < \infty$, for any state $x \in E$. This measure is unique up to a multiplicative constant.

To prove the above statement, notice that for any fixed state $x^\circ \in E$, one can always construct an invariant measure, say $q^\circ$, with the property $q^\circ(x^\circ) = 1$. For example, one can set

$$q^\circ(x) = \mathsf{E}_{x^\circ}\left[\sum_{k=0}^{\sigma_{x^\circ}-1} I_{\{x\}}(X_k)\right], \quad x \in E,$$

where $\sigma_{x^\circ} = \inf\{n \geq 1 : X_n = x^\circ\}$. In order to show that the above measure is indeed invariant (and that therefore *invariant measures exist*), it would be enough to show that for any function $f \in \mathscr{L}^0(E, \mathscr{E}; \mathbb{R}_+)$ one has

$$\langle q^\circ \mathbb{P}, f \rangle = \langle q^\circ, f \rangle.$$

In conjunction with the strong Markov property established in Problem 8.2.13, the last relation follows from the following chain of identities:

$$\langle q^\circ \mathbb{P}, f \rangle = \langle q^\circ, \mathbb{P} f \rangle = \mathsf{E}_{x^\circ}\left[\sum_{k=0}^{\sigma_{x^\circ}-1}(\mathbb{P}f)(X_k)\right] = \mathsf{E}_{x^\circ}\left[\sum_{k=0}^{\sigma_{x^\circ}-1}\mathsf{E}_{X_k} f(X_1)\right]$$

$$= \sum_{k \geq 0}\mathsf{E}_{x^\circ}\left[I_{\{k<\sigma_{x^\circ}\}}\mathsf{E}_{X_k} f(X_1)\right] = \sum_{k \geq 0}\mathsf{E}_{x^\circ}\{I_{\{k<\sigma_{x^\circ}\}}\mathsf{E}_{x^\circ}[f \circ \theta_k \mid \mathscr{F}_k]\}$$

$$= \sum_{k \geq 0}\mathsf{E}_{x^\circ}\{\mathsf{E}_{x^\circ}[I_{\{k<\sigma_{x^\circ}\}}f \circ \theta_k \mid \mathscr{F}_k]\} = \sum_{k \geq 0}\mathsf{E}_{x^\circ}\left[I_{\{k<\sigma_{x^\circ}\}}f \circ \theta_k\right]$$

$$= \mathsf{E}_{x^\circ}\sum_{k \geq 0} I_{\{k<\sigma_{x^\circ}\}}f(X_{k+1}) = \mathsf{E}_{x^\circ}\sum_{l=1}^{\sigma_{x^\circ}} f(X_l) = \mathsf{E}_{x^\circ}\sum_{k=0}^{\sigma_{x^\circ}-1} f(X_k)$$

$$= \langle q^\circ, f \rangle.$$

In addition to the normalization $q^\circ(x^\circ) = 1$, the measure $q^\circ$ constructed above also has the property $0 < q^\circ(x) < \infty$, for all $x \in E$. This last property follows from the following simple fact about excessive measures.

Suppose that the underlying Markov chain is irreducible and that the measure $q \in \mathscr{M}_+$ is excessive, i.e., $q\mathbb{P} \leq q$. If there is a state $x^\circ \in E$ for which $q(x^\circ) = 0$ (note that $q(x^\circ) < \infty$), then for any $x \in E$ one must have $q(x) = 0$ (note that $q(x) < \infty$). To see why this claim can be made, observe that for any $x \neq x^\circ$ one can find an integer $n \geq 1$, for which $p_{x,x^\circ}^{(n)} > 0$. As a result, the relations

$$0 = q(x^\circ) \geq \sum_{y \in E} q(y) p_{y,x^\circ}^{(n)} \geq q(x) p_{x,x^\circ}^{(n)},$$

imply that $q(x) = 0$.

We will now show that, up to a positive multiplicative constant, $q^\circ$ is the only non-trivial invariant measure. For that purpose, suppose that $q$ is some invariant (and, therefore, also excessive) measure with $0 < q(x) < \infty$, for all $x \in E$. Set

$$f(x) = \frac{q(x)}{q^\circ(x)}, \quad x \in E,$$

and define the (dual of $p_{xy}$) function $\widehat{p}_{xy} = \frac{q^\circ(y)}{q^\circ(x)} p_{yx}$. Since for every fixed $x \in E$ one has

$$\sum_{y \in E} \widehat{p}_{xy} = \frac{1}{q^\circ(x)} \sum_{y \in E} q^\circ(x) p_{yx} = \frac{q^\circ(x)}{q^\circ(x)} = 1,$$

the matrix $\widehat{P} = \|\widehat{p}_{xy}\|$ can be treated as a transition probability matrix and one can write

$$\widehat{P} f(x) = \sum_{y \in E} \widehat{p}_{xy} f(y) = \sum_{y \in E} \widehat{p}_{xy} \frac{q(y)}{q^\circ(y)} = \sum_{y \in E} \frac{q^\circ(y)}{q^\circ(x)} p_{xy} \frac{q(y)}{q^\circ(y)}$$

$$= \frac{1}{q^\circ(x)} \sum_{y \in E} p_{xy} q(y) = \frac{q(x)}{q^\circ(x)} = f(x).$$

The function $f = f(x)$ is therefore $\widehat{P}$-harmonic. Since $\widehat{p}_{xy} = \frac{q^\circ(y)}{q^\circ(x)} p_{yx}$ by definition, for every $n \geq 1$ one must have

$$\widehat{p}_{xy}^{(n)} = \frac{q^\circ(y)}{q^\circ(x)} p_{yx}^{(n)},$$

which entails the following relation between the respective Green functions

$$\widehat{G}(x, y) = \frac{q^\circ(y)}{q^\circ(x)} G(y, x).$$

The last two relations imply that if the Markov chain $X = (X_n)_{n \geq 0}$, with operator $\mathbb{P}$, happens to be irreducible and recurrent, then the dual chain $X = (X_n)_{n \geq 0}$, with operator $\widehat{\mathbb{P}}$, must be irreducible and recurrent, too. However, if $f = f(x)$ is any (automatically non-negative) excessive function (in particular, if $f = f(x)$ is harmonic), then the sequence $(f(X_n))_{n \geq 0}$ must be a non-negative supermartingale relative to the measure $\widehat{\mathbb{P}}_\pi$, for any initial distribution $\pi$. This property was mentioned earlier. We also noted that for sequences of that form the limit $\lim_n X_n$ ($\equiv X_\infty$) exists $\widehat{\mathbb{P}}_\pi$-almost everywhere, and, therefore the limit $\lim_n f(X_n)$ ($\equiv Z$) must exist $\widehat{\mathbb{P}}_\pi$-almost everywhere, too. It is easy to see that if the chain is irreducible and recurrent, then for any two states, $x$ and $y \neq x$, one can claim that $X_n$ visits infinitely many times both $x$ and $y$. In particular, this shows that $f(x) = f(y)$, for every $x, y \in E$, so that $f(x) \equiv$ const.

We have thus established that any other invariant distribution $q$, such that $0 < q(x) < \infty$, $x \in E$, must be a multiple, with some positive constant factor, of the measure $q°$.

The result that we just established entails the following feature of all irreducible and recurrent Markov chains, which was mentioned earlier: the only invariant probability distribution $q° = (q°(x), x \in E)$ is given by $q°(x) = [E_x \sigma_x]^{-1}$, $x \in E$.

• We now turn to certain ergodic theorems for Markov chains with countable state spaces, i.e., theorems about *convergence almost surely* as $n \to \infty$ of quantities of the form $\frac{1}{n} \sum_{k=0}^{n-1} f(X_k)$, or, more generally, of the form

$$\sum_{k=0}^{n-1} f(X_k) \Big/ \sum_{k=0}^{n} g(X_k),$$

for certain classes of functions $f$ and $g$. We will again rely on the technique of "regeneration cycles."

Let $X = (X_n)_{n \geq 0}$ be some irreducible and recurrent Markov chain with countable state space $E$ and invariant measure $q°(x)$, such that $0 < q(x) < \infty$ for all $x \in E$ and $q°(x°) = 1$ for some fixed state $x°$.

Next, suppose that $f = f(x)$ and $g = g(x)$ are two function from the class $L^1(q°)$, i.e., $f = f(x)$ and $g = g(x)$ are two function on $E$ chosen so that $\sum_{x \in E} |f(x)| q°(x) < \infty$ and $\sum_{x \in E} |g(x)| q°(x) < \infty$, and set

$$Y_0 = \sum_{k=0}^{\sigma_{x°}^1 - 1} f(X_k) \quad \text{and} \quad Y_m = \sum_{k=\sigma_{x°}^m}^{\sigma_{x°}^{m+1} - 1} f(X_k) \quad (= Y_0 \circ \theta_{\sigma_{x°}^m}).$$

By the very definition of the invariant distribution $q°$ we have

$$E_x Y_0 = \langle q°, f \rangle,$$

and, due to the Markov property, for any initial distribution $\pi$ one must have

$$E_\pi Y_m = E_\pi \left[ E_{X_{\sigma_{x°}^m}} (Y_0) \right] = E_{x°} Y_0 = \langle q°, f \rangle.$$

Thus, relative to the measure $P_\pi$, the random variables $Y_1, Y_2, \ldots$ are independent and identically distributed, and, furthermore, have the property $E_\pi Y_m = \langle q°, f \rangle$ $(<\infty)$, $m \geq 1$. The strong law of large numbers now implies that for every initial distribution $\pi$ one must have $(P_\pi\text{-a. e.})$

$$\frac{1}{n} \sum_{k=0}^{\sigma_{x°}^n} f(X_k) = \frac{Y_0}{n} + \frac{1}{n}(Y_1 + \ldots + Y_{n-1}) + \frac{f(x°)}{n} \to \langle q°, f \rangle \quad \text{as } n \to \infty,$$

and, assuming that $\langle q^\circ, g \rangle \neq 0$, one must have (again, for every $\pi$)

$$\frac{\sum\limits_{k=0}^{\sigma_{x^\circ}^n} f(X_k)}{\sum\limits_{k=0}^{\sigma_{x^\circ}^n} g(X_k)} \to \frac{\langle q^\circ, f \rangle}{\langle q^\circ, g \rangle} \quad \text{as } n \to \infty \quad (P_\pi\text{-a. e.}).$$

Next, let $v_{x^\circ}^n = \sum\limits_{k=1}^{n} I(X_k = x^\circ)$, and notice that, since the chain is recurrent, one can claim that $v_{x^\circ}^n \to \infty$ as $n \to \infty$ ($P_\pi$-a. e.). Since $\sigma_{x^\circ}^{v_{x^\circ}^n} \le n < \sigma_{x^\circ}^{v_{x^\circ}^n + 1}$, the above convergence entails *the ergodic theorem for ratios*:

$$\frac{\sum\limits_{k=0}^{n} f(X_k)}{\sum\limits_{k=0}^{n} g(X_k)} \to \frac{\langle q^\circ, f \rangle}{\langle q^\circ, g \rangle} \quad \text{as } n \to \infty \quad (P_\pi\text{-a. e.}).$$

Finally, suppose that the Markov chain under consideration is irreducible and positive recurrent. In this case one can replace the measure $q^\circ$ with the probability distribution $\pi^\circ = (\pi^\circ(x), x \in E)$, chosen so that $\pi^\circ = 1/(E_x \sigma_x)$, and, as a result, arrive at the following ergodic theorem (for irreducible and positive recurrent Markov chains):

$$\frac{1}{n} \sum\limits_{k=0}^{n} f(X_k) \to \langle \pi^\circ, f \rangle \quad \text{as } n \to \infty \quad (P_\pi\text{-a. e.}),$$

for any initial distribution $\pi$ (in particular, for $\pi = \pi^\circ$).

# References

1. Aldous, D.J.: Exchangeability and related topics. In: École d'Été St Flour 1983.Lecture Notes in Mathematics, vol. 1117. Springer, Berlin (1985)
2. Arnol'd, V.I.: Tsepnye drobi. MCNMO, Moscow (2001)
3. Arnol'd, V.I.: Chto takoe matematika. MCNMO, Moscow (2004)
4. Baranov, V.I., Stechkin, B.S.: Èkstremal'nye kombinatornye zadachi i ikh prilozheniya. Fizmatlit, Moscow (2004)
5. Bentkus, V.: On Hoeffding's inequalities. Ann. Probab. **32**(2), 1650–1673 (2004)
6. Berger, M.A.: An Introduction to Probability and Stochastic Processes. Springer, New York (1993)
7. Bertrand, J.: Calcul des Probabilités. Gauthier-Villars, Paris (1889)
8. Bhattacharya, R.N., Waymire, E.C.: Stochastic Processes with Applications. Wiley, New York (1990)
9. Billingsley, P.: Probability and Measure, 3rd edn. Wiley, New York (1995)
10. Billingsley, P.: Convergence of Probability Measures, 2nd edn. Wiley, New York (1999)
11. Boldin, M.V., Simonova, G.I., Tyurin, Yu. N.: Znakovyǐ statisticheskiǐ analiz lineǐnykh modeleǐ. Nauka (Science), Moscow (1997)
12. Bol'shev, L.N., Smirnov, N.V.: Tablicy matematicheskoǐ statistiki. Nauka (Science), Moscow (1983)
13. Bradley, R.C.: The central limit question under $\rho$-mixing. Rocky Mt. J. Math. **17**(1), 95–114 (1987); corrections: ibid. (4), 891
14. Brémaud, P.: An Introduction to Probabilistic Modeling. Springer, New York (1988)
15. Brémaud, P.: Markov Chains: Gibbs Fields, Monte Carlo Simulation, and Queues. Springer, New York (1999)
16. Bulinskiǐ, A.V.: Predel'nye teoremy v usloviyah slaboǐ zavisimosti. Izdatel'stvo Moskovskogo Universiteta. Moscow University Press, Moscow (1989)
17. Bulinskiǐ, A.V., Shiryaev, A.N.: Teoriya sluchaǐnyh processov. Fizmatlit, Moscow (2003)
18. Burkholder, D.L.: Sufficiency in the undominated case. Ann. Math. Stat. **32**(4), 1191–1200 (1961)
19. Cacoullos, T.: Exercises in Probability. Springer, New York (1989)
20. Cameron, P.J.: Combinatorics: Topics, Techniques, Algorithms. Cambridge University Press, Cambridge (1994)
21. Capiński, M., Zastawniak, T.: Probability Through Problems. Springer, New York (2001)
22. Chandrasekhar, S.: Stochastic problems in physics and astronomy. Rev. Mod. Phys. **15**, 1–89 (1943)
23. Chaumont, L., Yor, M.: Exercises in Probability. Cambridge University Press, Cambridge (2003)

24. Cherubini, D., Luciano, E., Vecchiato, W.: Copula Methods in Finance. Wiley, Chichester (2004)

25. Chibisov, D.M., Pagurova, V.I.: Zadachi po matematicheskoĭ statistike. Izdatel'stvo Moskovskogo universiteta (Moscow University Press), Moscow (1990)

26. Chung, K.L.: Markov Chains with Stationary Transition Probabilities, 2nd edn. Springer, New York (1967)

27. Constantine, G.M.: Combinatorial Theory and Statistical Design. Wiley, New York (1987)

28. David, H.: Ordered Statistics, 2nd edn. Wiley, New York (1981)

29. de Finetti, B.: Theory of Probability: A Critical Introductory Treatment, vols. 1 and 2. Wiley, London (1974, 1975)

30. De Groot, M.: Optimal'nye statisticheskie resheniya. Mir, Moscow (1974)

31. Devroye, L.: A Course in Density Estimation. Birkhäuser, Boston (1987)

32. Dorogovcev, A.Ya., Sil'vestrov, D.S., Skorohod, A.V., Yadrenko, M.I.: Teoriya veroyatnosteĭ (sbornik zadach). Vishcha shkola, Kiev (1980)

33. D'yachenko, M.I., Ul'yanov, P.L.: Mera i integral. Faktorial, Moscow (2002)

34. Durrett, R.: Probability: Theory and Examples. Wadsworth and Brooks/Cole, Pacific Grove (1991)

35. Emel'yanov, G.V., Skitovich, V.P.: Zadachnik po teorii veroyatnosteĭ i matematicheskoĭ statistike. Izdatel'stvo Leningradskogo universiteta, Leningrad (1967)

36. Erdësh, P., Spenser, J.: Probabilistic Methods of Combinatorics. Academic, New York (1974)

37. Erdësh, P., Rényi, A.: On Cantor's Series with Divergent $\sum 1/q_n$. Academic, New York (1974)

38. Everitt, B.S.: Chance Rules: An Informal Guide to Probability, Risk, and Statistics. Springer, New York (1999)

39. Feller, W.: An Introduction to Probability Theory and Its Applications, vols. I and II, 3rd edn. Wiley, New York (1968)

40. Fristedt, B., Gray, L.: A Modern Approach to Probability Theory. Birkhäuser, Boston (1997)

41. Gmurman, V.E.: Rukovodstvo k resheniyu zadach po teorii veroyatnosteĭ i matematicheskoĭ statistike. Vysshaya Shkola, Moscow (1979)

42. Gnedenko, B.V.: Kurs teorii veroyatnosteĭ. URSS, Moscow (2001)

43. Goncharov, V.L.: Teoriya veroyatnosteĭ. Gos. izdatel'stvo oboronnoĭ promyshl, Moscow (1939)

44. Grimmett, G.R., Stirzaker, D.R.: One Thousand Exercises in Probability (Companion to Probability and Random Processes). Oxford University Press, Oxford (2001)

45. Gut, A.: Stopped Random Walks. Springer, New York (1988)

46. Hall, M.: Combinatorial Theory. Wiley, New York (1986)

47. Hazewinkel, M.(ed.): Encyclopaedia of Mathematics. Springer, Berlin (2002)

48. Hoffmann-Jørgensen, J.: Probability with a View Toward Statistics, vols. I and II. Chapman and Hall, New York (1994)

49. Ibragimov, I.A., Rosanov, Yu.A.: Gaussovskie sluchaĭnye processy. Nauka (Science), Moscow (1970)

50. Ikeda, N., Watanabe, S.: Stochastic Differential Equations and Diffusion Processes, 2nd edn. Kodansha, Tokyo (1989)

51. Imhof, J.-P.: Introduction au Calcul des Probabilités. Gauthier-Villars, Paris (1969)

52. Ivchenko, G.I., Medvedev, Yu.I.: Matematicheskaya statistika. Vysshaya shkola, Moscow (1984)

53. Jacod, J., Protter, Ph.: Probability Essentials. Springer, Berlin (2003)

54. Kannan, D.: An Introduction to Stochastic Processes. North-Holland, New York/Oxford (1979)

55. Karlin, S., Taylor, H.M.: A First Course in Stochastic Processes. Academic, New York/London (1975)

56. Karlin, S., Taylor, H.M.: A Second Course in Stochastic Processes. Academic, New York/London (1981)

57. Karr, A.F.: Probability. Springer, Berlin (1993)

58. Kawata, T.: Fourier Analysis in Probability Theory. Academic, New York/London (1972)
59. Kemeny, J.G., Snell, J.L.: Finite Markov Chains. Van Nostrand, Princeton (1960)
60. Kemeny, J.G., Snell, J.L., Thompson, G.L.: Introduction to Finite Mathematics. Prentice-Hall, Englewood Cliffs (1963)
61. Kendall, M., Moran, P.: Geometric Probability. Griffin, London (1963)
62. Khrennikov, A.: Interpretations of Probability. VSP, Utrecht (1999)
63. Kingman, J.F.C.: Random variables with unsymmetrical regression. Math. Proc. Camb. Philos. Soc. **98**(2), 355–365 (1985)
64. Kochen, S., Stone, Ch.: A note on the Borel-Cantelli lemma. Ill. J. Math. **8**, 248–251 (1964)
65. Kolchin, V.F.: Sluchaǐnye otobrazheniya. Nauka (Science), Moscow (1984)
66. Kolmogorov, A.N.: Foundations of the Theory of Probability. Chelsea, New York (1956); Osnovnye ponyatiya teorii veroyatnosteǐ, 3ᵉ izd.. FAZIS, Moscow (1998)
67. Kopp, P.E.: Martingales and Stochastic Integrals. Cambridge University Press, Cambridge (1984)
68. Kornfel′d, I.P., Sinaǐ, Ya.G., Fomin, S.V. Èrgodicheskaya teoriya. Nauka (Science), Moscow (1980)
69. Kotz, S., Nadarajah, S.: Exreme Value Distributions: Theory and Applications. Imperial College Press, London (2000)
70. Kovalenko, I.N., Sarmanov, O.V.: Kratkiǐ kurs teorii sluchaiǐnyh processov. Vishcha shkola, Kiev (1978)
71. Kozlov, M.V.: Èlementy teorii veroyatnosteǐ v primerakh i zadachakh. Izdatel′stvo Moskovskogo universiteta. Moscow University Press, Moscow (1990)
72. Kozlov, M.V., Prokhorov, A.V.: Vvedenie v matematichskuyu statistiku. Izdatel′stvo Moskovskogo universiteta. Moscow University Press, Moscow (1987)
73. Krief, A., Levy, S.: Calcul des Probabilités: Exercises. Hermann, Paris (1972)
74. Lando, S.K.: Lekcii o proizvodyashchih funkciyah. MCNMO, Moscow (2002)
75. Lawler, G.F.: Intersections of Random Walks. Birkhäuser, Boston (1991)
76. Lehmann, E.L.: Testing Statistical Hypothesis, 5th edn. Wiley, New York (1970)
77. Letac, G.: Problèmes de Probabilité. Presses Universitaires de France, Paris (1970)
78. Lindgren, B.W., McElrath, G.W.: Introduction to Probability and Statistics. Macmillan, London/New York (1966)
79. Liptser, R.Sh., Shiryaev, A.N.: Theory of Martingales. Kluwer, Dordrecht (1989)
80. Liptser, R.Sh., Shiryaev, A.N.: Statistics of Random Processes I and II, 2nd edn. Springer, New York (2000)
81. Long, R.L.: Martingale Spaces and Inequalities. Peking University Press, Beijing (1993)
82. Lovász, L., Pelicán, J., Vesztergombi, K.: Discrete Mathemtics: Elementary and Beyond. Springer, Berlin (2003)
83. Lukacs, E.: Characteristics Functions. Griffin, London (1972)
84. Lukacs, E.: Developments in Characteristic Function Theory. Macmillan, New York (1983)
85. Mazliak, L., Priouret, P., Baldi, P.: Martingales et chaînes de Markov. Hermann, Paris (1998)
86. Meester, R.: A Natural Introduction to Probability Theory. Birkhäuser, Basel (2003)
87. Mikosch, T.: Non-life Insurance Mathematics. Springer, Berlin (2004)
88. Moran, P.A.P.: An Introduction to Probability Theory. Clarendon Press, Oxford (1968)
89. Mosteller, F., Rourke, R.E.K., Thomas, G.B.: Probability with Statistical Applications. Addison-Wesley, Reading (1970)
90. Nevzorov, V.B.: Rekordy: Matematicheskaya teoriya. FAZIS, Moscow (2000)
91. Norris, J.R.: Markov Chains. Cambridge University Press, Cambridge (1999)
92. Ortega, J., Wschebor, M.: On the sequence of partial maxima of some random sequences. Stoch. Process. Appl. **16**(1), 85–98 (1984)
93. Peskir, G., Shiryaev, A.N.: Neravenstva Hinchina i martingal′noe rasshirenie sfery ih deǐstviya. Uspehi matematicheskih nauk. **50**(5), 3–62 (1995)
94. Petrov, V.V.: Predel′nye teoremy dlya summ nezavisimyh sluchaǐnyh velichin. Nauka (Science), Moscow (1987)

95. Petrov, V.V.: A generalization of the Borel–Cantelli lemma. Stat. Probab. Lett. **67**(3), 233–239 (2004)

96. Petrov, V., Mordecki, E.: Theoría de Probabilidades (Con más de 200 ejercicios). Editorial URSS, Moscow (2002)

97. Poincaré, H.: Calcul des probabilités., 2 éd., Paris (1912); Russian translation: Puankare A.: Teoriya veroyatnostei. Regulyarnaya i haoticheskaya dinamika, Izhevsk (1999)

98. Port, S.C.: Theoretical Probability for Applications. Wiley, New York (1994)

99. Prohorov, A.V., Ushakov, V.G., Ushakov, N.G.: Zadachi po teorii veroyatnostei: Osnovnie ponyatiya. Predel'nye teoremy. Sluchainye protsessy. Nauka (Science), Moscow (1986)

100. Révész, P.: The Law of Large Numbers. Academic, New York/London (1968)

101. Révész, P.: Random Walk in Random and Non-random Environments. World Scientific, Teaneck (1990)

102. Révész, P., Tóth, B. (Eds.): Random Walks. Bolyai Society Mathematical Studies, vol. 9. János Bolyai Mathematical Society, Budapest (1999)

103. Revuz, D., Yor, M.: Continuous Martingales and Brownian Motion, 3rd edn. Springer, Berlin (1999)

104. Romano, J.P., Siegel, A.F.: Counterexamples in Probability and Statistics. Wadsworth and Brooks/Cole, Monterey (1986)

105. Rosenthal, J.S.: A First Look at Rigorous Probability Theory. World Scientific, River Edge (2000)

106. Ross, S.M.: Introduction to Probability Models, 8th edn. Academic, Boston (2003)

107. Ross, S.M.: Stochastic Processes, 2nd edn. Wiley, New York (1996)

108. Ruzsa, I.Z., Székely, G.J.: Algebraic Probability Theory. Wiley, Chichester (1988)

109. Ryzhik, I.S, Gradshteyn, I.M.: Table of Integrals, Series and Products, 7th edn. Elsevier, Burlington (2007)

110. Sachkov, V.N.: Kombinatornye metody diskretnoi matematiki. Nauka (Science), Moscow (1977)

111. Sachkov, V.N.: Veroyatnostnye metody v kombinatornom analize. Nauka (Science), Moscow (1978)

112. Schervish, M.J.: Theory of Statistics. Springer, New York (1995)

113. Seshadri, V.: The Inverse Gaussian Distribution. Clarendon, New York (1993)

114. Sevast'yanov, B.A.: Kurs teorii veroyatnostei i matematicheskoi statistiki. Institut komp'yuternyh issledovanii, Moskva-Izhevsk (2004)

115. Sevast'yanov, B.A., Chistyakov, V.P., Zubkov, A.M.: Sbornik zadach po teorii veroyatnostei. Nauka (Science), Moscow (1980)

116. Shafer, G., Vovk, V.: Probability and Finance. Wiley, New York (2001)

117. Shiryaev, A.N.: Optimal Stopping Rules. Springer, Berlin (1978)

118. Shiryaev, A.N.: Probability (Russian editions). Nauka (Science), Moscow (1980, 1989)

119. Shiryaev, A.N.: Probability (English editions). Springer, New York (1984, 1990)

120. Shiryaev, A.N.: Essentials of Stochastic Finance: Facts, Models, Theory. World Scientific, Singapore (1999)

121. Shiryaev, A.N.: Probability-1 and Probability-2 (Russian editions). MCCME, Moscow (2004, 2007)

122. Sklar, A.: Fonctions de répartition à n dimensions et leurs marges. Publications de l'Institut de Statistichue de l'Université de Paris, **8**, 229–231 (1959)

123. Sklar, A.: Random variables, distribution functions, and copulas: a personal look backward and forward. IMS Lecture Notes in Monograph Series. **28**, 1–14 (1996)

124. Skorohod, A.V.: Veroyatnost' vokrug nas. Naukova dumka, Kiev (1980)

125. Spitzer, F.: Principles of Random Walk, 2nd edn. Springer, New York (2001)

126. Stoyanov, J., Mirazchiiski, I., Ignatov, Z., Tanushev, M.: Exercise Manual in Probability Theory. Kluwer, Dordrecht (1989)

127. Stoyanov, J.: Counterexamples in Probability, 2nd edn. Wiley, Chichester (1997)

128. Sveshnikov, A.A. (red.): Sbornik zadach po teorii veroyatnostei, matematicheskoi statistike i teorii sluchainyh funkcii. Nauka (Science), Moscow (1965)

129. Székely, G.: Paradoxes in Probability Theory and Mathematical Statistics. Kluwer, Norwell (1986)
130. Tutubalin V.N.: Teoriya veroyatnosteĭ i sluchaĭnyh processov. Izdatel'stvo Moskovskogo universiteta. Moscow University Press, Moscow (1992)
131. Volodin, V.G., Ganin, M.P., Diner, I.Ya., Komarov, L.B., Sveshnikov, A.A., Starobin, K.B.: Sbornik zadach po teorii veroyatnosteĭ, matematicheskoĭ statistike i teorii sluchaĭnyh funkciĭ (pod red. A.A. Sveshnikova). Nauka (Science), Moscow (1965)
132. Wall, C.R.: Terminating decimals in the Cantor ternary set. Fibonacci Q. **28**(2), 98–101 (1990)
133. Williams, D.: Some basic theorems on harnesses. In: Kendall, D.G., Harding, E.F. (eds.) Stochastic Analysis, pp. 349–363. Wiley, London (1973)
134. Williams, D.: Probability with Martingales. Cambridge University Press, Cambridge (1991)
135. Williams, D.: Weighing the Odds: a Course in Probability and Statistics. Cambridge University Press, Cambridge (2001)
136. Wise, G.L., Hall, E.B.: Counterexamples in Probability and Real Analysis. Oxford University Press, New York (1993)

# Author Index

A.N. Shiryaev, *Problems in Probability*, Problem Books in Mathematics,     413
DOI 10.1007/978-1-4614-3688-1, © Springer Science+Business Media New York 2012

# Subject Index

## Symbols

$(B, S)$-market
  arbitrage-free in strong sense, 323
  arbitrage-free in weak sense, 323
$A$-integral, 193
$\sigma$-algebra:
  countably generated, 65
  separable, 65
"Kurtosis" parameter, 135
"Skewness" parameter, 135

## A

Absent-minded secretary problem, 10
Absolute continuity:
  of the Lebesgue integral, 88
Almost invariant:
  random variable, 261
Almost uniform convergence, 145
Amount of information, 45
André's reflection principle, 348
Appell relations, 383
Arbitrage, 323
Arbitrage-free $(B, S)$-market:
  in strong sense, 323
  in weak sense, 323
Arcsine law, 50, 51
Atom, 79
Atomic measure, 79
Atoms:
  of partition, 360
Auto-regression model, 268
Autoregressive model:
  first order, 272

## B

Balance equation, 332
Bell numbers, 362
Bentkus inequality, 303
Beppo Levi's theorem, 93
Bernoulli number, 379
Bernoulli polynomials, 379
Bernoulli scheme:, 330
  with random probability for success, 133
Bernoulli shifts, 265
Bernoulli transformation, 265
Bernstein inequality, 255
Bernstein polynomials, 67
Bernstein's criterion, 157
Bernstein's estimate, 35
Bernstein's theorem, 165
Berry-Esseen inequality, 41
Bertrand's paradox, 77
Bessel function, 115, 372
Beta-function:
  incomplete, 135
Bijection, 20
Binary bracketing, 2
Binary operation, 331
Binomial coefficient, 1
Binomial identity, 13
Binomial moment, 127, 379
Birkhoff-Khinchin theorem, 385
Bit:, 44
  of information, 44
Black–Scholes formula, 325
Bochner-Khinchin theorem, 154
Boole inequality, 63
Borel normal numbers theorem, 266